CULTURAL GEOGRAPHY FORM AND PROCESS

(Essays in Honour of Prof. A.B. Mukherji)

Edited by

NEELAM GROVER

Prof. of Geography
Panjab University
Chandigarh

KASHI NATH SINGH

Prof. of Geography
University of Addis Ababa
Ethiopia

CONCEPT PUBLISHING COMPANY PVT. LTD.
NEW DELHI-110059

ISBN: 978-81-8069-074-7

First Published 2004

Published and Printed by

Concept
Since 1974

Concept Publishing Company Pvt. Ltd.
A/15-16, Commercial Block, Mohan Garden,
New Delhi-110059 (INDIA)
T: +91-11-4110 1460
E: publishing@conceptpub.com; *W*: www.conceptpub.com

CULTURAL GEOGRAPHY
FORM AND PROCESS
(Essays in Honour of Prof. A.B. Mukherji)

About the Editors

Neelam Grover, a student of Professor A.B. Mukerji is currently a Professor in Geography, Panjab University, Chandigarh. She has had a brilliant academic record at Panjab University from where she received her M.A. and PhD. degrees. Her research in the field of Rural Settlement Geography within the perspective of Cultural Geography and as an element of cultural landscape provided a new and major thrust in the field of Rural Settlement Geography. The focus of her interest is in the cultural-geographical interpretation of rural settlements of different cultural groups in different ecologial settings, the processes of settling and the landscape results of those processes as integral to ethnic Rural Settlement Geography.

Professor Kashi Nath Singh, an eminent Cultural Geographer of India, retired head of the department, Department of Geography, Benaras Hindu University, Varanasi in 1993. Currently he is Professor, University of Addis Ababa, Ethiopia. Professor Singh received his M.A. and PhD. degrees from Benaras Hindu University and also did post-doctoral research at the Rutgers University, U.S.A. He has lectured at American, British, German and Japanese Universities. His research in the field of Historical Settlement Geography (*Historisch Siedlungsgeographie*) has been acclaimed by Cultural Geographers, scholars of Social and Regional History, and Cultural Anthropology.

CONTENTS

SECTION-III

CULTURE PROCESS AND CHANGE

SECTIO-IV
CULTURE TRAITS AND DISTRIBUTION

SECTION-V
CULTURAL ECOLOGY

LIST OF FIGURES

LIST OF TABLES

Prof. A.B. Mukerji

PROFILE OF A.B. MUKERJI

Professor A.B. Mukerji was born on November 7, 1929 at Jhansi (India) where his father S.C. Mukerji taught Geography as assistant master at the Government Intermediate College. After the premature death of his father, his family moved to Allahabad and settled there in the house of his grandparents.

Professor A.B. Mukerji had his early schooling in Allahabad, Gazipur, Barabanki and again in Allahabad. He completed his intermediate education from Anglo-Bengali Intermediate College, Allahabad. Through all the years in this college he was a student of Science (Physics, Chemistry and Mathematics) and also studied Geography upto High School. At home he learnt Bengali language and literature from his grandfather. It was at his insistence that he also had to study formally both Urdu and Sanskrit. He graduated from the University of Allahabad with Geography, Economics and English literature. He received his Master of Arts in Geography in 1951 from the University of Allahabad. As an undergraduate and post-graduate student he was taught Geography by Professor R.N. Dubey, who had received his D. Litt. at the University of Paris. It was Professor Dubey who first inspired the young Mukerji to become a geographer.

After obtaining his M.A. in Geography A.B. Mukerji taught for five years in a college in a *Mofussil* town. During the five years of his stay at Modinagar he carried out extensive field observation, collected *patwari* maps, visited villages of different castes including those of Muslims and gathered a large amount of information about all the aspects of their life and economy. As he moved from one village to the other, he gradually realised that the study of rural settlements could become his principal research vocation. He was aware that he was self taught and needed guidance for a better and deeper understanding of the rural settlements. He submitted his first paper on the Jat community in 1955 to Professor N.K. Bose, then editor of the *Geographical Review of India*. Professor Bose was not only very happy to publish it in this journal, but also wrote a long letter of encouragement to the novitiate.

Professor A.B. Mukerji started meeting Professor N.K. Bose whenever there was an opportunity and learnt his first lessons on cultural landscape from him. A.B. Mukerji remembers the profound and inspiring lessons that he received from him in informal meetings. He also remembers with deep gratitude the moral strength and appreciation that he received from Professor Bose, Professor S.P. Chatterjee and Professor P.K. Sarcar, all of whom became editor of the Calcutta based *Geographical Review of India*. Two of Mukerji's comprehensive papers on the Jat rural settlements and their house types were published by Professor V.L.S. Prakasa Rao in the Indian Geographical Journal, Madras. Self taught as he was, he began to evolve his own simple methodology for field work and for formulating broad generalisations. Even to this day he remembers how alone on a bicycle and an Alfa camera hanging from his shoulder and thick notebook he rode to specific villages and engaged the unwilling farmers to willing conversations and in their presence wrote some words and drew diagrams. He was fascinated by the local vernacular terms on the one hand, and the morphology of the village settlements on the other.

In July 1956 Mukerji was appointed a lecturer in Geography and Field Work Study at the Rural Institute, Udaipur, India. Between 1956-57 he carried out detailed field work on the settlement and land-use pattern of a Bhil village and some years later published a paper on this theme. His stay in Udaipur proved to be providential. He came to know of Professor Fred B. Kniffen and wrote to him about his field-work on the Jat settlements and enquired of the possibility of a scholarship at Louisiana State University, U.S.A. Professor Kniffen offered Mukerji a teaching assistantship for two years to work for a PhD. degree on a theme of Cultural Geography.

He went to Louisiana State University and joined the Department of Geography - Anthropology in September 1957. He took courses in Cultural Anthropology, Cultural Geography and, in addition, on those courses in Physical Geography which were relevant to his programme of study. He read courses with Professor F.B. Kniffen, R.C. West, and R.J. Russel. The lasting influence on his thoughts was exerted by Professors Kniffen and West. Professor Mukerji remains indebted to these two masters for the little success that he has had in his University teaching career and research.

After he completed his PhD. he returned to India and joined the Osmania University, Hyderabad in October 1960. Very soon he started going to the Telengana settlements and conducting intensive field work. Almost immediately after joining Osmania University he started publishing his research on the Telengana house types, village settlement morphology and field patterns. Mukerji considers his study at Osmania as a very fruitful period of field-work in an area completely unknown to him. His published research brought him acceptance and recognition by his own teachers at Louisiana State University and by David Sopher, easily the most prominent American specialist on the Cultural Geography of India.

Professor Mukerji joined the Department of Geography, Panjab University Chandigarh at the beginning of 1963 and retired there in 1994. For two years thereafter, beginning 1995 to end 1996, Professor Mukerji was Professor of Geography at the University of Poona. During the period he was at Chandigarh he published a large number of research papers in the journals of Europe and England and in books printed in U.S.A. and Germany. Professor Mukerji became interested in the Siwalik Hills adjoining Chandigarh and the Dun that spreads from Pinjore to Nalagarh. In this area he worked particularly on the Gujar settlements and their rural landscapes.

Throughout his long research and teaching career he has applied consistently the cultural-geographical methods and princples. In recent years he has become interested in the cultural geographical interpretation of late 19th and early 20th century historical records.

An interesting feature of Professor Mukerji's research is the sustained search for elements in the rural landscape which have had historical roots that can be discovered by observations and by quiet conversation. He never used a formal structured questionnaire while conducting field work. All his substantive research on rural landscape in based on encounters in the field and never pre-meditated. The final results of his analysis as they appeared in published papers were all validated by what existed in the field.

Professor Mukerji received many honours from the University Grants Commission during his teaching career. He was appointed UGC National Lectureship, UGC Emeritus Fellowship at Chandigarh, Jaipur, Pune, Shillong, Guwahati and Bhagalpur. He delivered special lectures on rural landscape at Varanasi, Allahabad, Aligarh, Ahmedabad, Baroda and Pune. He was appointed Distinguished Visiting Professor of Geography at the University of Missouri and delivered special lectures at Akron, Kent State, Southern Illinois, Whittington, Syracuse and Bowling Green Universities, U.S.A. He started lecturing at the Universities of Switzerland and Germany from 1977 onwards as Visiting Professor. He gave lectures at the Universities of Bern, Basel, Zurich and St. Gallen and E.T.H. in Switzerland and at Goettingen, Wuppertal, Friburg, Hiedelberg and Ruhr-Bochum in Germany and at the Open University in England.

He continues publishing research papers often on themes which have large canvas, but whenever there is an opportunity he goes to the field and engages himself in discussion on some selected themes and sometimes intriguing questions.

PUBLICATIONS OF PROFESSOR A.B. MUKERJI IN CULTURAL GEOGRAPHY AND RELATED THEMES

1953

1. 'Modinagar : A Study in Urban Geography', *Geographical Review of India*, Vol.15, No. 4, Dec., pp. 8-21.

1954

2. 'The Jats : A Study in Human Geography', *Geographical Review of India*, Vol. 16, No. 2, June, pp. 12-32.

1955

3. Kashmir : A Study in Political Geography — Eternal Relations', *Geographical Review of India*, Vol. 17, No. 1, March, pp. 19-33.

1956

4. 'Kashmir : A Study in Political Geography — Internal Patterns', *Geographical Review of India*, Vol. 19, No. 1, March, pp. 15-34.
5. 'Jat Settlements and Habitations : Spatial Patterns', *Geographical Review of India*, Vol. 31, Nos. 3 and 4, July-Dec., pp. 102-116.
6. 'Review of Geographical Factors in Indian History', by K.M. Panikkar, Bharatiya Vidya Bhawan, Bombay, 1955 in *The Indian Geographer*, Vol.1, No.1, August 1956, pp. 65-68.

1957

7. 'Jat Settlements and Habitations : Cultural Landscape', *Geographical Review of India*, Vol. 32. Nos. 1 and 2, Jan-June, pp. 13-28.
8. 'Nomeclature of the Jat Villages', *The Geographer*, Vols. 8 and 9, pp. 19-26.
9. 'Agricultural Geography of the Upper Ganga-Jamuna Doab', *The Indian Geographer*, Vols. 1 and 2, Nos. 2 and 1, August, pp. 122-164.
10. 'The Bi-Weekly Market of Modinagar', *The Indian Geographer*, Vol. 2, No. 2, pp. 271-294.
11. 'Review of Indian Village', by S.C. Dube, Routledge and Kegan Paul, London, 1955, in *The Indian Geographer*, Vol. 2, No. 2, December, pp. 331-332.
12. 'Review of the Kamars', by S.C. Dube, Lucknow, The Universal Publications Ltd., 1951, in *The Indian Geographer*, 1957, Vols.1 and 2, Nos. 2 and 1, pp.125 - 189.
13. 'Review of the Adivasis by the Publications Division', New Delhi, Ministry of Information and Broadcasting, 1955, in *Indian Geographer*, Vols. 1 and 2, Nos. 2 and 1, 1957, pp. 183-185.

1961

14. 'Sea Transport Among the Indians of South America', *Indian Geographical Journal*, Vol. 36, No. 1, Jan-March, pp. 28-45.
15. 'The Migration of the Jats : A Study in Historical Geography', *The Indian Geographer*, Vol. 6, No.1, August, pp. 41-54.
16. 'Field Patterns in the Upper Doab Jat Villages', *Journal of Geography*, Vol. 3, No. 1, pp. 1-8.
17. 'Review of Studies in the Geography of Ancient and Medieval India', by D.C. Sircar, Motilal Banarsidas, Delhi, 1960, in *The Indian Geographer*, August, 1961, pp. 119-120.
18. 'Review of Nepal : A Physical and Cultural Geography' by P.P. Karan, University of Kentucky Press, Lexington, 1960, in *The Indian Geographer*, December 1961, pp. 224-227.
19. 'Review of the Individuality of Portugal', by Dan Stanislawski, University of Texas Press, Austin, 1959, in *The Indian Geographer*, December, 1961, pp. 227-228.
20. 'Review of Peasant Life in India', (ed.) N.K. Bose, Anthropological Survey of India, Calcutta, 1961 in *Geographical Review of India*, March 1961, pp. 60-61.

1962

21. 'Bhawanti House Type in Andhra Pradesh', *Geographical Outlook*, Vol. 3, No.1, pp. 16-24.
22. 'Field Patterns in a Telengana Village', *Indian Geographical Journal*, Vol. 37, No.1, Oct.-Dec., pp. 153-163.
23. 'Agricultural Regions and Geographic Planning for Indian Agriculture', *National Geographer*, Vol. 5, pp. 40-48.
24. 'The Umland of Modinagar', *National Geographical Journal of India*, Vol. 8, Nos. 3 and 4, pp. 250-269.
25. 'Jat House Type', *Geografia*, Vol.1, Winter, pp. 27-37.
26. 'Review of Geographical Essays in Memory of Alan G. Ogilvie', (ed.) R. Miller and J.W. Watson, Thomas Nelson and Sons, Edinburg, 1959 in *Geographical Review of India*, March, 1962, pp. 61-62.
27. 'Review of Pre-history and Protohistory of Eastern India', by A.H. Dani, K.L. Mukhopadhyaya, Calcutta, 1960 in *Geographical Review of India*, Vol. 24, Nos. 3 & 4, 1962, pp. 61-63.

1963

28. 'Land Use Pattern in a Mewar Village, Rajasthan', *Geographical Review of India*, Vol. 25, No. 1, March, pp.11-34.
29. 'Cultural Geography of Wells in Telengana Reddi Villages', *The Indian Geographer*, Vol. 8, Dec. pp. 77-84.

30. 'Review of Eat Not This Flesh', by Frederick J. Simoons, Wisconsin University Press, Madison, 1961, in *The Indian Geographer*, December, 1963, pp. 128-130.
31. 'Review of the Geography of the Bible : A Study in Historical Geography', by Danis Baly, Harper and Brothers, New York, 1957 in *Geographical Review of India*, Vol. 25, No. 3, June, 1963, pp. 146-148.
32. 'Review of Land and People of the Andamans', by P.K. Sen, Postgraduate Book Mart, Calcutta, 1962, in *The Indian Geographer*, December, 1963, pp. 131-141.
33. 'Review of India's Urban Future', by Roy Turner, Oxford University Press, Bombay, 1962, in *Geografia*, Summer, 1963, pp. 60-65.
34. 'Review of Diffusion and Adoption of Agricultural Practices', by A. Rahim, Pakistan Academy of Village Development, Commila, 1961, in *Geografia*, Winter, 1963, pp. 137-139.
35. 'Review of Economic Development and Social Change in South India', by T.S. Epstein, Oxford University Press, London, 1962, in *Deccan Geographer*, Vol. 2, No.1, July, 1963, pp. 77-81.
36. 'Review of The Greater Bombay : A Study in Suburban Ecology', by C. Rajagopalan, Allied Publishers, Bombay, 1962, in *Geografia*, Summer, 1963, pp. 58-60.

1964

37. 'Cultural Landscape of a Telengana Village : A Morphogenetic Study', *Geografia*, Vol. 3, No.1, Summer, pp. 11-42.
38. 'Succession of Cultural Lanscape in Telegana Reddi Villages', *Indian Geographical Journal*, Vol. 39, Nos. 1 and 2, Jan-June, pp. 42-58.
39. 'Review of a Hundred Years of Geography', by T.W. Freeman, Gerald Duckworth, London, 1961, in *Deccan Geographer*, January, 1964, pp. 178-183.

1965

40. 'Review of Culture Worlds', by R.J. Russell and F.B. Kniffen, Mc Millan & Co., New York, 1960, in *Deccan Geographer*, January 1965, pp. 76-78.

1966

41. 'Review of Tarai Region of Uttar Pradesh', by L.R. Singh, Ram Narain Lal Beni Prasad, Allahabad, 1965, in *Deccan Geographer*, July, 1966, pp. 82-83.
42. 'Review of Neolithic Cattle-Keepers of South India', by F.R. Allchin, Cambridge University Press, Cambridge, 1963, in *Deccan Geographer*, July, 1966, p. 83.

1967

43. 'Muslim House Type in Meerut Rural Area', *Folklore*, Vol. 8, No. 4, April, pp. 116-130.
44. 'The Origins of the Jats : An Ethnogeographic Interpretation', *Geographical Review of India*, Vol.29, No. 2, June, pp. 75-90.
45. 'Review of Diffusion of the Decision to Irrigate', by L.W. Bowden, Chicago University Press, Chicago, 1965, in *Deccan Geographer*, Jan.-Dec., 1967, pp. 118-120.
46. 'Review of Villages, Towns, and Secular Buildings in Ancient India', by Amita Roy, K.L. Mukhopadhyaya, Calcutta, 1964, in *Deccan Geographer*, Jan.-Dec. 1967, pp. 112-114.

1968

47. 'The Jats of the Upper Ganga-Yamuna Doab, Some Ethnogeographic Considerations', *Human Events*, Vol. 2, No. 3, March, pp.113-141.
48. 'The Jats of the Upper Ganga-Yamuna Doab, Origin, Diffusion, Distribution', *Deccan Geographer*, Vol.6, No. 1, January, pp. 22-51.
49. 'Spatial Pattern of Literacy in Andhra Pradesh, India', *Selected Papers, 21st International Geographical Congress*, Vol. 3, pp. 181-186.

1969

50. 'Rural Settlements in the Jatakas', *Geographical Review of India*, Vol. 31, No. 2, June, pp. 17-32.
51. 'Spacing of Rural Settlements in Andhra Pradesh : A Spatial Interpretation', *Geographical Outlook*, Vol. 6, pp. 1-19.
52. 'Climatological Lore in the Yajur Veda,' *Indian Geographical Journal*, Vol. 54, Nos. 1 and 2, Jan.-June, pp. 1-8.
53. 'Review of Environmental Perception and Behaviour', (ed.) David Lowenthal, Chicago University Press, Chicago, 1967, in *Deccan Geographer* in Jan.-June, 1969, pp. 72-74.
54. 'Review of The Changing Cultural Landscape of Nadia', by B. Banerjee and N.M. Bowers, India Publications, Calcutta, 1965, in *The Indian Journal of Geography*, Vol. 4-6, 1969-1971, pp. 78-79.
55. 'Review of Middle America : Its Land and Peoples', by R.C. West and J.F. Angelli, Prentice-Hall, Englewood, 1966, in *Indian Journal of Geography*, Vol. 4-6, 1969-1971, pp. 75-77.

1970

56. 'Changes in Nutritional Density of Population in Uttar Pradesh, 1951-1961 : A Spatial Interpretation', *Indian Journal of Geography*,' Vols. 4, 5 and 6, No.1, pp. 12-20.
57. 'Spacing of Rural Settlements in Rajasthan : A Spatial Analysis', *Geographical Outlook*, Vol. 1, No. 1, April, pp. 1-19.

58. 'The Religious Composition of India's Population : A Spatial Analysis', *Tijdschrift voor Economische en Sociale Geografie*, Vol. 6, No. 2, March-April, pp. 31-100 (with G.S. Gosal).

1971

59. *Scheduled Castes of Andhra Pradesh*, Calcutta: Indian Publication, 1971
60. 'Female Participation in Agricultural Labour in Uttar Pradesh : Spatial Variations', *National Geographer*, Vol. 6, pp. 13-19.
61. 'Urbanisation in the Districts Surrounding the Union Territory of Delhi : 1951-1961', *Manpower Journal*,' Vol. 7, No.3, Oct.-Dec., pp. 64-78 (with R.C. Chandna and P.K. Bali.)
62. 'Regional Contrasts in Distribution, Density, and Relative Strength of the Scheduled Caste Population in Andhra Pradesh', *Indian Geographical Journal*, Vol. 46, Nos. 2-4, July-Dec,. pp. 23-49.
63. 'Geographic Patterns of Changes in Agricultural Wastelands in Uttar Pradesh : 1951-1971', *Geographical Review of India*, Vol. 33, No. 2, June, pp. 120-132.

1972

64. 'Rural Settlements in the Siwalik Hills Tract of Punjab (India) : An Analysis of Empirical Attributes', *National Geographical Journal of India*, Vol. 18, No. 2, June, pp. 57-63.
65. 'Distributive and Relative Concentration of the Scheduled Caste Population in India', *Census Centenary Monograph No.7*, *Census of India*, pp. 473-487.

1973

66. 'The Muslim Population of Uttar Pradesh, India, A Spatial Interpretation', *Islamic Culture*, Vol. 47, No. 3, July, pp. 213-230.

1974

67. 'Levels of Urbanisation in Uttar Pradesh, 1961', *Geographical Review of India*, Vol. 35. No. 1. March, pp. 31-43.
68. 'Female Participation in Rural Agricultural Labour in Andhra Pradesh : A Study in Population Geography', *Manpower Journal*, Vol. 9, No.1, April-June, pp. 64-106.
69. 'Small Town of Uttar Pradesh : An Exploratory Analysis', *Manpower Journal*, Vol. 9, No. 3, Oct.-Dec., pp. 82-98.
70. 'Road Transportation Network and Levels of Urbanisation in Rajasthan', *Transport Communication, No. 258*, May, pp. 14-26.
71. 'Spacing of Rural Settlements in Upper Ganga-Yamuna Doab', *Geographical Review of India*, Vol. 36, No. 2, June, pp. 155-165.
72. 'Morphogenetic Analysis of Rural Settlements of the Chandigarh

Siwalik Hills', *Panjab University Research Bulletin, Arts*, Vol. 5, No. 2, October, pp. 59-93.

73. 'Urbanisation in Awadh Region', *The Awadh Geographer*, Vol.1, pp. 13-29.

1975

74. 'Female Participation in Rural Agricultural Labour in India : Patterns and Processes', *Tijdschrift voor Economische en Sociale Geografie*, Vol. 66, No. 2, pp. 103-108.

1976

75. 'Chandigarh Siwalik Hills (India) : Some Aspects of Regional Rural Development', *S.M. Ali Memorial Volume on Theoretical and Applied Geography*, Sagar University of Sagar, pp. 302-315.
76. 'Rural Settlements of the Chandigarh Siwalik Hills : A Morphogenetic Analysis', *Geografiska Annaler*, Vol.58, Ser. B, No.2, pp. 95-115.
77. 'Patterns of Rural Settlements of the Siwalik Hills : An Application of Multiple Collinear Diffusion Model', *Geographical Dimensions of Rural Settlements*, Varanasi, pp. 54-67.
78. 'Review of Geography of Transport and Market Centres : A Case Study of Hadaoti Plateau', by H.M. Saxena, Chand and Company, New Delhi, 1975, in *Deccan Geographer*, Vol. 14, No. 2, pp. 177-178.

1977

79. *Physical Geography*, New Delhi: N.C.E.R.T., 1977.
80. 'Geographical Perspective on Rural Settlements : Recent Indian Efforts', *The Geographical Viewpoint*, Vol. 8, Nos. 1 and 2, April-Oct., pp. 1-19.

1979

81. 'Tribe', *A Survey of Research in Geography, 1969-1972*, (ed.) Moonis Raza, New Delhi: Allied Publishers Private Limited, pp. 97-108.
82. 'Religion', *A Survey of Research in Geography, 1969-1972*, (ed.) Moonis Raza, New Delhi: Allied Publishers Private Limited, pp. 108-109.
83. 'Social Geography : Some Neglected Aspects', *A Survey of Research in Geography, 1969-1972*, (ed.) Moonis Raza, New Delhi: Allied Publishers Private Limited, pp. 121-126.
84. 'Geography of Rural Settlements', *A Survey of Research in Geography 1969-1972*, (ed.) Moonis Raza, New Delhi: Allied Publishers Private Limited, pp. 177-190.
85. 'The Early Man's Habitat in the Siwalik Hills : Some Statements on the Geomorphological Setting', *Early Man in Northwestern India*, Delhi: Allied Publishers, pp. 98-107.

1980

86. *The Chamars of Uttar Pradesh—A Social Geography*, New Delhi: Inter-IndiaPublications, 1980.
87. 'Disease Ecology of a Small cul-de-sac : Chandigarh Dun', *Social Science and Medicine, British Journal of Medical Geography*, Vol. 14-D, No. 3, Sept., pp. 331-337.

1981

88. 'Review of Agrarian Egalitarianism', (eds.) M. Rahman and D. Bagchi, Kendall Hunt Publishing Company, Dubuque, 1981, in *Geojournal*, Issue 2, 1981, pp. 112-113.

1982

89. 'Cultural Ecology of Rural Settlements in the Chandigarh Dun Region', in *India, Cultural Patterns and Processes* ,(eds.), Allen Noble and Ashok Dutt, Boulder : Westview Press, pp. 337-355.

1983

90. 'Dispersal and Resettling of Refugees in Independent India : A Cultural-Ecological Model', *Indian Geographical Journal*, Vol. 58, No. 1, June, pp. 1-29.
91. 'Economic Degradation and Population Migration : Himachal Pradesh', *National Geographer*, Vol.18, No.2, pp. 153-165.

1984

92. 'Reflections on the Teaching of Environment in Geography', *Hill Geographer*, Vol. 3, No. 2, pp. 49-54.
93. 'Progress in Rural Settlement Geography', *A Survey of Research in Geography, 1972-75*, (ed.) S. Manzoor Alam, New Delhi: Concept Publishing Company, pp. 119-129.

1985

94. 'A Cultural-Ecological Appraisal of Refugee Resettlement in Independent India', *Population Redistribution and Development in South Asia,* (ed.) L.A. Kosinski, Boston: D. Reidel Publishing Company, pp. 89-111.
95. 'Foreword' to *Urban Development in the Metropolitan Shadow* by K. D. Sharma, New Delhi, Inter-India Publications, 1985, pp. 5-6.
96. 'Foreword' to *Rural Settlements : A Cultural - Geographical Analysis* by Dr. Neelam Grover, New Delhi, Inter-India Publications, 1985, pp. 9-12.

1986

97. 'Landscape and Man : Development and Destruction in Western

Himalaya', *Spectrum of Modern Geography*, (eds.) M. Shafi and M. Raza, Delhi: Concept Publishing, pp. 397-429.

98. 'The Region of the Seven Sisters : A Distant View', *Indian Geographical Journal*, Vol. 61, No. 1, June, pp. 1-12.
99. 'Foregleam', *Geography and Environment*, Delhi: Concept Publishing, pp. 339-346.
100. 'Review of Evolution and Spatial Organisation of Clan Settlements', by S.H. Ansari, Concept Publications, New Delhi, 1986, in *Maharshi Dayanad University Research Journal (Arts)*, Vol.1, No.1, April, 1986, pp. 117-118.
101. 'Review of Environmental Perception : The Human Factors', in *Urban Planning*, by Anjana Desai, Ashish Publishing House, New Delhi, 1985, pp. 71-72.
102. 'Review of Central Places in Western India', by Jaymala Diddee, University of Poona, Pune 1984, in *Transactions, Institute of Indian Geographers*, January, 1986, pp. 72-74.
103. 'Foreword' to *Periphery and Flux : Changing Chandigarh Villages,* by Dr. Ulrich Bichsel, Bern, Geographica Bernensia, 1986, pp. 5-14.
104. 'Foreword' to *Rural Settlements : A Cultural - Ecological Perspective,* by Dr. S. Sharma, New Delhi: Inter-India Publications, 1986, pp. 7-11.
105. 'Foreword' to *The Gujjar Settlements,* by D.S. Manku, New Delhi: Inter-India Publications, 1986, pp. 7-12.

1988

106. 'Periodic Markets of India : Characteristics and Developmental Implications', *Commercial Activities and Rural Development in South Asia*, Delhi: Concept Publishing, pp. 13-39.

1989

107. 'On the Relevance of Sankalia to the Cultural Geography of India' in *Panjab University, Research Bulletin (Arts)*, Vol. 20, No. 1, pp. 141-166.

1991

108. 'Review of Averting the Apocalypse : Social Movements in India Today', by Arthur Banner, Duke University Press, Durham, 1990 in *Panjab University Research Bulletin - Arts*, Vol. 22, No. 2, October, 1991, pp. 315-316.

1994

109. 'The Greening of the Siwalik Himalayas', *Himalayan Environment: Pressure, Problems, Processes*, (eds.) B. Mesurli, T. Hofer, and S. Wymann, Bern: Geographica Bernensia, 1994, pp. 63-77.

1996

110. 'Traditional Domestic Fuel in Rural Himachal Himalaya: A Cultural Ecological Understanding', *Umwelt Mensch Gebirge*, (eds.) Martin Grosjean, Hans Hurni, Hans Kienholz, Heinz Wanner, Urs Wiesmann, Bern: Geographisches Gesselschaft, pp. 49-58.

2000

111. 'Territorial Expressions of the Dominant Castes in Early 20th Century Haryana', *Geographic Planning Research Themes for the New Millenium*, (eds.) Baleshwar Thakur, Allen G. Noble, A.B. Mukerji, and Frank J. Costa, Delhi: Vikas Publishing House Private Limited, pp. 381-397.

2002

112. 'Geographic Personality: Cultural Geographical Interpretation' in *Allen Noble's Festschrift*, (eds.) Ashok K. Dutt *etal.* (In press)
113. 'Enlarging the Understanding of Geographic Personality : Evidences From Creative Literature and Painting', in *Allen K. Noble's Festschrift*, (eds.) Ashok K. Dutt *et al.* (In press)

INTRODUCTION

Cultural Geography has its ancestory in German Scholarship particularly in Ratzel's Anthropogeographie and the works of Schluter, Richthofen, Meitzen and Gradman, to name a few. In Germany it attained a degree of specialisation where prominent universities instituted and developed separate Institute of *Kulturgeographie* or *abteilungs* (departments) of *Kulturgeographie* in the Institutes of Geography. It was in Germany that studies in *Kulturlandschaft* were pursued with vigour and commitment and continue to be so.

What was formalized as Cultural Geography in Europe was developed in the English speaking world by Carl O. Sauer, and became a major thrust of geographic study in the United States especially at Berkeley. Breaking away from the environmentalist wranglings of the early twentieth century, Sauer saw on the landscape an evident rapport between man and nature concretised as creations on the landscape which reflected not only mans' perception of his micro-cosm but also an inherited value system and tradition intercepted by his technological know-how. Focusing on simpler, folk cultures, his methodology stemmed from the problem in the field itself, and observation alongwith explanation sought not only in the physical settlement but in the inhabiting groups' perception of the space in nature it transformed, remained fundamental. Carl O. Sauer was indeed a stalwart of Cultural Geography.

Interestingly however, Cultural Geography has never been formally defined in terms universally acceptable, with each country redefining it for their own situation. Practitioners of Cultural Geography have had the tendency to avoid defining it in strict terms although there is a general consensus that it deals with the expression of culture in areas in a large variety of ways both material and non-material—a settlement, religious structure, periodic market or even language, traditions and value system. In the words of Mikesell, Cultural Geography 'was more a shared curiosity and set of preferences than an explicit programme or doctrine'.[1]

Cultural Geography as practised by cultural geographers has at all times been identified with certain distinctive characteristics. Deeply rooted in Cultural Anthropology, Cultural Geography essentially focuses on the rural context and rural content. Simpler societies have been the prime original interest and it is only recently that Cultural Geography has branched on to study cultural symbols and meanings in urban areas. The abiding theme, however, has always been and is human occupance, the tenancy of the earth. Man in the process of adapting himself changes the earth in a way that the results of the changes introduced by man reflect or conform to his own value system and needs. In other words, man acts in and upon nature and not only in response to the dictates of nature. The results of these changes as expressions of culture on the landscape (rather than the invisible features) have been the prime focus of concentration in Cultural Geography since its inception.

Cultural Geography, has however, for a while stood at cross-roads as many of its characteristics have been and are undergoing change slowly as a new brand of Cultural Geography, different from the traditional Cultural Geography evolves. The new concerns hinge on urban structures and urban societies and their value systems as problems. But what emerges supreme in the earlier and new variety of Cultural Geography is the continuity of the concept of culture as originally enunciated by Edward Tyler when defining culture as, 'that complex whole which includes knowledge, belief, art, morals, law, customs and any other capabilities and habits acquired by man as a member of society'.[2]

Comprehension and understanding of the diversity in both material and non-material features of the earth remains at the core of cultural geographical enquiry as does the historical genetic approach so emphatically followed at Berkeley.

Man's impress on the landscape from perhaps a clearing made to meet the requirements of his *genre de vie*, or a stiucture created to fulfil his personal, secular, economic or religious needs, a pathway perhaps providing the fundamental link to create an interaction leading to the formation and evolution of a kinship, clan or culture group—all deserve the avid attention of cultural geographers. As such there is no limit to themes. While each such impress is anchored in space, the temporal factor cannot be ignored either. When it was created, how and why, are questions that link any attempted interpretation of the present day impressions of man with the past and with the future as also with the areal/spatial context in which they occur. As Mikesell states, 'Any sign of human action in a landscape implies a culture, recalls a history and demands an ecological interpretation; the history of any people evokes its setting in a landscape, its ecological problems and its cultural concomitants; and the recognition of a culture calls for the discovery of traces it has left on the earth'.[3]

The scope of Cultural Geography, therefore, is unlimited. However, its essence remains the conscious application of the concept of culture, its description, analysis and understanding. An analysis of the themes of research in Cultural Geography suggests that there have been clusters related to culture, culture history, culture ecology, culture area and cultural landscape. Cultural

Geographical research has been conducted on one or the other theme of these enmeshed clusters. After all a cultural landscape is located within a culture area, its creation the result of the interaction of culture with the ecology and its subsequent transformation through culture history. These conform to the four different dimensions of Cultural Geography.

In India, Cultural Geography is a relatively recent field of study. Only a handful of the graduate departments of Geography in the Indian Universities offer one or the other courses in Cultural Geography. Few teachers of the graduate programme in Geography and fewer students are familiar with Cultural Geography and what it stands for. The cultural landscape of India reflects the value system, the technological baggage and the nexus between culture and economy. India represents also cultural integration as integration of innumerable material and non-material traits can be observed. Therefore, it would require no strong and comprehensive arguments to suggest that for a country as ancient as India, with its complex historical fabric and immense diversities of culture groups and cultural traditions, it is not merely desirable but also *sine qua non* to develop an understanding of the country's rich cultural geography to develop a more meaningful understanding of India as a civilisation.

Exploring cultural variations in the Indian space has attracted little attention of Indian geographers, while most of the contributions of lasting value have come from American geographers, two among them—David Sopher and Fredrick Simoons being students of Carl O. Sauer who, as mentioned earlier, was unquestionably the greatest practitioner of American Cultural Geography and indeed its founding father. A whole large group of American and Indian students all trained in the U.S.A. emerged thereafter engaged in the task of investigating India's Cultural Geography. Deryck Lodrick, Robert Stoddard, William A. Noble, Allen G. Noble, Charles J. Bennett, Michael J. Libbee, Thomas J. Wilbanks, James Duncan, Nancy Duncan, Brian Murton, Richard Palmiere, Robert Haffpauir, Surendra M. Bhardwaj, Ramesh C. Dhussa, Bharat Bhatt, Surendra Aggarwal, Satish Devgan, Kiran Banga, H.N. Sharma, M. Hasan, Ashok Dutt, P.P. Karan and A.B. Mukerji, to name some.

From among them it was A.B. Mukerji who pioneered the systematic study of Cultural Geography in India. It is not so that others before him were never interested in such studies, but research in the field was limited. Professor Nirmal Kumar Bose for one, was deeply into teaching landscape studies and proved an extremely potent force of influence in the development of A.B. Mukerji's interest in landscape observation and analysis, an interest which bloomed and was groomed as he pursued his studies in the United States under Fred B. Kniffen, R.C. West, Richard Russel and Carl O. Sauer, the stalwarts of Cultural Geogrpahy.

When A.B. Mukerji returned to India to take up the job of Assistant Professor in Geography at Osmania University, Hyderabad, neither his colleagues nor the geographers of India cared much about Cultural Geography. Many considered it as an invention of American geographers and some thought of it as another name for the nascent Social Geography. Not to be discouraged or frustrated by a general atmosphere of apathy, Mukerji with single minded commitment devoted all his time and energy in conducting field work in the Telengana cultural landscape and succeeded in publishing papers on house types, field patterns, village landscape and evolution of cultural landscape, all of which brought him recognition and appreciation from his American and European teachers and professional friends. He soon moved to Panjab University at Chandigarh and continued to pursue his interests there.

Professor Mukerji thus introduced Cultural Geography to Indian Geography. His interest in archival material, Cultural Anthropology, in which he was formally trained, in problems and questions rooted on Indian soil, Indian ethos and Indian tradition, leading to conclusions Indian in essence and a deep commitment to field work, led him to evolve themes which came to him while he observed the landscape. He has been very interested in the historical genesis of landscape and landscape elements, their origin, change, diffusion and culture groups whichever area may have been the focus of his interest, Telengana, western Uttar Pradesh or the Chandigarh Siwalik Hills and Dun. However, his discussion has never been a theoretical or statistical model but historical. Mukerji never mentally separated the essential similarity of the methodology of Geomorphology and Cultural Geography because both involve processes and historical evolutionary analysis. In fact, he researched historical evolution and then discovered correspondence of areal landscape.

Authentication of any printed data for him was essential. For his studies basic methodology was field work, keen observation of and encounter with the landscape accompanied by sketches in the mind and explanations sought in the oral tradition and specificities of culture which he always emphasised, and archival material of which he was deeply conscious. Collection of data for him was like collecting building material, but with no pre-meditated design or method of construction. They according to him came from speculation, meditation.

A.B. Mukerji has practised in India a brand of Cultural Geography which is home grown, home bred and home spun. His contributions to the indigenous development of Rural Settlement Geography in the context of Cultural Geography significantly include the social functional divisions of the ground plan, especially into the broad divisions of social space or inter-family space and private or intra-family space, the application of principles of segregation, dominant clan and lineage in the understanding of the micro-morphology of a settlement, and application of the concept of succession of cultural landscapes in India. Not only did he introduce the idea of Cultural Geography as a systematic sub-discipline, he also applied Cultural Geography to Rural Settlements and initiated studies of the Cultural Geography of Rural Settlements in India.

As a teacher he communicated his learning through his teaching becoming the inspiration to delve further into problems for all his students. His strength and charisma lay in trasporting his students from the drab classroom to the vistas of the Himalaya, Siwalik Hills, Dun, Upper Ganga-Yamuna Doab, Mewar to Telenagana, which came alive as the repositories of endless research themes and areas identified with cultural groups who in turn lent identity to their habitat as they created their settlements interacting with the environs through the medium of culture. Thus, each aspect of the rural settlements especially the house and the fields come alive with cultural meaning to be traced and understood in the context of history, Indian ideology, religion, value system, tradition and within the triology of culture, history and ecology.

Notes and References

1. Wagner, Philip L. and Mikesell, Marvin W. (Eds.) (1962). *Readings in Cultural Geography*, Chicago: University of Chicago Press, p. 1
2. Friedl, John and Pfeiffer John (1977). *Anthropology : The Study of People*, New York: Harper and Row, p. 284.
3. Mikesell, Marvin W. (1978). 'Tradition and Innovation in Cultural Geography', *Annals of the Association of American Geographers*, Vol. 68, No. 1, pp. 1-16, ref. p. 4.

SECTION-I

METHODOLOGICAL STATEMENTS

1

THE CULTURAL GEOGRAPHER'S INTEREST IN REGIONS

ROBERT H. STODDARD

Introduction

Regions are important to cultural geographers. This is because regions help in understanding the spatial distributions of social phenomena. Like any classification scheme, a set of regions provides organisation to a large body of data. Although occasionally phenomena being studied may already be grouped into regions, this is not usually the case. Consequently, cultural geographers, whether engaged in research or teaching are frequently faced with the task of defining and delineating a set of meaningful regions. The purpose of this paper, therefore, is to discuss some of the methodological issues that confront geographers when they organize their data into regions.

The Task of Creating Regions

The fundamental objective in regionalisation is to match the patterns displayed by a map of regions with the actual areal variations as they occur in the real world. To accomplish this objective, the geographer must select meaningful criteria, operationalise them, collect the appropriate data, display the data spatially, and convert the mapped patterns into regions.

The initial task — the selection of meaningful criteria — is highly dependent on the topic being studied, that is, it is primarily just a matter of definitions. It only requires the geographer to select features, population characteristics, or other criteria that truly represent the phenomenon being regionalized.

The next step, however, is critical because it directly affects the validity of the work. Validity is achieved only to the degree that the conceptual definitions of the regionalising criteria coincide their operationalization. For example, a scholar who wishes to regionalise dwelling types must commence with an operational definition of a dwelling. Even though the concept of 'dwelling' is well understood, the operational task of distinguishing between what are truly dwellings from those structures that are not to be regarded as dwellings is difficult to accomplish objectively.

Despite the fact that the next tasks, that of collecting data and mapping it for each phenomenon, are basic parts of most geographic endeavours, they are not always achieved easily, especially when working with cultural topics. Attitudes, belief systems, behavioural patterns, and the many other attributes that constitute culture are not measured nor observed without considerable difficulty. Nevertheless, these steps are not discussed further here because the focus of this paper is on the last task: the conversion of mapped data into meaningful regions.

What are the major issues facing a cultural geographer who attempts to convert areal data into meaningful regions? The issues discussed here concern scale, precision and contiguity.

The issue of scale, of course, is a component of all geographic work because distributions and spatial relationships are heavily influenced by the size of the minimum areal unit and the size of the study area. Whether the minimum areal unit for the collection of data is a hectare, a city block, a district, or a state, influences the degree to which phenomena are identified with a particular location. Furthermore, the size of the minimum areal unit in comparison to the size of the total study area partially affects the complexity of the distributional patterns.

The issue of precision affects the formation of regions in the same way it affects other classification systems. In general, the greater the number of classes, the more precise is the differentiation of the elements being categorized and the higher the level of homogeneity among the elements in each class. For example, an area divided into seven regions will reflect the spatial variations more precisely than if the area were partitioned into only two regions.

Another aspect of precision, that of areal generalisation, arises when regionalising phenomena, but is not a problem in classification. This is because the characteristics of a small area may be dissimilar from the surrounding area, but the size of the dissimilar place is too tiny to delineate as a separate region. For instance, a scholar regionalizing a country into land use regions may decide to classify a fairly large area as 'an agricultural region', even though it is known that a town exists within the boundaries of this delineated region.

The issue of contiguity plagues all who attempt to classify two-dimensional data. To group linear data only requires finding meaningful breaks along a continuum of values, which means the classification of non-spatial data is much easier than regionalising an area. In contrast, the task of grouping areal data (*i.e.*, the task of regionalising) involves finding meaningful breaks in two dimensions and then reconciling the results with the goals of areal generalisation. The reconciliation, of course, depends on the purpose for the regionalization. If the goal is to emphasize the spatial variations of phenomena irrespective of how they are arranged, then the need for contiguity is diminished. For example, when regionalising climatic differences in the world, the wide separation of the regions with Mediterranean type climate presents no problem in interpretation or in utilisation. However, to form a state with non-contiguous parts would be regarded as a severe handicap to the viability of that political unit. Therefore, delineating the boundaries of this kind of region almost requires that all parts of the state (region) are connected.

An Illustration in Sri Lanka

The way these issues confront the cultural geographer can be illustrated by examining a current political controversy in Sri Lanka. That is, what is sometimes called 'the ethnic conflict' can serve as a suitable example for the kind of academic problems that continually face cultural geographers as they attempt to regionalise other phenomena in other settings.

The conflict in Sri Lanka results from two opposing objectives. One objective, held by the Tamils who identity themselves as culturally distinct, is to gain greater political control over 'their' territory. The second objective, the position held by the government, is to oppose any challenge to the integrity of the state. In spatial terms, the controversy concerns whether or not Tamils should achieve greater autonomy (even political independence) over a specified administrative area. The aspect of the controversy that is examined here, however, concerns only the delineation of territory that might form a more autonomous region. Therefore, the background for the conflict, the justification for the different positions, the sequence of developing events, and many other aspects of the controversy are omitted here. Furthermore, it must be made clear to the reader that the discussion about potential regional divisions should not be interpreted as either for or against such a division of land. Instead, the focus is on the kind of real-world decisions confronting a geographer who attempts to create meaningful regions. The Sri Lankan setting is merely used to demonstrate the spatial components of those decisions.

The first three tasks outlined above, the selection of meaningful criteria, their operationalisation, and the collection of appropriate data, are accomplished here by using government statistics. The collection of census data is by ethnic population for various levels of administrative sub-divisions. According to the 1981 census, the population of Sri Lanka consisted of the following ethnic groups: Sinhalese (74.0%), Sri Lankan Tamil (12.6%), Indian Tamil (5.6%), Moor (7.1%), and Others (0.7%).[1] With these data, then, the task is to explore issues associated with regionalising the distribution of the Tamil population in Sri Lanka.

Even though census definitions are accepted here, the task of selecting meaningful criteria is complicated slightly by the fact that the 'Tamil population' might be regarded as either just the Sri Lankan Tamils or both the Sri Lankan Tamils and Indian Tamils. Most scholars familiar with the controversy would probably regard just the Sri Lankan Tamils as the appropriate population. This is partly because the Indian Tamils have been less politically active, their citizenship status is problematic, and caste differences have tended to separate the two Tamil groups. Therefore, for most of the following discussion, the population that forms the basis of regionalisation attempts consists of only the Sri Lankan Tamils. However, for illustrative purposes, the Indian Tamils will sometimes be combined to see the effects of a different operational definition.

The definition and operationalization of the criteria being used for regionalization in this political example is also a function of what constitutes 'Tamil territory'. In a political system that recognises the wishes of the populace, is the pertinent level of support the majority or only the plurality? In most administrative sub-divisions of Sri Lanka, one ethnic group has a large majority, so this potential controversy occurs only infrequently. Nevertheless, in the very areas where regionalisation is the most difficult to solve, a majority does not always exist. The political answer to this question remains so complex that it is avoided here and both the patterns of majority and plurality population are examined.

The issues of one, the size of the areal units used for identifying and regionalising the population, and two, the persistent geographic problem of contiguity are evidenced by various scales. Both issues, along with the different definitions of the Tamil population and two levels of ethnic dominance (*i.e.*, the majority and plurality levels), are demonstrated by the maps that follow.

The first attempt to explore whether the spatial distribution of the Tamil population is arranged so that it can be meaningfully regionalised is at the scale of provinces. As can be seen by the table of listing the percentage of ethnic populations in each province (Table 1.1), seven of the nine provinces have a Sinhalese majority, one has a Tamil majority, and one, the Eastern Province, has no ethnic majority. In terms of a plurality measure, the Eastern Province is characterised by 40 per cent being Sri Lankan Tamils.

Locationally these two non-Sinhalese provinces are contiguous (Fig.1.1). Therefore, if a Tamil region were to be defined by only a plurality and if areal units were grouped at this scale, the Northern and Eastern Provinces do, indeed, form a contiguous Tamil region. However, if regions were to be defined by requiring an over-fifty-percent majority, then the Tamil region would be restricted to just the Northern Province. Incidentally, even if both Tamil populations (*i.e.*, those defined by the census as Sri Lankan Tamils and those as

Table 1.1 : Ethnic Groups by Provinces, 1981

(Data in Percentages)

Province	*Sinhalese*	*Sri Lankan Tamil*	*Indian Tamil*	*Moor*
Western	84.9**	5.6	1.6	6.2
Central	65.7**	7.3	18.7	7.6
Southern	95.1**	0.6	1.3	2.5
Northern	3.0	86.4**	5.7	4.7
Eastern	25.7	40.3*	1.2	32.2
North Western	90.0**	2.7	0.6	6.4
North Central	91.7**	1.4	0.1	6.5
Uva	75.9**	4.5	15.7	3.5
Sabaragamuva	85.4**	2.2	8.9	3.3

Note : Percentages for 'Others' are not included here.
* Plurality.
** Majority.

Indian Tamils) were to be combined into one definitional category for the purpose of regionalisation, there would be no change in the resulting regions.

The segregation of ethnic groups is also evident at the district level (Fig. 1.2). For example, in over half of the 24 districts, the percentage of the majority group exceeds 80, (Table 1.2), and in only three districts does the percentage drop below 50. Furthermore, the spatial distribution of districts having a majority of a particular ethnic group is such that the cores of regions are easily spotted, with a Tamil region in the north and a Sinhalese region over most of the rest of the country.

However, the arrangement of the districts having a majority ethnic population does not produce entirely contiguous regions. The four districts comprising the Northern Province do form the core of a Tamil region, but the other district with a Tamil majority (Batticaloa) is not contiguous with this core. In fact, the intervening district (Trincomalee) does not even have a plurality of Tamils, but instead, has a plurality of Sinhalese, which makes it impossible to form a single, contiguous Tamil region according to the criteria being discussed here.

The two remaining districts do not have majority populations and their pluralities are not Sinhalese. One of these is the Amparai District (which, in addition to Trincomalee and Batticaloa, forms the Eastern Province) where the Moor population constitutes the plurality. The other exception is Nuwara Eliya District, which has a plurality of Indian Tamils. If, when forming regions based on the criterion of plurality, the Indian Tamil population is regarded as distinct from the Sri Lankan Tamil ethnic group, then four regions result: the Sinhalese region consisting of 17 contiguous districts, the Moor region of the Amparai District, the Indian Tamil region of Nuwara Eliya District, and the Sri Lankan Tamil region, which includes the four districts of the north and the Batticaloa exclave.

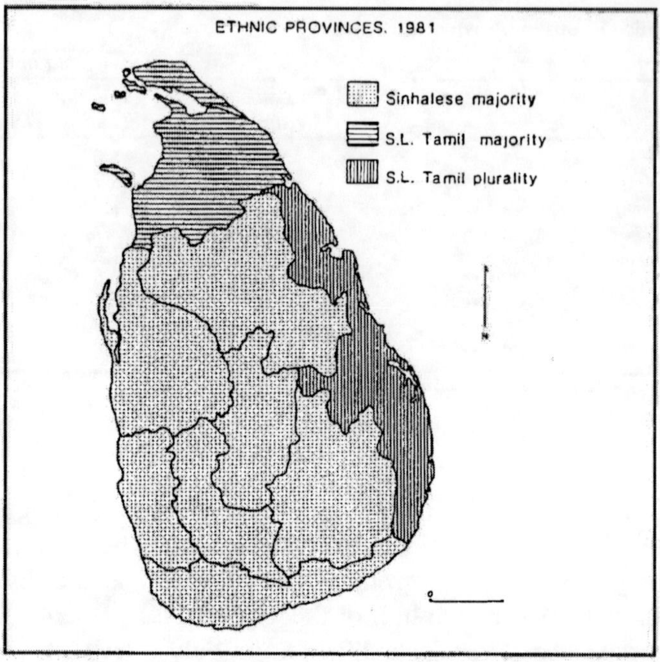

Fig. 1.1 : Provinces of Sri Lanka : According to Majority or Plurality Ethnic Population, 1981

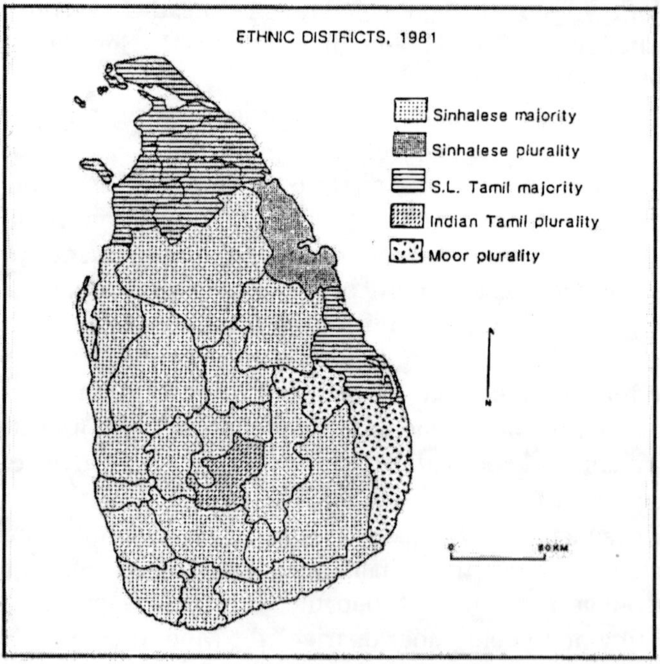

Fig. 1.2 : Districts of Sri Lanka : According to Majority or Plurality Ethnic Population, 1981

Table 1.2 : Ethnic Groups by Districts, 1981

(Data in Percentages)

Province	*Sinhalese*	*Sri Lankan Tamil*	*Indian Tamil*	*Moor*
Colombo	77.9**	9.8	1.3	8.3
Gampaha	92.2**	3.3	0.4	2.8
Kalutara	87.3**	1.0	4.1	7.5
Kandy	74.3**	4.9	9.3	10.7
Matale	79.9*	5.9	6.7	7.2
Nuwara Eliya	42.2	12.5	42.4*	2.5
Galle	94.4**	0.7	1.4	3.2
Matara	94.6**	0.6	2.2	2.4
Hambantota	97.4**	0.4	0.1	1.1
Jaffna	0.6	95.3**	2.4	1.7
Mannar	8.1	50.6**	13.2	26.6
Vavuniya	16.7	56.9**	19.4	6.9
Mullaitivu	5.1	76.0**	13.9	4.9
Batticaloa	3.2	70.8**	1.2	24.0
Amparai	37.6	20.1	0.4	41.5*
Trincomalee	35.8*	32.0	2.5	28.8
Kurunegala	93.1**	1.1	0.5	5.1
Puttalam	82.6**	6.7	0.6	5.1
Anuradhapura	92.1**	1.0	0.1	6.5
Polonnaruwa	90.9**	2.2	0.1	6.5
Badulla	68.5**	5.7	21.1	4.2
Moneragala	92.9**	1.8	3.3	1.9
Ratnapura	84.7**	2.3	11.1	1.7
Kegalle	86.3**	2.1	6.4	5.1

Note : Percentages for Others are not included here.
* Plurality
** Majority

If the relevant population to be grouped into a 'Tamil' region were to be defined as including both the Sri Lankan Tamils and the Indian Tamils, then the results would become even more spatially fragmented. This is because, when the two groups are combined, the total percentage in the Nuwara Eliya District exceeds 50 but the totals do not alter the tripartite composition of the Eastern Province. Thus, an all-Tamil region would consist of three widely separated, non-contiguous areas: the four-district area of the north, the east coast district of Batticaloa, and the central district of Nuwara Eliya.

The next smaller areal unit is what is commonly called an AGA in Sri Lanka. Although the AGA is too small a unit for high level government, this does not mean that district or provincial boundaries could not be redrawn to coincide with ethnic patterns produced at this areal scale. Therefore, it is informative to examine the arrangement of ethnic concentrations generated at this scale. As revealed by Fig. 1.3, the spatial fragmentation of the Tamils persists at this degree of detail. Not only are the Sri Lankan Tamils not dominant

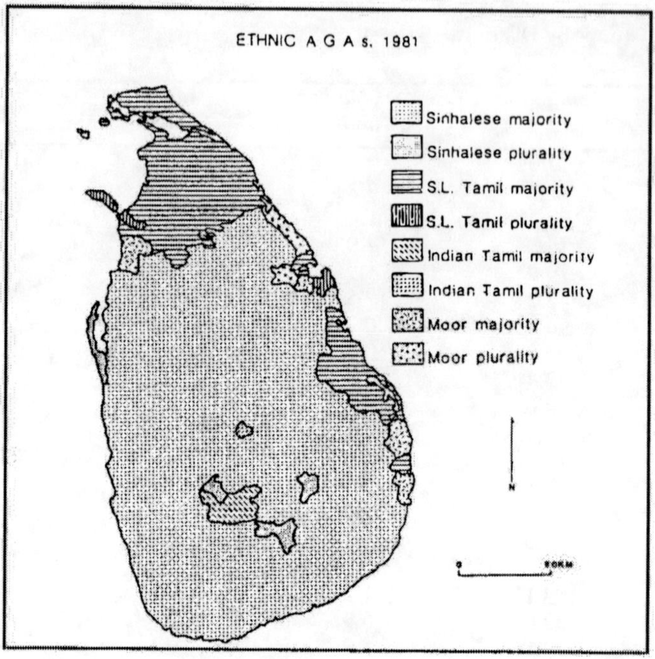

Fig. 1.3 : AGAs of Sri Lanka : According to Majority or Plurality Ethnic Population, 1981

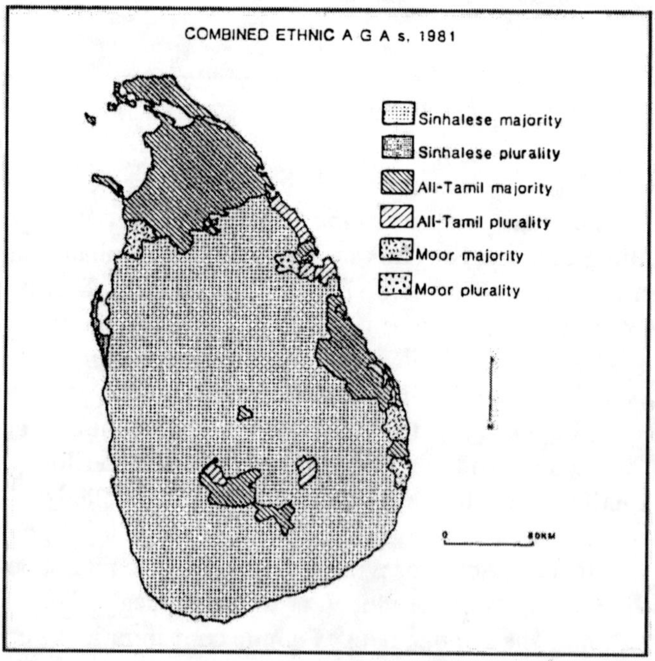

Fig. 1.4 : AGAs of Sri Lanka : According to Majority or Plurality Ethnic Population, with the Sri Lankan Tamils and Indian Tamils Combined, 1981

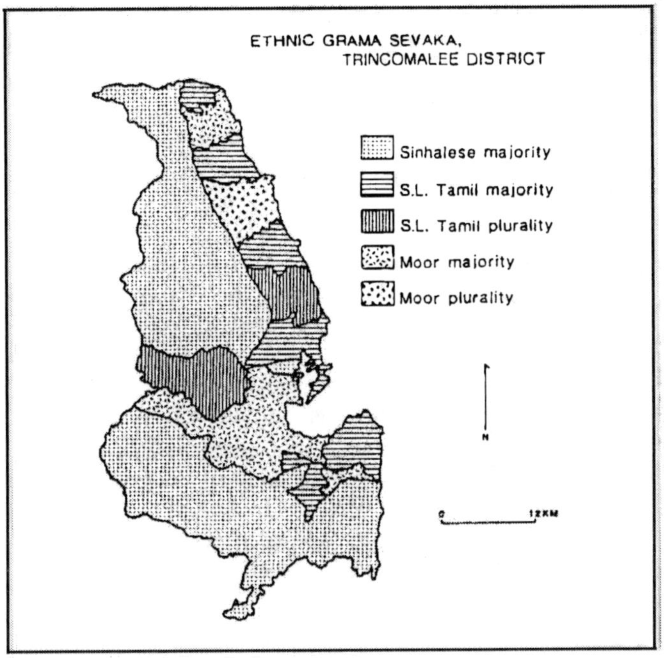

Fig. 1.5 : Grama Sevakas of Trincomalee District, Sri Lanka : According to Majority or Plurality Ethnic Population, 1981

in the Trincomalee District (as already noted), but within this district they form the majority in only one AGA, as well as a plurality in another. Both are isolated from Sri Lankan Tamil regions and from each other. Likewise, the Sri Lankan Tamils in the Amparai District are concentrated in one AGA, which is also separated from other Tamil areas.

At the geographic scale of the AGA, the Indian Tamils form a majority in a region consisting of two AGAs, and they constitute the plurality in another. If, for definitional purposes, the two Tamil groups were combined (Fig. 1.4), the task of forming a contiguous Tamil region remains impossible, even when only a plurality criterion is used. Under these conditions, the AGAs in the Trincomalee District are actually connected and, furthermore, they are linked with the northern core region. But, the Tamils in the Batticaloa District, in the isolated AGA in the Amparai District, and in the central portion of the country are all separated from each other and from the Tamil core in the north.

It is informative to examine the ethnic distributions in even greater detail because they continue to expose the complex mosaic of settlement patterns in the Eastern Province, especially in the northern district where the lack of contiguity presents a major regionalization barrier. Unfortunately, the data used here are from the 1971 census and the map (from a secondary source) may contain some errors.[2] Irrespective of these data limitations, the discontinuity of Tamil concentrations is revealed by mapping ethnic percentages in the critical district of Trincomalee (Fig. 1.5). The distribution of the Sri Lankan Tamil population at this areal scale (*i.e.*, using Grama Sevaka) vividly demonstrates the non-contiguity of ethnic groups along the coast and the impossibility of delimiting a contiguous Tamil region.

Conclusion

This examination of the distribution of populations at various scales reveals that, over much of Sri Lanka, one ethnic community tends to dominate in each administrative unit, so it is fairly easy to associate a particular population group with a particular territory or administrative unit. Furthermore, the areal units dominated by the main ethnic groups are generally clustered together, thus forming *de facto* ethnic regions.

However, in the Eastern Province, especially in the Trincomalee District, the population is arranged in a manner that does not permit the formation of contiguous regions based on an ethnic majority at any scale. In other words, it is impossible to create a Tamil region under the constraints of both contiguity and the higher degree of homogeneity associated with a majority.

One possible modification to the requirements of regionalization would be to relax the requirement that all parts of the Tamil region must be connected. As mentioned above, this might be a satisfactory solution when the goal of regionalisation is to show the distribution of climatic types, or even to display the patterns of ethnic groups for merely informational purposes. However, this relaxation of the contiguity constraint does not seem to be a viable solution when the objective is to form political entities that must interact with the population within a specified territory. Exclaves have always created severe problems for governments, and the difficulties in governing political outliers are especially great when the exclaves are surrounded by the territory of a hostile country.

Another alternative is to maintain the contiguity requirement but to group the areal units by using the criterion of only a plurality. This would produce a single region by joining the existing Northern and Eastern Provinces. Although such a Tamil region would achieve the contiguity desired, it would reduce the homogeneity of the resulting region, particularly in its peninsular extension along the east coast. Concurrently, such would diminish the level of areal precision because, as has been shown above, several small areal units within the Eastern Province contain non-Tamil majorities. In other words, at the provincial scale of measurement, the Eastern Province may be defined as 'Tamil' in terms of a plurality ethnic population; but this level of precision obscures the fact that non-Tamil majorities reside within several smaller areas within this region.

The discussion here is concerned with the issues facing cultural geographers as they attempt to delineate meaningful regions, so the various outcomes have been stated in terms of increasing the level of heterogeneity. In the case of Sri Lankan politics, however, the issue concerns the extent to which minority populations would be created by the drawing of various political boundaries. And, in turn, the effects of being a minority population depends largely on the political system and the ways the rights of all citizens, including those of a minority group, are protected. The feasibility of this regionalisation solution, therefore, is beyond the scope of this study.

In a political sense, a third option for creating regions that are contiguous and highly homogeneous might be to engage in what is known as 'ethnic cleansing', which is the forceful explusion of people of a particular ethnic group from certain areas. By altering the composition of the resident population, new percentages of ethnic groups are produced, which are then used to justify new territorial control. Besides, the fact that this solution involves inhumane tactics that most people do not approve of, this 'option' is not seriously examined here because the cultural geographer normally does not possess the power to shift people around. Even though moving cultural groups and features around would create a less complex set of regions, the geographer does not have this option.

In fact, because many cultural phenomena are not neatly clustered into obvious areal groupings, it becomes especially important that geographers undertake the task of organizing areal data into meaningful regions. The regionalization task, however, is not easy, especially because of the changing effects of scale and the difficulties of achieving contiguity. It is certainly appropriate to recognize the contribution of those cultural geographers who have wrestled with these problems and have produced meaningful results.

References and Notes

1. *Census of Population and Housing, 1981* (1981). Colombo: Department of Census and Statistics, Sri Lanka.
2. Hennayake, S.K. (1985). *Geographical Perspective on Ethnicity in Sri Lanka*, Unpublished M.A. thesis, Department of Geography, University of Peradeniya, Sri Lanka.

2

THE DEVELOPMENT OF INDIAN GEOGRAPHY

Allen G. Noble

Introduction

Although the roots of Indian academic geography lie in the dim past, the discipline is only now achieving maturity. Its development may be viewed as contained within a series of sequential phases. This study reviews the paradigm shifts which have taken place in the discipline and the major figures who shaped its direction.

Classical Indian Geography

Although geography was not then developed as a formal discipline, early Indian scholars had a well-developed geographical sense and clearly understood spatial relationships. As P.P. Karan argues, 'India developed indigenous geographical ideas long before the Greeks were developing the basic concepts of the Western world'.[1] The two best known modern scholarly works which deal with particular aspects of this very long period are those by Ali[2] and by Dubey[3] although modern academic interest in the ancient texts reaches all the way back to the late 19th century.[4] At about the same time that Dubey and Ali were producing their studies, A.B. Mukerji contributed an excellent study of geographical material on those settlements appearing in the *Jataka* text.[5]

Beginning in the 8th century, India was exposed to Islamic geographical concepts and ideas. Gradually, these materials came to be better known and more widely accepted, and Muslim geographers began to take their place beside Hindu scholars in contributing to the maturing of geographical study in India.

The Colonial Period

The arrival of the British and other European colonial powers in the seventeenth century forced an adjustment in Indian intellectual circles. Under the leadership of Ram Mohan Roy and other Indian scholars and social reformers, higher education became more modern in orientation. These leaders recognized that a reapproachment between Indian classical culture and thought, on the one hand, and modern European attitudes and practices, on the other, was required and they set about to lay the ground work for it.

Initially in the colonial period, the emphasis in Indian geographical study was on the collection and presentation of information to illuminate the various regions of the country, and to a much lesser extent, the continuing geographical analysis of classical Indian literature. Several observers have commented that the regional cataloguing of information was later encouraged by British colonial administrators' perceived need for detailed information on areas within their responsibility. The magnificent collection of Indian district and other gazetteers was an outgrowth of such efforts. These reference books are as useful today as they were over a century ago when they first appeared, although the direction of that utility has shifted. Nevertheless, these works remain primarily encyclopaedic in nature, with virtually no methodological or conceptual basis. Similarly, the Indian Census reports have made an excellent contribution to the advancement of Indian geographical knowledge and study.

The outlines of a more professional and academic approach to the discipline date from the 1920s. The first formal geography courses were introduced in 1920 in Punjab University, Lahore (now in Pakistan), followed shortly afterwards by courses at Aligarh Muslim University, which was also the first Indian University to introduce graduate courses in geography (1931). In the same decade, Patna University and Madras University also began instruction in geography. In the 1930s, Agra, Allahabad and Calcutta universities followed suit by establishing geography programmes. It is interesting to speculate that the establishment of the academic programmes in geography in colonial India coincides largely with the growing awareness of intellectual Indians in the future likelihood of an independent India and the consequent necessity to expand knowledge which would be critically useful for a new country.

Beginning in the mid-1930s and extending up to the time of independence in 1947, Indian geography fell under the influence of three individuals trained in French universities — R.N. Dubey, M.H. Rahman and S.P. Chatterjee. These scholars, especially Chatterjee, gave Indian geography its first particular methodological orientation. Man-land relationships expressed through the possibilist philosophy of the great French geographers, became a significant aspect of the newly-forming Indian geography. German geographic philosophy and ideas correlating physical elements with cultural elements to form 'landscape', also were introduced, largely by N.R. Kar and K.R. Dixit. Nevertheless, the outlines of Indian geography were most strongly influenced by then current British geographical orientations. British geographical study was in a period of considerable ferment with historical, political and economic directions, increasingly influencing basic geographical philosophy. Those most responsible for transmitting British geographical ideas to India in the early period, both regional and topical approaches, include M. Shafi, R. L. Singh, C. D. Despande, P. Dayal and George Kuriyan.

Each of these individuals, together with S.P. Chatterjee, R.N. Dubey and M.H. Rahman, all discussed above, became closely identified with university geography departments which evolved as the leading centers for the growth and development of the discipline. Their scholarly contributions have been summarized elsewhere.[6]

The Immediate Post-Independence Period

After Indian independence, a consideration for national, regional, and even local, development began to dominate the field, as academic geographers cast about to justify both their academic careers and their discipline of choice in a society and country with scarce resources, widespread poverty, and competing and conflicting demands.

Beginning in 1955, urban geography came to occupy a steadily more important segment of all geographic enquiry. Initially these studies emanated from Banares Hindu University, but as students of R.L. Singh took up leading positions in universities and colleges in other parts of India, urban geography studies proliferated across the entire country.

The situation in India was not unlike that in Western and Western-oriented countries generally. Urban geography, because of the increasing numbers of urban dwellers, the increasing concentration of power in large cities, and the continuing rapid growth of urban centers, began to be perhaps the most popular sub-field of geographical study after World War II. For example, the first urban geography produced in the West was published as late as 1937, even though it did not contain the words 'urban geography' in its title.[7] That designation was not used until 1949[8] and even that book was not truly a modern urban geography. Only in the late 1950s did such texts begin to appear when the sub-field began to assume its mature orientation. In India, the first comprehensive urban geography text did not appear until the early 1980s.[9]

Another major thrust of Indian geography was an emphasis on local or district land use and population studies. The popularization of such research stemmed from the importance of agriculture and the essentially rural nature of India. The inspiration probably was drawn from the great British geographer and student of Asia, L. Dudley Stamp. As had been the case earlier, most studies by geographers still emphasized description much more than analysis.

The early French and German influence was not long sustained (German longer than French, however) and most Indian geographers, up to the 1960s, received their advanced training in the United Kingdom. The prestige of receiving a foreign degree, especially from an institution located in the lands of the former colonial masters, helped ensure that returning geographers received positions of leadership in the rapidly growing numbers of geography departments in Indian universities and colleges. Even as early as 1950s, eight universities offered advanced study in geography[10] and by 1960, 24 institutions taught geography courses.[11]

The 1960s and American Influence

In the 1960s, a dramatic shift occurred in the training of Indian geographers.[12] Support opportunities for Indian graduate students in the United Kingdom began to decrease as British national economic difficulties mounted. Paradoxically, university expansion at the same time in both Canada and the United States offered the possibility of graduate study in North America. Professor V. Nath was the first of many Indian geographers to receive a Ph.D. in the United States. A steadily increasing number of students took advantage of these opportunities, and a substantial number remained in North America after completing their graduate work. By contrast, few Indian geographers at any period remained in Britain after completing graduate study. Those who returned from North America to India found themselves in situations similar to their UK-returned colleagues of a generation or so earlier. The major difference was that fewer leading positions were then available because the UK-trained professors already occupied them.

Nevertheless, the influence of American-oriented geography continued to grow. As a result, physical geography began to lose influence, field work was less stressed and much greater specialization became acceptable. Thus, in the late 1970s it was possible to identify urban geographers, medical geographers, agricultural geographers, political geographers and so on, each identified primarily by his/her subfield of specialisation.

Not unimportant to the process was the function of those Indian geographers who remained in the United States (or a smaller number who returned to India for a short time, but then migrated to North America to continue the rest of their academic careers). This movement was partly a response to the higher compensation, greater research opportunities and generally more sympathetic academic environment in North American colleges and universities than in India at the time. Even though they remained in North America they developed research interests on Indian topics, published in scholarly journals widely read in India, and maintained close and often very fertile contacts with Indian academic colleagues. Among the most productive of these scholars may be mentioned P.P. Karan, A.K. Chakravarti, A.K. Dutt, Ranjit Tirtha, and Beru

Sukhwal. A few Indian geographers were attracted to opportunities in other English speaking countries. Edison Dayal in Australia is probably the best known of such scholars. A significant cross-fertilisation was maintained which benefitted both sides. At the same time, Indian geographers were finding their own way, although not without difficulties.

Regional Rivalries, Societies and Journals

One of the most distinctive characteristics of Indian geography has been the coalescence of academic departments around the personality of a single individual. In part this is a persistence of the earlier British tradition of departments having only one professor, with all other faculty occupying lower academic ranks in a strongly hierarchical system. By the 1970s this tradition was gradually beginning to be replaced by the system prevalent in North America, where several professional positions exist in a single department. The very large size of India, the great distances which separated Indian university centers, and the difficulty of communication and travel (at least until very recently) also partly explains the tendency to independent orientation. One result has been the establishment of essentially regional geography journals (although often incorporating the words 'India' or 'National' in their titles). Partly to support these publishing outlets, professional geography societies were formed in these departments. Alternatively, journals were initiated to strengthen already existing local geographical societies. The Madras University geography department under the leadership of George Kuriyan, produced the first journal, the Indian Geographical Journal in 1926. Ten years later the Geographical Review of India was begun in Calcutta by S.P. Chatterjee and in 1948 Prof. M. Shafi started the Geographer at Aligarh Muslim University.[13] These three departments, journals and geographers became the foundation for the development of Indian geography as Independence dawned in 1947.

Some foreign observers have been sharply critical of Indian geographical journals terming them 'virtually unreferred, poorly edited, and badly printed'.[14] Such comments ignore the difficulties under which the journals have been produced, the stage of their development, and the real utility which most of them have had for Indian students and scholars. A recent review of one journal, *The National Geographical Journal of India* which may be taken as representative of Indian geography journals suggests that major changes are occurring. At the beginning of the 1980s, the number of pages in that journal devoted to major research articles averaged between 150 and 200 pages per volume, but by the close of the decade the average had climbed to between 225 and 300 pages. Furthermore, the number of professional reviews early in the decade occupied between four and five pages, but beginning in 1984 'the pages devoted to reviews increased dramatically'.[15]

One of the most serious problems facing academic departments of geography in India is the widely spread practice of hiring ones own graduates. Such a practice stifles independent thought, encourages provincialism and fosters inbreeding which inhibits the cross-fertilization produced by the introduction of diversity of opinion and training.

Only very recently, have attempts to hold truly national professional meetings, to organize a national society, and to publish a 'national' journal been successful. These activities, together with a better awareness of scientifically designed research techniques, and the improving quality of journals confirm the growing maturity of Indian geography.

The Maturing of Indian Geography

Only from the end of the 1980s, were Indian geographers sufficiently confident to question approaches, methodologies and terminology, which stemmed mostly from British and North American sources. A brief and largely ineffective infatuation with techniques of Marxist methodology occurred in the 1970s and 1980s and remained concentrated largely around Jawaharlal Nehru University in Delhi. By 1992 over 60 institutions of higher education offered geography courses.[16]

For the most part, the leading departments of geography in India today are those of long standing. Among such institutions may be listed (in no particular order) the departments at the University of Madras, University of Calcutta, Banaras Hindu University, Aligarh Muslim University, University of Delhi, Jawaharlal Nehru University (Regional Studies), Panjab University Chandigarh, University of Bombay and Pune University.

Among the individual scholars who are currently active leaders in the profession may be cited the following with their particularly important specialisation: Attur Ramesh (medical), Rana P.B. Singh (cultural), Baleswar Thakur (urban), R.P. Misra (regional planning), Rais Akhtar (medical), C. Ramachandran (urban), Sudesh Nangia (urban social), R.S. Dixit (marketing), R.D. Dikshit (political), B.K. Roy (population), Onkar Singh (environmental), K.V. Sundaram (regional planning), Gopal Krishan (administrative), C.S. Yadav (urban), D.N. Singh (quantitative methods), S.C. Bose (soils), N.B.K. Reddy (regional analysis), and Jasbir Singh (agriculture). Professor A.B. Mukerji, although retired, is widely acknowledged as the leading cultural geographer of India and one of the foremost such in the world.

A number of other important figures in geography, although retired have remained active and continue to exercise an important influence in the discipline of geography. Among these elder statesmen are R.L. Singh, G.S. Gosal, B.N. Sinha, C.D. Despande, Mohammed Shafi, K. Bagchi, N.R. Kar, E. Ahmed,

K. Sita, S. Manzoor Alam. The successful efforts of recently deceased Moonis Raza to introduce geography into the school curriculum were of enormous significance to the discipline.

Today's Problems and the Future

Clearly, Indian geography is today at an important turning point. The foundation laid down after Independence by geographers who are now in retirement is being challenged by newly developed or introduced methodologies or research techniques such as remote sensing, quantitative analysis, G.I.S. and cultural geography. At the same time, Indian geographers are just now beginning to look beyond their own regional boundaries, indeed even to the rest of the world outside India for research subjects.

In a brilliant valedictory article, A. B. Mukerji laid out an analysis of the major problems which have restricted full development of Indian geography in the past and which still remain in some measure to be overcome in future.[17] While one may question the extent to which each of these difficulties still affects the discipline, one must still respect the initiative and courage of Professor Mukerji's analysis which hopefully will force introspection by Indian geographers and ultimately produce continually improving professional attitudes. All of these developments and critiques should ensure, not only a more viable geography discipline in India, but in future, a greater integration of Indian geographical studies with those of other countries.

References and Notes

1. Karan, P.P. (1992). 'Development of Geographic Though in India', *National Geographical Journal of India*, Vol. 38, pp. 179-193.
2. Ali, S.M. (1966). *Geography of the Puranas*. Delhi : Peoples Publishing House.
3. Dubey, B. (1967). *Geographical Concepts in Ancient India*, Varanasi : National Geographical Society of India.
4. Das, N. (1894). 'Geography of Asia Based on the Ramayana, *Journal of the Royal Asiatic Society*.
5. Mukerji, A.B. (1969). 'Rural Settlements in the Jatakas', *Geographical Review of India*, Vol. 31, No. 2, pp. 17-32.
6. *See* ref. 1.
7. Van Cleef, Eugene. (1937). *Trade Centres and Trade Routes*, New York : D. Appleton Century.
8. Taylor, Griffith. (1949). *Urban Geography*, London : Methuen.
9. Prakasa Rao, V. L. S. (1983). *Urbanization in India*, New Delhi : Concept Pulishing Company.
10. *See* ref. 1.
11. Singh, R.L. and Rana P.B. Singh (1992). *The Roots of Indian Geography : Search and Research*, Varanasi : National Geographical Society of India.

12. Mukerji, A.B. (1992). 'What Ails Indian Geography? Some Statements', *National Geographical Journal of India*, Vol. 38, pp. 205-218.
13. Sinha, B.N. ed. (1986). *Trends in Geographical Research in India*, Bhubaneswar: Indian Council of Geographers.
14. Schwartzberg, Joseph E. (1983). 'The State of South Asian Geography', *Progress in Human Geography*, Vol. 7, pp. 69-72.
15. Noble, Allen G. (1993). *Analysis of Articles Published in the National Geographical Journal of India, 1980-1989*, Paper read at the Association of American Geographers Annual Meeting.
16. *See* ref. 11.
17. *See* ref. 12.

3

CIVILIZATION AND DEVELOPMENT

Some Critical Reflections on the Understanding of Terms : Case of India

DIRK BRONGER

Introduction

We owe to A.B. Mukerji one of the finest lessons in applied Social Geography: his comprehensive study of the *Chamars* in Uttar Pradesh. This contribution should always be a benchmark when one talks about Social Geography especially in connection with India : 'It is obvious that the social group which owns the means of production, mainly, the agricultural land, also commands the ownership, control, and utilisation of all the resources and facilities needed in the processes of production and socio-economic sustenance. In almost all the regions of Uttar Pradesh in India, this group comprises upper, higher, clean, and *dwija* (twice-born) castes. They too have problems but these do not and should not inspire an urgency in their analysis and solution. At the other extremity of scale is located the enormous volume of landless agricultural labourers drawn mainly, if not entirely, from the lower, unclean, polluted (*antyaj*) untouchable castes, the *shudras* as well as the economically backward among the artisan and service castes. They are the exploited and backward communities. If Social Geography has to attain a meaningful status and significance both in theory and practice, then it should give up its value-free stance and focus attention on these backward and exploited castes and areas'.[1]

His conclusion regarding the present situation of this single largest 'scheduled caste' of India is also thought-provoking especially in conjunction with the term 'Indian civilization'. To quote from this work,'The Chamars are the exploited, low caste, untouchable, landless agricultural labourers mostly concentrated in the rural areas. They are the victims of institutional mechanisms, social and economic. These institutional arrangements have denied the Chamars an access to the ownership of land, which is the principal means of production in the rural areas where they are the most highly concentrated. With this fundamental disability imposed upon them by the high caste land-owing classes is coupled the set of sanctions prescribed against them in the ritual purity-pollution principle governing the international behaviour of the four distinct classes which comprise the rural society. The Chamars, therefore, are either discouraged through subtle suggestions of words and actions or actually debarred from making use of the social, cultural, and economic facilities, amenities and production structural and infrastructural elements.

........'The Chamars form an integral part of the exploitative *jajmani* system and of the *zamindari* mode of production. Both display a facade of benevolence behind which they victimise the Chamars socially, culturally and economically by denying them entry to schools, dispensaries, better-wage jobs, irrigation, and indispensable farming inputs. The illiterate tenant cultivators, thus, regress into landless agricultural labourers, a state of semi-bonded slaves, in practice, if not legally. The Chamar is not an independent *kameen* supplying leather goods to his *Brahmin, Rajput, Thakur, Ahri*, and *Tyagi jajmans*, he is entirely dependent on them for his daily survival which he can ensure for himself very largely by selling his labour for agricultural purposes for a pittance. The Chamar helps his *jajman* to create and accumulate surplus capital which helps in the purchase of more land. The Chamar is further, and to an increasing extent, alienated from his land and driven to a state of despair symbolised by illiteracy and ill-health, reduced longevity, overwhelming dependence on agricultural labour in the rural areas and on menial jobs in the urban areas, high sex ratio, burden of dependency and enormous base of child population. Since the sources of development and modernisation have been deliberately located in the urban and the Chamars are overwhelmingly rural, they have been deprived of all opportunities of sharing the fruits of development'.[2]

This bitter accusation is also to be seen when we think about the interdependencies between 'civilization' and 'democracy' on one side and caste system on the other. To cite O.H.K. Spate in his well-known standard work on Indian geography, 'It is very difficult indeed to see how any really democratic society can co-exist with such an avowed, not to say violent, assertation of human inequality — not the natural inequality of individuals, but the automatic inferiority of whole classes of men, utterly irrespective of any individual or virtue'.[3]

I

It is a well-known fact, that ancient India together with Mesopotamia and Egypt is considered to be the first region of the earth in which a high stage of civilization had been achieved already five thousand years ago. To quote V. Gordon Childe (1897-1957), archaeologist and director of the Institute of Archaeology of the University of London, 'India confronts Egypt and Babylonia by the 3rd millenium with a thoroughly individual and independent civilization of her own, technically the peer of the rest. And plainly it is deeply rooted in Indian soil. The Indus civilization represents a very perfect adjustment of human life to a specific environment. And it has endured; it is already specifically Indian and forms the basis of modern Indian culture'.[4] Later, the Mauryan Empire as well as the Gupta period and, thousand years later, the reign of Akbar the Great (1556-1605 A.D.) are simultaneously considered as 'great periods', 'golden ages of Indian history'. It was during this time, that India not only began to regain its position among the nations of Asia and Europe but it actually became one of the richest and most powerful countries of the world.

Unfortunately India today belongs to the so-called underdeveloped or developingcountries in which, according to a recent World Bank study, 'India's poor account for an extremely large segment of the country's population. Along with many countries in Africa (*i.e.* including Egypt and, at least, parts of former Mesopotamia, D.B.), India has the highest incidence of poverty in the developing world'.[5] See Table 3.1.

Table 3.1 : Poverty in the Developing World, 1985 and 1990

Country groups	*Higher poverty line*[6]				*Lower poverty line*[6]			
	Number of poor (millions)		*Headcount index (percentage)*		*Number of poor (millions)*		*Headcount index (percentage)*	
	1985	1990	1985	1990	1985	1990	1985	1990
Aggregate	1,052	1,113	30.5	29.7	614	644	17.8	16.9
South Asia	532	562	51.8	49.0	308	287	30.0	25.0
of which India	421	448	55.0	52.8	250	226	32.7	26.6
East Asia	182	169	13.2	11.3	80	74	5.8	5.0
of which China	108	128	10.3	11.3	41	63	3.9	5.6
Sub-Saharan Africa	184	216	47.6	47.8	123	152	31.8	33.6
Middle East and North Africa	60	73	30.6	33.1	42	53	21.3	24.0
Eastern Europe	5	5	7.1	7.1	3	3	3.9	3.5
Latin America and the Caribbean	87	108	22.4	25.2	59	76	15.1	17.8

Source : World Bank 1992 : 78.

Taking this surprising fact, not to say contradiction, into consideration we have to question at least some of the causes of the apparent decline. In this line we first have to define the central terms 'civilization' and 'development'. In the next step one needs to question the character of 'civilization' and 'development' in India's 'golden ages' of history.

II

In English linguistic usage the term 'civilization' firstly comprises all stages of civilizations—in this connection the term 'culture' is synonymously used. In viewing 'civilization'/'culture' in this broad sense, one distinguishes between 'primitive civilization' or 'barbarian cultures' on one side and of 'highly civilized cultures', 'advanced stage of civilization' or 'highly developed civilization' on the other. Secondly, in a more restricted sense the term 'civilization' is equated with the latter meaning. So we speak of 'Indus Civilization' (not 'Indus Culture') in the sense that this civilization is characterized by large urban centres, surplus wealth, the use of metals and the development of writing including a numerical system. All in all this means, that the term 'civilization', however, is not generally used. It is employed more or less as a description of a situation or phase in the above mentioned broad and restricted sense as well.

In German nomenclature the 'advanced stage of civilization' is an established term and concept, named '*Hochkultur*' which is used in contrast to '*Primitivkultur*' ('primitive civilization'). '*Hochkultur*', as distinct from '*Primitivkultur*' is defined as follows by Hirschberg '(i) more efficient technology, *e.g.* usage of ploughs and draught animals, storage of surplus production in warehouses; (ii) collection and storing of information, usage of writing and of a numerical system; (iii) economically a functional differentiation in conjunction with the differentiation of the population into social strata (priests, noblemen, farmers, pastoralists, traders, craftsmen, artisans, scribes, serfs and slaves); (iv)social integration through the establishment and organization of a market system including money economy together with the creation of a differentiated political system with a professional bureaucracy and army; (v) urbanization, *i.e.* establishment of a hierarchical system of urban centres and thus, the formation of an urban-rural dichotomy; and (vi) formation and systematisation of especially theological and judicial knowledge and, later on, distinction between the ruling, and ruled (controlled) strata including their different cultural knowledge and behaviour'.[7]

All these factors and elements are interdependent; it is still controversial, which criteria out of these are more important and, at the same time, valid for which period. Explicitly the Indus Civilization is named as '*Hochkultur*'; the same applies to the Maurya and Gupta empire.

No complete agreement has yet been reached upon the concept of 'development'. The various sciences turn their attention to different objectives and aims of 'development', each based on its own ideas of development. However, general agreement in development research now exists on the following, namely, that the concept of 'growth', which contains only the idea of quantitative change, must not be equated with the concept of 'development', which also aims at qualitative changes, *i.e.* of the economic (and consequently social) bases. In other words, while the principal objective of development operations is economic the operations themselves are bound to extend to the entire culture.[8] Thus the concept of 'development' covers the combined effect of socio-cultural, political and economic changes or, in brief, the maximum sweeping changes in the social structure for the benefit of mankind. In more concrete terms, we can say that it is impossible to speak of 'development' until it can be ascertained that the entire economic and social structure, including its qualitative aspects, is adapted to the situation which has to be changed, *i.e.* when the structures and the attitudes and behaviour evolve towards optimum adjustment to and utilization of the new situation—and until the process is guided in that direction by the political system supported by a majority of the population.[9]

Later on two other German authors, Nohlen and Nuscheler have named the following five interdependent aspects and targets as essential to achieve 'development', ' (i) work/employment, (ii) economic growth, (iii) social justice/ structural change, (iv) participation, (v) economic and political independence'.[10] In other words, economic growth is incorporated in a network of development targets in which as much as possible all social strata participate on all kinds of improvements, whether economic, social and/or political. To quote the Encyclopedia Britannica in this connection, ' A civilization (as ... the Civilization of India) is that kind of culture which includes the use of writing, the presence of cities and of wide political organization and the development of specialisation... usually maintained by a large population (stressed by the author) over a considerable period, which has these elements'.[11]

However, it is not merely an interesting fact but it has to be emphasized, especially regarding the aspect of participation of the different social strata, that in the discussion/interpretation of the terms 'culture' and 'civilization', a marked distinction also quite usually is made between the 'primitive' and the 'modern civilization'. To quote the leading article on 'The Concept and Components of Culture' in the New Encyclopedia Britannica,'One of the most important, as well as characteristic, features of the economic life of preliterate societies, as contrasted with modern civilizations, is this: no individual and no class or group in tribal society was denied access to the resources of nature; all were free to exploit them. This is, of course, in sharp contrast to civil society in which private ownership by some, or a class, is the means of excluding others—slaves, serfs, a proletariat - from the exploitation and enjoyment of the resources of nature. It is this freedom of access, the freedom to exploit and to enjoy the resources of nature, that has given primitive society its characteristics of freedom and equality. And, being based upon kinship ties, it had fraternity, as well'.[12]

One has to (repeat and to) realize that not only the differentiation of mankind into different social strata but also the different *i.e.* limited 'access to the resources of nature' is named a hallmark of '*Hochkultur*'/'modern civilization'!

However, has 'civilization'/'*Hochkultur*' to be paralleled with 'development' as in this connection of 'highly developed civilizations'? Are there any doubts or even contradictions between these two (basic) terms?

If we treat both terms as equal *i.e.* 'civilization' as the story of human achievement and development, as has usually been done in science and literature and where 'participation' is considered as an integral part of 'development' if not its key indicator, we have to ask: Who and how many of which social strata really participated in the 'highly developed civilization' ('*Hochkultur*')?

III

Till now there has been relatively little empirical study on this qualitative aspect, *i.e.* to what extent the (majority of the) people have participated in 'civilization' as a pre-condition for 'develpoment'. Lets turn back to the early Indian 'civilizations' (*Hochkulturen*). As far as the Indus Civilization is concerned because of scarcity of data, we cannot answer our question. Therefore, we have to turn to the next period, considered as '*Hochkultur*', the Maurya dynasty, which laid the foundation of the first Indian empire (321-185 B.C.). The Maurya empire showed many of the aforementioned characteristics of an '*Hochkultur*': a strong political system with a highly differentiated as well as efficient working administration (especially the machinery for collecting different kinds of revenue), existence of a hierarchical system of urban centres, headed by the capital of the empire, Pataliputra, which was at that time not only one of the biggest (twice the size of imperial Rome) but most developed cities of the world. Further a differentiated economy (with agriculture as its backbone) and society, a powerful professional army—all headed by an emperor, who, in the case of king Asoka (268-233 B.C.), saw his role as essentially paternal ('all men are my children'). He issued edicts defining the idea and practice of dharma with its main principles of 'non-violence, tolerance of all sects and opinions, obedience to parents, respect for the Brahmins, liberality toward friends, human treatment of servants, and generosity toward all'.[13]

But we know little about the reality, the life of the people, the actual development, *i.e.* the participation of the different social strata in this 'glorious period'.

A bright spot in this historical darkness forms the well-known *Artha-sastra* of Kautilya. The date of origin of the *Artha-sastra,* however, remains debatable, with suggested dates ranging from Maurya up to the Gupta period, another 'classical age of Indian civilization'. Most authorities now agree, that the kernel of the book was originally written, at least initiated by Kautilya. He was prime minister to Chandragupta Maurya (321-293 B.C.). the founder of this empire. The *Artha-sastra* offers not only a profound view of the structure of the Mauryan government but it contains also a list and detailed enumeration of salaries given to the occupational groups. This list allows not only a deep insight into the hierarchy of authorities, but reveals highly pronounced income disparities as one will seldom find even today.[14]

The *Artha-sastra*, a unique source for our problem contains no data regarding the quantitative participation of the different strata with respect to the income groups. Let us therefore have a view of the social structure during the Maurya period. How it was organised and how the different social strata were assigned numerically?

Table 3.2 : Income Disparities in the Mauryan Empire

(Income per years in panas)

Income Group	*Income per year*	*Index*	*Social-strata—occupational group*
1	48,000	100	sacrificer, king's instructor, chief minister, royal priests, commander-in-chief, crown-prince, king's mother, king's chief wife (1 person each)
2	24,000	50	royal steward, chief tax-collector, chief treasurer etc. (1 person each)
3	12,000	25	princes, prince mothers, cabinet council etc.
4	8,000	16,7	corporation's chairmen, chief supervisors of the elephants, horses and war-chariots, judges
5	4,000	8,3	cavalry, infantry, war-chariot and elephant-troop supervisors
6	2,000	4,2	war-chariot fighters, elephant commanders, horse trainers etc.
7	1,000	2,1	astrologers, story-teller, war-chariot drivers, bards, teachers and scholars, spys and agents
8	500	1	infantry, accountants, writers etc.
9	250	0,5	actors
10	120	0,25	craftsmen
11	60	0,13	subordinate supervisors and employees
12	15†	0,03	agricultural labourers, coolies, pastoralists, slaves (government's land)

† and kind

Source : *Artha-sastra* of Kautilya, cited in : EMBREE/WILHELM 1967: 76f (compilations by the author).

Regarding the first, *i.e.* the qualitative aspect, besides the *Artha-sastra* another valuable source of exactly the same time; is the report of Megasthenes, the emissary of the diadochus Seleukos Nikator to the court of the same Chandragupta Maurya in Pataliputra from 303 to 292 B.C. He left his observations in the form of a book, the Indica. Megasthenes used the following words to describe the Indian caste system, 'It is not permitted to contract marriage with a person of another caste, nor to change from one profession to another, nor for the same person to undertake more than one, except if he is of the caste of philosophers, when permission is given on[15] account of his dignity'. As the German sociologist Kantowsky rightly pointed out, the notable characteristics of the caste system—limited freedom of choice in the professional realm, prohibition of intermarriage, privileges of a ritual upper class— were all observed with surprising precision by Megasthenes.[16] The character, if not to say rigidty of the caste system, obvious already during this time was correctly emphasised by Sir Herbert Risley when he characterized the distinction to the guilds of medieval Europe, 'Several writers have laid stress on the analogy between Indian caste and the trade guilds of medieval Europe. The comparison is misleading. In the first place the guild was never endogamous in the sense that a caste is; there was nothing to prevent a man of one guild from marrying a girl of another guild. Secondly, there was no bar to the admission of outsiders who had learned the business; the guild recruited smart apprentices just as the Baloch and Brahui open their ranks to a fighting man who has proved his worth. The common occupation was a real tie, a source of strength in the long struggle against nobles and kings, not a symbol of disunion and weakness like caste in India. If the guild has been a caste, bound by rigid rules as to food, marriage, and social intercourse, and split up into a dozen divisions which cannot eat together or intermarry, the wandering apprentice who was bound to travel for a year from town to town to learn the secrets of his art and who survives, a belated but romantic figure, even at the present day, could hardly have managed to exist, still less could he have discharged, like Quintin Matsys and a host of less famous craftsmen, the traditional duty of marrying his master's daughter. It seems indeed possible that the decadence and sterility of Indian art at the present day may be due in some measures to the trammels by which the caste system has checked its natural growth. A guild may expand and develop; it gives free play to artistic endeavour; and it was the union of the guilds that gave birth to the Free Cities of the Middle Ages. A caste is an organism of a lower type; it grows by fission, and each step in its growth detracts from its power to advance or even to preserve the art which it professes to practice'.[17]

Of course the priests including the religious teachers were the most respected and, what is also important, not only handsomely paid but, additionally, exempted from taxation. Also the soldiers were fairly paid, especially when compared to those engaged in agriculture (Table 3.2). But regarding the numerical assignment and thus, the participation of the different groups, it passed on from that period that the *Brahman varna* formed the smallest group in number (Kautilya himself was a *Brahmin*). If Pliny's figures regarding the strength of the Mauryan army are correct, it consists of 30,000 cavalry (income group 5-7, partly 4 in Table 3.2) and 600,000 infantry (income group 8). Both together amounted for 1.26 per cent of a suggested population of a total 50 million during Asoka's time. If we further consider the character and strictness of the caste system already during this time and take into account that changing of caste was (and is) extremely limited, then a quantitative assignment (estimate) can be deduced even from the present time: the share of the 'clean *jatis*' amounts to at most 10 per cent of the total population. That means : it is very much likely that only a small minority participated economically and socially in this 'classical age of civilization' - so much the more as the peasants and pastoralists, not included in this list, were more or less all dependent farmers because the state, *i.e.* the king, was the sole owner of the land. In addition to this it is safe to assume that the remaining social strata, *i.e.* the traders, craftsmen, serfs and slaves would have hardly received income higher than their caste members listed in Table 3.2.

On the whole we can conclude that the vast majority of the people, say income group 9 and below, *i.e.* 90 per cent of the total population, just managed to survive during these 'golden ages of Indian civilization'.

IV

Further let us turn to another prime period of Indian history, the Mughal Empire (1526-1761), more precisely to the greatest in the entire history of India, Akbar the Great (1556-1605). Indeed 'Akbar's idealism, natural gifts, and force of character as well as his concrete achievements entitle him to a high place among the rulers of mankind'.[18]

Akbar's historical merit was his own deep conviction of the king as God's representative on earth and the impartial ruler of all his subjects, irrespective of their religion. His comparative study of religions convinced him that there was truth in all religions but that no one possessed absolute truth. Therefore, he must be absolutely tolerant to every creed, establish universal peace in his dominions, and work ceaselessly for the welfare of all classes of his people. (He even disestablished Islam as the religion of the state.) But was the time ripe for the successful implementation of an idealistic programme of that type?

As land control is the key in a Predominantly agrarian society answering our question regarding the participation of the different social strata, we have to examine who actualy controlled the land in Mughal India during the reign of Akbar?

This information is proivded by the *Ain-i-Akbari*, a description and handbook of Akbar's kingdom. It contains (*inter alia*) a list of the caste groups from which revenue was obtained, *i.e.* a picture of the distribution of the caste-wise revenue payment can be obtained. Table 3.3 shows these interdependencies in a limited region in the middle Ganga valley, more precisely in the 4 districts of Varanasi, Jaunpur, Ghazipur and Ballia.[19]

Although some minor castes who also paid revenue were probably omitted in the *Ain-i-Akbari,* it is a fact that the three highest ranking *jatis* - Brahmin, Bhumihar and Rajput—were responsible for the payment of 90 per cent (together with the Muslim-Rajput even 94 per cent) of the total revenue demand. At least

Table 3.3 : Castes Paying Land Revenue According to the Ain-i-Akbari, ca. 1596

Caste	*Jaunpur*		*Ghazipur-Ballia*		*Banaras*		*all districts*	
	Revenue	*Per cent*	*Revenue*	*Per cent*	*Revenue*	*Per cent*	*Revenue*	*Per cent*
Rajput	445,000	64	112,000	39	30,000	16	587,000	50
Brahman-Bhumihar	43,000	6	170,000	59	151,000	79	364,000	30
Brahman-Rajput	122,000	18			11,000	5	133,000	11
Muslim	35,000	5					35,000	3
Muslim-Rajput	43,000	6					43,000	4
Other	6,000	1					6,000	1
Unknown			7,000	2			7,000	1

Source : COHN 1969: 56.

a fairly good idea of their numerical strength, not given in the *Ain-i-Akbari,* can be obtained from the Census 1931. The percentage of these three *jatis* to the total population of the four districts at that time amounted to 20.3 per cent (22.2 per cent of the Hindu population).

Despite the fact that the designation of the rural society in Mughal India as an undifferentiated mass of pauperized peasants is clearly too simple, the tax burden (as well as the rights of the land) undoubtedly had been distributed very unequally among the different strata of the peasantry and this fact being far away from the idea of the unchanging 'village republic' articulated *inter alia* by Sir Charles Metcalfe,'In much as it was convenient for the revenue authorities to treat the village as a unit for collection (and even assessment), it was natural for them to rely upon the headman or a small stratum of upper peasants. This dominant group, then, collected the tax at rates fixed by themselves from every peasant, putting the collection in a pool (*fotal*), with its accountant, the *patwari*. From this pool the land revenue would be paid, so also the fees and perquisites of certain officials and certain common expenses of the village. ...

Those who controlled the pool usually evaded paying their due share of revenue. Lower rates were also levied upon some favoured elements, the *khwud-kasht* peasants in Northern India, *gharuhalas* in Rajasthan and *murasdars* in Maharashtra. Such a favoured category of peasants was found in almost every part of the country. The smaller peasants, *reza ri' aya* (*maltis* in Rajasthan, *kunbis* in Maharashtra) forming the bulk of the peasantry, were thus called upon to pay more than their due share of the revenue in order to make up the total. It was common in Mughal administrative literature to bemoan such exploitation of the smaller peasantry by the 'dominant elements' within the village.[20]

More sharply Hasan has pointed out that 'the *zamindar* class played a vital role in the political, economic, and cultural life of medieval India. During the Mughal period its importance increased ... In spite of the constant struggle

Table 3.4 : Size of Land Ownership by Caste in 6 Deccan Villages

Socio economic strata	*Caste*	*No. of households Total*	*%*	*Size of Holding (in acres)* >100	>5	>20	>10	>5	>1	<1	Σ	*Without land*
Priests/Landlords	Brahmin											
Traders	Vaishyas	67	2.8	4	5	11	2	4	3	1	30	37
Bigger landowners	Reddi, Kapu	169	7.1	11	21	64	36	17	16	2	167	2
Smaller landowners	Telaga, Mushti Muttarasi, Boya	587	24.6	—	6	31	55	40	73	9	214	373
Craft and Services (Potter, Smith,Carpenter, Shepherd, Weaver, Toddytapper, Barber Washermen, Stone-crusher)	Kummara, Ousala, Kammara, Vodla, Golla, Kuruba, Padmashali, Goundla Mangala Dhobi, Voddara	1,013	42.5	—	3	21	32	47	83	48	234	779
Services (inferior)	Mala, Madiga (S.C.)	547	23.0	—	—	6	15	11	37	36	105	442
Total		2,383	100.0	15	35	133	140	119	212	96	750	1.633

Source : Own investigations (1967-1970) in : BRONGER 1976 : 220.

between the imperial government and the *zamindars* for a greater share of the produce, the two became partners in the process of economic exploitation'.[21]

Thus, we can state a considerable economic differentiation not only among the peasantry but also among the people living in rural (as well as urban) areas as a whole. There was a sharp contrast between a small group of the ruling class living a carefree, luxurious life on one side and the miserable standard of living of the masses — most of the peasants, the artisans and the domestic attendants, on the other. There is hardly any doubt that the ordinary rural folk used to live in mud huts thatched with straw.[22] We can summarize that the vast majority of the people in Mughal India participated very little in the economic and social development during this 'glorious period of civilization'.

Conclusion

It is definitely true that 'the sharp contrast between the standards of living of the ruling classes and the common people was, of course, not peculiar to India; it existed in a greater or lesser degree everywhere, including Europe'.[23] Taking up the present picture regarding the distribution of the resources, in terms of the ownership of the means of production, mainly the agricultural land which is also depicted for Uttar Pradesh by A.B. Mukerji, shows obviously very little change since then. Up to now the members of the *dwija* and the dominant castes, forming together merely 10.25 per cent of the rural population control 60.85 per cent of the arable land. These interdependencies are proved by a large number of studies covering almost all parts of India including the author's investigations of 6 Deccan villages[24] (Table 3.4).

In addition the Indian caste system itself in its principles and strictness has even strengthened its position since Megasthenes. To quote Dube regarding its present role for Indian life and thus development, 'Social scientists in India—and also abroad—criticize their fellow social anthropologists and sociologists for their caste fetish. To them their concern with caste is excessive. But this view under rates the pervasive role played by caste in India. Like an octopus caste has its tentacles in every aspect of Indian life. It bedevils carefully drawn plans of economic development. It defeats legislative effort to bring about social reform. It assumes a dominant role in power processes and imparts its distinctive flavour to Indian politics. Even the administrative and the academic elites are not free from its over-powering influence. So how can it be ignored as a social force?[25]

Our brief discussion and historical reflections of the interdependencies between 'civilization' and 'development' reveals that the often used term 'civilization'/'advanced civilization' ('*Hochkultur*') and also the phrases 'glorious times', 'golden ages' etc. should be thought over much more, differentiated and thus, used much more carefully. Such a critical reflection of the reality helps (all of) us for a better understanding of 'development' including the developmental problems. And this should validate, of course, not only to India.

References and Notes

1. Mukerji, A.B. (1980). *The Chamars of Uttar Pradesh : A Study in Social Geography* Delhi p.155.
2. *See* ref. 1, pp. 114-115.
3. Spate, O.H.K and A.T.A. Learmonth. (1967). *India and Pakistan : A General and Regional Geography*, London : Methuen & Co. Ltd; p. 877.
4. *New Encyclopedia Britannica*. (1986). *India*, Vol. 21, pp. 1-71.
5. World Bank (Ed.) (1992). *India : Stabilizing and Reforming the Economy*, Washington, pp. 184.
6. 'Higher and lower poverty line is based on an international poverty line set at 1985 US$ 31.23 per person per month in 1985 constant purchasing parity prices (PPP) across countries. These prices refer to those of a common basket of goods and services across countries. At these prices, India's per capita of US $ 340 in 1990 poverty line translates into US $ 9 per person per month - the cost of maintaining a minimum calorie intake at Indian prices and exchange rate.' *World Bank* 1992, pp. 78.
7. Hirschberg, W. (Ed). (1988). 'Hotchkulturen', in *Neues Worterbuck der Volkerkunde*, Berlin, pp. 215.
8. Behrendt, R.F. (1961). 'Entwicklungslander; Soziologische Problematik', in *Handbuch der Sozialwissenschaften*, Vol. 3, Stuttgart pp. 230-242.
9. Behrendt, R.F. (1968). Geselischaftliche Aspekte der Entwicklungsforderung in Fritsch, B. (Ed). *Entwicklungslander*, Koln, Berlin.
10. Nohlen, D. Nuscheler, F. (1974). 'Entwicklungstheorien und Entwicklun-gsbegriff', in *Handbuch der Dritten Welt*, Nohlen, Nuscheler (Eds) Hamburg, pp. 13-33.
11. 'Civilization and Culture' in *Encyclopedia Britannica*, Vol. 5, 1968 London, pp. 824-833.
12. 'The Concepts and Components of Culture', in *The New Encyclopedia Britannica* Vol 16. Chicago, 1986, pp. 925-932.
13. *The New Encyclopedia Britannica* (1986). Chicago, Vol. 2, p. 38.
14. Embree, A.T. Wilhelm (1967): 'Arthashastra of Kautilya' cited in *Indien*, Vol. 17, Frankfurt Fischer, Weitgesthichte.
15. Wilson, J (1877). *Indian Caste*, 2 Vols, London.
16. Kantowsky, D. (1970). 'Dorfentwicklung and Dorfdemokratie in *Indien*, Bielefeld, pp. 171.
17. Risley, H. (1931). 'Caste, Tribe and Race' in *Census of India 1931*, Calcutta, pp. 489-557.
18. *See* ref. 4, pp. 67.
19. Cohn, B.S. (1969). 'Structural Change in Indian Rural Society 1596-1885' in Frykenberg, R.E (Ed.) *Land Control and Social Structure in Indian History*, Madison, pp 53-121.
20. Habib, I. (1982). 'Agrarian Relations and Land Revenue. North' in Rayachaudhuri T. and Habib I. (Eds). *The Cambridge Economic History of India*, Vol. 1 Cambridge, pp. 235-249.
21. Hasan, S.N. (1969), 'Zamindars under the Mughals', in Frykenberg, R.E. (Ed). *Land Control and Social Structure in Indian History*, Madison, pp. 17-31.
22. Chandra, B. (1982). 'Standard of Living : Mughal India', in Rayachaudhuri, T. and Habib I.

(Eds.) *The Cambridge Economic History of India* Vol.1, Cambridge, pp. 458-471; Fukazawa, H. (1982). 'Standard of Living: Maharashtra and the Deccan', in Rayachaudhuri, T. and Habib I. (Eds.) *The Cambridge Economic History of India* Vol.1, Cambridge, pp. 471-477.

23. Chandra, S. (1982) 'Standard of Living : Mughal India,' in Rayachaudhuri, T. and Habib, I. (Eds.) *The Cambridge Economic History of India*, Cambridge Vol.1, p. 458.
24. Bronger, D. (1976). 'Formen raumilcher Verflechtung von Regionen in Andhra Pradesh / Indien als, Grundlage einer Eutwicklungsplanning', in *Beitrag dar Angewandten Geographie zur Entwicklungslanderforschung* Paderborn, p267.
25. Dube, S.C. (1968). 'Foreword' in Atal, Y., *The Changing Frontiers of Caste,* Delhi, pp.V-IX.

4

A CULTURAL GEOGRAPHICAL ANALYSIS OF AN INDIAN VILLAGE

Some Conceptual and Methodological Statements

NEELAM GROVER

Studies of individual villages have a firm and old tradition in the growing corpus of both Western Geography and Indian Geography. However, the use to which these studies have been put has been notably different. Rural sociologists, economists, and anthropologists have contributed much in this direction. Towards the middle of the present century, studies of individual villages were conducted widely by the American and British anthropologists under the sponsorship of Foreign Area Studies financed by the Office of Naval Research and the Department of State in U.S.A. and the Social Science Research Council in U.K.[1] Several studies were financed by various fellowships such as the Guggenheim Fellowship and also those awarded by the American Institute of Indian Studies and the Indian Institute of American Studies. Indians too followed the pattern set by their American and British counterparts and the focus of several doctoral dissertations were single village studies. The best known among these studies were produced by M.N. Srinivas[2] and S.C. Dubey.[3]

A further strong impetus to these kinds of studies was provided by the Social Studies Division of the Census of India in their compilation of village Monographs. Providing the most useful, wide-ranging, and comprehensive inventory of all aspects of individual villages they, however, failed to provide any theoretical leads.

In fact individual village studies received much attention from Indian geographers since the late fifties and through the sixties. The emphasis of these studies was, however, consistently on different aspects of the attributes of rural settlements,[4] which alone were highlighted and very often the physical factors were given predominance in explanation. The role of history and culture was practically completely ignored. These investigations lacked any conceptual framework within which they were conducted. Their conclusions, therefore, were mere summaries of their Individual works which contributed nothing by

Note : Reprinted from *Punjab University Research Bulletin* (Arts), Vol. XXI, No. 1, April, 1990.

way of generalisations or strengthening any existing conceptual outline. These studies were essentially inventories lacking internal cohesion or a well defined focus. Thus, studies of individual villages in geography have been highly fragmented and, therefore, limited in scope. They have dealt with individual villages not in a holistic manner but as if each settlement represents a broken part of the spectrum.

On the other hand, individual village studies in sociology and anthropology have tended to be holistic. Those by anthropologists particularly, reflect a recognition and consideration of the concept of culture. To mention a few, studies by McKim Marriott, Srinivas, and others harbour an underlying thread of recognition of a paradigm of something Pan-Indian, an Indian Village. Hence, detailed studies of several individual villages would lead to drawing generalizations which could hold good for the whole. However, sociologists and anthropologists did not concern themselves with satisfying the norm of regional representation. Although the villages selected by them for study did reflect the regional character, they were not studied as representative of the region. Geographers on the other hand, did recognise the village as representative of a region, as also a region associated with villages having common empirical attributes. Their studies, however, revealed a persistent, underlying lack of concern with the recognition of, or a conscious, deliberate application of the concept of culture.

A pre-occupation with the study of the elements of Rural Settlement Geography and the empirical attributes of the rural settlements led to scant attention being paid to villages as units in a cultural landscape. Pioneer studies by Mukerji[5] of rural settlements and their attributes in the essential context of Cultural Geography with an attempt to discover the cultural meanings, processes involved in the gradual evolution of the attributes of rural settlements, and to the rural cultural landscape, gave new direction to studies in Rural Settlement Geography. Subsequently, while studies in empirical attributes, functions, and elements of rural settlements continued, a leaning towards their analysis in the context of Cultural Geography became evident. The significance of these studies can at no time be undermined, but it cannot be denied that studies of individual villages recognising their constituent settlement and field landscapes and their evolution through time and space; to recognise the regional representative as part of a greater whole and thereby linking them to the entire spectrum of the rural settlement fabric, have not been common.

Rural settlements need the special attention of geographers not merely to expedite their supreme significance as the incipient units of the settlement fabric, but also as the hearths of culture. In fact, most of the ancient cultures in the Oriental Civilization, have been and continue to be rural in origin and character. A rural settlement itself is really a corollary of the operation of cultural processes in a given environment through time. It is in fact a cultural artefact having been moulded from and contributing to the surrounding area, being the expression of the culture of the group that inhabits it. Each rural settlement, therefore, is the outcome of certain settlement processes, each controlled in its expression by the culture of the group that inhabits it.

In this context, an indepth study of one individual village in India has much more meaning in the understanding of the Indian rural scene in its social and cultural matrix, than perhaps a study of thousands of villages through sampling. 'The study of a single village is productive of much more than knowledge about a single village'.[6]

Having for long remained the domain of rural sociologists and anthropologists, the significance of such studies in Cultural Geography, especially in Rural Settlement Geography, is implicit. No doubt indepth studies of individual rural settlements will have an intimate relationship with and relevance to the work of sociologists and anthro pologists in the same manner in which the latter have contributed to the understanding of rural settlements.

Any rural settlement is the microcosm reflecting the macrocosm of which it is the nucleus. Studies undertaken by social scientists of individual villages belonging to different culture groups and in different regions reveal, rather distil of something Pan-Indian, while at the same time being conscious of regional variations. In fact, two different character clusters interplay with each other, that of the regional tradition and that of the Pan-Indian tradition. Individual settlements are to a large extent region and ecology specific, each maintaining its identity but contributing to the Pan-Indian principle. The regional tradition is a modification of, in fact, a reflection of the Pan-Indian tradition. The latter is, however, so deeply entrenched that differentiation between the two becomes almost impossible. As seen in Fig. 4.1, the constituent villages within the different regions, R_1, R_2, R_3, and R_4, will have closer affinity with each other than perhaps with the villages of the other regions. thus, V_1, V_2, V_3, and V_4 of R_1 are more similar and culturally closer to each other than to V_1, V_2, V_3, and V_4 of R_2, R_3, or R_4. Although each village is a distinctive and unique entity, it definitely partakes of the regional culture, which it reflects and with which it can be immediately identified. Each region as a whole partakes something of the Pan-Indian village, but at the same time retains and maintains its own distinctive character, which distinguishes it from the other regions. The underlying thread of the paradigm of the Indian Village, however, links the different regions, R_1, R_2, R_3 and R_4. Thus, R_1, R_2, R_3, and R_4 act as a media reflecting onto their constituent villages their own respective, distinctive character as also something of the holistic image of the Indian Village, thereby building up the entire spectrum of the rural fabric. The prime significance of individual village studies lies in that it contributes to the paradigm of the Indian Village and thus, provides

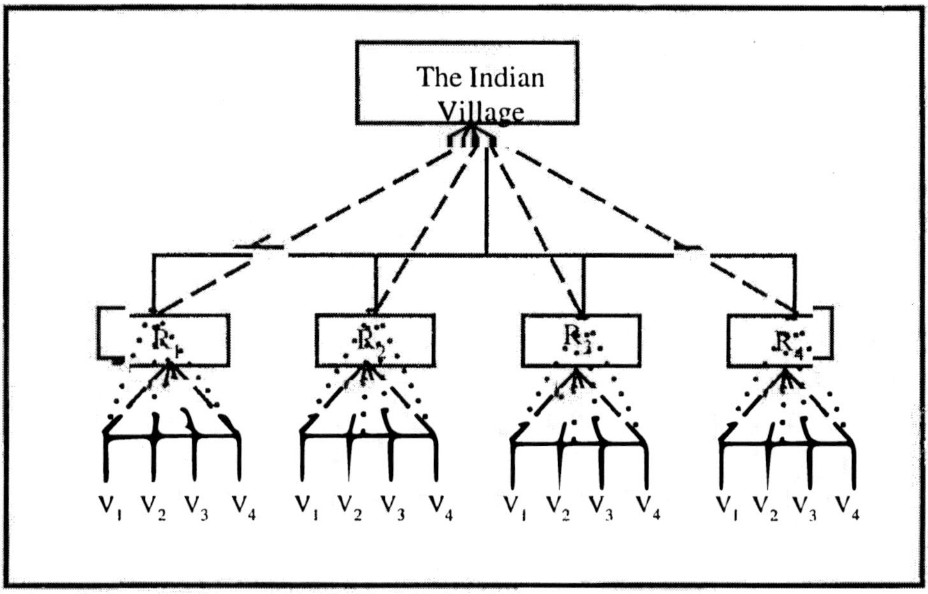

Fig. 4.1 : The Indian Village

towards the additional validation of the principle of a Pan-Indian tradition which is arching over everything. At no time, however, can the Pan-Indian tradition be regarded as a summation or aggregation of regional tradition.

In individual village studies we are on much surer ground with specificities of site, situation, size, type, caste, and culture. In fact, they provide an empirical illumination and expansion of the deductive generalizations of Pan-Indian nature. This involves no attempt at statistical jugglery. Region and ecology specific, the study of individual settlements is and should continue to generate interest and be important till such time that the variations in the Indian rural fabric have been fully documented and analysed.

Individual settlement studies could be either of a multi-caste or a uni-caste settlement. The emphasis need not be on the caste structure alone, but its roie in the social and cultural functions of the settlement. This would provide an ideal opportunity to study its landscape expressions both in the settlement and field landscapes, and the hierarchical *jajmani* system[7] which links it to the fundamental social structure of the Indian rural fabric. On the other hand, the uni-caste or the uni-clan settlement becomes important in providing a different dimension, wherein it becomes possible to study in greatest detail the intricacies and landscape expressions of the intra-family and inter-family relationships in the social, private, and domestic realms and their cultural uniqueness. In the broader framework too, the individual settlement is also a link in the horizontal *jajmani* system, as result of which it thrives and which it in turn supports.

Where the individual settlement belongs to a single caste or culture group, a study of the processes of settling and their landscape expressions become vital tools in the understanding of the culture group, its *genre de vie* and its place in the matrix of the Indian rural scene. The study of a single settlement provides insight into the two essential components of the vernacular cultural landscape, the settlement landscape and the field landscape. Contrary to the belief that, 'India's ancient and traditional villages are unchanging or slowly changing social phenomena',[8] the evolution of the settlement and field landscapes reflects the subtle dynamism inherent in each rural settlement. In fact, the existing landscapes provide only a link in the evolutionary process and an analysis of the landscape expressions of the material and non-material elements of culture lead to the discovery of relict features, which although endowed with new meanings and perhaps functions, are the key to its cultural history, helping to reconstruct the succession of cultural landscapes and hence, the processes of settling.

Such a study involving the settlement landscape necessitates first an understanding of the culture of the group inhabiting the settlement, its *genre de vie*, traditions, norms, value systems, and religion. A study of the field landscape initiates a study of the land tenure system, the initial acquisition and subsequent division of land resulting in the evolution of the field pattern, basic traditions and norms adhered to in this respect, level of technological know-how, land utilisation by way of the crops grown, the irrigation methods practised.

In the wider context of culture, history, and ecology, a single settlement leads to the study of its own intra-relationships within this triology and its inter-relationships with the settlements of the surrounding region and through them to the entire rural settlement fabric of India. An expression of culture on the landscape, a settlement is essentially the creation of the interaction of the inhabiting culture groups to the ecological context through time. Each one of its elements therefore, can lead to an endless amount of discovery.

An attempt in this study is being made to illustrate the importance and significance of the study of any single rural settlement in the context of Cultural Geography to an understanding of a people, a region as perceived, exploited, and transformed into one deriving its identity from and lending its identity to the inhabitants. As Mikesell observed, 'any sign of human action in a landscape implies a culture, recalls a history, and demands an ecological interpretation, the history of any people evokes its setting in a landscape, its ecological problems, and its cultural concommitants, and the recognition of a culture calls for the discovery of traces it has left on the earth'.[9]

Kniffen emphasised that the material details of the landscape are the geographer's lore.[10] In this context the study of individual settlements gains primary significance as it has a vertical structure and also a spatial structure or horizontal extensions. Thus, if several hundred settlements were to be studied, a detailed portrayal of the vertical structure, which by nature is individual and not amenable to horizontal, spatial translation, would be rendered impossible. In such a case an analysis of how the vertical structure is generated, sustained and how it undergoes modification would not be possible. This can be done only by the indepth study of a single settlement. For such a study any rural settlement can be selected.

Lost in the vast anonimity of the rural landscape, each rural settlement is the home or microcosm of either one or more culture groups. Unique in its identity, a settlement is intricately linked to the regional rural fabric necessitating a look back into its culture, culture history, its culture ecology, the interplay of which has resulted in the creation of its cultural landscape, making it one of a vaster culture area. A cultural-geographical analysis of any individual settlement must, therefore, essentially follow a similar sequential order to study in detail all its aspects individually as also one of a larger spectrum of rural settlements of India. Mikesell defined the personality of Cultural Geography to include, '(a) historical orientation; (b) a stress of man's role as an agent of environmental modifications; (c) a preoccupation with material culture; (d) a bias in favour of rural areas in this country and non-Western or pre-industrial societies abroad; (e) a tendency to seek support from anthropology; (f) a commitment to substantive research and a consequent attitude of extreme individualism; and (g) a preference for fieldwork rather than "armchair geography"'.[11] The methodology adopted should attempt at illustrating and validating these proposed theoretical formulations. The essence of such studies should, therefore, essentially stem from fieldwork. Analysis of relevant topographical maps, published and unpublished records, and oral and recorded history must also be used but primarily as tools to substantiate information collected through fieldwork wherein keen observation of the landscape is the most fundamental and pivotal task. Drawing of sketches, taking notes, interviews and photographs would become secondary but corroboratory to an observation of details. Beginning with the observation of the general rural landscape, a bird's eye view, a nesting hierarchical structuring of the manner and details of observation should first be worked out. This only reiterates the suggestions first put forth by Suess and Bruhnes and later supported by Kniffen.[12] Thus, what is involved is the collection of information at different levels of observation (Fig. 4.2).

An individual settlement being the focus of study, as a first step, it becomes important to extensively traverse and keenly observe the surrounding regional rural landscape in an effort to discover and understand the economic, social and cultural regional milieu which has created and sustained the settlement. The evolutionary processes involved in the evolution of any one settlement do

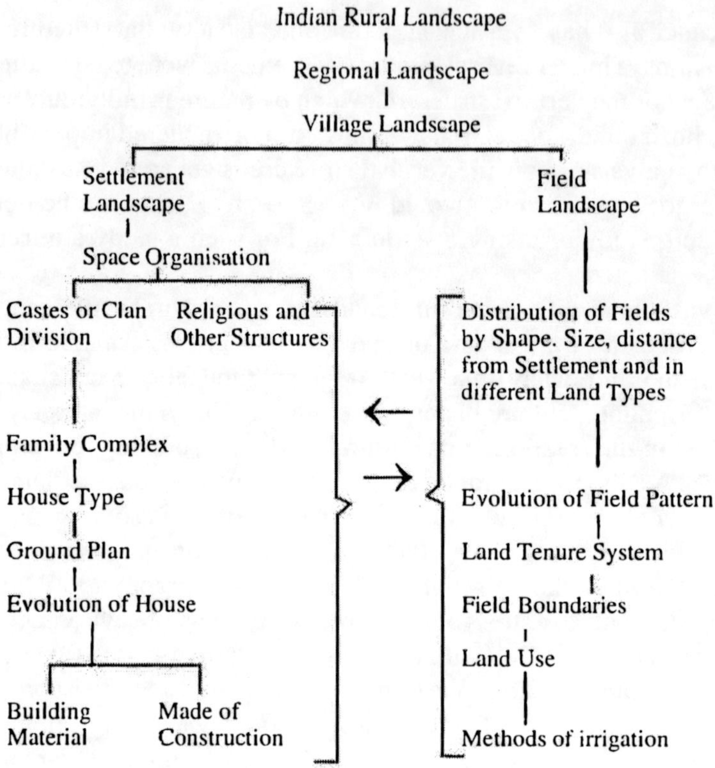

Fig. 4.2 : Indian Rural Landscape

partake something of the evolutionary processes operative in the region as a whole. Subsequently, at a larger scale of observation, coming to the village landscape; the focus would shift to the constituent settlement landscape, the field landscape, the house types, and the field pattern, all forming the core elements of Cultural Geography. A study and observation of the village landscape would involve frequent visits to the village under study and an observation of the physical endowments of its site, the surrounding ecology, the exploitation of which provides sustenance to the inhabitants. Also involved would be an observation of the extent of the village territory, its outer limits, whether formed by natural or man-made boundaries, the expression on the landscape of the boundary or *hadbast*. For any later reference to official records it is vital at this stage to note the *hadbast* number of the village. Within the *mauza* (village territory), the location of the *majra* (settlement proper), whether on the periphery, in the centre or at any prominent point too, should be noted. As mentioned earlier, the core elements, the settlement landscape, the field landscape, the house type, and field pattern, should then be taken up in detail.

Within the purview of the settlement landscape, a careful observation of the organisation of space should be the prime focus. Fieldwork at this stage involves drawing the ground plan of the entire settlement in an attempt to discover any underlying principles of space organisation, particularly the relationship of one house to another, of a house with the street, and of one street to another street. An identification of any division of the settlement area on caste, clan or family lines, subtly revealing the operation of the ritual-purity pollution principle[13] should be attempted. This must be done simultaneously with collecting information from the village elders about the establishment of the settlement, its name, any cultural, ecological, or historical associations. Also should be noted information regarding the location of the oldest house, the gradual evolution of an individual house, phases, if any, in the evolution of the individual house, and the processes, whether additive or divisive, involved in its evolution. Similarly, stages in the evolution of the family complex and ultimately the settlement as a whole should also be observed and noted. Collection of such information must necessarily be accompanied with sketches. Composition of the society, its caste and clan composition, specific cultural dictates and their expressions on the landscape, must be keenly observed. Religious and other buildings on the landscape, their symbolism, their function, and their significance in the mode of living of the inhabiting group should be found out.

Further, the house itself should be the focus of study, being the most vital expression on the landscape of both the culture of the inhabiting culture group and its *genre de vie*, by which the group sustains itself. A detailed diagram of the ground-plan or partitioning of the house space must first be drawn, and the constituent components of the house, be clearly marked. A close observation of the house components, the living rooms, rooms for cattle, rooms for storing grain or agricultural implements, the courtyard, verandahs, kitchen hearth area, and any other structure within the precincts of the house, and their arrangement in the context of the culture of the people as also the available ecological context, would be fundamental to the understanding and study of any individual rural settlement in the context of Cultural Geography. An attempt should be made to understand the perception and recognition of the components of the house as social, domestic, and private realms, offering different levels of familiarity in inter-family and intra-family interactions. This would highlight the functional, cultural, and religious symbolism of the house. Variations in the arrangement of the house components in different houses must also be noted as also the reasons for doing so. Information regarding the local vernacular names given to the different rooms and other integral parts of the house, their functions, the use to which they are actually put, and religious symbolism attached, if any, should be accompanied by relevant sketches and photographs of the components of the individual house. Finally, information about the building material used and the mode of construction, the architectural design and symbolism attached to it, must also be gathered.

Any study of the rural landscape would be incomplete without the analysis of the field pattern. The field landscape is essentially an expression of the community decision through their culture and mode of living. A study of the field landscape, therefore, entails not merely a description of the fields in terms of their size, shape, distance from the settlement, their location in the different land types, and their land use, but also an attempt at analysing the evolution of the field pattern through time and in space. The field pattern is not a static phenomenon, nor is the present day field landscape a spontaneous creation. It has evolved gradually over a period of time, each phase being closely related to, and in fact, the product of, the then existing political conditions, land tenure system, technological know-how, the culture group and their *genre de vie*. As Nitz observed, 'it is, therefore, obvious that it is not only important to describe the various field patterns, but also, as far as possible, to trace the origin and evolution of the patterns in order to find out the socio-economic and political factors in operation at the time of the settlement'.[14]

Such a study of the field landscape involves first securing from the village *patwari* in the *tehsil* or district headquarter, the village map showing the field boundaries. Before proceeding to the field, it is important to first trace the field map and also note the number of each field. What is important at this stage is to secure access to all the unpublished records pertaining to the village under study. These are generally available with the *kanungo* at the district headquarter in a bundle known as the *basta* and indentified by the village *hadbast* number. This contains, in addition to village maps showing the fields, their boundaries and identification numbers, prepared at different intervals of time, the *khasra khatauni* or the record of field ownership, the *khasra girdauri* or field-wise record of the land use, all documents and agreements pertaining to the village territory, and exchange, surrender in lieu of debt or otherwise, or acquisition of new fields, and the *sajra nasab* or genealogical tree for the individual settlement, which usually goes back to the founder of the village. While these records are fundamental to the study of the field pattern, they are also vital to the reconstruction of the evolution of the village territory and the field pattern. In reconstructing the evolution of the field landscape it becomes important to prepare a diagrammatic representation of the genealogical tree of the settlement under study with the help of the *sajra nasab* and the *malkiyatnama* or record of ownership wherein, in addition to the names of the owners of each field, the names of their fathers too are recorded. This helps in preparing backwards the genealogical tree to the very founder or founders of the settlement under study. As a next step, the present day ownership of the fields should be plotted on the latest available map showing the field boundaries. Subsequently, by re-tracing the generations with the help of the names of the fathers of the present owners, and a regrouping of the fields as per their present ownership to an earlier generation, the ownership landscape of the fields as it then existed is created. Simultaneously, on the map is revealed the distribution of the fields by their shape, size, distance from the settlement, and in the different land types. Thus, is recreated one phase in the evolution of the present field landscape. Similarly, maps pertaining to different periods and generations revealing the manner in which fields have been divided, sub-divided and sometimes even regrouped, can be prepared. An analysis of these maps in the context of the existing land tenure system during that period shall help in the discovery of any underlying principles and cultural dictates in the division of fields among colatterals of a single generation, and the sub-division of fields down the line of generations, contributing to the evolution of the field pattern and the field landscape. The maps shall also reveal the gradual extension, if any, of the land under cultivation, reduction in the extent of area under *shamilat* or common land, and all changes in the shape and size of the fields. Thus, the present day field landscape as it exists today becomes a key to the understanding of the processes of its evolution through time.

On the ground, the type of field boundaries, the methods of irrigation used and the irrigation channels, the crops grown, any perceived divisions of the fields meaningful to the farmers alone, should all be closely observed as also information regarding them should be collected by interview.

Maps of the fields should also be prepared according to their size, in fact fields should be classed into different categories of size, and a map showing these related to the map showing the different land types should be carefully studied and analysed to discover the underlying relationship of the field size to the distance from the settlement and to the different land types.

At no stage should the evolution of the settlement and field landscapes be considered to be independent of each other. In fact, they are simultaneously operating processes, for generally a division of the extended family into its nuclear components alone initiates a division of the fields owned by that family, resulting in an increase both in the number of households within the settlement landscape and in the number of fields and perhaps a reduction in their size, within the field landscape.

The methodology, therefore, aims at a portrayal of the hierarchy of contents of the landscape. The Indian Village though an abstract entity, has in fact, been abstracted from reality.

Thus, the study of an individual settlement within the framework of Cultural Geography aims at formulating the concept of an Indian Village.

References and Notes

1. Literally every state of India was covered by anthropologists and several of these works were published. See Haimendorf, E. Von Furer (1958). *An Anthropological Bibliography of South Asia*; Paris : Mouton & Co. To mention a few : Marriott, McKim (1952). 'Social Structure and Change in a U.P. Village', *The Economic Weekly*, 25, August. pp. 869-874; Opler, M.E. and Singh, R.D. (1952). 'Two Villages of Eastern Uttar Pradesh, India : An Analysis of Similarities and Differences', *American Anthropologists*, Vol. 54, No 2, Part 1; Marriott, McKim (ed.) (1955). *Village India*, Chicago : University of Chicago Press; Lewis Oscar (1958). *Village Life in Northern India*, Urbana : University of Illinois Press; Epstein, T.S. (1962). *Economic Development and Social Change in South India*, Manchester : Manchester University Press; Nair Kusum (1962). *Blossoms in the Dust*, London : Gerald Duckworth & Co. Ltd.; Karve, I. and Dante. Y.B. (1963). *Group Relations in Village Community*, Poona : Deccan College Monograph Series 24; Patnaik, N. (1964). *Caste and Social Changes: An Anthropological Study of Three Orissa Villages*, Hyderabad; National Institute of Community Development; Fraser Jr., T.M. (1968). *Culture and Change in India : The Barpali Experiment*, Amherest : University of Massachusetts Press.
2. Srinivas, M.N. (1960). *India's Villages*, Bombay: Asia Publishing Co.; Srinivas, M.N. (1966). *Social Change in Modern India*, Berkeley : University of California; Srinivas, M.N. (1976). *The Remembered Village*, Delhi : Oxford University Press.
3. Dubey, S.C. (1955). *Indian Village*, London : Routledge and Kegan Paul; Dubey, S.C. (1958). *India's Changing Villages*, London : Routledge and Kegan Paul.
4. I.C.S.S.R. (1972). *A Survey of Research in Geography*, Bombay : Popular Prakashan, pp. 188-202; Raza, Moonis (1979). *A Survey of Research in Geography, 1969-1972*, New Delhi Allied Publishers Private Limited, pp. 177- 189.
5. To name a few: Mukerji, A.B. (1956). 'Jat Settlements and Habitations, Part-I', *Indian Geographical Journal*, Vol. XXXI, Nos. 3 and 4, pp. 102-116; Mukerji, A.B. (1957). 'Jat Settlements and Habitations, Part II', *Indian Geographical Journal*, Vol. XXXII, Nos. 1 and 2, pp. 13-28; Mukerji, A.B. (1962). 'Bhawanti House Type in Andhra Pradesh', *Geographical Outlook* Vol. III, No.1, pp. 16- 25; Mukerji, A.B. (1962). 'Field Pattern in a Telengana Village', *The Indian Geographical Journal*, Vol. XXXVII, No.4, pp. 153-164; Mukerji, A.B. (1964). 'Succession of Cultural Landscapes in Telengana Reddi Villages', *Indian Geographical Journal*, Vol. XXXIX, Nos. 1 and 2, pp. 42-58.
6. Srinivas, M.N. (1977). *Village Studies and Their Significance, Anthropology in the Development Process*, Hari Mohan Mathur (ed.), New Delhi : Vikas Publishing House, Private Limited, pp. 329-345, ref. p. 330.
7. The *jajmani system* may be defined as a system of service relationship that exists within the hierarchical caste and class structure into which the Hindu rural society is divided. See Darling, M.L. (1934). *Wisdom and Waste in the Punjab Village*, London; Wiser, W. (1936). *The Hindu Jajmani System : A Socio-Economic System Inter-relating Members of a Hindu Village Community in Service*, Lucknow; Gould, H. (1958). 'The Hindu Jajmani System : A case of Economic Particularism', *Southwestern Journal of Anthropology*, 14, pp. 428-437; Lewis, O. (1958). *Village Life in Northern India : Studies in a Delhi Village*, Urbana; Marriott, M. (1960). 'Caste Ranking and Community Structure in Five Regions of India and Pakistan', *Deccan College Monograph Series*, 23, Poona; Rao, M.S.A. (1961). 'The Jajmani System', *The Economic Weekly*, 13, pp. 877-878.
8. Bichsel, Ulrich (1986). *Periphery and Flux : Changing Chandigarh Villages*, Berne : Geographisches Institut der Universitat Berne, p. 5.
9. Wagner, P.L. and Marvin W. Mikesell (1962). *Readings in Cultural Geography*, Chicago : University of Chicago Press, p. 23.
10. Kniffen, F.B. (1974). 'Material Culture in the Geographic Interpretation of the Landscape', *The Human Mirror*, Miles Richardson, ed., Baton Rouge: Louisiana State University Press, pp. 252-267, ref. p. 252.

11. Mikesell, M.W. (1978). 'Tradition and Innovation in Cultural Geography', *Annals of the Association of American Geographers*, Vol. 68, No.1, pp. 1-16, ref. p.4.
12. Kniffen, F.B. (1960). To Know the Land and Its People'. *Landscape*, Vol. 9, p.22.
13. The ritual purity-pollution principle is applicable only to the Indian rural society which is fragmented along caste lines. In the traditional analysis we have the upper castes, *Dvijya* or the twice born and the lower castes, *Antyaj* or the once born. This division is valid as per the Manusmriti and accordingly they must be placed at a social distance. Thus, in order to prevent the violation of the Purity Pollution Principle, whereby the upper castes lose their purity or become polluted by associating with those belonging to the lower castes, in a settlement, the two have separate sectors or *muhallas* separated by the streets, which represent a symbolic distance. Mukerji, A.B. Technical Communication.
14. Nitz, H.J. (1966). 'Geographical Studies in the Field Patterns of Northern India and Germany', *The Geographer*, Vol. XIII, pp. 61-74.

SECTION-II

IMPRESSIONS OF CULTURE ON LANDSCAPE

5

EVOLUTION OF FOLK HOUSE TYPES IN MAHARASHTRA

JAYMALA DIDDEE

Introduction

Philosphy of a house or *Ghar* as it is referred to in Hindi, emerges from the idea of a roof and in the context of rural areas, in India particularly, the implication is that a man attaches himself to a piece of land called his own and hence the neccessity to construct a house. The rural folk strongly believe that one builds a house not for oneself but for one's progeny and for posterity. A man stabilises himself on his territory, forges a permanent connection with it through the house. Hence unlike the west, in India the philosophical mould in which the house is set is vital to our understanding of the different house forms. Attention here is focused on the Hindu house type or *Ghar* because it is the basal group and other groups such as the Muslims, Sikhs or Jains are the sub-strata of the same.

Works on early Indian archietecture[1] have explicitly stated that a house in India cannot be considered apart from the Hindu religion because that, in the first instance, prescribes the rules and the basic forms of the building of a house. Hence, religion is the base on which the town's super structure is built. Dutt clearly mentions that the social stratification in Hindu society was clearly reflected in houses prescribed for each caste.[2] For the Brahmins or the priests, the *chatusala*, the Kshatriyas or the warriors the *trisala,* the Vaishyas or the traders the *dvisala* and the Sudras or the untouchables the *eksala* meaning houses with four, three, two, and one structure respectively.

Concept of a Folk House

Before analysing the folk house types of Maharashtra, some conceptual clarification as to what is meant by a folk house will be relevant to the ensuing discussion.

Folk houses are generally of low scale, being humble dwellings of farming households in rural areas.[3] Folk houses in general reflect vernacular architecutre, which is neither ornamental nor monumental.[4] These houses are constructed with materials from the local ecological niche. There is no formal plan or blueprint as the house form is simple and it is constructed with family and community labour. In any regional type of a folk house, there is a strong tendency for the ground plan, vertical plan and the roof form to repeat itself, regardless of where it occurs within the region. This means that the tradition of the form of the house is preserved in the traditional consciousness of the people, and this design is repeated house after house and owner after owner, which results in strengthening the continuity of the traditional design. Rural folk houses thus, strongly resist change since the rural house type is more a cultural tradition and demonstrates that it is influenced more by tradition and culture and less by environmental deterministic principles.

Methodology and Approach

To study the Traditional Rural House Types of Maharashtra (TRHT) as the folk houses are commonly referred to, field observations were the principal method adopted. Field work to areas in East Maharashtra could not be undertaken and hence Vidharbha, which is one of the five culture-historic regions of Maharashtra, is not included in this discussion.

Maharashtra is traditionally divided into five regions : Konkan, Desh, Marathwada, Vidharbha, and Maval. Maharashtra as a state is unique in respect to its location. Lying mid way between the Aryan north and the Dravadian south, it is in many ways a zone in transition between the two cultural regions. There exists a tremendous diversity in material and non-material culture traits between the North and South and the folk house type is one such material trait, which has distinct regional distribution. But despite this vast extent of diversity in material culture traits, there is overarching unity (pan Indian) which is governed by the Hindu religious traditions.

Therefore, before analysing the major house types of Maharashtra, it would be useful to put in proper perspective the place the folk house type occupies in the larger Indian tradition. The folk house, as the cultural geographer views it, does not reflect the great Indian tradition but it emphasises the great differences in the Little Traditions of different historical cultural regions each characterized by its own ecological resources and modes of living.

The Evolution of the Folk House Type

The antecedents of the folk house type must lie in the early community settlement of pastoral groups, which in due course of time, broke into individual houses.

The *Kraal* of the Masais or the *Corral* of the Tepi Indians can be cited as examples of community dwellings. Today the different forms of folk houses observed on the cultural landscape may have evolved in response to successive phases of sedentarisation in human society. The following sequence is visualized in the climax form of a folk house.

1. Pastoral Nomadism
2. Pastoral Nomadism + Sedentary Farming
3. Sedentary Farming + Pastoral Nomadism
4. Sedentary Farming (remanent cattle culture)

In the context of Maharasthra it needs to be reiterated that agriculture has never been very prosperous with the result that the rural society has been largely egalitarian. In the absence of much economic disparity, the *Balutedari* system as the *Jajmani* system in Maharashtra is called, was not very rigid and though the caste groups in the form of a pyramid, like the all India model do exist, the regional structure is not that rigidly defined as in the Aryan north or Dravadian south. One can even now, with a little effort, discover some vestiges of the original tribal mode of living and simultaneously, a near absence of many of the rites and rituals of the Great Indian Tradition that overlies and partly even obscures the basal diversities of the multitude of Little Traditions.[5] For example, one of the culture traits that rightly belongs to the great Indian Tradition is fire worship, but the *homa* or *havana* is not a must at the time of many of the religious ceremonies all over Maharashtra. Also, all the *rites de passage* that are integral to the Maharashtrian society need not be performed in the open court-yard —which is a must in North Indian society. Such rites even when performed by the Brahmin castes in Maharashtra can be conducted within a room. Thus, both the religious as well as the social significance of the open court-yard centered house type is virtually eliminated in Maharashtra. This is in contrast to the importance of the court-yard in the evolution of folk or rural house types in all of South Asia which has been time and again emphasised by many researchers.[6] All of them have emphasised that the traditional Indian house is essentially a court-yard dwelling. Outside India, studies by Kniffen, Stone and Jordan[7] have all dealt with rural settlements and demonstrate how folk house types are an integral part of the cultural landscape of a region and this material trait is an important attribute to distinguish one culture region from another. Strangely the court-yard in all its guises, though present all over India from Bengal in the east to Haryana, Punjab, and Rajasthan in the west and from Uttar Pradesh, Bihar, Madhya Pradesh and down south in Andhra Pradesh, Kerala and Tamil Nadu, it is in Maharashtra where it ceases to be the starting point of a rural house type, marking the centre as well as the boundaries of the house. The space of the court-yard, except for the Marathwada region, shrinks to the small open space (sometimes covered with a lean-to roof) either in front or at the back of the house. This is a significant difference which sets the folk house of Maharashtra in a class apart in the classification scheme of folk houses. In the case of Maharashtra, with the sole exception of the Marathwada region, the court-yard centered house type, irrespective of the region, is associated with the rural elite or overlords or the village chieftains.

Based on ground plan and roof type, folk houses of Maharashtra can be categorised into four broad types with strong regional distribution (Fig. 5.1).

Beginning with the coastal littoral, the west coastal plain of Maharasthra has the typical four gable roof with walls made of laterite stone. This Konkan house type is referred to as *chaupakee*, meaning a roof with four parts (Fig. 5.2). In south Konkan, the pagoda type compound roof, the smaller four gabled roof sitting atop the lower four gabled roof, has the attic which is used as a living or storage space. The ground plan of the Konkan house has three bays. The front covered porch is called the *oota*, this is the social space where interaction with outsiders takes place and could be designated as public space. All and sundry visitors, irrespective of caste and status affiliation, are allowed entry here. The inner central bay or *varai*, is the family space, here the image of Ganesha is painted on the wall and the altar of family deities is kept here. All members of the extended or joint family have access to this area. There is no sub-division of the *oota* or *varai* even with the partioning of the house. As the family expands the house grows by accretion and rooms are added on the side, with separate entrances. Each household maintains separate hearths. This is personal or private space for intra-family interaction. The *padvi* or cattle shed may be found along the side wall. In the Konkan, the rice fields are small plots and agricultural operations do not require drought animals.

The next characteristic region is the Western Ghats area and can be identified with the Maval house type (Fig. 5.1). Here the house can have a single gable sloping roof with a lean-to roof at the front and back of the house. The walls are made of rough basalt stone or even wattle and mud plaster. The two gable roof is more common with bamboo framework supported by wooden beams. The ground plan, depending on the economic condition of the peasant, will have one *sopa* or room or two *sopas*. Three and four *sopa* houses are less common. In a house with two *sopas*, the first is the *oosari* and is used for sitting and has the bed, while the family cow or bullock is also tied in one corner. Thus man and beast live together in many Maval houses due to poverty. The middle room is the *mazgar* and is used for storing grain, as a kitchen and a small niche in the corner is reserved for bathing. If a third bay is present it is generally at a lower level and is used for cooking, storing fuel wood and here the water pots are also kept. Most houses have lofts for storage in the *mazgar*. In one corner of the

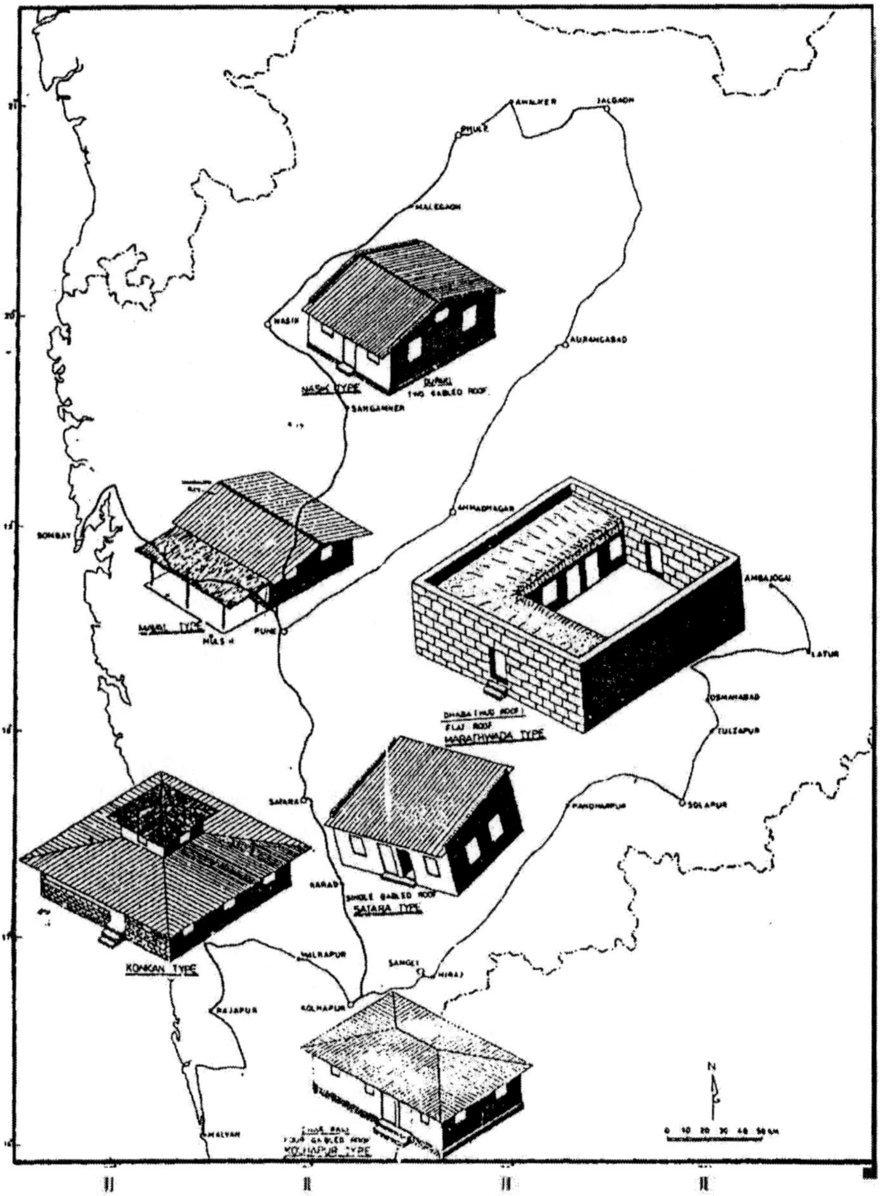

Fig. 5.1 : House Types of Maharashtra

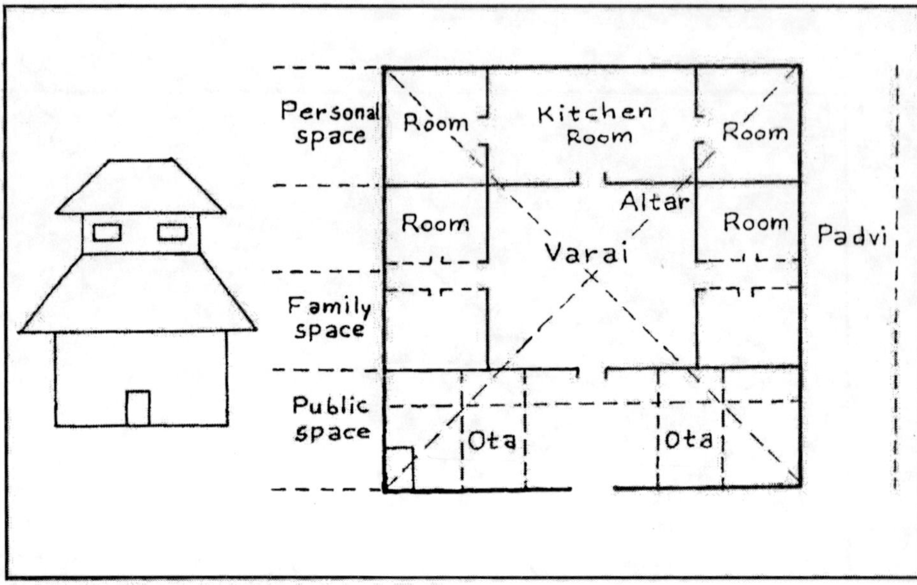

Fig. 5.2 : Konkan House Type

kitchen, the family altar is placed, while the *oosari* has gaudy pictures of the gods and filmstars. It is not un-common to find a television and a clothes line, which acts as a wardrobe, with jeans and t-shirts hanging! There are niches in the wall which also function as cupboards. There is very little movable furniture, except an iron bed in the house. In a number of areas in the Maval tract the houses are enlarged by extending the roof and sub-dividing the rooms by partitioning them. The *oosari i.e.* the porch and the *angan* which is the open space in front of the house, where the sacred basil plant is kept, is rarely sub-divided and is shared by the joint family. The family house resembles an 'l' or long house.

The third house type is the Desh (Fig.5.3) type and this region lies to the east on the leeward side of the Western Ghats. Here, the Pune and surrounding region has a two gable sloping tiled roof with walls made of stone. The poor have mud walls and corrugated iron sheets for roofs. The internal arrangement or ground plan is quite similar to the Maval house type except that the cattle are always penned at the side in a separate enclosure. The roof is commonly made of tiles and the houses have at least three bays or rooms. In southern Desh, and in the north Nashik and Khandesh area with more prosperous agriculture, the houses are sometimes of four gables and some may be of the *wada* type. The *wada* is a mansion like house with high walls made of dressed basalt stone with

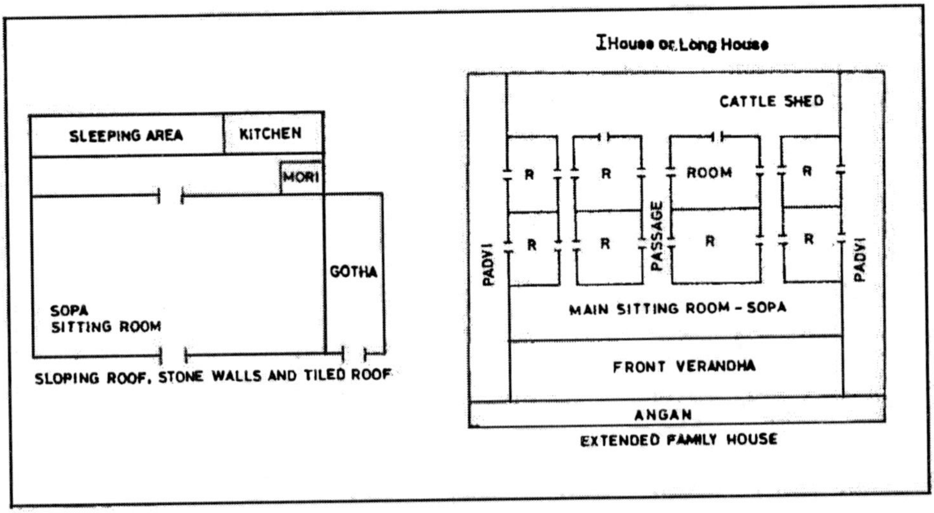

Fig. 5.3 : Desh Type (Shot Gun House)

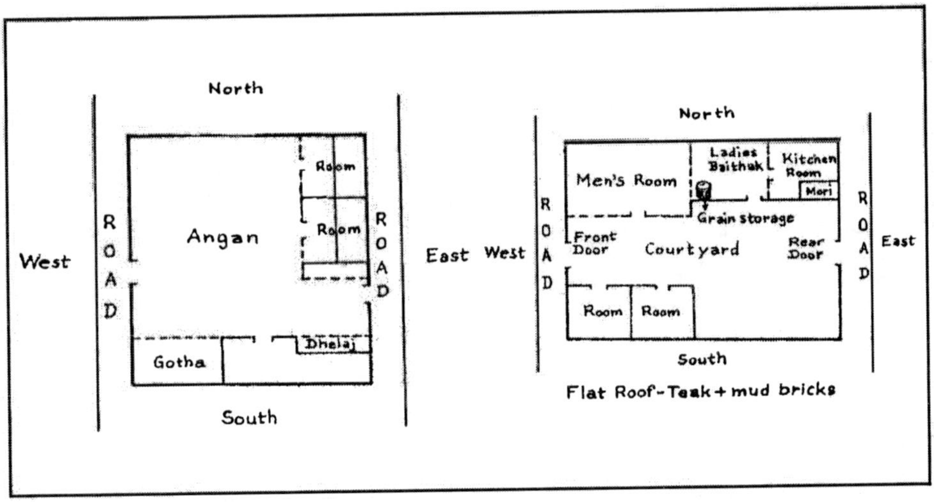

Fig. 5.4 : Marathwada House Type North

imposing teak doors, with carved eaves. There is generally a central court-yard surrounded by several rooms on all four sides. These *wadas* are generally two storeyed structures and sometimes they can accomodate several families of the same clan.

The fourth type of the folk house is the Marathwada type which is totally distinct both in the ground plan as well as the roof form (Fig. 5.4). This area lies in the south central dry core of Maharashtra. It has been an area of constant friction - a marchland between the Marathas on the west and the Bahamani kings in the east. Later this region formed part of the Nizam's territory and was included in the erstwhile princely state of Hyderabad. Here the house forms undergo a drastic transformation. The court-yard centered house appears on the landscape here with a flat roof which is the dominant house form. It would be tempting to suggest here that the flat mud or *dhaba* roof and the central court-yard are related to the arid conditions prevalent in these parts. This on close scrutiny appears far too simplistic an explanation, because historically the entire Marathwada region, until recently had been part of the Nizam's domain and subject to deep and pervasive muslim influence, epxressed in the form of *purdah* (veil) and seclusion of females is an important consideration here which is clearly reflected in the ground plan of this regional house type. Gender-wise separation of space is very clearly marked here in contrast to the Desh, Konkan and Maval house types, where only social and personal spaces are defined. The widespread presence of the enclosed court-yard in the Marathwada folk house type appears to be associated, if not altogether explained with the marchland conditions prevailing here till very late in the history of the Mughal period and even later.

References and Notes

1. Acharya, P.K. (1927). *Indian Architecture According to Manasara Silpasastra*, London: Oxford University Press.
2. Dutt B.B. (1977). *Town Planning in Ancient India*, Delhi: New Asian Publishers.
3. Kniffen Fred B. (1965). 'Folk Housing : Key to Diffusion', *Annals, Association of American Geographers*, Vol. 55, No. 4, December, pp. 549-577.
4. Rapoport, Amos (1969). *House Form and Culture*, Englewood Cliffs, N.J. : Prentice Hall, Inc.
5. Redfield, Robert (1955). *The Little Community*, Chicago: Chicago University Press.
6. Acharya, P.K. (1927). *Indian Architecture According to Manasara Silpasastra*, London: Oxford University Press; Singh, R.L. (1957). 'Typical Rural Dwelling in the Umland of Benaras (India)', *National Geographic Journal of India*, Vol. 3, No. 2, June, pp. 51-64; Mukerji, A.B. (1962). 'Jat House Types', *Geografia*, Vol.1, No.1, pp.27-37; Mukerji, A.B. (1962). 'Bhawanti House Types in Andhra Pradesh', *Geographical Outlook*, Vol. 3, No.1, May, pp. 16-25; Mukerji, A.B. (1967). 'Musalman House Types in Meerut Rural Area', *Folklore*, Vol. 8, No. 4, April, pp. 116-130; Mukerji, A.B. (1976). 'Rural Settlements of the Chandigarh Siwalik Hills: A Morphogenetic Analysis', *Geografiska Annaler*, Vol. 58, Ser. B, No. 2, pp. 95-115; Sinha, V.N.P. (1969). 'Houses and House Types of Chota Nagpur Plateau', *The Geographical Knowledge*, Kanpur, Vol. 2, No. 2, July, pp.91- 101; Mitra, A. (1966). *Report on House Types and Village Settlement Patterns in India*, Delhi: Manager of Publications
7. Stone, Kirk H. (1965). 'The Development of a Focus for the Geography of Rural Settlements', *Economic Geography*, Vol. 41, Oct., pp. 346-355; Kniffen, Fred B. (1936). 'Lousiana House Types', *Annals, Association of American Geographers*, Vol. 26, pp. 179-193; Kniffen, Fred B. (1965). 'Folk Housing : Key to Diffusion', *Annals, Association of American Geographers*, Vol. 55, No.4, Dec., pp. 549-577; Jordan, Terry G. (1966). ' On the Nature of Settlement

Geography', *Professional Geographer*, Vol. 18 (1), pp. 26-28; Jordan, Terry G. (1978). *Texas Log Building: A Folk Architecture*, Austin: University of Texas Press; West, Robert C. (1974). 'The Flat-Roofed Folk Dwelling in Rural Mexico', *Geoscience and Man*, Vol. 5, June, pp. 111-132; Winberry, J.J. (1977). 'Ecology and Etymology of the Rural Folk House: Example of the Log House in Mexico', in R.C. Edit, K.N. Singh, and R.P.B. Singh (eds.), *Man, Culture and Settlement*, New Delhi: Kalyani Publishers, pp. 170-175.

6

THREE RURAL HOUSE TYPES OF THE RAVI RIVER VALLEY

A Culturo-Geographical Analysis

BHUPINDER SINGH MARH

Introduction

The Rural House Type

A rural settlement comprises a number of visible cultural artefacts such as field patterns, cluster of houses, streets, non-residential structures, and house types which are very intimately related to both the culture of the people occupying the area and also the physical environment of the area. One component of the cultural landscape of rural settlements which can be considered the best indicator of both the culture of the group of people occupying the area as well as the physical environment of the area is the traditional rural (or folk) house.[1] Such a house has a place of primacy in the settlement structure and therefore, should be given special attention in a cultural—geographical analysis of rural settlements. The settlement is a collection of houses,[2] and a traditional rural house expresses'the distinctive social organisation of its inhabitants'.[3] Thus, the analysis of the house types does lead to the discovery of the social organisation of its occupants.[4]

Genre de vie or the mode of living of the people is the element of culture which has great impact on house types. Max Sorre expresses the relationship between *genre de vie* and cultural landscape as follows... 'As assemblages of techniques, *genres de vie* are active expressions of the adaptations of the human groups to their natural settings. The specialisations and stability of this adaptation depends on the specialisation, stability and durability of the *genre de vie*, and the local changes in the former are expressed in the variations in the latter'.[5] By the same logic the house types are an expression of the interaction between the *milieu* (or environment) and *genre de vie* of the people. The rural house is an important key to clarifying a settlement structure. The roof and plan of the rural house vary from place to place as they reflect the physical environment as well as the specific way of living of the people of an area. The rural house type, thus, can be considered as a regional indicator.

House type 'reflects the cultural heritage, current fashion, functional needs, and positive and negative aspects of non-cultural environments'.[6] Kirk H. Stone has given so much importance to the study of house types in settlement geography that he has defined settlement geography as ' ...the description and analysis of the distribution of the buildings by which people attach themselves to land ...'[7] Although the houses as such are never permanent and they keep on being mutilated and rebuilt simultaneously yet nevertheless they maintain their permanent features[8] through cultural persistence. Thus, the houses keep on reflecting the intitial conditions under which they were developed and of course their present form also reflects different waves of cultural change by way of consequent incorporations. The house types '.... do not always, but in most cases, reflect precisely the particular environment in which the particular dwellings are situated'.[9] This might be the case when the inhabitants of a particular area have migrated from a different culture ecological zone. The original culture persists in some measure in the new area of occupance, despite the changed *milieu*. A study of house types is, thus, also a study of the people, their origin, the cultural hearth from which they dispersed, the routes of dispersal, the region of ultimate settlement, and their adaptations to the new environment.[10]

The dwellings referred to as house types are not the elegant mansions made very carefully by architects and engineers. These are the ' habitations of rural areas which are inhabited by human groups dominantly engaged in the arts and occupations of agriculture and animal husbandry'.[11] They are rural in origin, function, and characteristics and do not need any architectural designing by an expert.[12] They are built by the owner either singly or with the help of men who constitute his social interaction group. The house is built according to a customary plan and does not require formal training. The idea of a simple architectural design is passed from generation to generation, and houses with basically the same attributes go on being built generation after generation.[13]

The term house type itself needs clarification. A house, the like of which is not found elsewhere, meaning thereby that it represents none but itself and is totally unique, cannot be designated as a house type. A house type is both spatially and temporally representative. It represents a region and also a historical period. A house type is distributed through an area where the cultural groups for which that house type is characteristic have settled themselves during different periods of time.

In the light of this general background to the study of rural house types, three house types of the Ravi river valley, *viz*. the Churahi house type, the Gaddi house type, and the Sahoo house type, are the focus of this study. These three house types are occupied by people who have migrated into the study area from the adjoining Punjab Plains during different periods. Though at present these house types are found in areas having comparable environmental characteristics, yet the house types are appreciably different from each other.

The Study Area

The area of study is the upper part of the Ravi river valley comprising the Bharmour, Chamba, Churah and Saluni tehsils of the Chamba district of Himachal Pradesh (Fig.6.1). The Ravi river and its sub-valleys *viz.* Budhil, Sal and Siul valleys have mostly remained isolated and sheltered from the outside influences due to inhospitable terrain making access difficult. But some waves of influence from the Dogri and Punjab Plains' culture areas have been able to reach this area specially during the periods when the adjoining plain areas experienced onslaught of foreign invaders forcing people to take refuge in the inhospitable mountainous areas.

The Ravi river valley is located sandwiched between the Dhaula Dhar in the south and the Pir Panjal range in the north. The whole of the valley, except a narrow northwest-southeast trending strip of conglomerates, is comprised of gneisses and granites. The western part of the valley, lying in the Churah and Saluni tehsils, is drained by a right-bank tributary of the Ravi known as Siul *Nadi.* This in turn, is fed by the Chanju, Bara, Bera, and Matin *Nalas*. In the Chamba tehsil, Ravi is fed by several tributaries on both the banks, the chief among them being the Sal *Nala* which meets the Ravi near Chamba town. In the eastern part of the Ravi basin, lying in the Bharmour tehsil, Ravi is fed by one main right bank tributary known as the Budhil *Nadi*. The valley of the Ravi river is fairly wide in its main part near Chamba town and local relief is considerable. At places steep gorges having precipitous valley walls are encountered. On valley slopes are found small settlements associated with small terraced fields.

Almost the whole of the study area is located in the middle mountain zone having elevations ranging between 3,500 and 10,000 feet above mean sea level. The climate is temperate with cold to severe winters. Snowfall is heavier in Bharmour than in Churah and Saluni tehsils. Chamba tehsil receives small amounts of snowfall. Transhumant groups, *viz.* Gaddis and Gujjars inhabit the different parts of the Ravi valley.[14] The Churah and Saluni tehsils are predominantly inhabited by Ranas and Rathies with Gujjars found in the higher parts. The main tribal group of Chamba tehsil is Gujjar, although appreciable numbers of Gaddis and other groups are also found. The Bharmour tehsil is inhabited by the Gaddis, although the Gujjars are also found in lower areas.

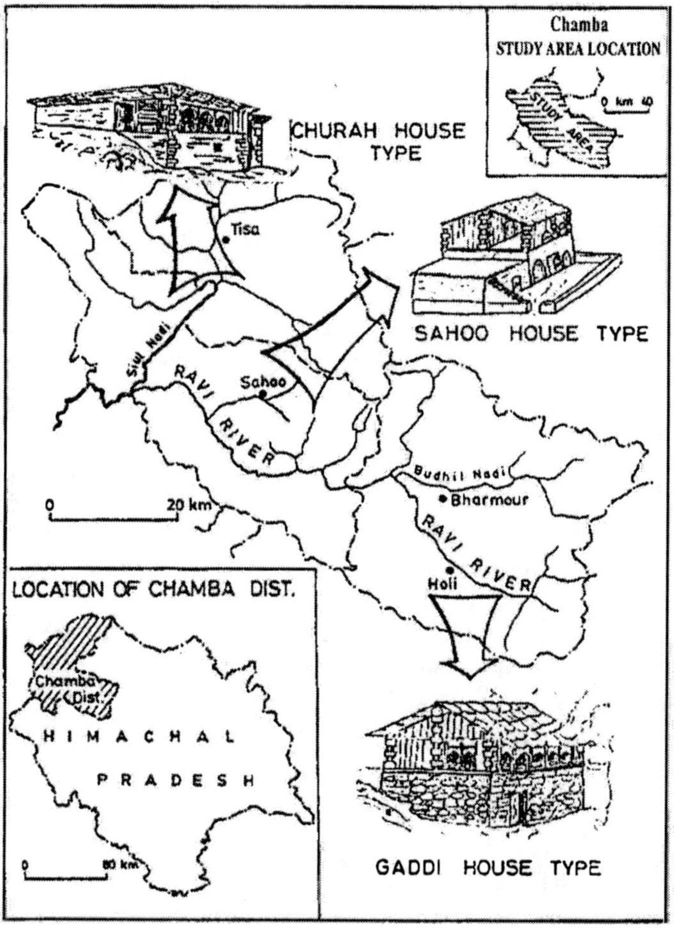

Fig. 6.1 : Ravi River Valley House Types

There are thus, two transhumant tribes, the Gujjars and Gaddis, inhabiting the study area. There are some noteworthy differences between the two groups. The Gujjar transhumance is based on buffalo raising without recourse to agriculture, while the Gaddi transhumance involves both permanent agriculture in their permanent villages in Bharmour tehsil and sheep herding. Gujjars rarely own land, Gaddis invariably do. Gujjars are Muslims, whereas the Gaddis are Hindus.

Data Collection

Most of the data and information used in the present study have been collected through field survey, though some information was also obtained from the series of Village Survey Monographs published by the Census of India (1961). Documentary sources about the mode of living of the people have been obtained from published articles.[15]

The information regarding house types was collected by making traverses through the study area, a method first adopted by Kniffen.[16] Houses were observed, measured, photographed and line sketches made of them. Ground plans were recorded and different compartments were labelled according to their use.

The peoples' ideas about the houses and the cultural antecedents of the houses were collected through interviews. This method was also used to obtain information about house construction methods.

Classification of House Types

One of the main problems encountered in the classification of house types is that, unlike classification in biological sciences, we cannot put any element of house type under a definite category. 'A biologist never finds the tail of a lion grafted to the body of a cow; the classifier of cultural forms has no such assurance'.[17] A typical European chimney may be found attached to an Indian folk house. A cultural geographer dealing with houses 'finds that what he suspects to be the minor variants of a theme have a common areal distribution, he may, with some assurance, group them together'.[18] Thus, while studying the house types of an area it is necessary to categorise them so that houses having similar characteristics are put under one category.

The roof form and its construction material are very important characteristics and help in the interpretation of house form. The roof form and its material are often given so much importance that house types have been distinguished only on this basis.[19] The following roof types have been distinguished

(I) Flat Roof:

(a) Covered with earth and thatch.
(b) Other materials.

(II) Sloping Roof:

(a) *Gable Roof:*
- (i) Slate shingles.
- (ii) Wooden shingles.
- (iii) Other materials.

(b) *Pyramidal Roof:*
  (i) Slate shingles.
  (ii) Wooden shingles
  (iii) Other materials.
(c) *Hip Roof:*
  (i) Slate shingles.
  (ii) Wooden shingles.
  (iii) Other materials.
(d) *Shed Roof:*
  (i) Slate shingles.
  (ii) Wooden shingles.
  (iii) Other materials.

An attempt has often been made to correlate the roof form with the amount and form of precipitation. Some correlation between the angle of the roof slope and the amount of precipitation has been suggested.[20]

The house is a real cosmos in every sense of the word—it is a micro-cosm within the macro-cosm in which we live. Every person tries to build his house in such a way that its form resembles the concept of the form of the universe held by him. The house is partitioned and different components arranged in a manner dictated by the complex of cultural and natural conditions of the area.[21] The partitioning of the private space within the house reflects the cultural beliefs of the people. The classification of the house types should, therefore, give due consideration to house form and plan and should try to trace the cultural antecedents of these traits.

Many scholars have classified houses on the basis of the predominant building materials used. This is justified by the fact that generally the building material used is available in the local area and can be manipulated and moulded conveniently. Thus, the building material reflects the regional character of an area.

The Ravi river valley contains house types which can be distinguished on the basis of roof types (Fig. 6.1). The Gaddi house type is a gable-roofed house using slate shingles. The Churahi house type is a flat-roofed house covered with mud and thatch. The Sahoo house type, the distribution of which forms a transition between the earlier two house types, is a house which is partly gabled and partly flat-roofed.

Churahi House Type

The Churahi house type is distributed in the Churah and Saluni tehsils of the Chamba district towards the western margin of the Ravi river valley. The houses of this type are flat-roofed and mostly single-storeyed. The houses appear like rectangular blocks inserted into the mountainside, making terraces. Often the flat roof of one house serves as a court-yard for the one just above it.

Flat-roofed houses have been mainly associated with areas of scanty precipitation, but this relationship does not hold in the case of the Churahi house type. In the area of distribution of this house type, heavy rainfall from July to September and heavy snowfall are experienced during winters. Thus, cultural antecedents of the flat roof need to be explored.

According to one theory the inhabitants of the area are derived from the Aryan culture group, who came into India from their Central Asian homelands in several waves during the Rigvedic period (1500 B.C.).[22] The sedentary farmers of the Central Asian region as well as those living in the adjacent Tibetan region make flat-roofed houses.[23]

According to another theory, suggested by Barnett[24] and supported by Hoernle and Grierson, the Aryans came from Persia. Some of them must have entered the geographic *cul-de-sac* of the Himalayan valleys such as the Ravi valley. Archaeological excavations in the West Asian countries as well as in the Punjab Plains have suggested that the houses made by the earliest Aryans had flat roofs. After having entered the Himalaya they became isolated and insulated from the external cultural influence. Their culture as well as the house form has changed little over the centuries. This point has been emphasised by Goetz thus, 'All the world over, high mountain ranges have served as refuges to the remnants of ancient peoples and civilisations in which, comparatively unmolested, they could preserve their identity and traditions, some till the present day.[25]'

The Churahi houses are very poorly ventilated and the entrance door is the only opening. A small circular hole for the exit of smoke is made in the roof. Absence of windows and other such openings could be due to very cold winters (minimum temperatures falling to a few degrees above zero centigrade). In order to protect themselves from the intrusion of the cold winds they either construct only a few windows, or do not construct them at all.[26]

The houses of the poor comprise only one multi-purpose room which is used both by the family members and their cattle. Area allocated to animals is separated by a low row of stones. These houses do not have any court-yard. The more elaborate houses of the well-to-do, in contrast, have separate rooms for cattle and storage, although it is a common practice to pen cattle in the living room during winters (Fig.6.2). This could be for the purpose of warmth as suggested by Rapoport.[27] A verandah with arched railing Fig.6.3(d) with provision for vertical sliding doors to be closed during nights and winters is a characteristic feature of these houses. A handloom is generally kept in the verandah. Court-yards are found only in a few houses.

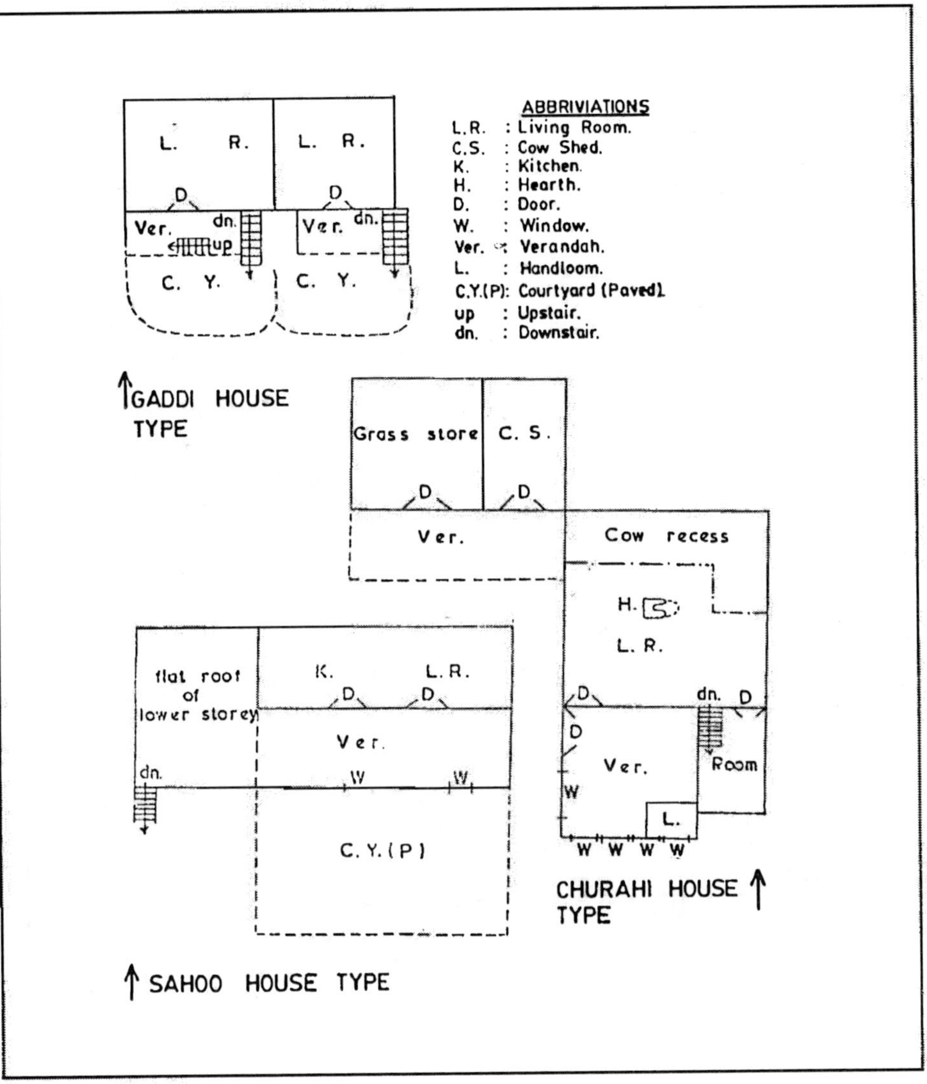

Fig. 6.2 : Ravi River Valley—Typical Floor Plans

Generally the roof of the main room of the house is supported in the centre by one or two wooden pillars. These pillars are finely carved in some well-to-do houses (Fig. 6.4) and not much decorated in the houses of the poor. The presence of a central post *(stambha)* supports the tree model of a house.[28] *Stambha* is a very important element of an Aryan house suggesting that the houses of the study area have antecedents in the Aryan culture.

The building material of these houses is derived from the local area. Locally available stone and wood are the principal building materials. Mutually riveted

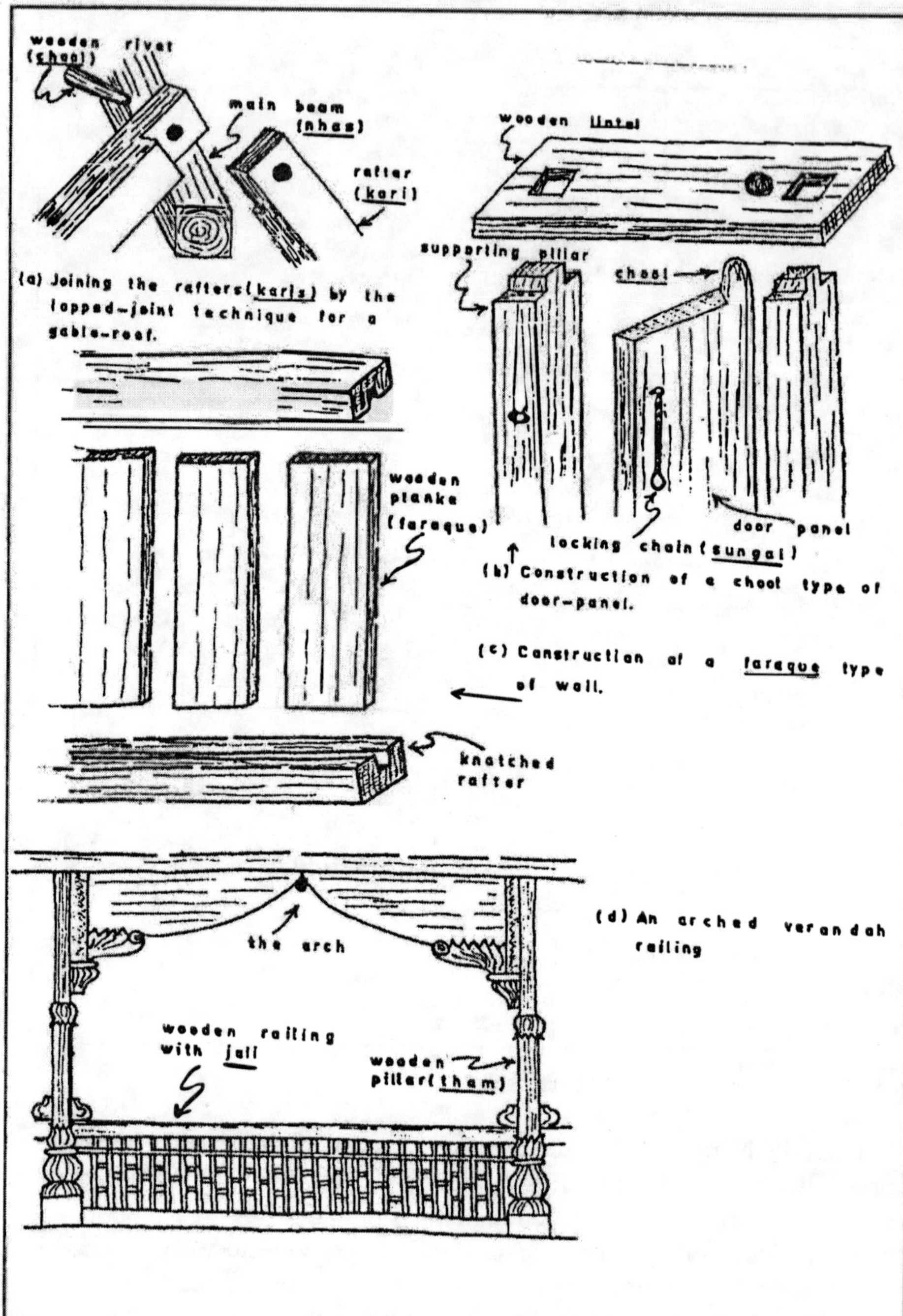

Fig. 6.3 : Elements of Construction

wooden planks are used as re-inforcement in the walls made of stone. At all the four corners of the house and other points of greater stress, hollow wooden pillars *(tholas)* filled with stone are erected (Fig. 6.5). For the roof too, the main building material is wooden. Successive layers of wooden rafters (*karis*), twigs (*chhaniatu*), birch bark (*bhojpatra*) and soil make a thick flat roof (Fig.6.5).

House construction in a Churahi village is a community affair which binds the whole village community together. At least one member from each household participates in the construction of any house in the village. Cash payment is made only to the blacksmith and carpenter.

No general rule is followed regarding the aspect of houses. They face downslope irrespective of the compass direction. Ease of construction, good view of fields located downslope, warming by valley breezes (*anabatic*) and prevention of cooling by mountain breezes (*katabatic*) could be some of the reasons for such orientation.

Gaddi House Type

The Gaddi house type is associated with the Gaddi tribe which is distributed in the Bharmour tehsil, Holi sub-tehsil and the adjacent areas of Chamba tehsil (Fig.6.1). The Gaddis are a pastoral, semi-nomadic, transhumant tribe who have remained isolated from the outside world for long and thus have been able to retain their cultural traits, including house types, in practically their original form.

The Gaddi houses are gabled and are covered with rough and heavy slate shingles. Most houses are three storeys high. A variation of the hip-roof in which the side gables are slightly lower than the main gables is found in Chhitrari village. The pitch angle of the roof is very low which, after Brunhes, could be associated with climate.[29] Brunhes suggests that the snow retained on such gentle roofs works as an insulator.

All Gaddi houses have a verandah on the first floor which is used to entertain guests and to store unprocessed agricultural produce (Fig.6.2). The interior of the house is approached through a narrow, low door. The rooms have low ceilings and very few, rather small windows are provided.

The court-yard (*angan*) is an integral part of a Gaddi house. It is paved with thick slates and is enclosed by a parapet wall about 2 to 3 ft high. The courtyard is used as a threshing ground and to tether cattle, sheep and goats in addition to its use as a place to do other family chores. The importance of a court-yard in a traditional Indian house is well-recognised. 'The central and most important fact in the traditional Indian house . . . . is not the house but the courtyard.'[30] It is the court-yard where the greater part of life of the occupants is spent. The ritual and secular functions of the courtyard in the North Indian houses have been described in detail by Mukerji.[31]

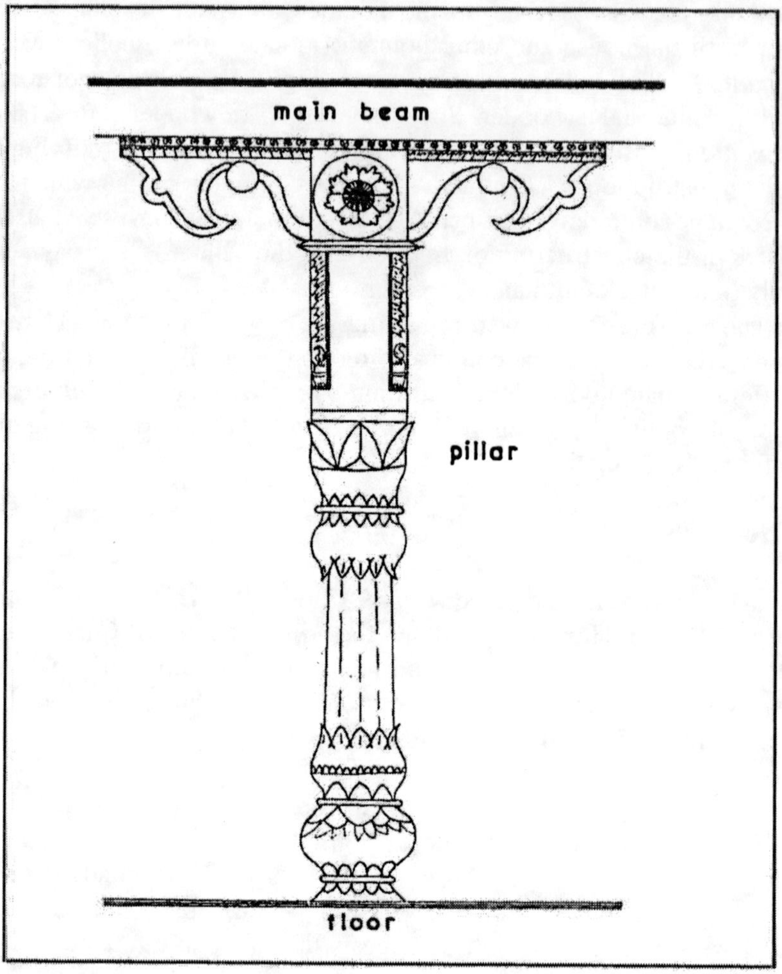

Fig. 6.4 : A Finely Carved Pillar in the Centre of a Dalan or Main Living Room

In a typical Gaddi house comprising three storeys, the ground floor is used to pen cattle and to store fodder and agricultural implements. The second floor is used as living quarters which comprises a room used as a kitchen and another room which is a bed-room. In some houses the kitchen is not a separate room, but an area comprising a raised platform is used for the purpose. This raised platform is considered sacred and is not approached by all. A hearth and a waste water outlet are provided in this area. The third storey is usually used as a place to store grains etc., but in some cases as that of extended families, this floor might be used as living quarters of the nuclear family just separated from the joint family.

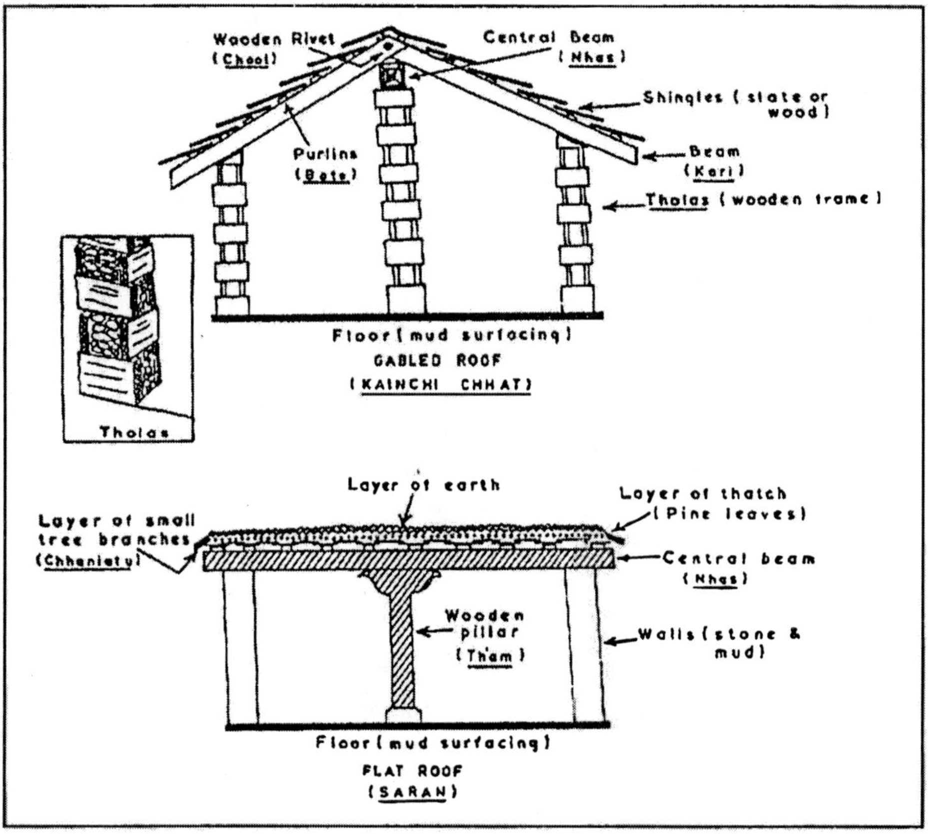

Fig. 6.5 : Elements of Construction

Locally available materials, stone and wood, are used in house construction. Except for the roof, the method of construction is the same as in the Churahi house. Hollow wooden pillars (*tholas*) are erected at corners and points of greater stress. The walls of the first and second floors are made with wooden planks (*faraque*) which are fixed between joints (Fig.6.3c). Rough, coarse and heavy slate-shingles are used for roofing (Fig.6.5). Slate shingles are fixed on one inch thick planks which rest on 2 inches by 3 inches purlins. The purlins are fixed on wooden rafters, one end of which rests on the wooden plates fixed on top of the walls, and the other rests on the central beam (*nhas*) which rests on the *tholas*. These wooden parts are riveted together with wooden rivets.

The origin of the Gaddi house type may be traced back to the 10th century A.D. Punjab Plains. The Gaddis are beleived to be descendents of the Rajputs who escaped the plains from the tyranny of Muslim rulers. The Gaddis developed the gabled roof, either as a response to the local environment, or acquired this technique en route through Kashmir, or from the Gaddis with the Sanskritic

Aryan culture, cannot be ruled out. Linguistic affinities to Sanskrit can be cited to support this belief. The presence of the courtyard, which is-a-well-established Aryan trait ,[32] in a Gaddi house suggests its Aryan origin. Although the rugged terrain and very steep slopes do not favour the presence of the courtyard, the trait, a legacy of the Aryan culture, is incorporated in every Gaddi house.

Sahoo House Type

The Sahoo house type, named after the village of this name, (Fig.6.1) has been distinguished on morphological basis. It combines the characteristics of the gabled Gaddi house type and the flat-roofed Churahi house type. It forms a transition between these two house types. This house type is distributed throughout the Chamba tehsil and the adjoining parts of other tehsils. The main, larger part of this house type is gabled, and a smaller portion with a flat roof is attached to it. The gabled portion is a double-storeyed structure similar to the Gaddi house type. This part is used exclusively for living purposes and comprises a verandah called *van* and a room *obri*. A wooden pillar (*stambha*) is invariably found in both the *van* and *obri* to support the roof. An arched railing with shutters which can close, are found on the front side of the *van*. The *van* is used as a meeting place and for storing unprocessed grain. *Obri*, which has only one door and no windows, is a living room. The upper floor of the gabled portion comprises an open verandah and a room used as a kitchen. This portion is approached through a staircase to the attached flat-roofed part.

The attached flat-roofed portion is only one storey high and is used exclusively as a cattle shed. It comprises one room which sometimes is partitioned by a wooden wall. The two parts are respectively used as cattle shed and store for fodder and implements.

The method of construction and the material used for the construction of the Sahoo house type are similar to the other house types of the study area. The predominance of wood as house building material is seen here too.

Conclusion

Among different traits of material culture, house types is the one which reflects both the physical characteristics of the region and the various aspects of culture and culture history of the people. House types, thus, function as an index of regional identity, and, if correctly interpreted may lead to a deeper understanding of the regional character and culture processes. The three house types of the Ravi river valley analysed in the present study confirm the belief that the house type is a regionalindicator being an index of the totality of environment, culture, and culture history. It has been found that the distribution of the gabled and flat-roofed types does not support the precipitation hypothesis. The areas of Churahi and Gaddi house types have more or less similar precipitation characteristics, but have contrasting house types. The explanation probably lies in the culture history of the occupants. The people of Churah are the descendents of the Aryans who came into the study area from the Punjab Plains during the first phase of the Aryan migration (1500 B.C.). These people who came from either Central Asia or the Middle East brought with them the technology of flat-roofed houses. This has persisted here due to relative inaccessibility from the outside world. The Gaddis, on the other hand, came into the mountains during the 10th century A.D. from the Punjab Plains. Their route of migration was through Kashmir. They acquired the technology of gable roof construction either *en route* in Kashmir or learnt it from the indigenous people in the study area. Both the Churahi and Gaddi house types, thus, reflect their historical antecedents. The use of local building materials has probably been learnt from the indigenous tribes. The technology for the construction of walls, which uses large amounts of wood in contrast to the houses in the Punjab Plains, was probably a local development or acquired from the indigenous people.

The courtyard, which is a well-established Aryan trait, has been observed in two out of the three house types of the study area, *i.e.* in the Gaddi and Sahoo house types. Although the central courtyard was noticed rarely, the open courtyard is a common sight. Despite the non-availability of adequate flat ground, the courtyard is made by making high retaining walls. The absence of the courtyard in the Churahi house type is associated with topographic limitations.

The Sahoo house type which combines the morphological characteristics of the other two house types is more likely to be a local development. This house type is distributed in the transition zone between the other two house types.

Thus, it has been found that the distribution of house types in the Ravi river valley is related to the physical factors only to a very small extent. Different elements of culture, culture history (origins), culture ecology (relation of culture to *milieu*), and the culture processes (diffusion and culture contacts) help in interpreting the form of the rural houses to a great extent.

References and Notes

1. Jordan, T.G. and L. Rowntree. (1990). *The Human Mosaic : A Thematic Introduction to Cultural Geography*, 5th ed. New York : Harper & Row Publishers.
2. Brunhes, J. (1952). *Human Geography*. London: George G. Harrap.
3. Houston, J.M. (1953, revised edition 1968). *A Social Geography of Europe, London:* New York: Praeger.

4. Wagstaff, J.M. (1965). 'House Types as an Index in Settlement Study : A Case Study from Greece', *Transactions Institute of British Geographers*, No. 37, Dec. 1965, pp. 69-75.
5. Sorre, M. (1962) 'The Concept of *Genre-de-vie*' in Wagner and Mikesell (eds.): *Readings in Cultural Geography*, Chicago: University of Chicago Press, pp. 399-401.
6. Kniffen, F.B. (1965). 'Folk Housing: Key to Diffusion', *Annals of the Association of American Geographers*, Dec. 1965, pp. 549-577.
7. Stone, K.H. (1966). 'Further Development of a Focus for the Geography of Settlement', *Professional Geographer*, Vol. 18, No.4 pp. 208-209.
8. Brunhes, J. (1952). *Human Geography*. London : George G. Harrap; Fuson, R.H. (1964). 'House Types of Central Panama', *Annals of the Association of American Geographers*, June 1964, pp. 190-208.
9. Fuson, R.H. (1964). 'House Types of Central Panama', in *Annals of the Association of American Geographers*, June 1964, pp. 190-208.
10. *See* ref. 9.
11. Mukerjee, R. (1968). *Man and His Habitation*, Bombay: Popular Publishers.
12. *See* ref. 1.
13. Rapoport A. (1969). *House Form and Culture*, Englewood Cliffs, N.J. : Prentice Hall Inc.; Raglan, Lord (1968). 'The Origin of Vernacular Architecture', in Forter and Alcock (eds.) '*Culture and Environment*,' London: pp. 373-387.
14. Randhawa M.S. and Nath (1959). *Farmers of India Vol. I, Punjab, Himachal Pradesh, and Jammu & Kashmir*, I.C.A.R. New Delhi.
15. Randhawa, M.S. (1954). 'Gaddi Shepherds of Great White Range', in *The Tribune*, June 24, 1954, Ambala ; Kayastha, S.L. (1957). 'Studies of Dhaula Dhar : A Case Study in Human Ecology', in *National Geographical Journal of India*, June, 1957, pp. 65-78 ; Newell, W.H. (1957). 'A Gaddi House in Goshen Village, Chamba State, North India', *Man*, Feb. 1957, pp. 24-55 ; Bose, S. (1967), 'Environment and Occupation in High Altitude', *Geographical Review of India*, December 1967, pp. 8-12 ; Sharma, J.P. (1973). 'Pastoral Economy and Transhumance in the Himalaya Ravi-Chenab Interrfluve', *Geographical Review of India* , September, 1973, pp. 232-247.
16. Kniffen, F.B. (1936). 'Louisiana House Types', *Annals of the Association of American Geographers*, Vol. 26, pp. 179-193.
17. *See* ref. 16.
18. *See* ref. 16.
19. *See* ref. 9.
20. *See* ref. 2.
21. Mukerji, A.B. (1961). 'Jat House Types', *Geographia* Vol. 1, No.1, pp. 27-34.
22. Grierson, G.A. (1916). *Linguistic Survey of India*, Vol. IX, Part IV, Calcutta; Goetz, H. (1969). *Studies in History and Art of Kashmir and the Indian Himalaya*, Wiesbaden; Upadhyaya, B.S. (1976). *The Ancient World*, Delhi : S. Chand and Co.
23. Stein, Sir Aural (1928). *Innermost Asia-Detailed Report of Exploration in Central Asia, Kan-Su and Eastern Iran*. London : Oxford, Clarendon Press; Huntington, E. (1945). *Mainsprings of Civilization*, New York : Wiley : New American Library of World Literature.
24. Barnett, L.D. (1913, reprinted 1964). *Antiquities of India*, Calcutta : Punthi Pustak.
25. Goetz, H. (1969). *Studies in History and Art of Kashmir and the Indian Himalaya*, Wiesbaden.
26. Rapoport A. (1969). *House Form and Culture*, England Cliff, N.J. : Prentice Hall Inc.
27. *See* ref. 26.
28. Shukla, D.N. (1960). *Hindu Science of Architecture.*
29. *See* ref. 26.
30. Mitra A. (1966). *Report on House Types and Village Settlement Patterns in India*, Delhi : Manager of Publications
31. Mukerji, A.B. (1961). 'Jat House Types', *Geographia*, Vol. 1, No.1, pp. 27-34; Mukerji, A.B. (1967). 'Mussalman House Types of Meerut Rural Area', *Folklore*, April 1967, pp. 116-130.
32. *See* ref. 31.

7

THE CHANGING RURAL HOUSE TYPE OF PUNJAB

An Expression of Socio-Economic and Cultural Change

MEHAR SINGH

Introduction

Most studies concerning rural house types have been conducted assuming that the characteristics of rural houses do not change over a fairly long period of time. Although implied in the methodology, no mention of this assumption is generally made, except in few cases where it is quite explicitly stated. For example, Mukerji states that the *Jat* houses in Upper Ganga-Jamuna Doab can be studied in their original form, function and structure because they have resisted change to a great extent.[1] Hart also endorses the same view when he states that 'most houses outlive the men who build them, and in a stable peasant society the house comes to be associated with ancestors. Its size and style tend to become sacrosanct, and its archaic symbols and meaningless details are repeated through the centuries'.[2] It is therefore not surprising that in most studies of rural houses little attention has focused on the changes in house types. It may be true that in most areas rural houses are resistant to change, but there are also areas where rural houses have experienced rapid changes. One such area is Punjab. The rural house type of this area has been experiencing notable changes in its form, function and structure, since Independence. In fact, within just five decades, the rural house type of Punjab has been transformed almost completely. Surprisingly, while evident to even a lay man in Punjab, these changes have escaped the attention of most scholars. Only a few researchers have tried to understand these changes. Krishan for one, studied the impact of population growth on rural housing and observed that because of mounting population pressure, regular adjustments in the physical make up of the house take place in the form of: (a) sub-division or gradual roofing of available space within the house, (b) extensions onto the adjoining space, and (c) shift to the land reserved for *baras*.[3] Manku too in his study of the foot-hill tract of Punjab identified some differences in the old and new Gujar houses.[4] Certain changes in the rural house type of Punjab were also observed by Gill[5] and Singh.[6] However, none of these studies provide a comprehensive picture of the changes which have taken place in the form, function and structure of the rural house of Punjab. The present study tries to fill this gap. The main objective of this study is to identify the changes that have taken place in the form, function and structure of the rural house type of Punjab since Independence and to explain them in the context of contemporary socio-economic and cultural changes. It is also expected that the study will draw the attention of researchers to a very important dimension of rural houses which has not been properly addressed so far. The main questions that this study seeks to answer are:

(i) What kind of changes have taken place in the rural house type of Punjab since Independence, and what will be its future form?
(ii) To what extent are these changes in the rural house type of Punjab associated with the contemporary socio-economic and cultural changes in the area?
(iii) Can changes in the rural house be used as reliable indicators of socio-economic and cultural change?

Methodology

The study is based largely on personal observation, and focuses primarily only on those changes in the rural house type of Punjab, which occurred after Independence. At the outset it shall be relevant to state that there exist in Punjab marked local and regional variations in the rural house type, and that it was not possible to cover all types in this study, which focuses its attention only on the dominant house type of the common farmers (middle and lower middle class). It has been observed, however, that the house types not covered in this study too have undergone changes similar to the selected house type. The changes enumerated in this study are based on the observations made mostly in the eastern Malwa tract comprising districts of Roopnagar, Patiala, Ludhiana and Fatehgarh Sahib.

The study period has been divided into two major phases, that is, the pre-Green Revolution phase upto the late 1960s, and the post-Green Revolution phase, from the 1960s to the present. Each of these phases is then further divided into sub-phases. It may also be stated that in this study the focus is changes which have been incorporated when building a new house at a new site or rebuilding a house after demolishing the previous one at the same site. The changes which are continuously taking place in the existing houses by way of sub-division or roofing of available open space inside or outside the house to meet the ever increasing need for housing in families are not covered in this study. The focus is mainly on the changes which have taken place in the building material, ground plan, vertical plan, and architecture. For a proper appreciation of these changes it was considered necessary to provide first a brief outline of the form, functions and structure of the traditional, that is, the pre-Independence house of the area.

The Pre-Independence House

The pre-Independence house or the traditional rural house of Punjab was a single storey, flat-roofed, *kutcha* structure. Since it was designed by the people themselves for their own use, it reflected their social and cultural values, their level of socio-economic development, their mode of living, their preferences and prejudices, their beliefs and disbeliefs, etc.

The Building Material

The building material used for the traditional *kutcha* rural house of Punjab was mud, thatch and wood. The walls of the house were built using sun dried mud blocks, unbacked bricks or mud in wet form itself. These walls were plastered with mud paste mixed with wheat chaff. The floor of the house too was *kutcha*. After levelling the ground a thick layer of mud mixed with wheat chaff was applied on it. Once this mud layer dried up, a thin coating of mud and cow dung paste was applied to give the finishing touch. The roof of the house was made of wood, thatch, dry earth and mud. Since the most common roof type of this area, is the flat roof, a very strong frame was required for its support. In the traditional house this frame was made of wood. It comprised of *shateers* (logs, about 3 metres in length and having a diameter of about 30 cm) and *balas* (rafters, 1.5 metres in length and 7.5 cm × 7.5 cm in thickness) or *karis* (twisted rafters of uneven thickness which are cheaper than *balas*). The *shateers* were placed on the walls in a horizontal position parallel to the walls along the width of the room at a spacing of about 1.5 metres. *Balas* or the *karis* were then arranged across the *shateers* at a spacing of about 15 cms from each other. For a large sized room this wooden frame comprised of three layers in which, for the lower most layer, wooden logs of greater thickness (about 45 cm diameter) called *latains*, were used. Usually two, and sometimes four *latains* were used for this type of roof. One end of these *latains* was placed on the walls and the other end on a pillar located in the centre of the room. The other two layers of this frame were constructed using *shateers* and *balas* in the same manner as described above. On the wooden frame was then placed a layer (about 15 cm thick) of thatch, usually of *sarkanda* grass (*Sachcharum sarkara*). The thatch was then covered with a layer of dry earth and finally by a thick coating of mud mixed with *toori* (wheat chaff) or chopped grass to prevent it from cracking up.

The Ground Plan

A farmer's house in Punjab (the house type under investigation in this study) usually comprises of two physically separated units. While one of these units is used for residential purposes by the family and is called the *vasughar*, the other is primarily meant for cattle and is known as the *bara* (enclosure). The *vasughar* is commonly located within the *abadi* (settlement), while the *bara* is invariably found on the periphery of the *abadi*. The ground plan of the traditional *vasughar* comprised an elongated room, called *deodi* in front of the *vasughar*, behind which lay the *verah* (courtyard) enclosed on two sides by high walls (Fig.7.1). Beyond the *verah* were the living quarters comprising a large sized living room, an elongated storage room behind the living room, and an elongated *chhapar* (a sort of a verandah with slanting thatch roof) in front of the living room. The roofless kitchen was located adjacent to the *chhapar* in one corner of the courtyard. Thus, while the *deodi* served as a social space, the unit, behind the *deodi* constituted the private space. The *deodi* had two doors, one of which opened onto the street and the other into the inner part of the house. However, these doors were positioned in such a manner that direct view into the inner part of the house from the street was completely blocked. The primary function of a *deodi* was to provide *parda* (privacy) to the private space in the house. Outsiders, especially the males, were not supposed to enter the house beyond this room, wherein lay the females' domain. In a way the *deodi* served as a *ghungat* (veil on the face of a woman) for the house as both perform the same function.

In the traditional house the *verah*, a most essential element of every rural house, was generally squarish or rectangular in shape and covered about one-third to two-thirds of the total area of the house. Most of the daily chores were performed in the *verah* by the women, which was also used for purposes of sleeping by the family for most part of the year. Although the *verah* in a traditional house constituted the private space yet on certain occasions such as birth, death or marriage, it served as a social space, as most of the rituals related with the *rites de passage* were performed here. In addition, as an open space, the *verah* served as the lungs in the body of the house.

As mentioned earlier the actual living area in the house was located beyond the *verah*. A large sized multifunctional room occupied a pivotal position within the living quarters. In the eastern Malwa tract of Punjab this room was also referred to as *latain* (probably because *latains* were used in the construction of its roof). It was generally squarish in shape and fairly large (about 6×6 metres or more) in size [Fig.7.1(a)]. The distinctive feature of a *latain* was a central pillar which provided the necessary support to the roof. *Latains* with two pillars were also not uncommon. However, in some areas of Punjab, such as the *Bist*

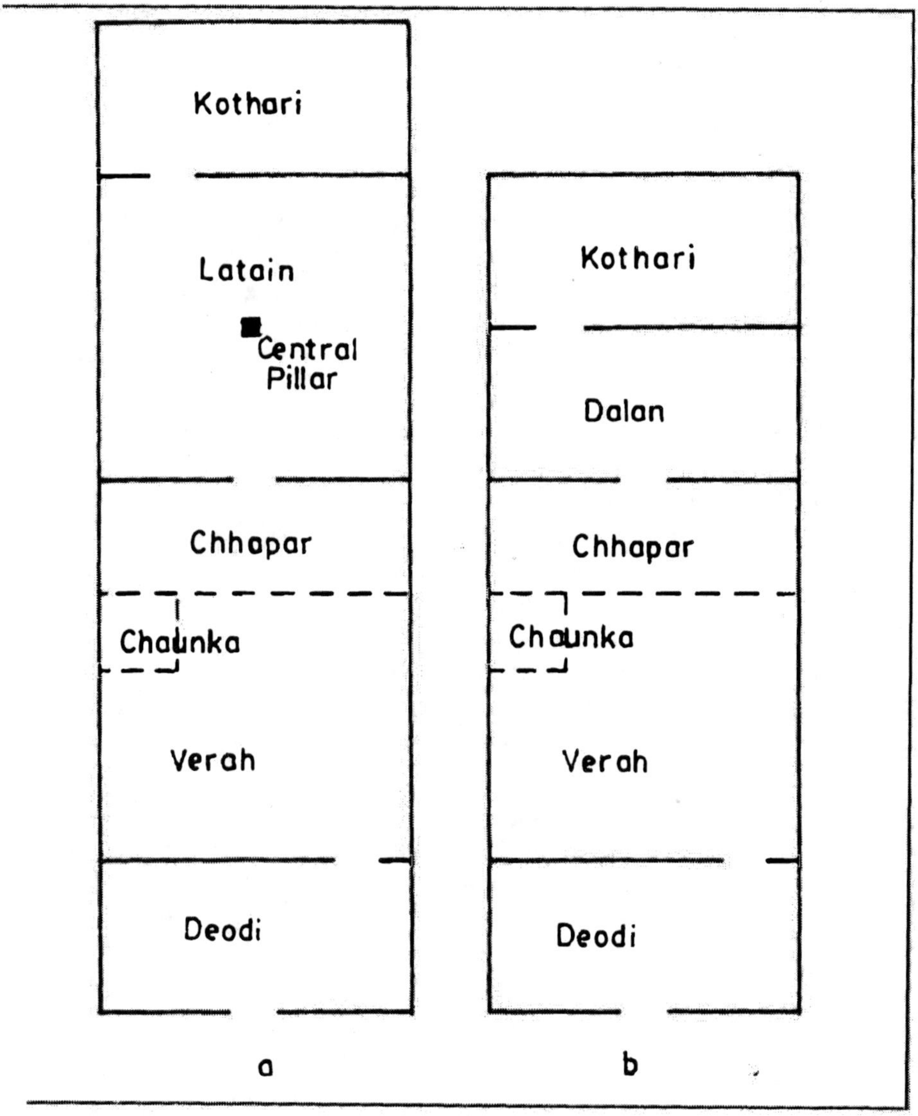

7.1 : Ground Plan of the Pre-Independence Vasughar

ıb (area between Satluj and Beas rivers), the *Majha* (area between Beas and 'i rivers), and in some parts of the *Malwa* tract (area south of river Satluj) an ıgated room of normal width (about 3 metres), known as *dalan* replaced fulfilled the same function as the *latain* [Fig.7.1(b)]. The length of this n, however, varied from house to house usually being around 6 metres, ough in certain cases it could be even more. A single but large sized (about metre) door opening onto the *chappar*, served as an entrance to both the *in* and *dalan*.

Similar to the *verah*, the *latain/dalan* too served a variety of functions, such as storage of grain, putting things of daily use, placement of large sized wooden boxes etc., but most significantly it served as a dormitory for the females of the family, all of whom, whether young or old, married or unmarried, slept in this room during winter. It may not be out of place here to mention that in a traditional rural house in Punjab there were no separate bed- rooms—not even for the married couples. Infact, till very recently, sleeping of husband and wife in a separate room to themselves or even in close proximity to each other, except in the case of newly married couples, was considered indecent and thus, undesirable. The husband and wife slept at a distance from each other, a distance which generally increased with advancing age. As such all the female members in a joint family, slept at one place, that is, the *latain/dalan* in winter, and the central courtyard during the summer. The male members, on the other hand, slept away from the females at different places, in the *bara*, the *deodi*, or on the roof top. Thus, the prevailing value system in society which segregated women even within the family influenced the ground plan of the house in that it necessitated the construction of a dormitory (*dalan/latain*) in the *vasughar*, but discouraged the construction of separate bed-rooms.

At the rear end of the *latain / dalan* was an elongated room, (of the same length as the *latain/dalan* and about 3 metres in width) known as the *kothari* or *pichhla ander* (rear room). Its rear position led this room to be considered as the most secured room, and thus most suitable for storing household valuables. It was also used for sleeping purposes in winter.

The front end of the *latain /dalan* was generally covered with an elongated *chhapar* which served the purpose of a verandah. It was just a slanting thatch roof which sloped forward from the front wall of the *latain /dalan*. While one edge of the *chhapar* rested on the front wall of the *latain /dalan,* the other edge was held in position on wooden or brick supports.

A characteristic feature of the traditional rural house in Punjab was the absence of any covered kitchen area. Cooking was done in the open, a practice still prevalent in a large number of rural houses in Punjab. Thus, instead of a *rasoi* (a proper covered kitchen), one generally finds an uncovered cooking space called *chaunka* in these houses. A *chaunka* is a slightly raised platform varying in size from 2×2 metres to 3×3 metres, with one or two fixed U-shaped *chulas* (hearths). An additional pit formed a circular hearth, known as *hara*, used for slow cooking of *dal* (pulses) and *sag* (green leafy vegetable) or for day long slow boiling of milk, may also be found. A *chaunka* is usually located in the open courtyard along some wall, but always near or adjacent to the main residential part of the house. A low wall (about 0.75 metre high) is usually erected both to demarcate it, as also to provide *parda.* The entrance to the *chaunka* never faces the inner door of the *deodi* but is normally accessed from either side of the living quarters. In the hot summer afternoons or during the rainy season when cooking in the open is either inconvenient or impossible, a temporary roof may be provided on the *chaunka* or else the *chula* is shifted temporarily under the *chhapar / verandah* or to some other sheltered place. It may be pointed out that the use of the word *rasoi* is not very common in rural Punjab. In day-to-day conversation, during rituals and ceremonies associated with religious beliefs and customs, especially at the time of birth, death and marriage, or in Punjabi folk songs and folk tales, the word *chaunka* rather than *rasoi* is used. This is indicative perhaps of the absence of a tradition of constructing a covered kitchen in rural Punjab. It may be interesting to probe this absence but that is beyond the scope of this study.

Further, the absence of a bathroom and toilet in the traditional rural house type of Punjab is another characteristic feature. Men, who spent most of their time in the fields or the *bara,* probably never felt the need of these two facilities in the house. It was only the women who faced inconvenience. They had no option but to bathe in the open courtyard by making a temporary enclosure with the help of cots and also go to the fields for purposes of defecation either before sunrise or after sunset.

Thus, the ground plan of the traditional *vasughar* reveals the strong impact of the status of women in rural Punjab. The presence of the *deodi* in front of the house, the entrance to the house through the *deodi,* offset positioning of the inner and outer doors of the *deodi*, the central location of the courtyard, and the provision of *dalan/latain* (an exclusive dormitory for the females), all indicate that women were restricted and kept strictly under *parda*. Similarly, the absence of a covered kitchen, bathroom, toilet and separate bed-rooms also reveals the insenstivity towards the needs of the women indicating thereby that women were accorded a very low status in the rural society of Punjab.

In traditional farming societies cattle are indispensable to man and thus, a farmer plans not only for the needs of his family but also cattle. A *bara*, which literally means an enclosure, in this context an enclosure for cattle, therefore, is as essential as the *vasughar* for farmers (Fig.7.2). A *bara* may be nothing more than an open space enclosed by a fence, although in most cases a *bara* includes some built-up structures. A large part of the *bara,* however, is kept as an open space used for penning cattle, stocking dry fodder, making cow-dung cakes, drying and storing them, threshing minor crops after harvesting, drying grains, dumping cow-dung etc. The commonly found built-up structures in the pre-Independence *bara* included a cattleshed, a fodder room, a *chhapar* and a *baithak* (sitting room). These structures were arranged within the *bara* in a very simple manner. The cattleshed and the fodder room were elongated and generally constructed adjacent to each other along any one side of the *bara* or in an L-shaped arrangement [Fig.7.2(a) and 2(b)]. Their size depended upon the number of cattle. The front of these rooms was covered by an elongated *chhapar*. Generally both rooms had flat roofs, although in certain cases the gabled thatch roof too was used. Such a structure then was known as *chhan*.

The cattleshed and the *chhapar* in front of it, were used for stall-feeding cattle. However, depending on the weather conditions, the cattle were sometimes kept in the cattleshed, while at other times under the *chhapar*, or in the open. The fodder room, also known as the *toori wala kotha* (wheat chaff room) was used, as indicated by the name itself, for the storage of *toori* or other dry fodder. The part of the *chhapar* in front of this room was used for functions such as installation of *toka* machine (fodder chopper), keeping of agricultural implements, or then parking of bullock cart during the rainy season.

Although meant primarily for cattle, the *bara* is also used by humans. In fact, man and cattle generally live together in the *bara*, with man's presence here being justified by the fact that cattle are the precious possession of every farmer and require careful tending However, another, and a more compelling reason of man's presence in the *bara* was the prevailing social milieu, which till very recently was highly influenced by the *parda* system. Elderly males in the family generally avoided going to the *vasughar*. They rarely crossed the inner door of the *deoḍi* because, as mentioned earlier, the area beyond was the domain of the females. Male guests too were not taken to the *vasughar*. This necessitated the partial use of the *bara* for human beings, particularly males. In

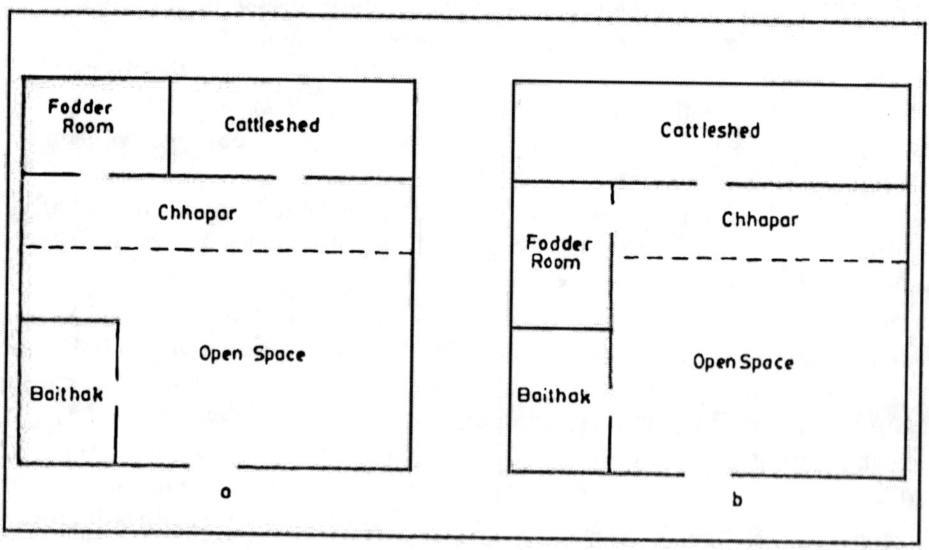

Fig. 7.2 : Ground Plan of the Bara

fact, for the elderly males, the *bara* was their virtual home. Even the other male members in the family spent most of their leisure time in the *bara*. Friends and male guests were generally received in the *bara*, which is why the females avoided visiting the *bara*, especially in the mornings and in the evenings when most of the males were present there. Surprisingly, not many structures were constructed in the *bara* for human use. Often a single, medium sized (about 3×4.5 metre) room known as the *baithak*, which functioned as male dormitory-cum-male guest room alone was constructed. Located not too close to the cattleshed, it was positioned in a manner that while sitting in the *baithak* proper vigil on the cattleshed and other structures could be maintained. A characteristic feature of the *baithak* was the presence of windows in front as also along the side walls, which made it airy and comfortable in summer.

The Vertical Plan

While the traditional rural house type of Punjab was and is a single storied structure, the two storey house was not unknown in the region. Even prior to Independence, such houses did exist in rural Punjab, but their number was very small and they were found only in large sized villages. Secondly, such houses did not belong to the common farmers. The second storey in the two storied house generally comprised a single room, known as the *chaubara*. It was viewed as a status symbol and served almost the same function as the *baithak*. It was also used as a bed-room by the newly married couples.

The Architecture of the House

As in other folk houses, the most striking feature of the architecture of the traditional rural house type of Punjab was its simplicity of design as well as of the material used for its construction. The said house type was basically a single storey flat-roofed structure, except in the case of the *chhapar* and *chhan* which had slanting roofs. The most important characteristic of the house, however, was that it was constructed using only the locally available materials. Mud was used for the construction of walls as well as for the laying of the floor. The roof was made using wood, thatch, dry earth and mud. The *chaukhats* (frame) and the *palas* (panels) of the doors and windows were made of locally available wood.

Most of the rooms in the house were rectangular or elongated in shape, with the exception of the *latain* which was generally squarish in shape. The width of the rooms was generally maintained between 3 to 3.5 metres. The size of the rooms was commonly expressed in terms of *khans* (spans), that is, the divisions between the adjacent *shateers*, which were placed generally about

1.5 metres apart. Although the size of rooms generally varied between 2 to 8 *khans*, the four *khan* room was most common. However, the *latain* generally comprised of 8 *khans* which were arranged in two rows.The thickness of mud walls in most cases was about 50 cm. The height of the flat roof from the floor level was around 2.75 metres. The width of most doors, which had two panels and a threshold, was about 1.2 metres.

As mentioned earlier the design of the house reflected the deep impact of the *parda* system and the low status accorded to women. This is further emphasised by the location of the sitting room for males, that is, the *baithak*, in the *bara*. Interestingly, whereas rooms in the unit where ladies lived were devoid of windows and ventilators, those which were exclusively used by males, such as, the *baithak* and the *chaubara* had windows in almost all the walls.

The Changes

The form, function, and structure of the rural house type of Punjab has experienced notable changes, although in phases, after Independence. The study period, therefore, has been divided into two major phases, (i) the pre-Green Revolution phase, and (ii) the post-Green Revolution phase, each of which has again been divided further into sub-phases. These phases and sub-phases of change in the rural house type of Punjab roughly coincide with the various phases of socio-economic and cultural change in Punjab, suggesting a close relationship between the two.

Changes During the Pre-Green Revolution Phase

The period from the time of Independence in 1947 to the onset of the Green Revolution in the late 1960s is characterised by a comparatively slow pace of socio-economic and cultural changes in Punjab. Even within this period, the pace of change during the first decade of Independence was much slower than during the second decade. Contemporaneously, there were also marked differences in the pace of change in the rural house type of Punjab during those decades. Thus, the pre-Green Revolution period can be conveniently divided into two sub-phases, (i) from the time of Independence to the late 1950s, and (ii) from the late 1950s to the time of the onset of the Green Revolution in the late 1960s.

During the first decade of Independence not much change took place in the social, cultural and economic life of the people of rural Punjab. Likewise very little change took place in the rural house type of Punjab. The most noticeable change during this period was the beginning of the trend to replace mud walls with brick walls. The kiln fired bricks were earlier used only by the rich but after Independence even common farmers started using *pucca* bricks for their houses. Initially bricks were used only for the construction of pillars erected along the load bearing walls at points where *shateers* were placed to take the load of the roof. In case of *latain* an additional pillar was also constructed in the centre of the room for the support of the roof. Soon the use of bricks was extended to the construction of some select walls, such as those of the *baithak*, the front wall of the *deodi*, or the load bearing walls of other rooms. Gradually bricks came to be used for the construction of all the walls of the house. However, the number of brick houses and walls remained limited till the late 1950s, and mud walls continued to dominate the rural scene. Moreover, even for the construction of brick walls, mud paste continued to be used as a cementing material. In many cases even the plastering of brick walls especially on the inner side, was done with mud paste. Use of cement till then was very limited. It was used only for covering the grooves between the brick layers or at the most plastering of walls from the outside. Apart from these changes in the building material no other significant change took place in the house till the late 1950s.

From around the late 1950s, however, significant social, cultural and economic changes began to take place in rural Punjab following the spread of education, increased rural-urban interaction, consolidation of land holdings, extension of irrigation, reclamation of wastelands, increase in net sown area and increase in total cropped area. Almost simultaneously the rural house of Punjab also underwent notable changes in the ground plan, vertical plan and its architecture, in addition to the changes in building material.

The beginning of this sub-phase almost marked the end of the age old practice of constructing mud walls in most areas of Punjab. From that time onward bricks became almost an exclusive material for the construction of walls although the use of mud-paste for laying bricks continued throughout this phase and even beyond. The use of cement for plastering of walls was, however, gradually increasing. Apart from the walls, the building material of the roof as also floor too began to change during this phase. The most significant change in the building material of the roof was the replacement of wooden *shateers* by steel girders, and the thatch layer by a tile layer. This change not only made the roof more sturdy and long lasting, but also elegant in appearance. Similarly, the *kutcha* mud floor was replaced by a semi *pucca* brick floor. The above mentioned changes not only improved the quality of the house and increased its life span considerably, but also enhanced the prestige of the family. By the late 1960s, a fairly large proportion of houses had been affected by these changes. In almost all the new houses constructed or in the renovated portions of the old houses, during the late 1950s and during the 1960s, bricks were used for the construction of walls and steel girders and tiles for the roof, especially in case of construction or renovation of the *baithak* and the *latain / dalan*. As far as the brick floor is concerned only a small proportion of houses experienced the change. Again the *baithak* and the *latain/dalan* were the most favoured areas for laying the brick floor. Thus, at the onset of the Green Revolution, houses, built partly with modern and partly with traditional building materials, dominated the rural scene.

The process of change in the ground plan of the *vasughar* and the *bara* started in the later part of the 1950s, and gathered momentum during the 1960s, when an effort was made to streamline the passage for entrance into the *vasughar*. The offset positioning of the two doors of the *deodi*, which provided the entrance to the *vasughar* rendered it impossible for the bullock cart or the tractor-trolley to be taken into the courtyard (located behind the *deodi*). Thus, arose the need to provide a separate passage from the *deodi* and hence, to fulfil this requirement a straight wide corridor was carved out of the *deodi*, as a result of which the *deodi* lost much of its intended purpose of blocking free entrance to the inner part of the house (Fig.7.3). The truncated *deodi* then came to be used for other purposes and in most cases was converted into a *baithak*, but without any door or window opening towards the inner side of the house. Entrance for the guests was provided from the outside. It had an additional door opening into the corridor but that was used only for serving tea or food to the guests. In other words, this *baithak* was designed in such a manner that the privacy of the inner part of the house was fully preserved.

This change in the ground plan of the *vasughar* although apparently a trivial one, affecting only a single element of the house, that is, the *deodi*, was indeed a very significant one. With this change the *deodi*, which till then was accorded a very important position in the house, lost much of its significance. The separation of the corridor from the *deodi* or the conversion of the truncated *deodi* into a *baithak* was not merely an act of re-alignment or replacement of one element by another, it was much more than that, it signified a change in the attitude of the people towards the *parda* system. The emergence of the *baithak* in the *vasughar* which was earlier located in the *bara* marked a major departure from the old value system. Another important change in the ground plan of the *vasughar* during the pre-Green Revolution period, was the replacement of the *chhapar* by a flat-roofed verandah. Viewed as a symbol of poverty, the *chhapar* was removed at a very early stage, a change associated with the improvement in the economic condition of the people. No major change took place in the ground plan of the *bara* during this phase.

The most significant change in the vertical plan of the house was the emergence of the *chaubara* in a fairly large number of houses. Viewed at that time as a status symbol, many farmers added *chaubaras* to their houses for the enhancement of their status in society.

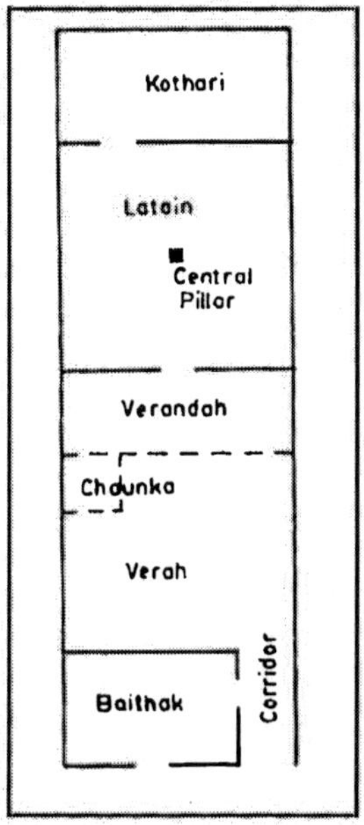

Fig. 7.3 : Ground Plan of Vasughar with Modified Deodi

The architecture of the house underwent no major change during this phase except that the thickness of the walls decreased to about 35 cm, and in many cases to even 23 cm., following the use of bricks. There was also an increase in the number of windows and ventilators in the house. However, the fact that no door, window or ventilator from the *baithak* of the *vasughar*/modified *deodi* opened into the central courtyard suggests that the design of the house continued to be influenced by traditional values, especially the observance of *parda*.

Changes During the Post-Green Revolution Phase

More significant and rapid changes in the rural house type of Punjab took place after the onset of the Green Revolution. The post-Green Revolution phase can be divided into three sub-phases, that is, (i) from the late 1960s to the middle of the 1980s; (ii) from the middle of the 1980s to the middle of the 1990s; and (iii) from the middle of the 1990s to the present. Each of these sub-phases coincides with the various phases of socio-economic development in Punjab. The first sub-period between the late 1960s to the middle of the 1980s was marked by rapid socio-economic development and a significant change in the status of women. This phase was followed by a decade during which conditions in Punjab remained disturbed due to militancy and terrorism, as a result of which the pace of development in the state slowed down considerably. With the return of peace in the state in the mid 1990s, the process of development once again picked up. The socio-economic and cultural changes which took place during these phases have left their own imprint on the rural house of Punjab.

The most dynamic phase of change of the house type in Punjab began after the onset of the Green Revolution in the late 1960s. In fact, the period from around 1970 till the mid 1980s was a most remarkable phase in the history of evolution of the rural house type of Punjab in many ways. Firstly, notable changes took place in the building material, ground plan, vertical plan and architecture of the house during this period. Secondly, construction activity in the state was pursued most vigorously during the same period. By the mid-1980s, a large number of new houses were constructed on the periphery of the villages, mostly on plots allotted for *baras* at the time of land consolidation. In addition, most of the old *kutcha* houses were renovated and converted into *pucca* houses. In fact, such transformation of the rural landscape in such a short span of time, was unparalleled in the history of Punjab.

The most significant change regarding the building material was that the difference between the material used for the construction of a rural house and an urban house almost disappeared. In the houses constructed during this phase, preference of the people for brick walls, RCC roofs and *pucca* cemented floors is clearly evident. The mud paste used earlier as mortar for brick laying and construction of walls came to be replaced by the preferred use of a mixture of sand, cement and water. Consequently, the use of cement and steel increased, with the use of cement necessitating the use of river bed sand and gravel/concrete. In addition, materials such as glass, cast iron pipes, plastic pipes, marble chips, glazed tiles, iron grills, wire mesh etc. too came to be used during this period.

Priorities in the Choice of Construction Material

However, no doubt modern construction materials came to be preferred for the construction of new houses or the renovation of the old ones, it has been observed that paucity of resources forced their use in a phased manner. In such a situation. priorities for their use were fixed. Top priority for the use of these materials was given to the construction of walls, followed by the roof. The laying of the floor was given the lowest priority. Such a fixing of priorities for the use of modern building materials is not difficult to understand, in that the component of the building, the construction material of which was replaceable most conveniently at a later date (*e.g.* the floor), was given the lowest priority. On the other hand, that component for which replacement of material was most cumbersome (as in the case of the walls) received the highest priority. Similarly, priorities for the use of modern building material for different constituents of the house were also fixed. As such the *baithak* received top priority, while the kitchen received the lowest priority. The practice of using semi-modern and even traditional materials along with the modern materials therefore, remained quite common not only during the pre-Green Revolution phase, but also during the early years of the post-Green Revolution phase. Thus during the 1960s and 1970s many houses were constructed partly using modern building materials and partly the semi-modern or traditional materials. However, the number of such houses began declining around the mid-1970s.

Although several factors led to the changes in the building material, the prime factor responsible for this transformation was the economic change which took place in Punjab after the onset of the Green Revolution. In addition to increased agricultural incomes following the Green Revolution, the state also experienced huge flow of remittances from foreign lands especially Gulf countries, U.K., Canada, and U.S.A., made by the NRI's. Also those who succeeded in getting jobs in government or private sectors also contributed significantly to rural incomes. A substantial portion of this money was spent on the construction of new houses or on the renovation of old ones.

In addition to economic change, the willingness on the part of the people to spend a sizeable proportion of their earnings on the use of modern building materials too has played an important role in bringing about this change. The reason for this willingness to opt for modern building material and incur the extra expenditure involved, lies in the fact that a house in rural Punjab is viewed not only as a shelter but also as a status symbol. There is a direct relationship between the kind of material used in the construction of the house and the socio-economic status of the family. Thus despite its higher cost, people most willingly opted for the use of modern building material because that significantly enhanced their status in the society. Moreover, it has been observed that in most villages of Punjab a very strong sense of competition between different families prevails. The adoption of any new thing especially that which is associated with affluence or modernity by one family generally leads to the adoption of that soon by others also. Thus, the use of modern building materials by a single or few selected families triggered the trend in favour of these materials in almost every village in Punjab.

Similarly, the beginning of the 1970s also witnessed significant changes in the ground plan of the *vasughar* and *bara*. Many elements of the old *vasughar* such as the *deodi,* the *dalan/latain*, and the *kothari* started losing out to newly incorporated elements such as separate bed-rooms, covered kitchen, a hand pump with *pucca* platform around it, and bathroom. The spatial location and size of some of the elements such as the *verah*, the *baithak*, and the verandah was shifted and changed, especially the verandah which was modified in both its size and shape. Thus, the arrangement of the various elements in the new ground plan of the *vasughar* was markedly different from that in the old.

The *deodi*, in its original or modified form no longer occupied the front of the house, just as the *verah* was moved from its central location to the front of the house [Fig.4(a)]. As a result now, one directly steps into the *verah* from the street which is separated from it simply by a wall. The built-up structures, excluding the verandah, in the new plan are positioned in a U shaped arrangement. Two rectangular or elongated rooms, placed side by side, form the base of the U. These rooms which are used as bed-rooms occupy the position which was occupied by the *kothari* in the old plan. A set of another two rectangular rooms, positioned at right angles to the rooms mentioned earlier, form the two vertical limbs of the U. One of these rooms is used as *baithak* and the other as a kitchen. In many cases the latter may be used for some other purpose, such as a store or a bed-room. In such cases, one of the vertical limbs of the U is extended further to accommodate the covered kitchen [Fig. 7.4(b)]. The space between the three limbs of the U is covered by the verandah. All the rooms open onto the verandah. In the new ground plan not only has the verandah been shifted, but its size and shape have also been modified. Enclosed within the three limbs of the U, the verandah in the new house is not only smaller in length but also less airy than in the old house.A staircase is usually provided from the open courtyard alongside one of the walls. Surprisingly one still finds an open kitchen in the new house, which is often located adjacent to the covered kitchen. In many houses, partly covered and partly open kitchens are found under the staircase. The hand pump, and the bathroom are generally located adjacent to each other, near or along the wall which separates the courtyard from the street.

Such changes in the ground plan of the *vasughar* clearly reflect a substantial decline in the *parda* system and a significant improvement in the status of women. The shifting of the *baithak* to that part of the house which till very recently was considered a female's domain is the most remarkable change which supports the above hypothesis. Also, from a time when elderly male members even from one's own family avoided stepping into the courtyard, to now when male guests are taken right into the living quarters without any inhibition reflects a major change in the attitude of the people. The elimination of the *dalan/latain* is also associated to a great degree with the decline in the *parda* system. It may be mentioned here that these changes, especially the disappearance of *deodi* from the ground plan of the house began around the same time when women in rural Punjab started discarding the veil from their faces. The emergence of the covered kitchen, bathroom, handpumps and especially toilet in the house indicate that the people have started caring for the needs and comforts of the women. Infact, the status of these elements in a house can serve as a fairly reliable indicator of the status of women in that house.

Certain changes in the ground plan of the new house are also associated with the breaking up of the joint family system. In a joint family, everything is

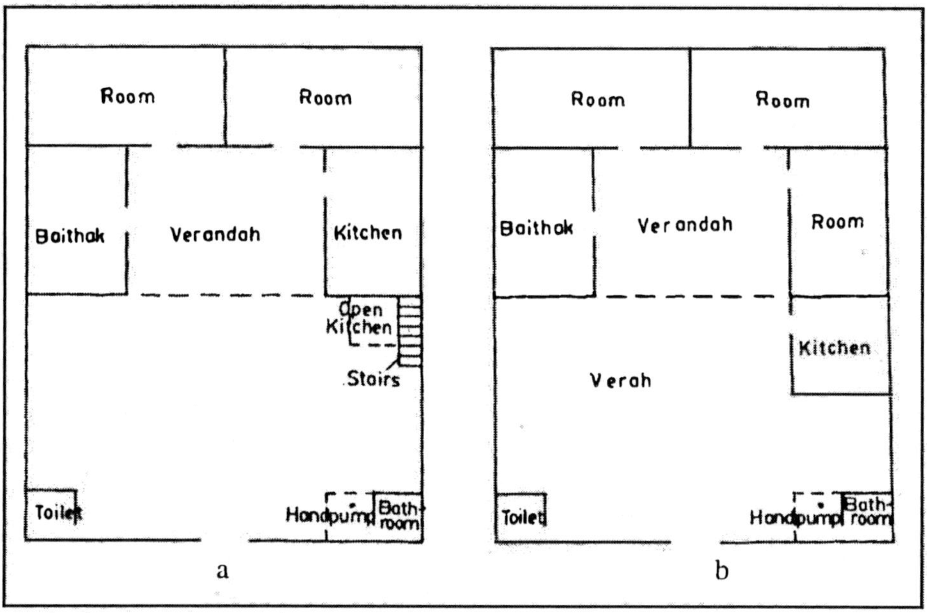

Fig. 7.4 : Ground Plan of the Post-Green Revolution *Vasugarh*

planned for the collective needs of the entire family, while needs of individual members generally get ignored. This is why in the traditional house there is a marked absence of any room which is exclusively meant for any individual. However, their presence in the new house is fairly common.The elimination of *'atainldalan* and the *kothari* are also associated, to some extent, with the breaking .up of the joint family, as these units were meant for the collective needs of the oint family.

Surprisingly the *bara* has undergone practically no change in its ground)lan except that a few more built-up structures to accommodate the new gricultural machinery, such as, tractor- trolley, cultivators, discs, drills, hreshers, etc., and for storage of grain, and fertiliser, have come up. However, ıe number of *baras* being used exclusively as *baras* declined very sharply uring this phase. Most of the former *baras* got converted into *bara*-cum-sidential houses. In the new arrangement, the ground plan of the unit used as *ıra* remained almost unchanged except that now there was no *baithak* in the *ıra*. The placement of the two units, that is, the *bara* and the *vasughar vis-ā-s* each other however differed from house to house. In some cases the built-) structure of the two units faced each other with the open courtyard in the iddle, [Fig.7.5(a)] while in others they were placed side-by-side with the urtyard in the front [Fig.7.5(b)]. This change in the *bara* is largely associated th the rapid increase in population and the break up of the joint family system.

The peripheral location of the *bara* which provided easy and direct access for the tractor and trolley also encouraged the people to shift to the *baras*.

It follows, therefore, that most of the changes in the rural house type of Punjab during the 1970s and the first half of the 1980s, were the result of the decline in the *parda* system, improvement in the status of women, break-up of the joint family system, rapid increase in population, and mechanisation of farming and the consequent change in the value system of the people and their housing needs. Accordingly, the relative importance of the various constituents of the house changed. While the importance of the *deodi*, the *latain/dalan* and *kothari* declined sharply, leading to their virtual disappearance from the ground plan of the new house, constituents such as the covered kitchen, bathroom and separate bed-rooms gained importance and thus, became regular features of the house.

The practice of constructing a *chaubara*, which started mainly in the 1960s and became more common after the onset of the Green Revolution, now came to be replaced by the construction of a pair of bed-rooms on the first floor above the rooms which formed the rearmost part of the house. Unlike the *chaubara*, these bed- rooms on the first floor were no longer viewed as a status symbol, but were largely linked with the changing housing needs of the people.

Whereas use of locally available building materials and emphasis on simplicity in design were the main characteristics of the traditional rural house type of Punjab, the new house type which emerged in the post-Green Revolution phase was built mostly with the factory made materials, such as cement, steel, crushed stones, kiln made bricks, glass, cast iron pipes, plastic pipes and with wood brought from outside the region, and displayed complexity in its design, which reflected considerable influence of modern architecture, especially that of Chandigarh style. Several features, generally associated with modern architecture, such as, the cantilever, fascia, lintel, tile or brick face work, mantelpiece, niches, stylist parapets on the roof edge, recessed rain water pipes, grills, wire-meshed doors, cupboards, etc., came to be incorporated in the new rural house. The rooms were now provided with windows and ventilators, as a result of which the house became airy and better lit. Thus, while features associated with the observance of *parda* and general prevalent poverty almost disappeared, those linked with modernisation and affluence became regular features. The casualty in the process was the folk element in the rural house architecture, although the process also indicated the end of the isolation of the rural society which was increasingly coming under the influence of urbanism and modernisation. The isolation of villages was further broken with the construction of village roads linking almost every village with the nearby urban centre, and there on with other urban centres. These changes also suggest a significant change in the attitude of the people and their value system. With

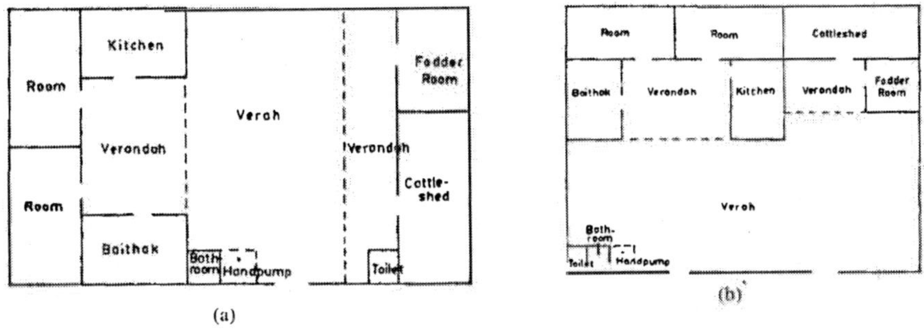

Fig. 7.5 : Ground Plan of the *Bara-cum-Vasugarh*

this, the house came to be viewed as a most reliable indicator of ones economic condition and status in society. Hence anything which looked traditional was avoided and that which appeared modern was incorporated in the design of the house.

After the mid-1980s, house construction activity in the state slowed down considerably, primarily because by then most farming families in Punjab had already built a new house or got the old one renovated. Secondly, conditions in Punjab during that period were highly disturbed which discouraged all type of construction activity. Hence, not many changes took place in the rural house of Punjab from the mid-1980s to around mid-1990s. The only noteworthy change which took place during this phase was the addition of a toilet in the house, a move initiated by incentives provided by the government as also the change in the attitude of the people. The disturbed conditions in the state led to high priority being given to the security dimension of the house, which led to an increase in the height of the boundary wall, putting high iron gates on the entrance to the house, and fixing of iron grills on the windows, all for security reasons. However, despite the disturbed conditions, the process of modernisation continued unabated. As a result gadgets, such as, refrigerator, television, electric fan, room cooler, washing machine and furniture items such as sofa set, dining set, box type beds, almirah and crockery and cutlery items, hitherto considered as the exclusive preserve of urban houses, made their appearance in the rural houses of Punjab. Entry of these goods into the rural houses attracted more costlier building material such as glazed tiles, modern sanitary fittings, plywood, marble chips and slabs, fancy electrical fittings, paints and varnishes etc. Greater emphasis on the look of house was another important development of this sub-phase.

With the restoration of normalcy in the state by the mid-1990s, construction activity picked up slowly once again. The house form which emerged during this phase displays greater similarity with the modern urban house. It is a bungalow type house with a courtyard in the front as well as at the back and sufficient open space along the other two sides. All built-up structures, which generally comprise bed-rooms with attached bathrooms, drawing room, dining room, kitchen, store, verandah etc. are organised in compact form, but they do not have any common ground plan. Most of these houses are two storied structures hardly displaying any characteristic of the folk architecture as they are not designed by the people who live in them, but by qualified urban architects. In fact, they are modern urban houses located in the rural area. Such houses, generally emerging either on the periphery of the village or on the farms belong to affluent and educated land owner farmers with fairly high incomes. However, the number of such houses as yet is not very large.

The Future Form

Although it is very difficult to visualise exactly the future form of the rural house type of Punjab, yet in the light of the changes taking place in the rural society of Punjab, some conjectures can be made. It seems that in many respects the design of the future rural house of Punjab will be closer to that of the urban house. However, it will still retain some of its distinctive features. The differences in the building material, ground plan, and architecture of the urban and residential unit of the rural house, are likely to narrow down further. Some of the constituents of the urban house, such as bed-rooms, drawing room, dining room, guest room and store, are likely to become regular features of the rural house. The relative position of the kitchen, bathroom, and toilet is bound to improve further. Thus use of glazed tiles, mosaic flooring, marble slabs, tapped water supply, and modern sanitary fittings etc., for these units will become common. The conventional modes of fuel, such as wood and cow-dung are likely to be replaced by modern fuel such as kerosene oil, bio-gas, L.P.G. and electricity. Consequently, the traditional *chula* may disappear from the rural kitchens in times to come. The present manner of cooking while squatting on the floor is likely to be replaced by standing type working system on a slab. The size of the house may become comparatively smaller, but there will be more judicious use of space. The standard architectural and construction norms are likely to be followed more seriously. Consequently, proper provision will be made in the new house for sunlight and air circulation.

However, so long as farming remains the main occupation of the rural people, and cattle continue to be their indispensable companions, the rural house will retain some of its distinctive features. As in the past the courtyard will continue to retain its prime position in the house. Its size is likely to remain large. Likewise the *bara* along with its constituents will remain an essential part of the rural house. The future rural house of Punjab shall most likely be a multi-storeyed structure in a composite *bara*-cum-residential form. However, the future model will be different from the present *bara*-cum-residential model, in that while the constituents of the *bara* are likely to occupy the ground floor the upper storeys will serve as the residential part. However, bungalow type houses, designed by architects and similar to the modern urban house may be preferred by the more educated and affluent, simply an urban, modern house in a rural environment.

Conclusion

It may be concluded that the rural house type of Punjab has experienced notable changes in its form, function, and structure after Independence. These changes have been the result of the socio-economic and cultural changes which this area has been experiencing since Independence. Whereas changes in the building material of the house are largely associated with economic change, those which have occurred in the ground plan of the house are mainly the manifestation of social and cultural changes. The changes in the vertical plan and architecture are linked with economic changes as also with the social and cultural changes.

The main factors responsible for the changes in the rural house of Punjab include, consolidation of land holdings, extension of irrigation, expansion of landunder cultivation, mechanisation of farming, marked increase in agricultural incomes after the onset of the Green Revolution, substantial enhancement in the volume of remittances by the NRI's and others, construction of rural roads, electrification of villages, spread of education—especially among the females, increased rural-urban interaction, exposure to modern urban life style through electronic and print media, invasion of rural houses by the goods and articles generally associated with modern, urban living, break up of the joint family system, marked increase in the status of women, decline in the *parda* system, notable changes in the values and priorities of the people, and the consequent changes in the housing needs of the people.

The transformation of the rural house in Punjab has taken place through different stages, each associated with some phase of socio-economic development in Punjab. A period which saw rapid socio-economic and cultural change coincided with more rapid and significant changes in the form, function, and structure, of the house. As compared to the pre-Green Revolution phase, the changes in the rural house type in Punjab took place more rapidly after the onset of the Green Revolution. The real transformation of the rural landscape of Punjab, especially that of the rural houses, took place mainly during 1970 to 1985 when most *kutcha* houses were converted into *pucca* houses and many new houses appeared on the periphery of the villages.

The study reveals an inverse relationship between the level of development and folk character of the house. As rural societies progress, the rural house begins to lose its folk character. Its dependence on locally available building materials, local designs and local construction technology start diminishing. On the other hand, the use of factory produced alien materials, urban designs and modern construction technology begin to increase. In fact, the degree to which the rural house type of an area has lost its folk character is an indication, in inverse proportion, of the level of socio-economic and cultural change which that area has undergone. However, so long as farming remains the main occupation of rural societies, and cattle continue to play an important role, the design of the rural house will continue to be influenced by the specific needs of farming and compulsions of providing shelter for cattle.

Although in Punjab, women are generally not involved in the designing of the house, yet their status in society leaves its own imprint on the design of the house, which, therefore, reflects the status accorded to women in different areas and in different times when the design was evolved.

Thus, changes in a rural house are highly reliable indicators of socio-economic and cultural change. While a rapidly changing house type in an area is a sign of dynamism and progress, the unchanging one indicates stagnation.

References and Notes

1. Mukerji, A.B. (1962), 'Jat House Types' *Geographia*, Vol. 1 (1), pp. 27-37.
2. Hart, John Fraser (1975), *The Look of the Land*, London: Prentice-Hall Inter- national, Inc.
3. Krishan, G. (1978), 'Impact of Population Growth on Rural Housing', *Geographical Review of India*, Vol. 40 (2), pp. 138-143.
4. Manku, D.S. (1986), *The Gujjar Settlements: A Case Study in Ethnic Geography*, New Delhi: Inter-India Publications.
5. Gill, M.S. (1991), 'Changing Rural Settlements of Punjab', *The Indian Geographical Journal*, Vol. 66 (1), pp. 1-5
6. Singh, Manjit (1991), *Pind Viyont ate Subhiyachar* (Rural Planning and Culture), Amritsar: Waris Shah Foundation.

8

GUJAR SETTLEMENT MORPHOLOGY IN THE BEAS- SUTLEJ KANDI

An Evolutionary Perspective

D.S. MANKU

The Beas-Sutlej Kandi comprising the piedmont region of Punjab between the Beas and Sutlej rivers and covering the districts of Hoshiarpur and Ropar, is a typical folk region (Fig.8.1). It includes the northwest-southeast running Siwalik Hills' tract flanked on both sides by adjacent uplands and dissected by numerous seasonal torrents (*choes*) (Fig.8.2). The region is characterized by a deep water table, soils which are sandy clay mixed with sand, and in some patches clayey, and natural vegetation of grass, bush and scattered trees.

The history of settlement of the region goes back to the Mughal period, when nomadic pastoral Gujars, coming from the area around Delhi settled here. Their mode of living was then centered on cattle culture, which has over time been replaced by farming and labour work (*paledari*) especially in the towns. This change, as also the extension of link roads and other services in the towns, Jat culture, and the urban culture of the plains, has greatly influenced the cultural landscape of this region.

Settlements, which are the vital expressions of culture on the landscape, in this region, lie along the hills and *choes* in a linear pattern and are nucleated. Each settlement consists of a number of clan *beras* interconnected by streets. Each house is a component of the clan *bera* and consists of one or two flat-roofed rooms and a kitchen for humans and a separate thatched two-gable room (*chhan*) for animals (*bara*).

The morphology of a settlement is characterised by its form, pattern, and type. 'Form implies the geometrical shape of the aggregate of buildings and streets. Pattern connotes the geometry of locational arrangement of a large number of settlements. Type indicates the house to house, house to street, and street to street relationship'.[1] The rural settlements of the folk region are amorphous without any perceptible internal or external design. At best, they can be called elongated quadrilaterals, the elongation resulting from the extension of the houses and *beras* (residential block) along the clan streets. The

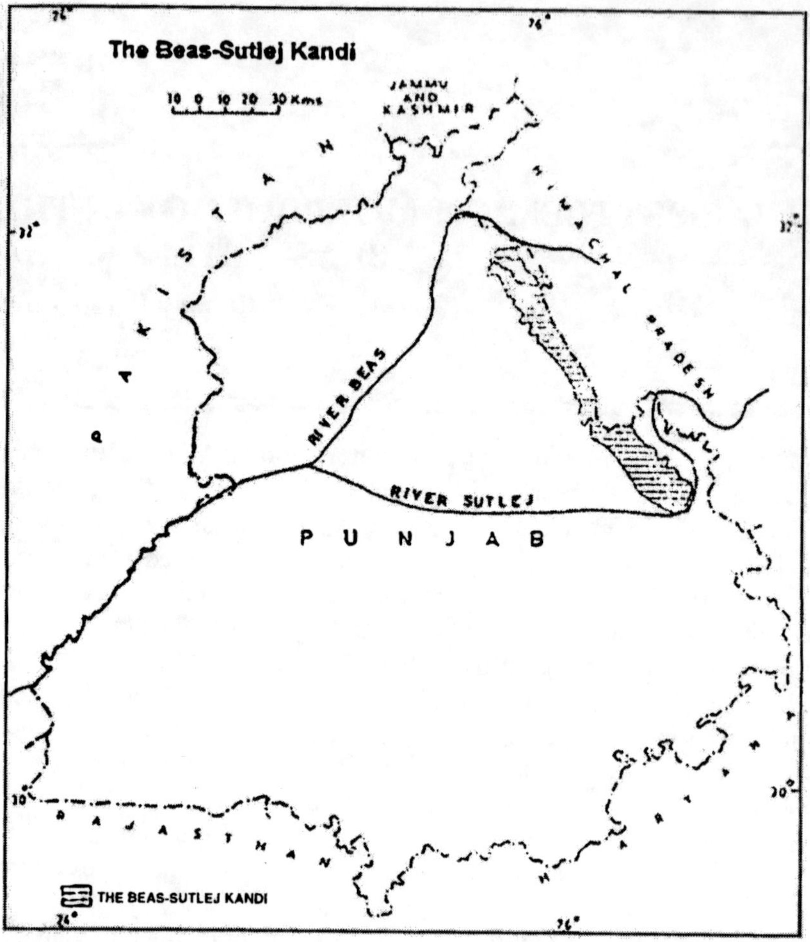

Fig. 8.1 : The Beas-Sutlej Kandi

phirni (circular road), created following the consolidation of the holdings, has given some shape to the settlements, although it need not always be the encompassing boundary. Within the *phirni* limits the settlements are more or less rectangular (Fig. 8.3) but houses have been constructed outside the *phirni* just as there are fields within the *phirni* limits. Essentially, the *phirni* encloses the settlement. Small settlements generally are rectangular.

Most of the Gujar settlements are located at the foot of the Siwalik Hills. A majority of these form two lines running from south to north along both the eastern and western flanks of the Siwalik Hills in the districts of Ropar (Nurpur Bedi block) and Hoshiarpur (Balachaur, Saroya, and Garhshankar blocks). A few, however are situated at some distance. Thus, the pattern of settlements is essentially linear, parallel to the hills (Fig.8.4).

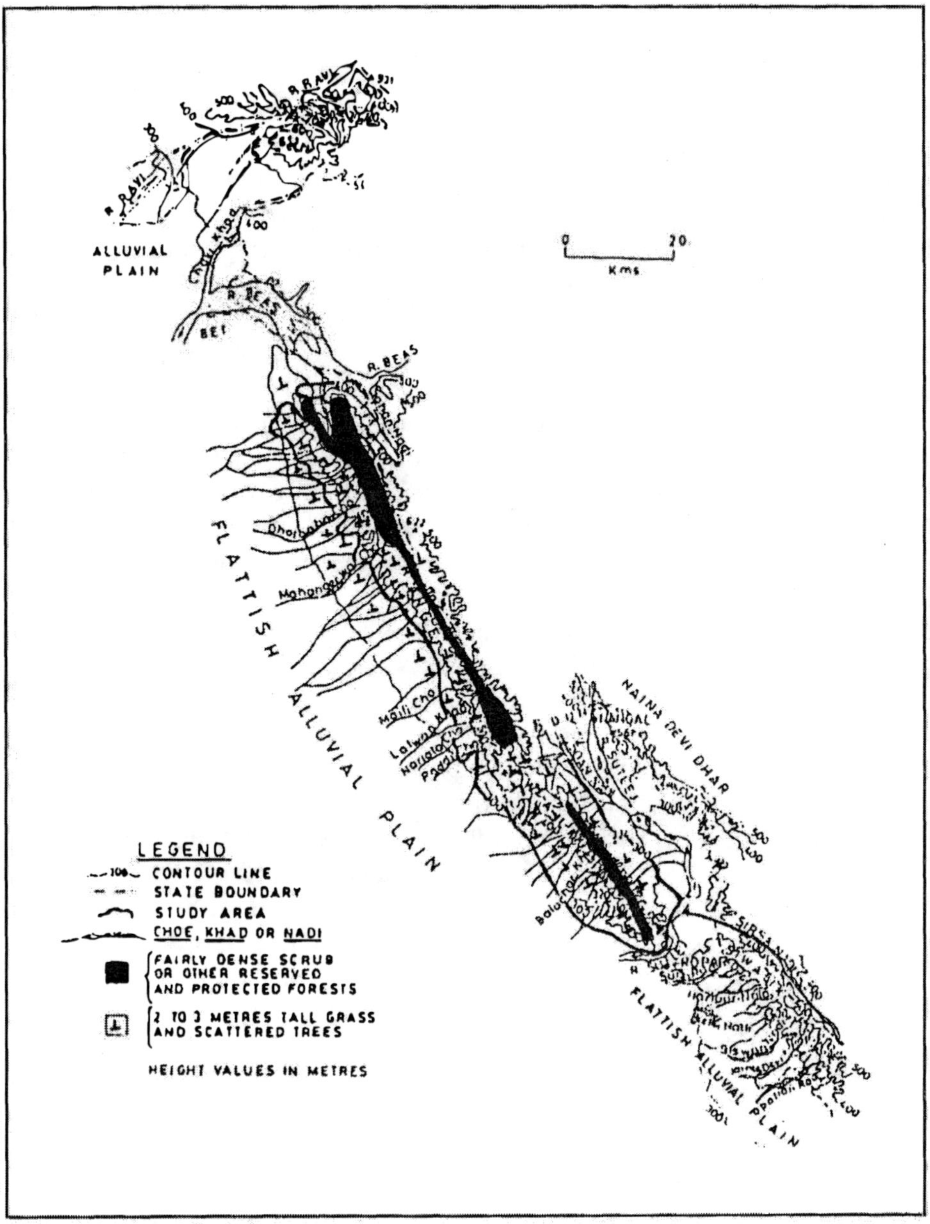

Fig. 8.2 : The Beas-Sutlej Kandi : Relief, Drainage and Natural Vegetation

The morphology of the settlements has been effectively influenced by site and also cultural attributes such as *genre de vie*, dominant clan, and clan organisation. In the past, the Gujars were primarily pastoral[2] and were roaming with their flocks of animals in search of grass.[3] Grass was and is abundant in the *bet* (new flood plain) area of the rivers and the hills. During historical periods, the Gujars formed the flanks of the Jat and Rajput movements. The latter occupied the fertile plains and the Gujars were pushed into the marginally productive areas,[4] typified by the Punjab *Kandi*. The pastoral Gujar families moved towards the hills and established their settlements on the flanks of the Siwalik Hills so that they could graze their animals in the hills and cultivate the adjoining less uneven lands. Moreover, the fear of forced conversion by the Mughal kings might also have forced them to take shelter in the Siwalik Hills.

The pattern of the Gujar settlements is determined to a large extent by the *genre de vie*, safety, tribal strength and organisation, and the lie of the Siwalik Hills. The Siwalik Hills form a 5 to 10 kilometres wide band of dry, dissected terrain, covered by open and scattered trees and scrub vegetation. The hills extend from southeast to northwest through the middle of the region (Fig 8.2). Unconsolidated and steeply dipping beds of sandstone and conglomerate, intense dissection and lack of level land for cultivation prevented the localisation of settlements on the hill slopes and tops. It is only in the *Bet* Manaswal of Garhshankar tehsil, where the Siwalik Hills assume the form of a plateau covered with productive soils, (Fig 8.4) that settlements are sited on the hills.

All the Gujar settlements are characteristically agglomerate. Aggglomeration has resulted from the extension of a clan family house into a clan *bera* complex (Fig 8.5) and single clan *bera(s)* into a multi-clan *bera* settlement. Agglomeration has been related to clan solidarity and organisation, less productive land, limited availability of water, and defence needs.

Though the individual house is the basic component of the settlement, it does not exist independently, rather it is an integral part of a lineage or clan *bera* complex (Fig 8.6). Every settlement is an aggregation of several *bera* complexes. Hence, the lineage *bera* is the most important component of the morphology of the settlement. A multi-clan settlement, which almost all the settlements of the folk region are, consists of a number of lineage and clan *bera* complexes.

Each lineage or clan *bera* complex develops along both sides of the central, open and elongated space which serves as *bera* (open courtyard) of the individual house and as street for the lineage or clan (Fig.8.6 and 8.7). The village has three grades of streets interlinking various parts, primary streets (*gohar*), secondary streets (clan), and tertiary (lineage or small). Most of them are irregular in width and length, alignment, form, and intersection (Fig.8.3).

The main street of the clan *bera* is termed as a clan street. It interlinks all

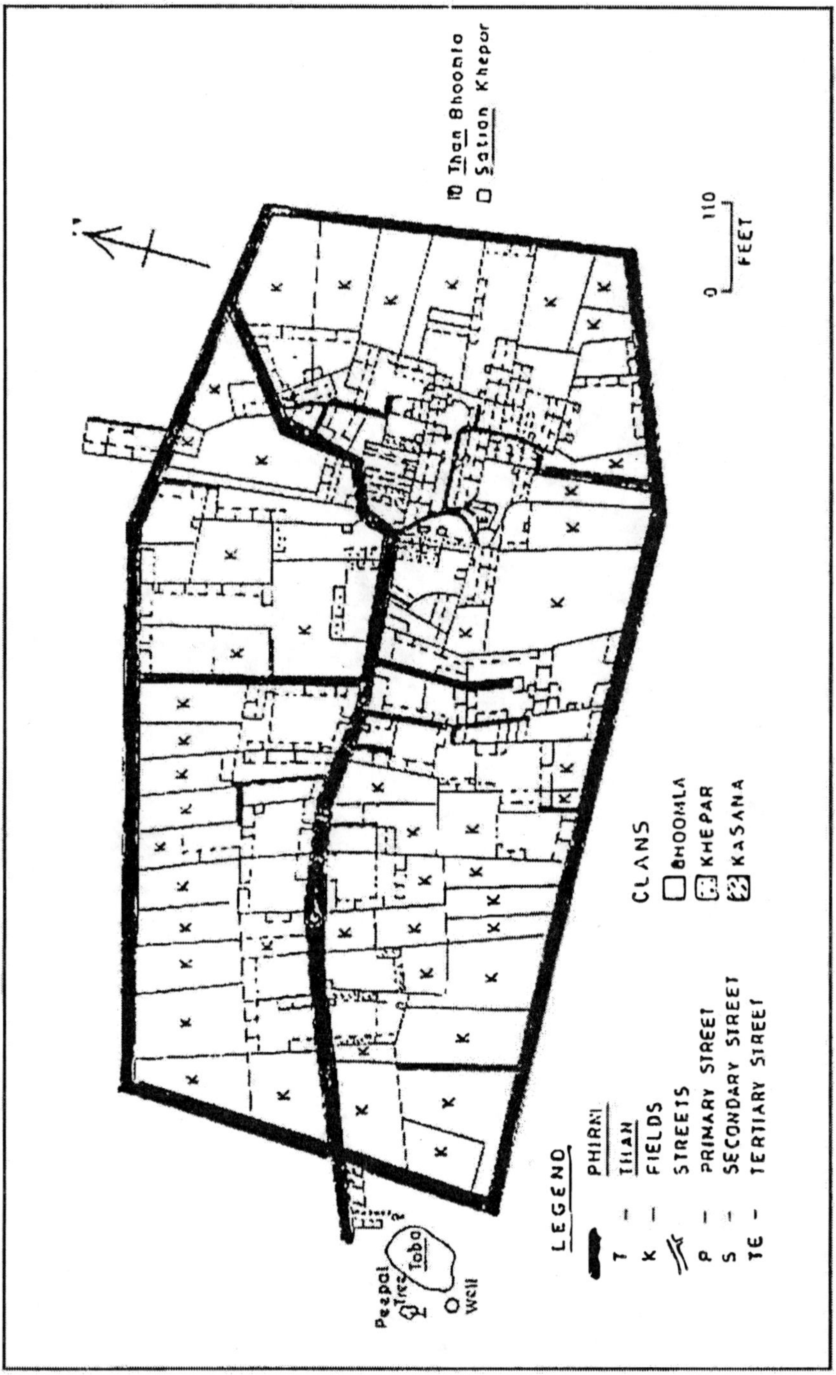

Fig. 8.3 : Village Kukarsuha : Morphology and Shape of the Settlement within Phirni Limits

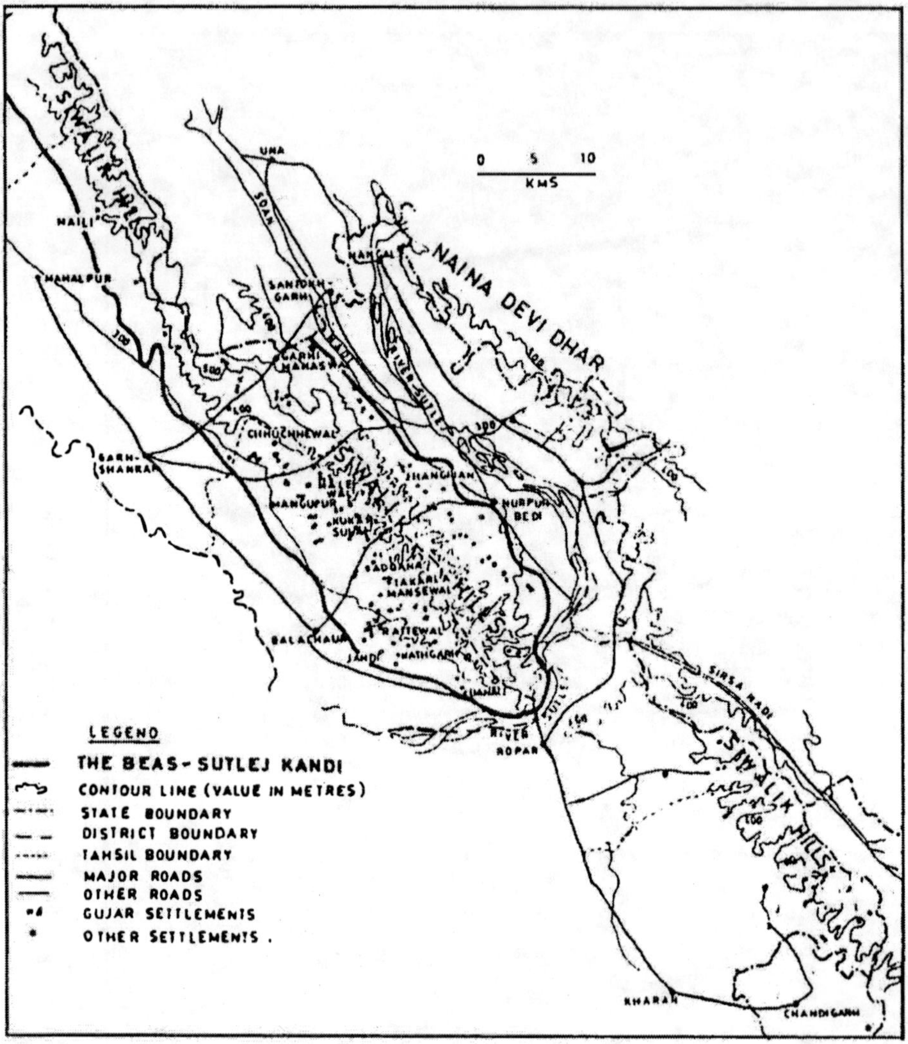

Fig. 8.4 : The Beas-Sutlej Kandi : Distribution of Gujar Settlements

the lineage streets of the clan and joins the clan streets of the other clans; thus, allowing inter and intra clan interactions. A multi-clan settlement has as many clan streets as is the number of clans. These are the main arteries of the body of the settlement. The clan streets can also be the clan street of the village termed *gohar*. The main street can also be the clan street of the dominant clan. It is much wider and heavily used by pedestrians, animals, and bullock carts. Further extension of clan or lineage *bera* evolves along the tertiary streets (small) stemming from the clan streets. Thus, the processes of agglomeration and extension of the clan family into nuclear families proceed simultaneously.

Clan and lineage *bera* complexes are inter-linked by tertiary, secondary, and primary street patterns which serve not only as social, economic, and religious links between various parts of the settlement but also function as channels for rain water. Because of their long and continued use by pedestrians and animals and the erosion caused by heavy monsoon showers, the streets are sunk (one to five feet below the *bera*, courtyard, level) (Fig 8.7). In many of the villages the streets have been paved with bricks, thus checking further downward erosion (Fig 8.7). Most of the streets meet at oblique angles showing unplanned growth of the settlement. In contrast streets and roads carved during *killabandi* procedure of consolidation are straight and uniform in width and have rectangular intersection with the *phirni* (Fig 8.3).

The Gujar settlements in the region have not evolved around some focal point of ancestor worship such as *khera* or *bhumiya*, though each clan has its separate place of worship, *pir sultan* termed *than*. This trait, having its origin in Islam became integrated with Sufism.[5] A mustard oil earthen lamp (*deeva*), is lighted on every Thursday at the *than*. A *than* usually lies near or in the house of the oldest family of the clan (Fig. 8.3). Every clan has its own *satian* (place of ancestor worship) in honour of some lady who performed *sati* or self-immolation by burning on the pyre of her husband. The *satian* (Fig 8.3) generally lies outside the settlement. Also some clans may have their *satian* in another village, a fact that suggests their migration.

Other important landmarks of the villages are the *toba* (water pond) and a *peepal* (*Ficus religiosa*). These usually lie at some distance from the settlement (Fig.8.3), fulfilling the needs of religion, building material, and water. With the excavation of mud for building purposes, the *toba* becomes large and deep and with the gradual expansion of the settlement the *toba* comes to lie near the *phirni* and in some cases is even enclosed by the buildings. The *toba* provides drinking water for animals and water for washing clothes, and bathing animals. Unlike the Jat villages, the *peepal* tree in Gujar villages has not developed into a community meeting place with a raised platform (*chauntha*) for the elders to sit and discuss community matters and where people could rest and enjoy the cool shade during summer. There are no trees in the old *beras* but in the later stage *beras*, and the present unified houses (*vassu*-cum-*bera*), some trees, especially *kikar* (*Acacia arabica*) or *neem* (*Azadirachta indica*) are found. In some villages, temples have also been constructed.

The evolution of the settlement from a clan family house unit into its present form has resulted from the extension and disintegration of clan and lineage families into nuclear families involving the expansion and extension of the *bera* complex and the individual house through the processes of internal division and addition (Fig. 8.5).

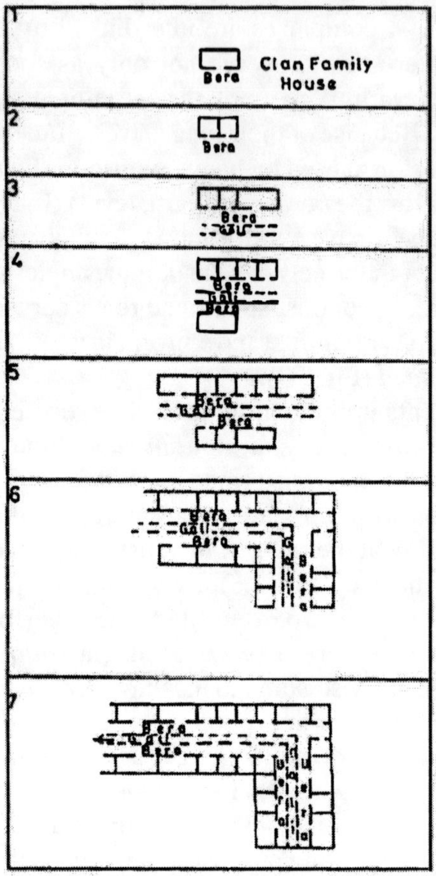

Fig. 8.5 : Sequential Evolution and Agglomeration of a Clan Family House into a Clan *Bera* Complex

In case of two sons in a family, nuclearisation, reflected in the internal division of the house follows the marriage of a son. In families having more than two sons, expansion of the family is manifest by addition, reflected through the construction of a new house, either in physical continuation of the existing house or, on a site physically separated from the existing house.

The evolution of a settlement can be better understood by the stages througl which it has passed from the initial stage to its present form. Four tentative stages have been postulated. They are, early or initial stage, second stage, pre consolidation stage, and finally, the present or the post-consolidation stage [Fig 8.8(a) and 8.8(b)]. Most of the settlements are more than 200 years old There has been, over time, a change in the (building material) as also the functions of space and buildings.

In the early stage (about 150 years ago), all the houses were made of thatcl roof and mud walls termed *chhan*. Every house contained one room whicl

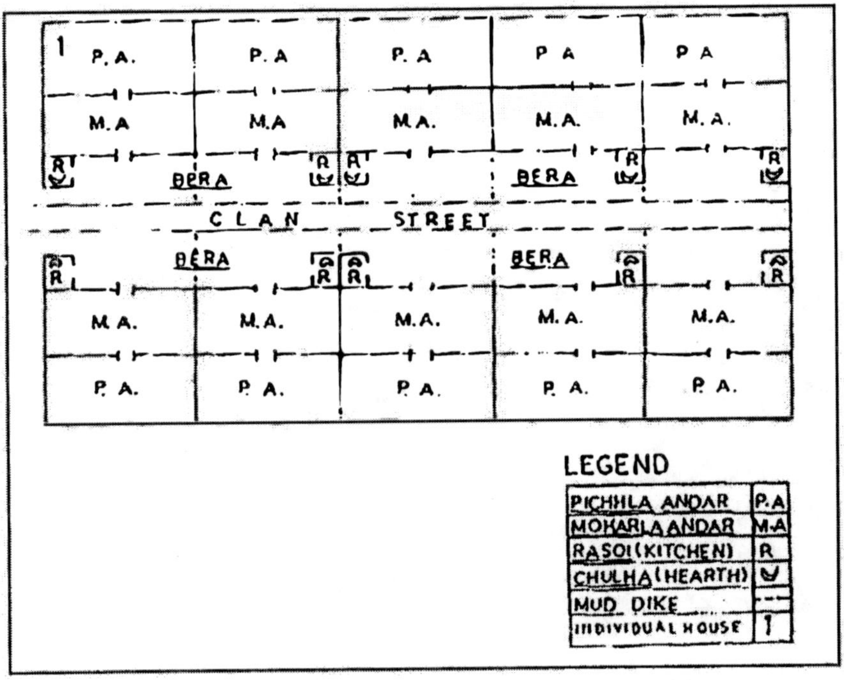

Fig. 8.6 : Lay Out of a Gujar Clan *Bera*

functioned as *vassu*-cum-*bara*. It served as living and sleeping space, kitchen and also as space for stalling the cattle. The open *bera* was used for all social, economic, and religious functions and also for stalling the animals during summer and day time.

In the early stage the number of houses was limited and the needs were few. There was extensive wild growth and a persistent fear of wild animals,[6] thieves, and marauders. For defence, fuller utilisation of the natural resources, and close tribal interaction and clan solidarity, the houses of different families were located in close proximity. But on account of different clan organisation and the difference in the status of the original families, invited families, and later migrating families, a common *bera* (open space) was not possible. Only their backwalls could be common [Fig 8.8(a)]. The initial settlement was agglomerate in type and comprised of three to four houses belonging to different clans [Fig 8.8(a)].

The site was selected near the choe and with reference to a water pond (*toba*) and a *peepal* tree. These elements fulfilled their religious and social requirements, as also needs of water and building material. In the initial stage each settlement did not have its own well. Many neighbouring villages would

Fig. 8.7 : Lineage Residential Block (Bera) The houses open on to the central rectangular space through the middle of which runs the sunken clan street. The Kitchens cut weges into the open space.

have a single common well. It was in the second or third stage that each settlement had its own drinking water well.

In the second stage the clan *bera* extended outward along the clan street through the processes of internal division and addition as the clan family disintegrated into many lineage families. In village Mansewal (Fig 8.4) there were four thatch-mud houses (*chhans*) in the early stage [Fig 8.9(a)]. Two houses belonged to the dominant Bhoomla clan after whose ancestor Mansa, the village was named, and one each to the Bajar and Chhabri clans. In the second stage (about 1900 A.D.) the houses of the Bhoomla clan extended towards the south along the Bhoomla clan street and the number of houses increased to six. This clan street towards the south leads to the fields belonging to the Bhoomla clan, *i.e.* the basis of their *genre de vie* and also to the common well dug in that period. The Bhoomla clan street, thus became the main street of the settlement. It upholds the importance of the dominant clan in the settlement morphology. The houses of the other clans extended along their respective clan streets. This form of growth of the village resulted in its agglomeration.

In the second stage a few tertiary streets also developed to connect new houses of the extended lineage families. Another important change effected

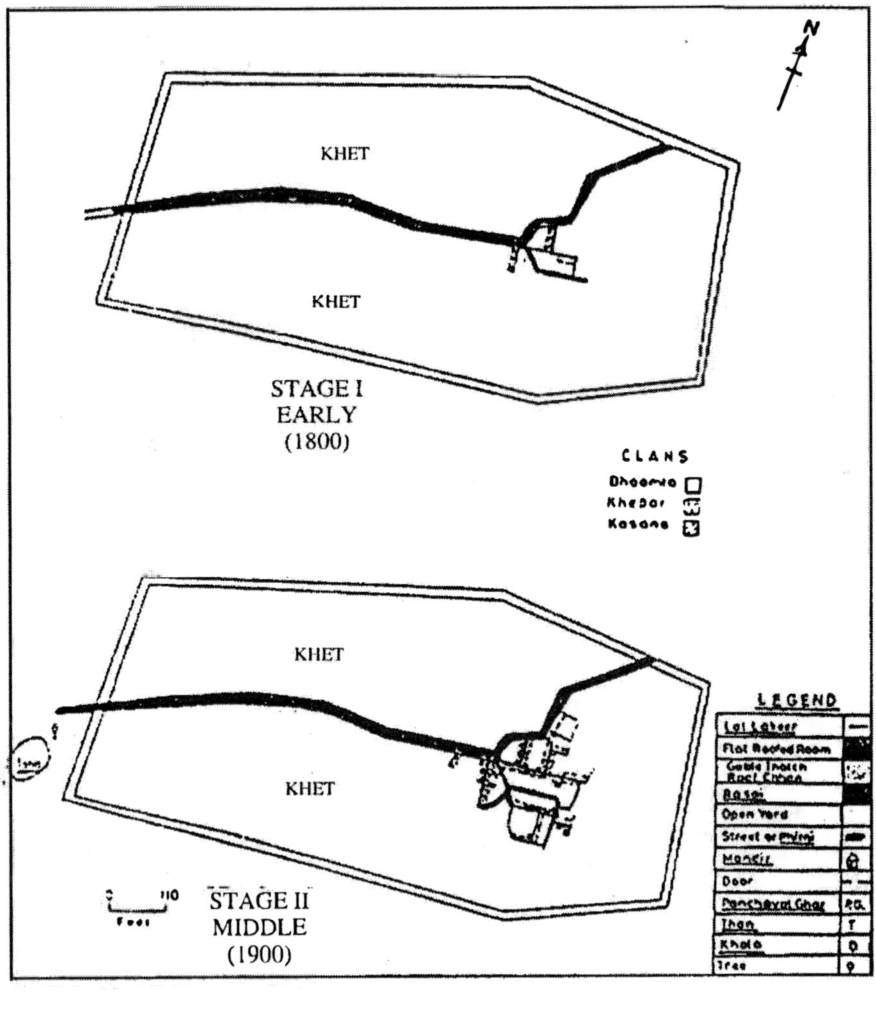

Fig. 8.8(a) : Village Kukarsuha : Evolution of the Morphology—Stage I and Stage II.

during the transition from the early stage to the second stage was the replacement of the gable thatch roof (*chhan*) houses with flat-roofed houses (*kothas*). This trait was adopted from the adjacent plain areas as a result of the interaction of Gujar culture with the plain areas culture, wherein flat mud roofs dominate. As the clan *bera* became more congested, animal stalling was also shifted from the all purpose initial room to a separate enclosure with a *chhan*, termed *bara*, lying physically separated from the all purpose room. The kitchen too was shifted from inside the room to the open yard, as the thatch-*cum*-mud flat roof

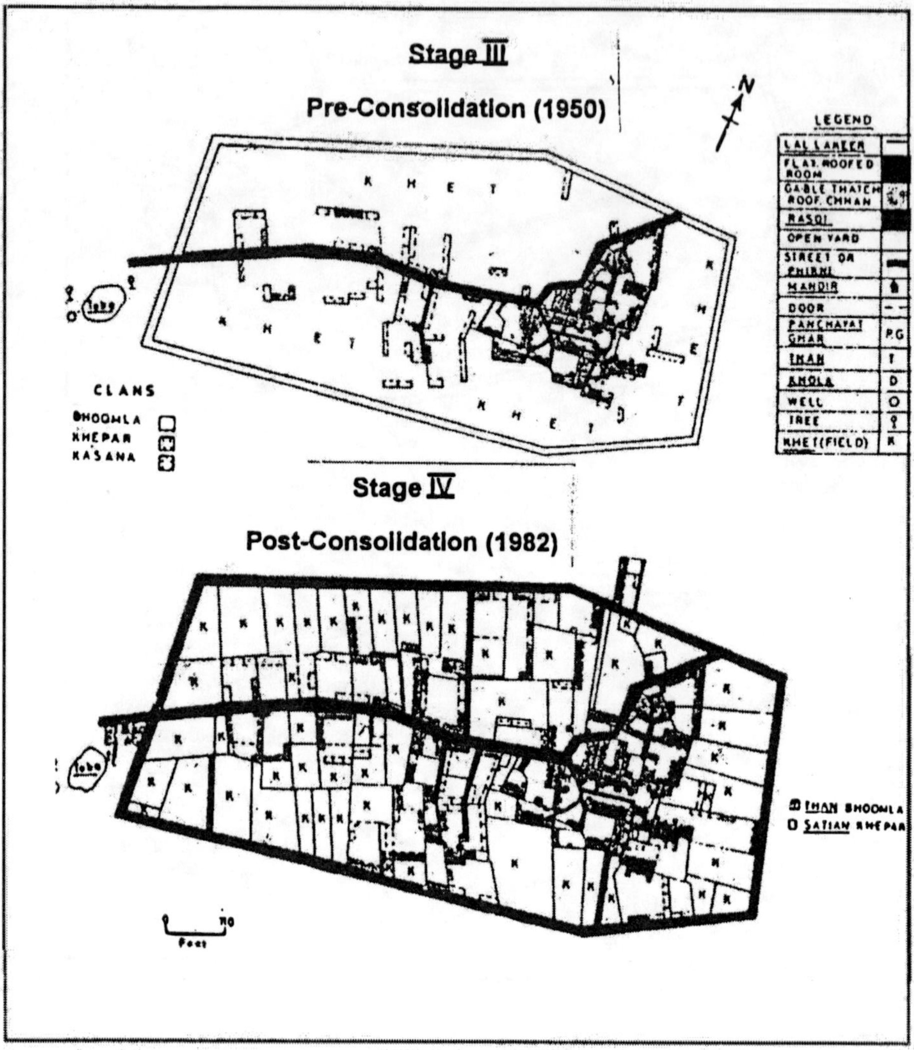

Fig. 8.8(b) : Village Kukarsuha : Evolution of the Morphology

did not permit the escape of smoke, as was possible in the gable thatch roof of the initial *chhan* house [Fig 8.9(a)]. All these changes were markedly gradual until the third stage was reached when changes became more pronounced.

The villages started growing and changing rapidly as new interactions developed following the participation of the Gujars in work in the grain markets of the towns of Punjab leading to increasing prosperity as also the inflow of more money and outside influence. In the third stage, which is termed as the pre-consolidation stage, as it existed before the consolidation of land holdings

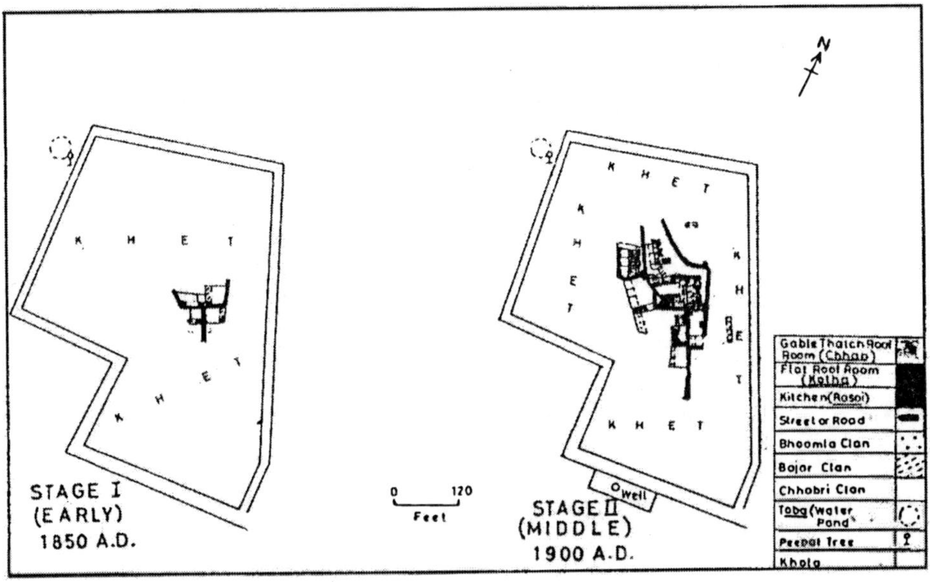

Fig. 8.9(a) : Village Manswal : Evolution of the Morphology

was effected, the number of houses increased, use of burnt bricks in the construction of walls and roofs started, and the settlement began expanding. In village Mansewal, the number of houses increased from nineteen in the second stage to forty-eight in the third [Fig. 8.9(b)]. In village Kukarsuha the increase was from fifteen to thirty-five.

With an increase in the number of houses there was an elaboration of tertiary streets along both sides of the clan street. In village Kukarsuha development took place along both sides of the village main street (*gohar*) which leads to the *toba*, the *peepal* tree, and the well lying in the west [Fig. 8.8(b)]. Because of congestion in the old residential block (*bera*) a number of new *chhans* were constructed on both sides of the *gohar* for animals, and to serve as sleeping place for the elder and unmarried male members of the family and also for the guests. These gable thatche structures (*chhans*) were the *baras* of various families.

The last phase, which is still continuing and can be termed post-consolidation stage, has been the most important in the growth of the settlement. It began with consolidation when a road, termed *phirni*, was constructed around the settlement. *Abadis* for future settlement were carved near the *phirni* (both inside and outside); new streets and roads were laid, space for school, *panchayat-ghar*, *janjghar* and *mandir* or *gurudwara* [Fig. 8.8(b)] was allocated, and in many cases the cremation ground was shifted to near the village boundary; the fields were made rectangular and consolidated at three or four sites. These changes affected the morphology of the settlements to a great extent. A look at

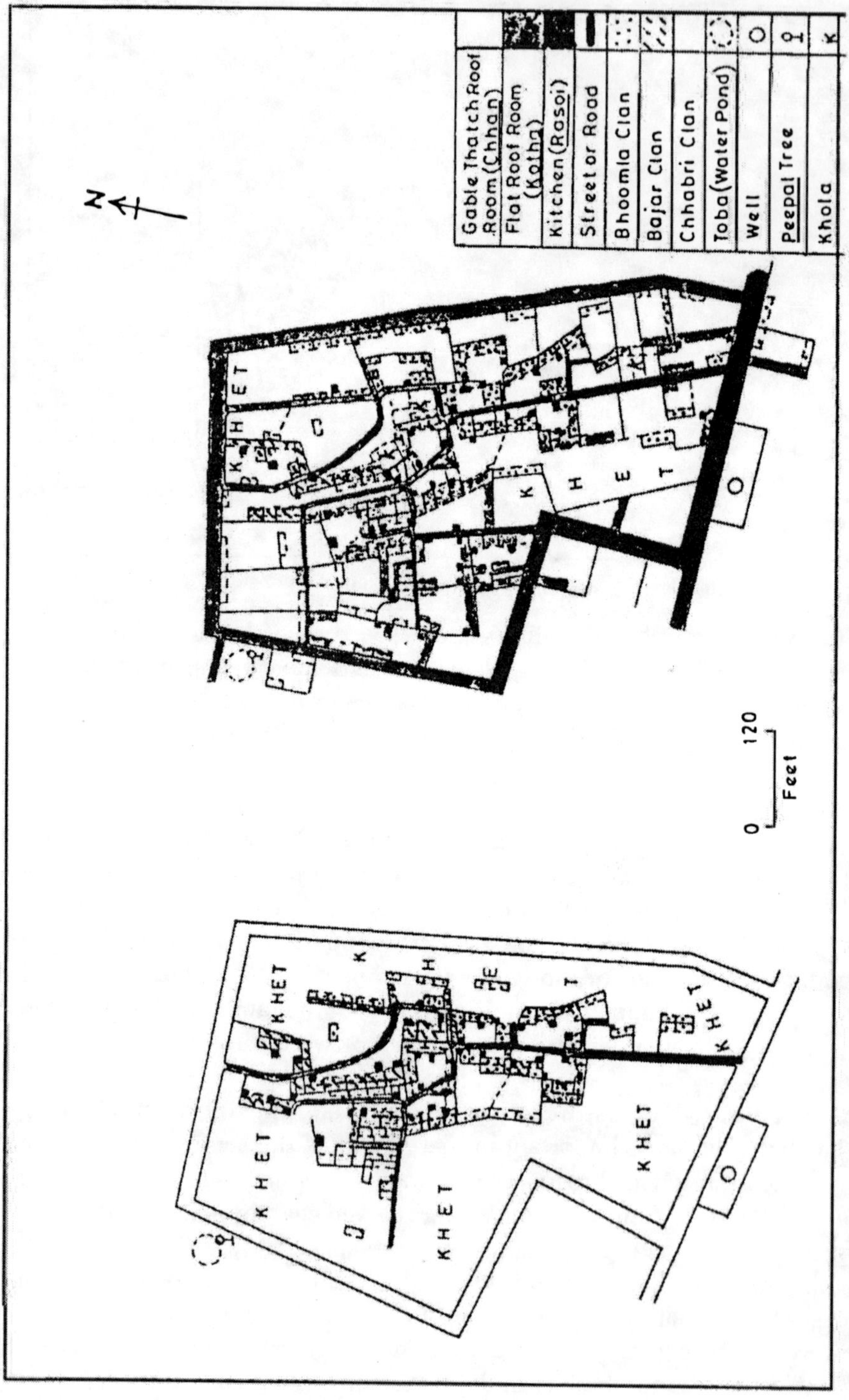

Fig. 8.9(b) : Village Manswal : Evolution of the Morphology Stage III Pre-consolidation 1950 A.D.

the pre-consolidation and post-consolidation notional maps of village Mansewal and Kukarsuha reveal the differences [Fig 8.8(b) and 8.9(b)]. Consolidation and *phirni* gave a definite shape to the settlement, which is more or less rectangular. People started abandoning the old houses in the core of the settlement and building *pucca* ones on or near the *phirni* or roads. This resulted in the creation of *kholas* (deserted houses) in the old settlement and many new houses on the periphery of the settlement [Fig 8.9(b)]. Construction activity continued along the streets or roads, similar to ribbon development in towns. Many old mud houses began to be replaced by burnt-brick houses with windows and ventilators.

Consolidation also resulted in the creation of the new Gujar house, which is a unified unit, comprising of the *vassughar* and *bara* units of the old Gujar house. Now both animals and human live in close proximity and the new house is more spacious, airy, open and accessible to the fields and the outside world. The newly carved streets and roads are straight and uniform in width, having rectangular intersections with the *phirni*.

References and Notes

1. Mukerji, A.B. (1976). 'Rural settlements of the Chandigarh Siwalik Hills (India) : A Morphogenetic Analysis', *Geografiska Annaler*, Vol. 58, Ser B., pp. 95- 115, ref.p.98.
2. Rose, H.A. (1970). *A Glossary of the Tribes and Castes of Punjab and North West Frontier Province*, Patiala : Languages Department, Punjab, Vol-II, pp.309-317.
3. Punjab Government (1872). *Shajra Nasab Village Rattewal*; Punjab Government (1872). *Shajra Nasab Village Malewal;* Punjab Government (1872). *Shajra Nasab Village Adoana;* Punjab Government (1872). *Shajra Nasab Village Maili.*
4. *See* ref. 1, p. 11.
5. Sufism is an Islamic mystical system akin to devotional mysticism of the Indian saints of medieval India. Sufis were the Muslim saints, many of whom were poets also, who preached and practised simplicity, abstention from the prohiited means of livelihood, tolerance, compassion, and human frailty. They brought Hindus and Muslims close to each other. With them the practice of adorning the memorials (tombs) of saints started.
6. Punjab Government (1872). *Sajra Nasab Village Rattewal.*

9

EVOLUTION OF FIELD PATTERN IN A PUNJAB VILLAGE

DALJIT SINGH

Introduction

The settlement, house types and field patterns are the three basic components of the rural cultural landscape, which is created through an interplay of culture, ecology and history. A cultural group through agriculture establishes a permanent relationship with the land and in this process creates field patterns which are displayed in the arrangement of parcels of land. As the landscape of the fields is created through interaction of culture with nature through time, a study of the field landscape is indicative of both the milieu and the processes of settling.

It has to be regrettably admitted that Indian geographers have made very few attempts to study in detail the field patterns of a small area having homogeneous ecological setting. In this study a detailed analysis of the evolution of the field patterns of a single village having homogeneous ecological setting has been attempted. The village under study is Dhenowali located in Jalandhar tehsil and district in Punjab (Fig 9.1 and Fig. 9.2). The study area, though small, reveals a diversity of field patterns. The reasons for such diversities can be found in the framework of an interacting system of historical, cultural, social, economic and revenue administrative elements. The focus of study are the field patterns and their essential attributes; size, shape, boundaries, distribution, scattering of the fields and their combinatorial arrangements. Several factors, such as social customs and cultural practices, the effect of the operation of Hindu law of inheritance, principle of equitable distribution of benefits and handicaps, principle of compensation, rights of ownership, caste, distance from the settlement, land types, irrigation methods, crops grown, technology used, market demand factors, methods of cultivation, land tenure systems, government policies, sale, mortgage or land gifts and the sub-division of fields and holdings, influence the attributes of the field pattern.

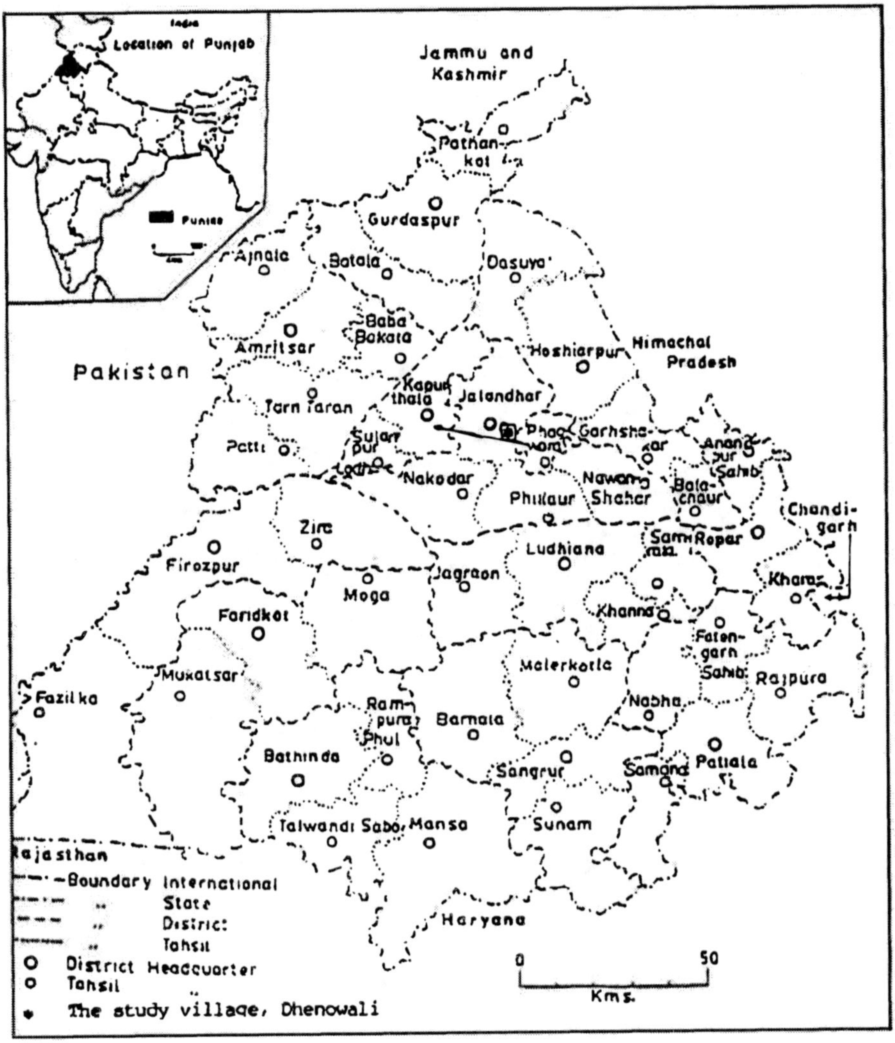

Source : Census of India, Punjab, 1981.

Fig. 9.1 : Location of Dhenowali (Punjab)

Evolution of the Field Pattern

The evolution of the field pattern reflects its dynamism. The contemporary field landscape provides only a link in the evolutionary process of fields and a key to the understanding of the processes of settling. It expresses the community decisions mediated by their culture, value system, technology and mode of living. 'The pattern, function and the evolution of fields can be understood well through an interpretation of the field patterns in an area which reflect all

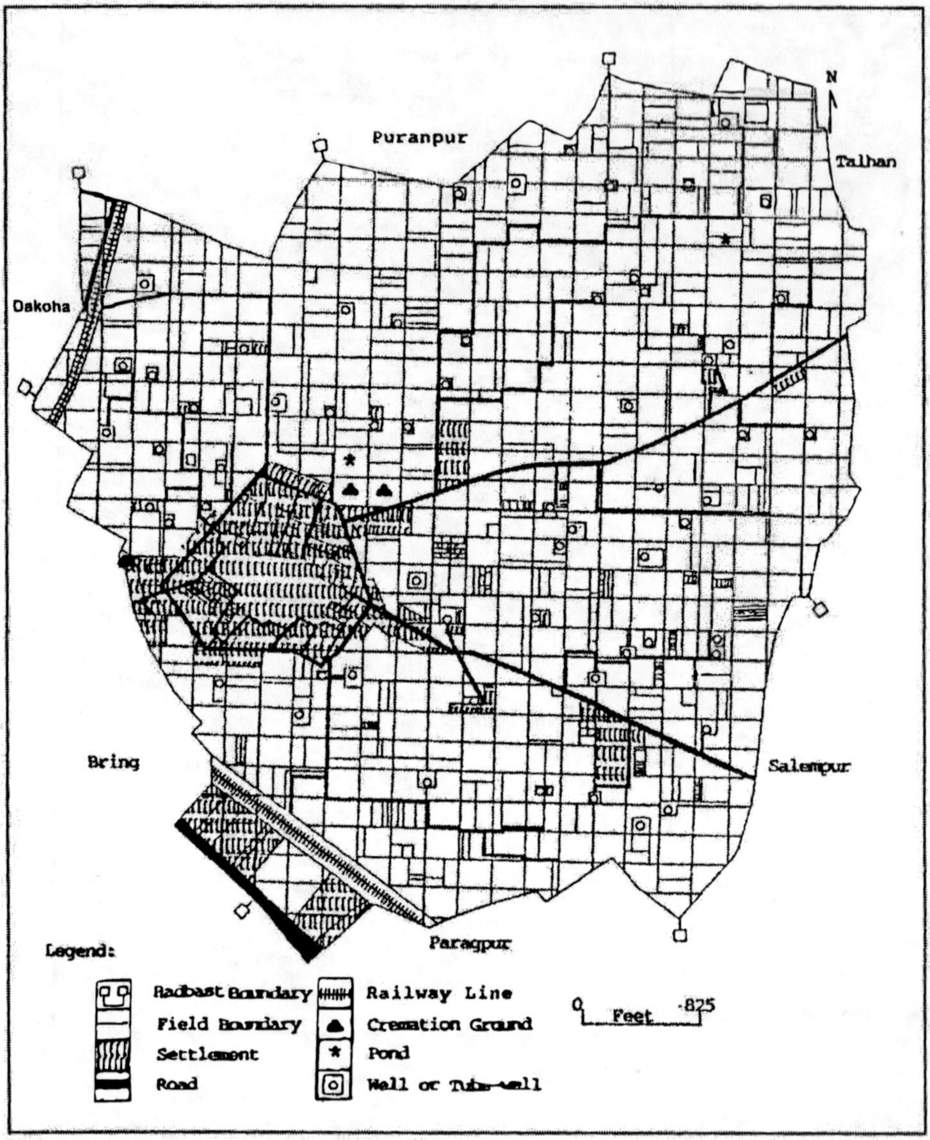

Fig. 9.2 : Dhenowali : Village Territory

these attributes and functions'[1]. The evolution of field patterns should be analysed through time and space. As Nitz has observed, '... that it is not only important to describe the various field patterns, but also, as far as possible, to trace the origin and evolution of the pattern in order to find out the socio-economic and political factors in operation at the time of settlement'[2]. The field pattern existing today has undergone various stages and is not a spontaneous creation. 'It has evolved gradually over a period of time, each phase being closely related to, and in fact, the product of the then existing political conditions, land tenure system, technological know-how, the culture group and their *genre de vie*'[3].

According to the *Shajra Nasab* (1884) the village Dhenowali was first owned by a Jat Randhawa family who migrated from Amritsar. The village began as a uni-religious, uni-caste, uni-clan settlement. The Jats acquired the village territory by removing forests and divided cultivable land equally among themselves. The boundaries of the village were marked by forests and uneven *tibbas* (sand ridges). The initial division of cultivated land led to the emergence of the initial field pattern. The expansion of the cultivated land continued till the Britishers defined the actual fields and village boundaries in 1846 and prepared the map showing field and village boundaries in 1884 after conducting actual field surveys (Fig. 9.3). The entire village territory was not brought under cultivation simultaneously, but was gradually extended in different phases. The land adjacent to the *abadi deh* (settlement) was the first to be brought under cultivation. In the first phase, cultivated land was extended to the north of the *abadi deh*. In the second phase, the extremely fertile land near the eastern and north-eastern territory of the village was brought under cultivation. In the subsequent stages cultivated land was extended towards the south and south-east of the *abadi deh* and then to other peripheral parts of the village. These phases are indicated by land types and smaller size of fields which reflects higher degree of sub-division (Fig. 9.4).

In the beginning the village territory was divided equally among the seven founder members of the Jat Randhawa family. Diwani was one of them. In the beginning of the nineteenth century either within Diwani's life time or soon after his death, the land belonging to him was divided among his three sons Harsa, Gorsa and Parsa (Fig. 9.5). The details of division regarding who got how much and where is not known because of non-availability of data and maps for that period. The division must have been made according to the then prevailing customs. In the third generation the land belonging to Harsa was divided among his six sons, Dhian Singh, Karam Singh, Gohinder Singh, Amar Singh, Bhag Singh and Nasib Singh. But Karam Singh and Nasib Singh died heirless and therefore, their share was also given to the remaining four brothers. Following the operation of the existing law of inheritance, the land would have been divided equally among all the sons following the Pan-Indian tradition, conceptualised by McKim Marriott as the Great Tradition. The discussions with the village elders reveal that it was at that time customary for the eldest son to take the charge of his father's property and divide it among all his brothers, in a spirit of mutual adjustment. The fourth generation field pattern and

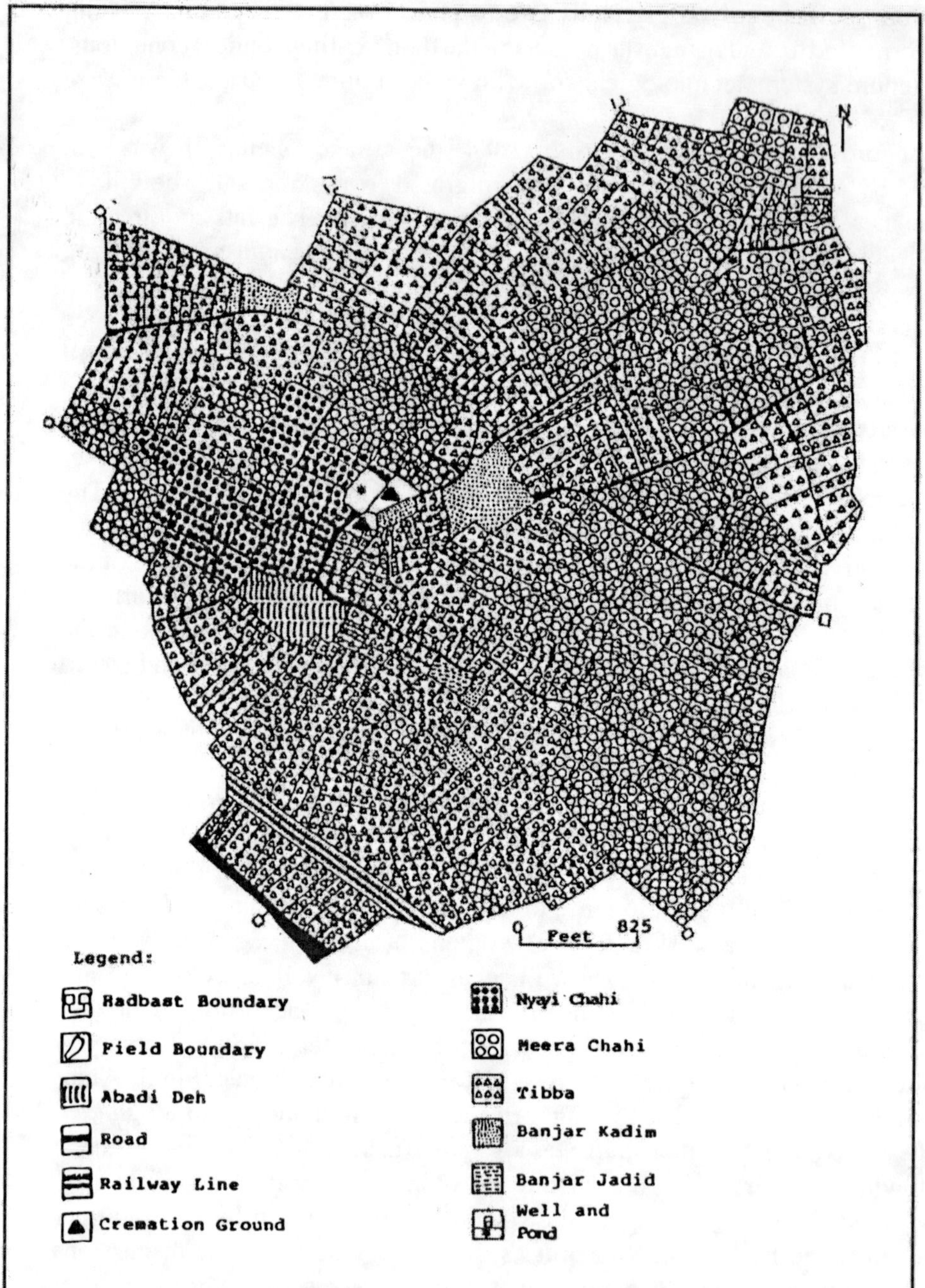

Fig. 9.3 : Village Dhenowali : Land Types and Field Patterns, 1884

Legend :

Hamir Singh and his sons (Gujar Singh, Toba Singh & Gulab Singh)

Hira Singh

Rann Singh, Kishan Singh, Hira Singh & Hamir Singh

Fig. 9.4 : Village Dhenowali : Ownership of Cultivated Land, (1884) (Fourth Generation of One Family)

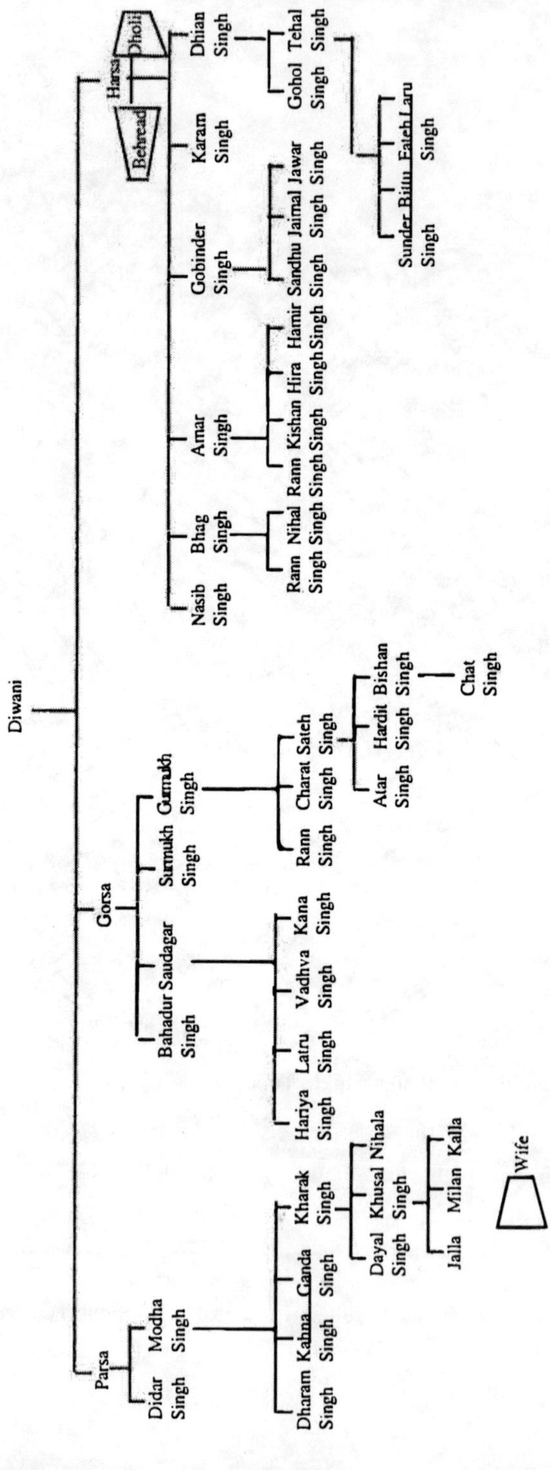

Fig. 9.5 : Village Dhenowali : A Section of Genealogical Tree (Shajra Nasab) of owners 1884

distribution of landed property among Amar Singh's four sons Hamir Singh, Hira Singh, Kishan Singh and Rann Singh as obtained around 1884 are seen in Fig.9.4. As the figure reveals it was the eldest son, Hamir Singh who got the largest share, about half of the total land (39 *kanals* and 6 *marlas*) of his father's property. Hamir Singh was joint owner with his three sons, Gujar Singh, Toba Singh and Gulab Singh. All of his individual fields were located only on one site in fertile land, *meera chahi*, irrigated by wells and relatively close to the *abadi deh*. Hira Singh who was younger to Hamir Singh had not only fewer fields but also more widely scattered in varying land types, such as *tibba* and *meera chahi*. His fields were largely located near the periphery. The rest of the fields belonging to Amar Singh were owned jointly by all the four sons. In fact, here a kind of gradation can be identified where the eldest son had the most and the best, the second eldest son had less than the eldest son and the remaining land was given to the youngest sons who had the minimum land. The size of land holdings and the number of fields owned by Gujar Singh, Toba Singh and Gulab Singh, sons of Hamir Singh, in the fifth generation are larger than those of Amar Singh's descendents of fourth generation. In the fourth generation (1884) Amar Singh's family owned 16 fields comprising a holding of 104 *kanals*-9 *marlas*. The number of fields in 1914 jumped to 51 and the size of the land-holding increased to 173 *kanals*-9 *marlas* (Fig. 9.6). This became possible either due to direct purchase of land or surrender of property in lieu of debt or return of the previous mortgaged land. The field map of 1958 reveals that the size of their compact holding further increased to about 250 *kanals*, where the number of fields remained almost the same as in 1914 (Fig. 9.7). The number of fields would not increase due to the process of consolidation of holdings by which most of the small fields of less than 8 *kanals* were converted into uniform size of one *killa* (198 feet to 220 feet or 8 *kanals*). All the family members hold land jointly either with their fathers, brothers or uncles or sons of their uncles.

In their eighth, present generation the fields are further sub-divided (Fig. 9.8 and Fig. 9.9). Sarvan Singh transferred his land to his only son Mohinder Singh. Karam Singh sub-divided his land between his two sons Santokh Singh and Gurpartap Singh. Bhagat Singh's property was divided among his four sons Kewal Singh, Balbir Singh, Chain Singh and Major Singh. The continuity of joint ownership of two or more generations in different phases is very evident from the study of the *Shajra nasab* and land ownership maps. Among the present generation joint ownership is slowly weakening on account of out migration of population from rural to urban areas mainly for employment. A man who is leaving the village for employment often wants to make official the individual status of his ownership of inherited land as separate from that of other members in order to avoid quarrels over the land. After officially securing

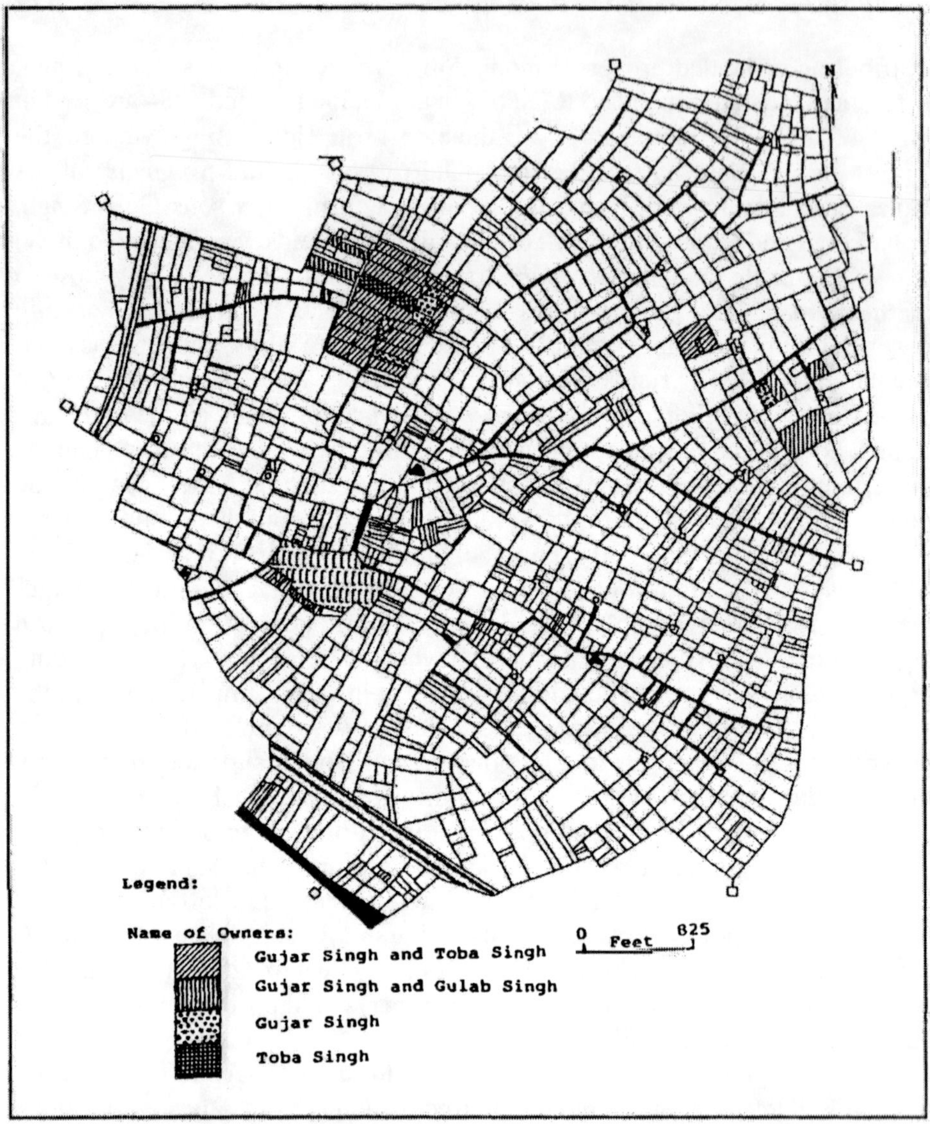

Fig. 9.6 : village Dhenowali : Ownership of Cultivated Land (1914) (Fifth Generation of One Family)

individual ownership of the land he sometimes rents or sells his land to his brothers or to someone else. Similarly, the generation to generation evolution of field pattern is almost the same for other families of this village, with sub-division as the basic governing process. Twentieth century onwards the evolution of the field pattern reveals continuous sub-division of the fields due to increase

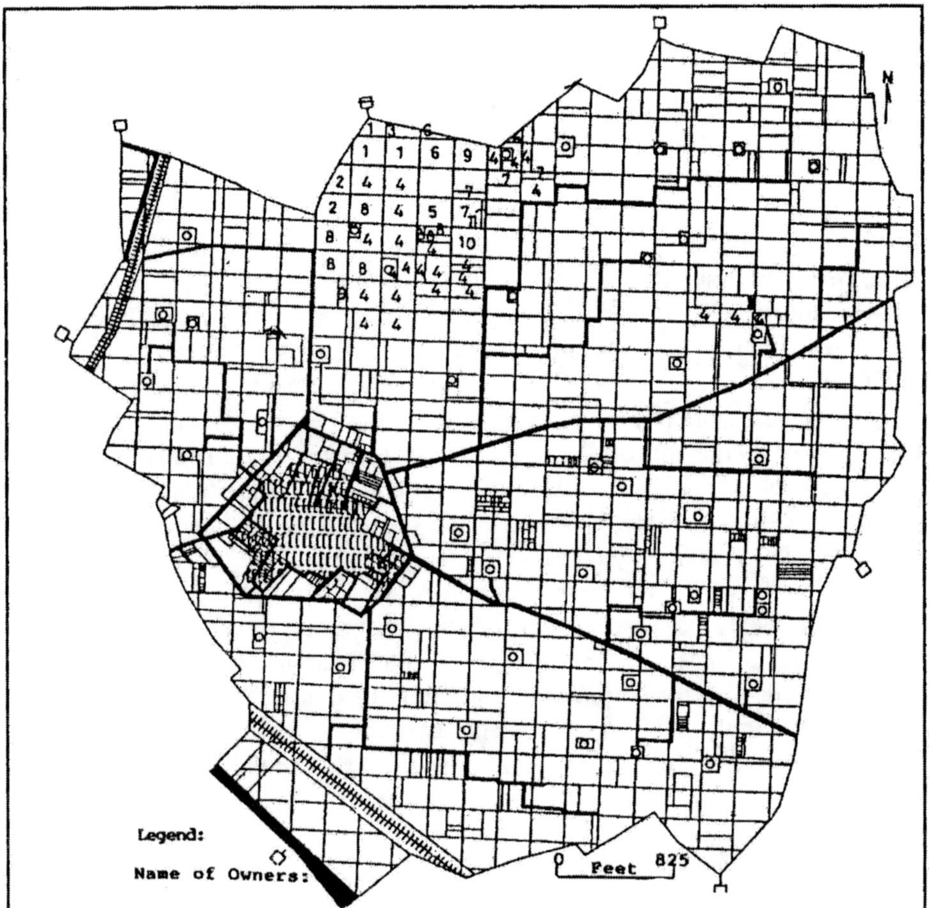

1 Bhola Singh Malla Singh, Jagat Singh Sons of Gujar Singh, Mehanga Singh, Gurdev Singh Sons of Arjan Singh

2 Bhola Singh, Malla Singh Sons of Gujar Singh, Mehanga Singh, Gurdev Singh Sons of Arjan Singh

3 Malla Singh S/o Gujar Singh, Mehanga Singh, Gurdev Singh Sons of Arjan Singh

4 Bhola Singh, Malla Singh Sons of Gujar Singh, Mehanga Singh Gurdev Singh Sons of Arjan Singh, Jagat Singh S/o Toba Singh

5 Mehanga Singh, Gurdev Singh Sons of Arjan Singh and Malla Singh S/o Gujar Singh

6 Mehanga Singh, Gurdev Singh Sons of Arjan Singh, Bhola Singh, Malla Singh Sons of Gujar Singh

7 Mehanga Singh, Gurdev Singh Sons of Arjan Singh, Jagat Singh S/o Gujar Singh

8 Mehanga Singh, Gurdev Singh Sons of Arjan Singh, Bhola Singh S/o Gujar Singh

9 Bhola Singh S/o Gujar Singh

10 Jagat Singh S/o Gujar Singh and adopted S/o Toba Singh

11 Malla Singh S/o Gujar Singh

Fig. 9.7 : Village Dhenowali : Ownership of Cultivated Land (1958) (Sixth and Seventh Generation of One Family)

in the number of male members of different families from generation to generation.

The principles which were discovered earlier with certain deviations and modifications, continue to operate in the present time and remain significant. In the village under study the operational law of inheritance is such that all the sons of a man will share equally in his land, regardless of their age with a fair distribution of advantages and handicaps. This has resulted in the fragmentation of holdings and sub-division of fields. At some places, the principle of compensation operates whereby the proportion of land owned by different sons in different land types varies.Anyone desirous of maximum land in the best land type is offered fewer fields which are also small and those with a limited share in the best land may have several fields of large size. The same principles were adopted by the consolidation officials at the time of land consolidation in 1958 and continue to operate. Due to the increasing population pressure and the entrance of female land owners made possible by the amendment of the Hindu Act which lays down that the daughters are entitled to share equally in their father's property, the per capita number of fields and size of holding has been continuously decreasing (Table 9.1). The per capita number of fields is reduced from 9 in 1884 to 3 in 1990. Similarly the per capita size of land-holding is reduced from 6 *killas* in 1884 to a little more than 1 *killa* in 1990. The cultivated land which was created in early phases is continuously decreasing due to encroachment of existing cultivated land by expanding settlement and road network.

Consolidation and Field Pattern

The historical evolution of the field pattern was disrupted due to the implementation of consolidation of land-holdings in 1958 in the study village. Till the middle of the last century the holdings of farmers comprised a number of fields located at different sites and in different directions of the village territory (Fig. 9.10). The fields constituting the holdings of different farmers were mostly intermingled. 'Dispersed and intermingled location of fields was one of the typical features of the field pattern in the pre-consolidation phase'.[5] The scattered fields and fragmented holdings ultimately resulted in low production. 'The small and fragmented fields retarded enterprise and created hinderances in smooth agricultural operation'.[6] Such a sub-division of the fields is essentially the result of the operation of the law of inheritance and equitable distribution of benefits and handicaps. To eliminate and reduce such problems, the government introduced the measure of consolidation of holdings by which the scattering of fields has been minimised by consolidating the irregular, small and inconveniently large fields and holdings into compact blocks of ractangular

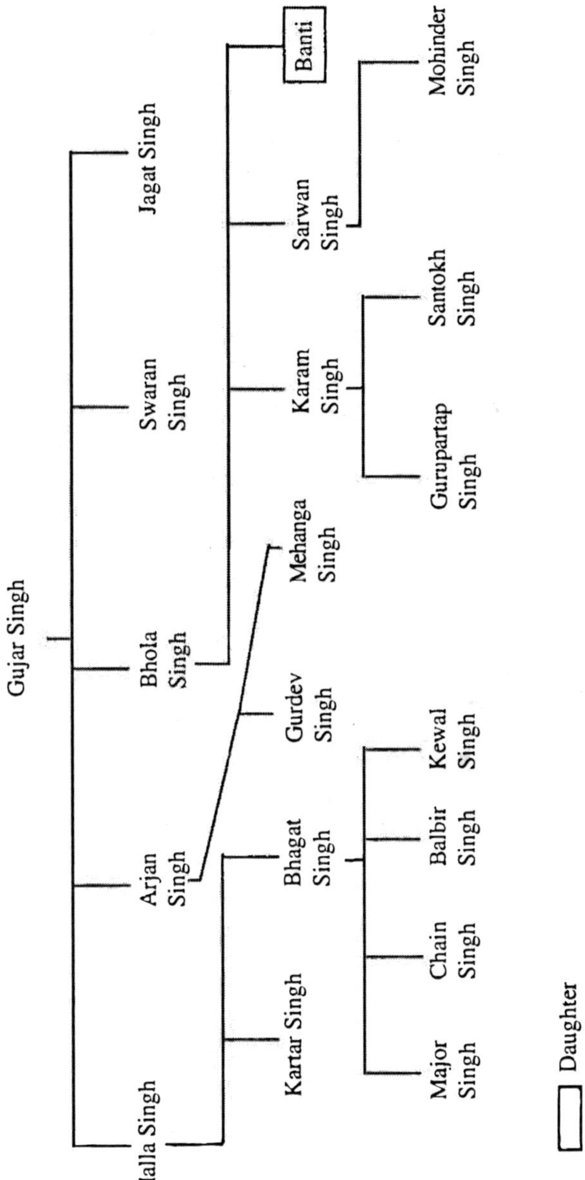

Fig. 9.8 : Village Dhenowali : A Section of Genealogical Tree (*Shajra Nasab*) of Owners (1990)

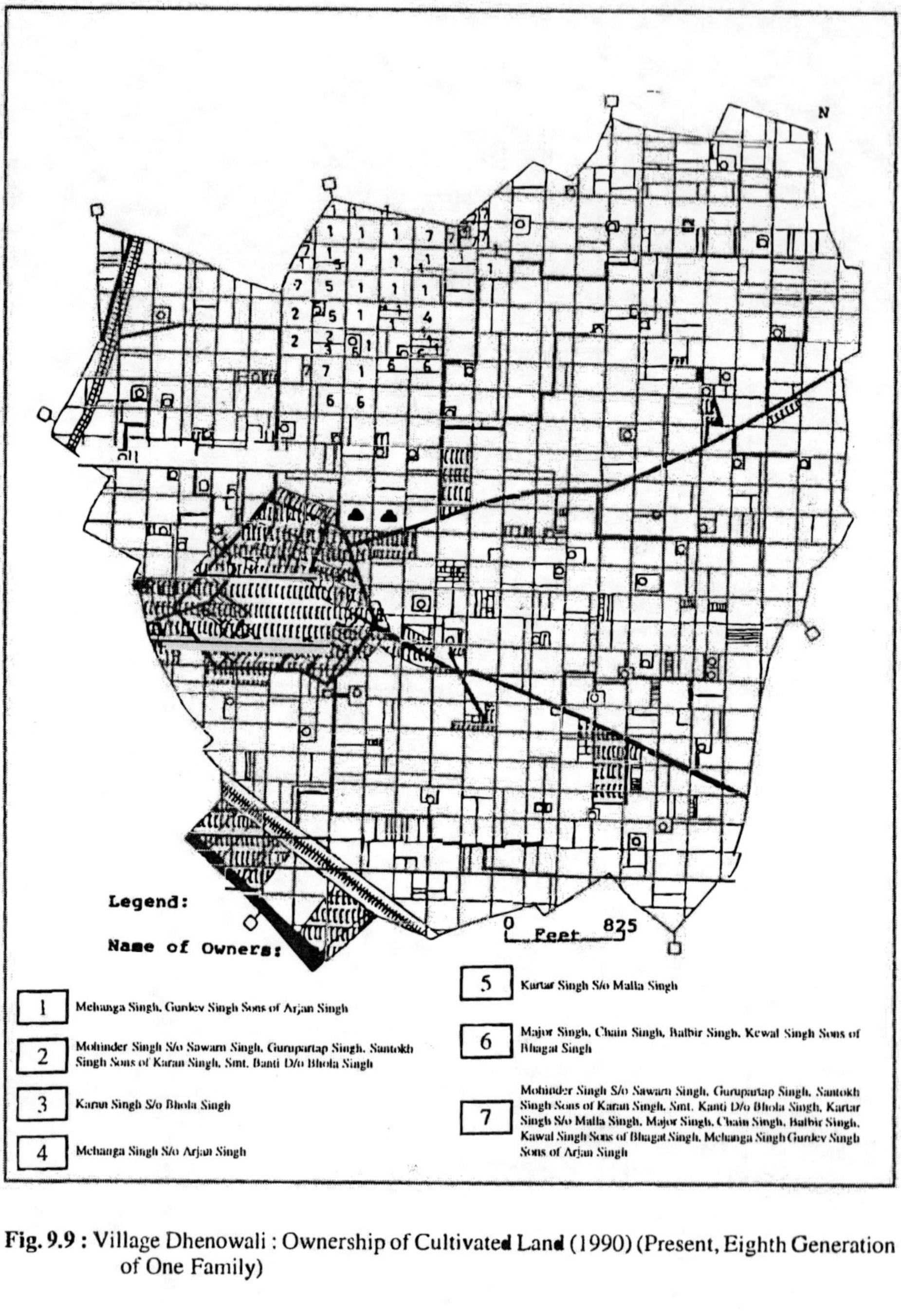

Fig. 9.9 : Village Dhenowali : Ownership of Cultivated Land (1990) (Present, Eighth Generation of One Family)

Table 9.1 : Village Dhenowali : Per Capita Number of Fields and Per Capita Size of Land-Holding

Year	*Per Capita Number of fields*	*Per Capita Size of Land-holdings*
1884	9	46 *kanals* (6 *killas*)
1914	8	29 *kanals* (3½ *killas*)
1958	4	12 *kanals* (1½ *killas*)
1990	3	9 *kanals* (1 *killa*)

Source : Jammabandis of Village Dhenowali, 1884, 1914, 1958 and 1990.

fields of uniform size of one *killa* (8 *kanals* or 198 feet × 200 feet) each. As a result the irregularities of the evolved landscape and fragmentation have disappeared. 'With the implementation of the consolidation of land-holdings there ushered a stage of a total and dramatic transformation of the field pattern and field landscape'.[7] The allocation of holdings was made subsequent to an evaluation of the plots of land, 'keeping in view the quality of the soil, the source of irrigation, the productivity of land and the distance from the *abadi*'.[8] Through the process of consolidation, the fields of a farmer which were found scattered at many sites were grouped into compact blocks located at not more than three sites adjacent to the largest part of his holding (Fig. 9.11). This resulted in the reduction in the number of sites in relation to the number of fields. An index of scattering can be computed by the formula suggested by Mukerji.

$$I_s = \frac{X}{Y} \times 100$$

where :

I_s is the Index of Scattering in percent.
X is the number of sites.
Y is the number of fields.

To study the change in the distribution and scattering of holdings between the pre-consolidation (1914) field landscape and the post-consolidation (1958) field landscape phases in the study village, an assessment of the land. holdings and fields belonging to three farmers (A, B and C) have been taken. The index of scattering of fields owned by farmer A has been reduced from 52 percent to 6.52 per cent, by farmer B from 27.58 per cent to 7.50 per cent and by farmer C from 27.77 per cent to 14.28 per cent in the pre-consolidation phase and post-consolidation phases due to grouping of scattered fields into compact blocks (Table 9.2). The value of the index of scattering of sites of fields may be the same for two farmers but in terms of distance of sites of the fields from the settlement, the farmer whose fields are located near the settlement enjoys more

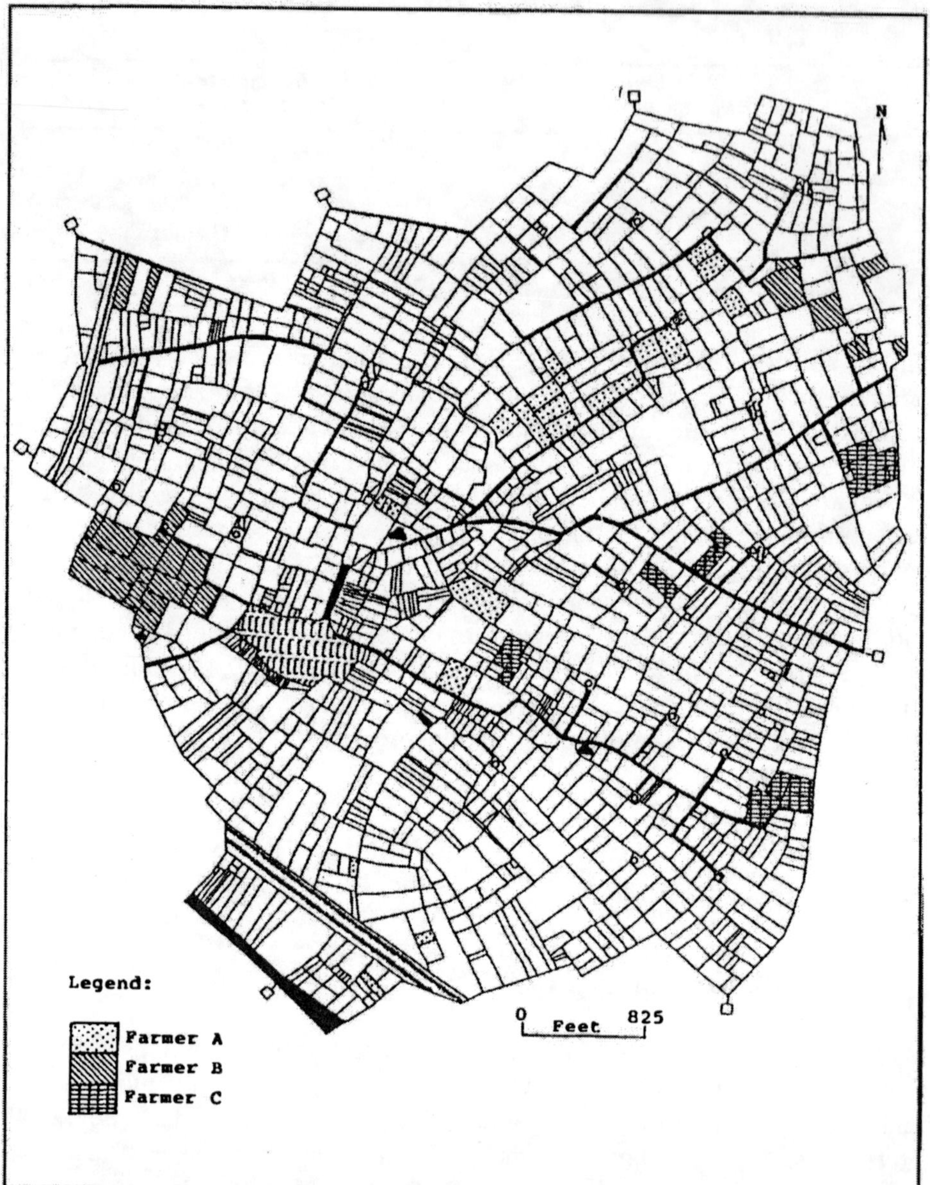

Fig. 9.10 : Village Dhenowali : Pre-Consolidation Scattering of Holdings, Field Forms and Field Patterns (1914)

benefits than the farmer whose fields are scattered at a great distance from the settlement. Therefore, the cumulative distance covered by the farmer from the *abadi deh* (settlement) to all the sites of the fields will make the index of

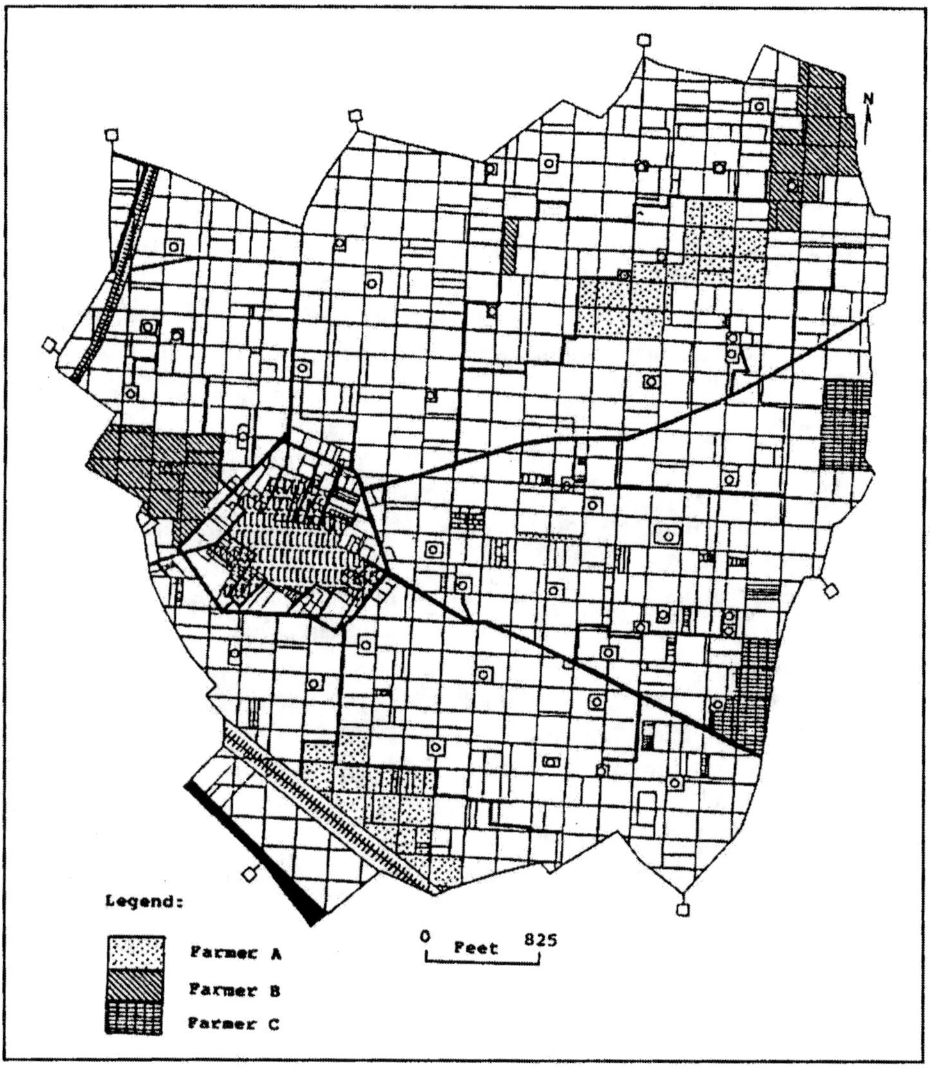

Fig. 9.11 : Village Dhenowali : Pre-Consolidation Scattering of Holdings, Field Forms and Field Patterns (1958)

scattering more operationally meaningul. Therefore, the index of scattering (as suggested by Mukerji) which takes into account the cumulative distance of all sites of fields from the settlement has been applied, *i.e.*

$$I_{SD} = \frac{I_S}{\Sigma D_{SF}}$$

where

I_{SD} = the index of scattering in terms of distance.

I_S = the value of index of scattering in percent.

ΣD_{SF} = the sum total or cumulative distance of sites of fields comprising a holding.

(ΣD_{SF} = the cumulative distance in inches on the map from the geometric centre of the *abadi deh* to all the geometric centres of sites of fields.)

There is an inverse relationship between the numerical value of I_{SD} and actual scattering of sites of fields in terms of distance (Fig. 9.12). Higher is the numerical value of I_{SD}, the lower will be the scattering of sites of fields in terms of distance.

In the pre-consolidation phase (1914) the value of I_S for farmers A, B and C was 52 percent, 27.58 percent and 27.77 percent respectively, whereas now the numerical value of I_{SD} is 1.818, 0.959 and 1.815 respectively. The I_S of farmer A is more than that of farmer C and B yet his fields in terms of distance are less scattered as his numerical value of I_{SD} is more than that of values of farmers C and B. In terms of I_S the fields of farmers B are less scattered than that of farmer A and C but in terms of distance his fields are most scattered as his numerical value of I_{SD} is the least.

In the post-consolidation phase there is a very evident change in the I_S and

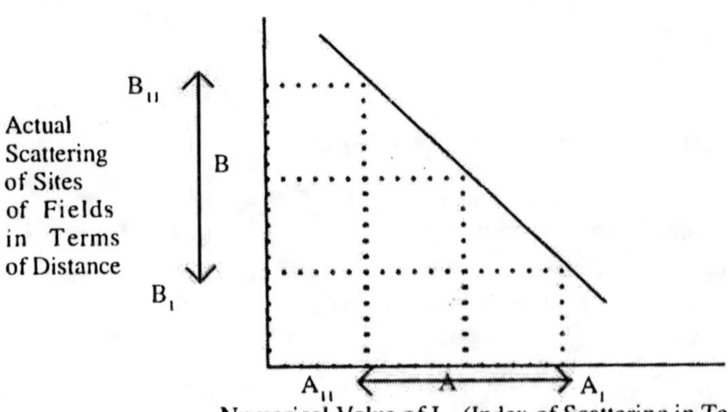

Fig. 9.12 : Relationship between Actual Scattering of Sites of Fields in Terms of Distance and Numerical Value of Index of Scattering in Terms of Distance

I_{SD}. The I_S is reduced for all the three farmers A,B and C, but I_{SD} has increased for all the three farmers as the numerical value of I_{SD} is reduced. This is due to the reduction in the I_{SD} whereas the actual $\sum D_{SF}$ is reduced for the three farmers. (Table 9.2). The scattering of fields in terms of distance tends to be almost the same for farmers A and B whereas it is the least for the farmer C although his I_S is maximum among them.

Size of the Fields

Size simply refers to the area of individual fields which comprise the holding of a farmer. The earliest (1884) available field map reveals that the fields near the *abadi-deh* were relatively smaller as a result of the higher frequency of subdivision of fields here in order to satisfy the desire of every member of the family to have a share in this land, not so much due to its productivity and fertility but due to its relative proximity and hence, the scope for building a house in future (Fig.9.3). The fields are larger where the incidence of subdivision is less. The fields at this period of time were mostly large ranging from blocks to elongated strips.

The consolidation of landholdings has had a great bearing on the size of the fields (Table 9.3). The number of fields have reduced from 1436 in the pre-consolidation phase (1914) to 1044 in the post-consolidation phase (1958). This reduction in the number of fields was related more to the reduction in area under cultivation from 5,040 *kanals* to 3,688 *kanals*, a reduction of 1,352 *kanals*. The cultivated land was encroached upon by the construction of new roads, houses and other *gairmumkin* (uncultivable) land causing reduction in the total number of fields in the post-consolidation phase. In some cases the number of fields of a farmer may have increased following the partitioning of joint holdings and reduction in the number of very large fields of above eight *kanals* in the post consolidation phase. In the pre-consolidation phase the number of such large fields covering an area of 1289 *kanals*-13 *marlas* was 102, comprising more than 25 percent of the total cultivated area, but in the post-consolidation phase, the number of such fields was reduced to only 3 (0.3 percent of total fields) covering an area of 28 *kanals* (less than 1 percent of the total cultivated area (Table 9.3). All of these three very large sized fields were located inside the *phirni* (circular road around *abadi deh*) where the fields had not been consolidated. The very large fields of above 8 *kanals* were converted into rectangular fields of uniform size of one *killa* (8 *kanals*) each. The medium sized fields of 1 to 4 *kanals* and the large fields of 4 to 8 *kanals* have also experienced a change in the post-consolidation phase. Both the number of fields and the area covered by medium sized fields have reduced in the post-consolidation phase. Such fields may be found in any part of the village territory

Table 9.2 : Dhenowali : Change in Degree of Scattering of Holdings from Pre-consolidation to Post-consolidation Phase

Pre-consolidation Phase (1914)						*Post-consolidation Phase* (1958)				
Owner	*Fields*	*Sites*	I_s	ΣD_{SF} (*in Inches*)	I_s	*Fields*	*Sites*	I_s	ΣD_{SF} (*in Inches*)	I_s
Farmer A	25	13	52.00	28.60	1.818	46	3	6.52	8.15	0.80
Farmer B	29	8	27.58	28.75	0.959	40	3	7.50	9.5	0.78
Farmer C	18	5	27.77	15.3	1.815	14	2	14.28	8.5	1.68

Source : Jammabandis of village Dhenowali, 1914 and 1958.
I_S Index of Scattering in percent.
ΣD_{SF} Cumulative Distance of sites of fields.
I_{SD} Index of Scattering in terms of distance.

Table 9.3 : Dhenowali : Field Size Analysis in the Pre-consolidation to Post-consolidation Phase

Pre-consolidation Phase (1914)					*Post-consolidation Phase* (1958)			
Field Size	*Number of fields*	*Percentage in total N.O.F.**	*Area covered (Kanal-merla)*	*Percentage in total C.A.C.**	*Number of fields*	*Percentage in total N.O.F.**	*Area covered (Kanal-merla)*	*Percentage in total C.A.C.**
Below 1 *Kanal*	290	20.2	142-5	2.8	235	22.5	72-10	2.0
1 *Kanal* to 4 *Kanals*	668	46.5	1559-3	30.9	381	36.5	848-14	23.0
4 *Kanals* to 8 *Kanals*	376	26.2	2049-5	40.6	425	40.7	2738-12	74.2
Above 8 *Kanals*	102	7.1	1289-13	25.6	3	0.3	28.04	0.8
Total	1436	100.0	5040-6	100.0	1044	100.0	3688-00	100.0

Source : Jammabandis of village Dhenowali, 1914 and 1958.
N.O.F.* Number of fields.
C.A.C.* Cultivated area covered.

but are more common near the *abadi deh*. Large fields are dominant both in terms of total cultivated area covered and number of cultivated fields in pre-consolidation and post-consolidation phases. These fields account for more than 40 per cent of the total fields and 74 per cent of the total cultivated area in the post-consolidation phase. Most of the large fields are rectangular and common everywhere in the village territory. Small fields of below one *kanal* remained practically unaffected in the post-consolidation phase. These fields though scattered in the entire village territory, are very common near the *abadi deh*. The fields near the *abadi deh* are small due to the greater degree of their sub- division in response to the desire of every member of the family to have a share in this land due to their proximity to the *abadi deh* and future scope for house making. Moreover, the continuous encroachment of cultivated land by *gairmumkin* land has further reduced the total number of fields from 1044 in

1958 to 1006 in 1990 (Table 9.4). The fields and holdings have become small in size due to continuous sub-division. As can be seen from the field map of 1990, small fields below 1 *kanal* comprise 24 percent of the total cultivated fields, but due to their small size they cover only 2 percent of the total cultivated land. The medium and large sized fields together comprise 76 per cent of the total number of fields and 98 per cent of the total cultivated area. All the categories of field size are found scattered everywhere in the village territory. The concept of distance and size of the fields is no more applicable.

Table 9.4 : Dhenowali : Field Size Analysis (1990)

Field Size	*Number of fields*	*Percentage in total number of fields*	*Area covered (Kanals-marlas)*	*Percentage in total cultivated area*
Below 1 *Kanal*	237	23.6	79-09	2.2
1 *Kanal* to 4 *Kanals*	365	36.3	868-06	24.5
4 *Kanals* to 8 *Kanals*	404	40.1	2603-04	73.3
Total	1006	100.0	3550-19	100.0

Source : Jammabandi of village Dhenowali, 1990.

At present in areas where tractors are used, fields are larger and accentuated rectangular to save both fuel and time as tractor needs less number of turns to plough the entire field. Bullock drawn fields are relatively smaller. An analysis of the existing cropping pattern in the village reveals that there exists no remarkable relationship between cropping pattern and field pattern.The *teenfasli* (three cops in a year) fields are relatively larger than *dofasli* (two crops in a year) fields. The rice grown fields are relatively smaller. Water can be kept in small fields with ease. Therefore paddy fields tend to be smaller. All other crops are found in all the fields irrespective of their size and shape tend to be smaller of their size and shape.

Shape of the Fields

Shape is an important and useful parameter for differentiating field patterns. The shape of the fields may be determined by property boundaries. In Fig.9.4, it can be seen that the fields are typically elongated, rectangular strips in shape (Fig.9.4). This shape is attributable perhaps to the ancient notion that the field should be of such a size that one plough team of farmers could plough the field in one day, and to the use of the wooden plough suited to a long furrow, 'so that there would be no wastage of time in turning the plough too frequently'.[9] The elongated rectangular and strip shape of the fields at that stage would have also been determined by micro- topography such as uneven *tibbas* (sandy tracks), poor landtypes, poor sources of irrigation, poor farm implements, the subsistence nature of agriculture, poor management of fields due to small family size and big holdings scattered at different sites. The local micro topography has played an effective role in the orientation of the long-side extension of the rectangles and strips.

Therefore, as evident from the map, the alignment of the fields varies in different parts of the village territory. For easy movement of human beings and cattle many of the elongated rectangular and strip fields comprising the original holdings of different farmers are oriented to the village and field roads (Fig. 9.4).

In the pre-consolidation phase (1914) the cultural landscape of the village was characterised by irregular fields (Fig. 9.10). Most of the fields were irregular rectangles, elongated and rectangular. These three classes together accounted for 91 per cent of the total fields. Irregular polygons, hexagons and triangular fields were also found (Table 9.5). In contrast the most characteristic feature of the post-consolidation (1958) field landscape is total regularity and uniformity. As a result of consolidation itermingled and irregular rectangular fields have changed into regular rectangular fields of uniform size of one *killa* each. Majority of the fields have become rectangular, regular strips and quadrilaterals (Table 9.5). Approximately half of the total fields have become rectangular in the post-consolidation phase whereas in the pre-consolidation phase only 10 per cent of the fields were rectangular. The proportion of irregular rectangular and elongated fields has reduced from 43 per cent to 10 per cent and from 38 percent to 27 per cent respectively. Fields, absolutely square in shape account for less than 1 per cent both in the pre-consolidation and post-consolidation phases.

Table 9.5 : Dhenowali : Field Shape Analysis in the Pre-consolidated and Post-consolidation Phases

Field Shape Class	*Pre-consolidation Phase*		*Post-consolidation Phase*	
	Number of fields	*Percentage in total N.O.F.**	*Number of fields*	*Percentage in total N.O.F.**
Three sided-Triangular	26	2.0	39	3.7
Four sided-Square 6	0.4	04	0.4	
Four sided-Rectangular	146	10.0	506	48.5
Four sided-Elongated or Strips	552	38.0	284	27.2
Four sided-with one or more sides Oblique irregular Rectangular or Quadrilateral	617	43.0	104	10.0
Five sided-Polygons15	1.0	24	2.3	
Six or more sided-Hexagons	74	5.0	83	8.0
Total	1436	100.0	1044	100.0

Source : *Sharja Kishtawar* of village Dhenowali, 1914 and 1958.

* N.O.F. Number of fields.

There is a slight increase in the percentage of triangular, polygon and hexagon fields in the post-consolidation phase. Such non- rectangular fields are found mostly along diagonal roads, railway lines, irregular village boundary, near the *abadi deh* and the sources of irrigation (wells and tubewells). The percentage of fields under different shape categories tends to be almost the same in the post-consolidation (1958) and existing (1990) fields maps (Table 9.6).

Table 9.6 : Village Dhenowali : Field Shape Analysis (1990)

Field Shape Class	*Number of fields*	*Percentage in the total number of fields*
Three sided—Triangular	37	3.7
Four sided—Square	2	0.2
Four sided—Rectangular	489	48.6
Four sided—Elongated or Strips	287	28.5
Four sided—With one or more sides Oblique irregular Rectangular or Quadrilateral	82	8.2
Five sided—Polygons	18	1.8
Six or more sided—Hexagons	91	9.0
Total	1006	100.0

Source : *Shajra Kishtawar* of village Dhenowali, 1990.

Field Boundaries

The field boundaries give shape to a field. ' The pattern formed by well defined field boundaries within the precincts of a village constitute an important element of the man-made landscape.'[10] On the village map, lines representing field boundaries are in fact the limits of property units each bearing a separate *Khasra* number (field identification number) and in the *Lal Kitab* (a record book of property units). On the ground a farmer divides each property unit into smaller parts for different purposes such as to suit requirement of irrigation or for growing different crops. These smaller units and property units are separated by low mud dikes known as *wats* and by *arrs* (narrow water channels). These *wats* and *arrs* constitute the smallest feature of the agricultural landscape. The *wats* are about eight inches high, one to two feet wide and have enoromous length. They are made of locally available wet soils. These *wats* are used as dams to retain water within the fields and as paths for movement of human beings and cattle. Some of the fields near the *abadi deh* and roads are enclosed by *wats* of *kande* (row of thorny branches) and a row of trees to protect fields from straying cattle and also children. Because of the process of consolidation the field boundaries have become regular in shape. At present (1990) on an average 85 per cent of the total fields have four boundaries and remaining 15 per cent of the fields have either less than four or more than four boundaries.

Table 9.7 : Dhenowali : Land Types of Cultivated Fields in the Pre-consolidation and Post-consolidation Phase

Pre-consolidation Phase (1914)					*Post-consolidation Phase* (1958)				
Land Type	*Number of fields*	*Percentage in total N.O.F.**	*Area covered (Kanal-marla)*	*Percentage in area covered*	*Land*	*Number of fields N.O.F.**	*Percentage in total*	*Area covered (Kanal-marla) covered*	*Percentage in area*
Chahi	772	53.8	2495-02	49.5	*Chahi*	708	67.8	2496	67.7
Banjar Kadim	74	5.1	147-17	2.9	*Banjar Kadim*	36	3.4	32	0.9
Banjar Jadid	45	3.1	27-00	0.5	*Banjar Jadid*	19	1.8	24	0.6
Barani Awal	34	2.4	129-11	2.6	*Barani*	281	26.9	1136	30.8
Barani Daum	439	30.6	2009-06	39.0					
Barani Saum	72	5.0	231-10	4.6					
Total	1734	100.0	5040-06	100.0	Total	1044	100.0	3688-00	100.0

Source : Jammabandi of village Dhenowali, 1914 and 1958.
N.O.F.* Number of fields.

The field boundaries run straight and most of them meet at right angles because of the regular field pattern (Fig. 9.13). The fields near the *abadi deh*, roads and source of irrigation (tube-wells) have irregular field boundaries.

Land Types and Field Patterns

Land types and field patterns are closely related. Land types have experienced a change both in space and time. The earlierst field map of 1884 reveals several types of land characterised by variations in inherent fertility and productivity (Fig. 9.13). The land types include *nyayi chahi, meera chahi, tibba banjar kadim* and *banjar jadid*. The *nyayi chahi* is the best quality, highly manured irrigated land adjacent to the *abadi deh*. The *nyayi chahi* land is found to the north of the *abadi deh* and has relatively small fields due to the higher degree of sub-division as every person wants to have a share in the irrigated and fertile land type. On the other hand, *meera chahi* is an irrigated land generally found away from the *abadi deh*. *Tibba* is an uneven sandy tract which cannot retain water for long and therefore is not well suited to agriculture. *Meera chahi* and *tibba*, the two predominant land types in1884 are characterised with large and irregular fields scattered all over the village territory. *Banjar kadim* and *banjar jadid* are terms used for inherently poor fallow land. While *banjar kadim* refers to the land which has been left fallow for atleast eight years or eight cropping seasons, *banjar jadid* refers to land which has been left fallow for three years or three cropping seasons. The fields of such land types are large in size and irregular in shape.

Pre-consolidation (1914) land types include *chahi, banjar kadim, banjar jadid and barani (awal, daum and saum)* (Fig. 9.14). The term *barani* denotes non-irrigated, sandy land having low inherent fertility and productivity. Notionally, *barani* land has further been divided into three types : *barani awal, barani daum* and *barani saum* on the basis of distance from the *abadi deh*, inherent fertility and use of manure. *Awal* means superior or first ranking, *daum* means inferior or second ranking and *saum* refers to the most inferior land type. *Barani awal* land has low sand content and can retain water for long. This type of land is found adjacent to the *abadi deh* and, therefore, is highly manured. *Barani daum* land has a higher proportion of sand and is generally found near the village territory but away from *barani awal* land. *Barani saum*, on the other hand, is sandy, cannot retain water and is the most inferior cultivated land. It is found near the village territory away from *barani daum*, although a few patches of *barani saum* are also found in the interior. The fields are relatively small near the *abadi deh* not due to their high productivity, because except in the north-west, the *abadi deh* is surrounded by land of low productivity, *banjar jadid, banjar kadim and barani (awal, daum, and saum)*

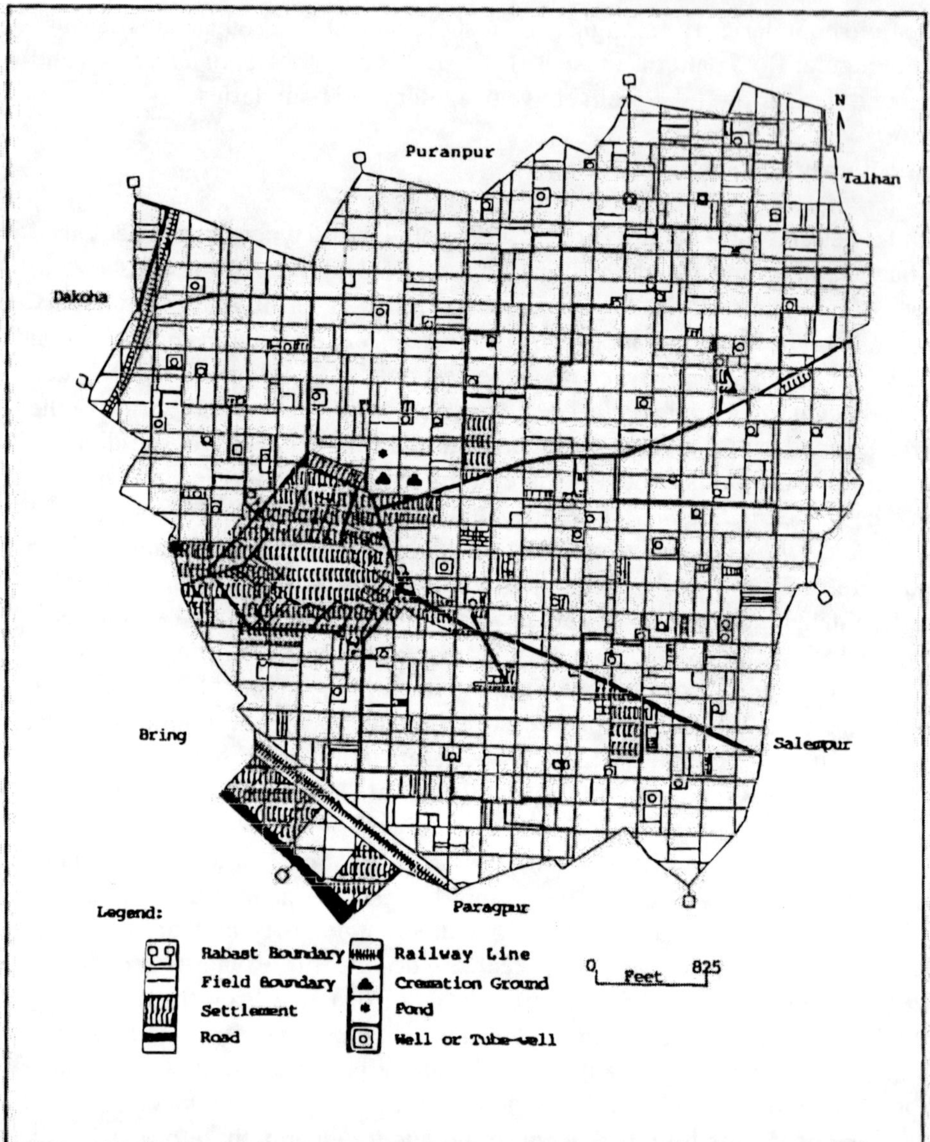

Fig. 9.13 : Village Dhenowali : Field Pattern (1990)

but as a result of the frequent sub-division of these lands on account of their relative nearness to the *abadi deh* and future scope for house construction. The pre-dominant land type in the pre-consolidation phase was *chahi* covering about half of the total cultivated land and more than half of the fields (Table 9.7). *Barani* land constitutes around 47 per cent of the total cultivated land and

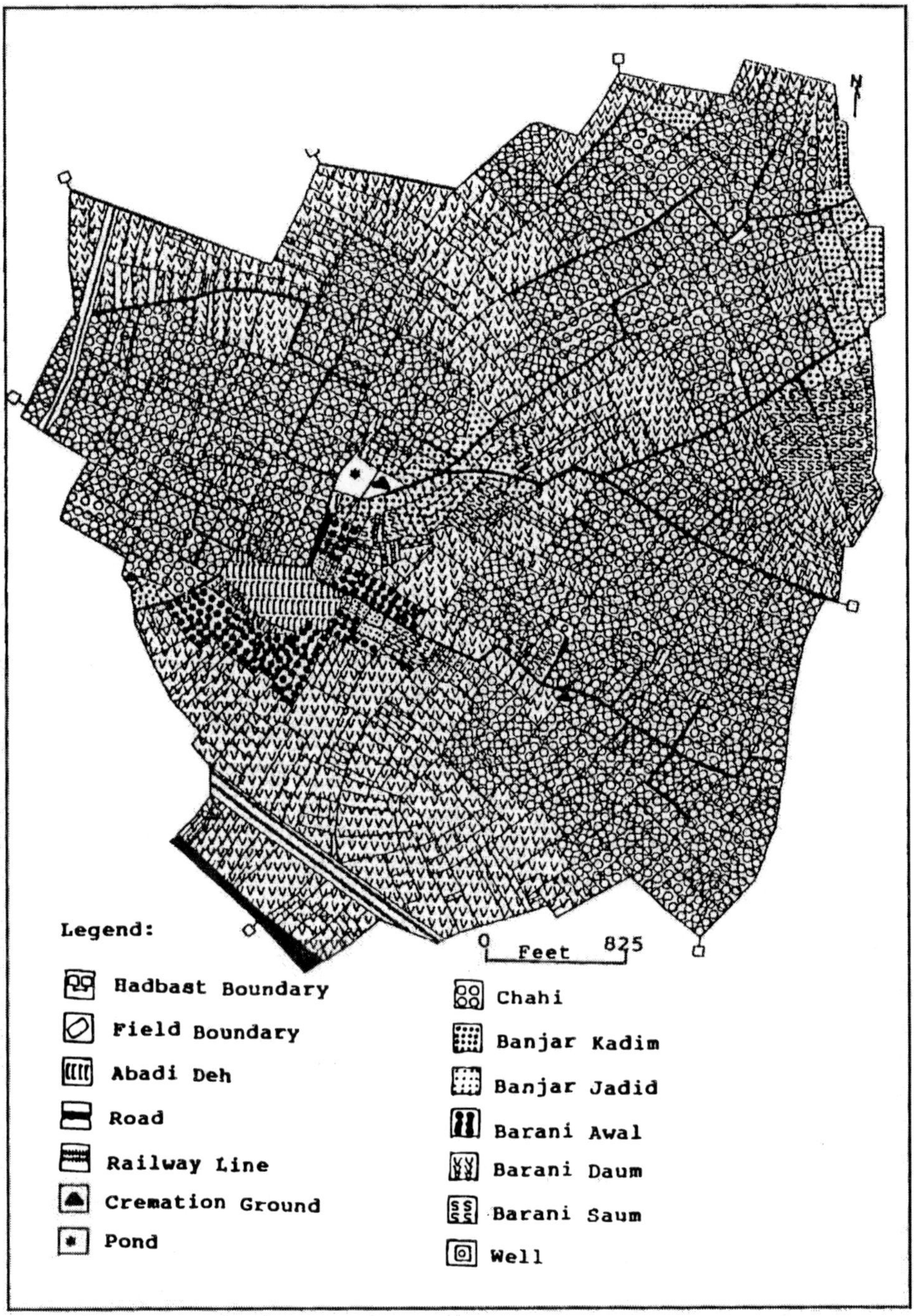

Fig. 9.14 : Village Dhenowali : Land Types and Field Patterns in the Pre-Consolidation Phase (1914)

38 per cent of the total fields. Among the *barani* land types, the *barani daum* covers the maximum accounting for 40 percent of the total cultivated land and 30 percent of the fields. Although fields of different sizes are found everywhere in the village territory, and in different land types yet on the basis of number of fields and area covered by these fields in each land type. The *chahi* fields are smaller than the *barani* fields.

Land types have improved in the post-consolidation phase (1958). Most of the *barani* fields have been converted into *chahi* fields (Fig.9.15). The percentage of *chahi* fields to the total fields and percentage of cultivated area under *chahi* fields both have increased from 53.8 per cent to 67.8 per cent and from 49.5 per cent to 67.7 per cent respectively (Table 9.7). The number of fields and cultivated area under *banjar kadim* and *banjar jadid* is reduced. Also the number of *barani* fields and the area under them has reduced greatly. This has become possible due to the increase in the number of wells from 36 to 46. The *banjar kadim* and *banjar jadid* fields are located near the *phirni* (circular road around the *abadi deh*). Most of the *barani* fields are found near the *abadi deh* while the *chahi* fields are found both in proximity as also at a distance from the *abadi deh*. Fields of different sizes are common in the entire village territory irrespective of land type but *chahi* fields are relatively smaller than the *barani* fields in the post-consolidation phase, although this association has gradually faded away with the adoption of Green Revolution technology in the study village. Almost the entire cultivated area and fields (99.80 percent) have been brought under irrigation with the help of tubewells (Fig. 9.16). Presently there exists no relationship between the distance from the *abadi deh* and prodctivity and fertility of land due to use of irrigation and fertilisers.

Land-owning Castes and Field Patterns

Ownership of land by different castes has a definite bearing on the field pattern. The extent of land owned by different castes display different field patterns. In general, fields and holdings belonging to the high castes are larger than those of the lower castes. The study village was first inhabited by a family of Jat Randhawa caste. At that time it was a uni-religious, uni-clan and uni-caste settlement. Later they offered and sold some land to other castes, such as Jats, Nagra, Khatri, *Tarkhan* and *Sunnar* to settle the village and practise agriculture on the existing fields (Fig. 9.17). Other low castes were invited to settle here and be employed as agricultural labour. They were not given cultivated land. The Jat Randhawas remained the dominant land owning caste, having more than 90 per cent of the total fields and cultivated area. Obviously the fields and holdings owned by them were larger and more regular than those belonging to other castes. The fields belonging to the Jat Randhawas and those of the other

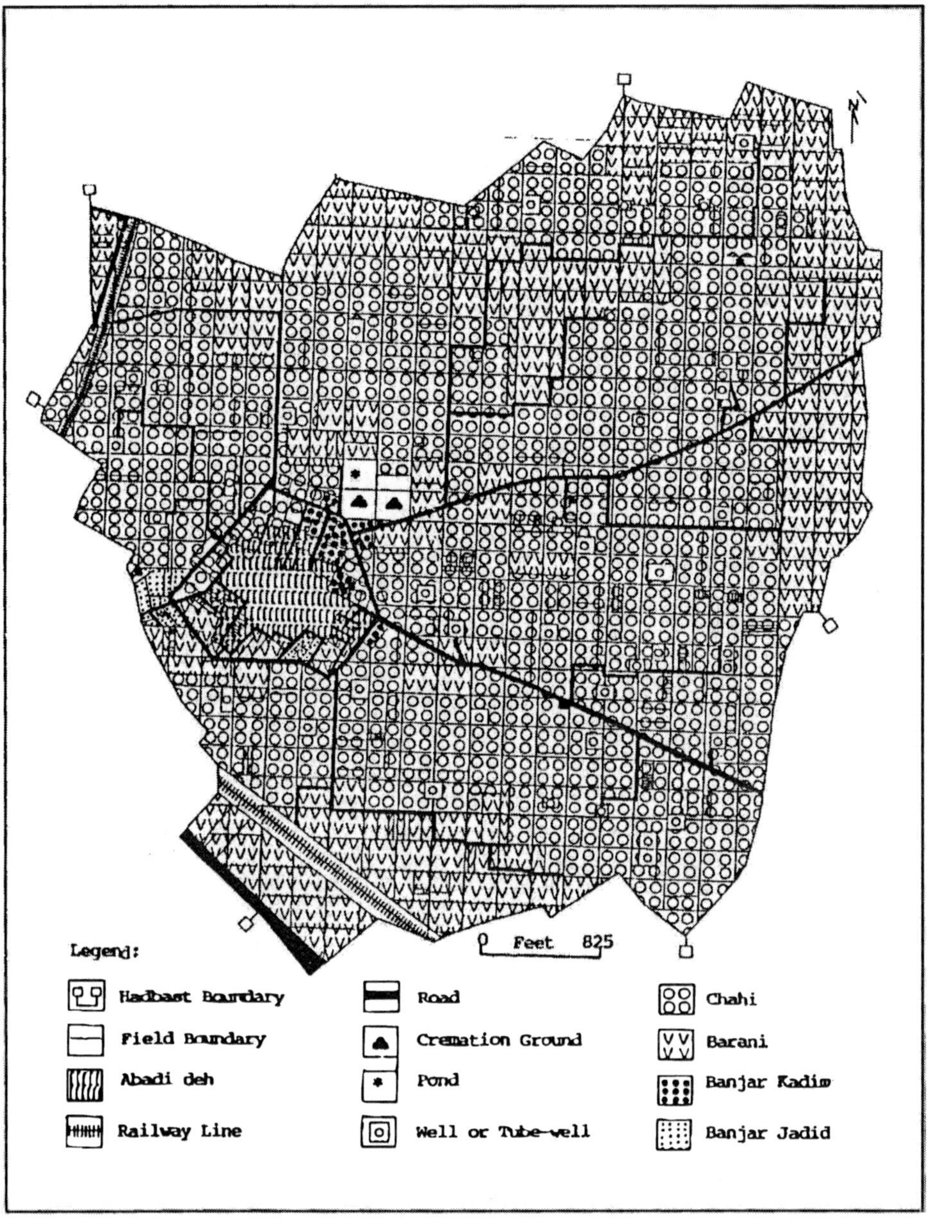

Fig. 9.15 : Village Dhenowali : Land Types and Field Patterns in the Post-Consolidation Phase (1958)

castes were found intermingled in the entire village territory, but more towards the north, north-east and north-west of the village.

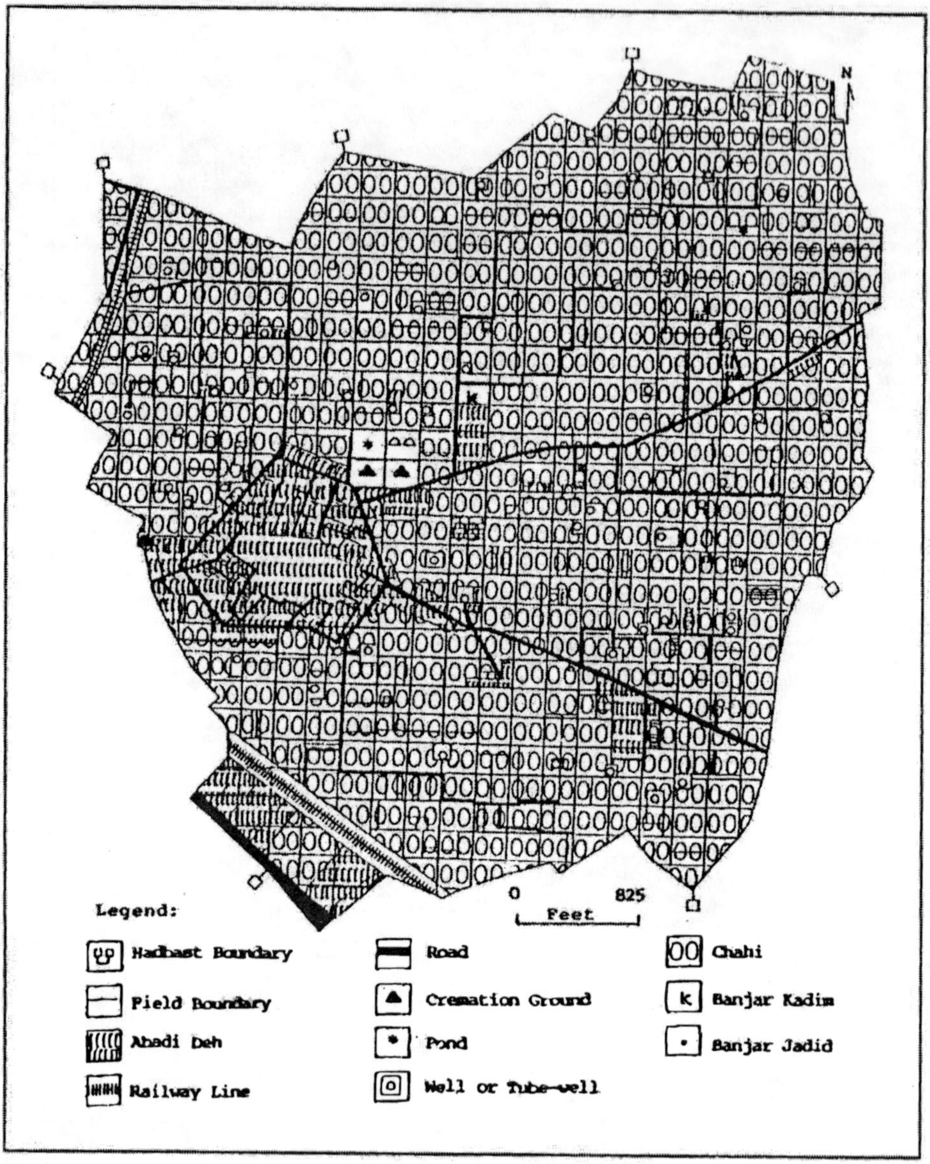

Fig. 9.16 : Village Dhenowali : Land Types and Field Patterns, 1990

Even today Jat Randhawas continue to be the dominant landowning caste. They have more than 93 percent of the total fields of the total cultivated area in their possession (Fig. 9.18). Their fields are large and regular rectangular. The fields and holdings of other existing castes (Jat Nagra, Jat Johal, Jat Malhi, Ramgarh Vadder, Ramgarhia Mander and *Tarkhan*) are relatively small. The

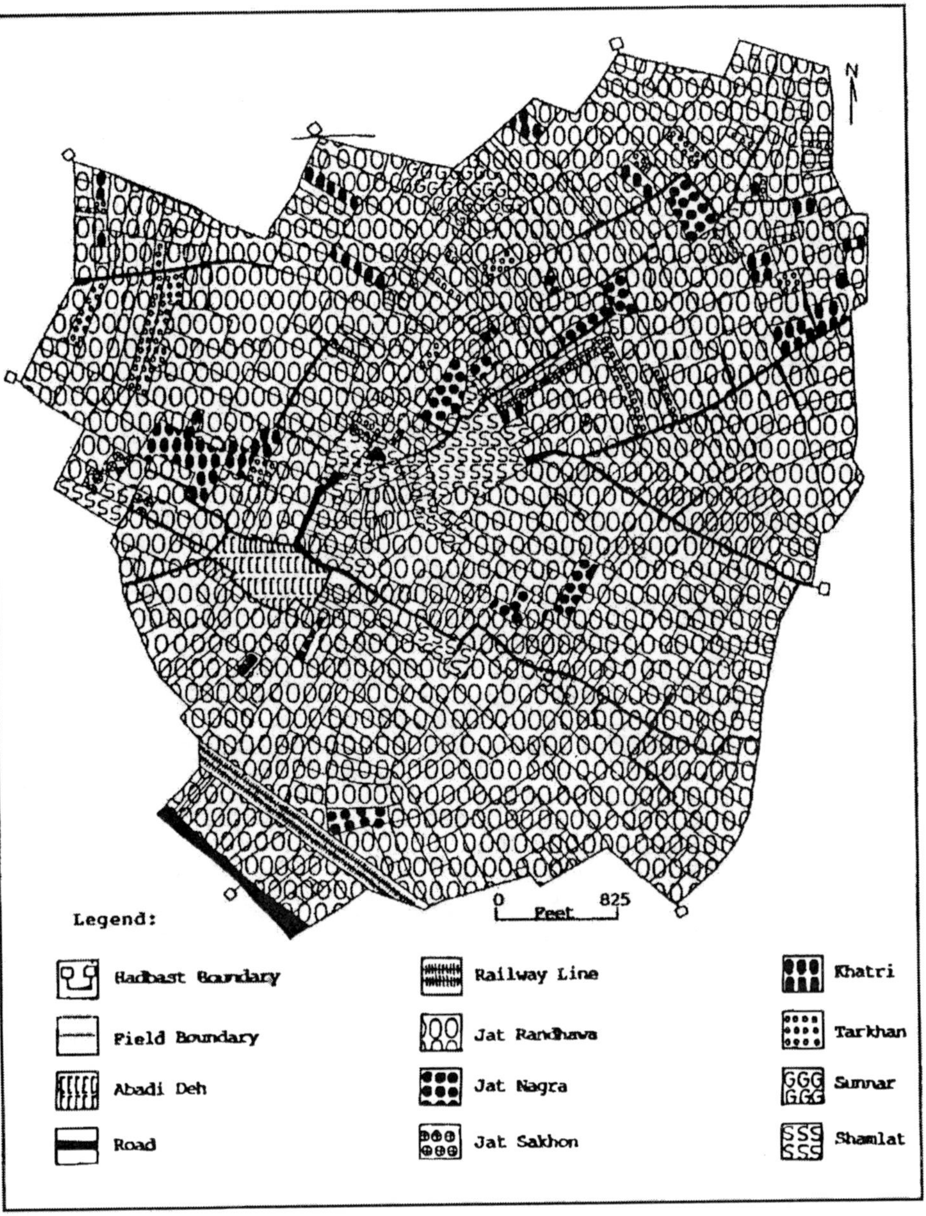

Fig. 9.17 : Village Dhenowali : Distribution of Field According to Land-Ownings Castes, 1884

holdings of these castes are intermingled with the holdings of Jat Randhawas. Their fields are distributed near the village periphery. No difference is found in terms of shape of the fields, source of irrigation, crops grown and farm mechnisation among the fields belonging to different castes. The low castes do

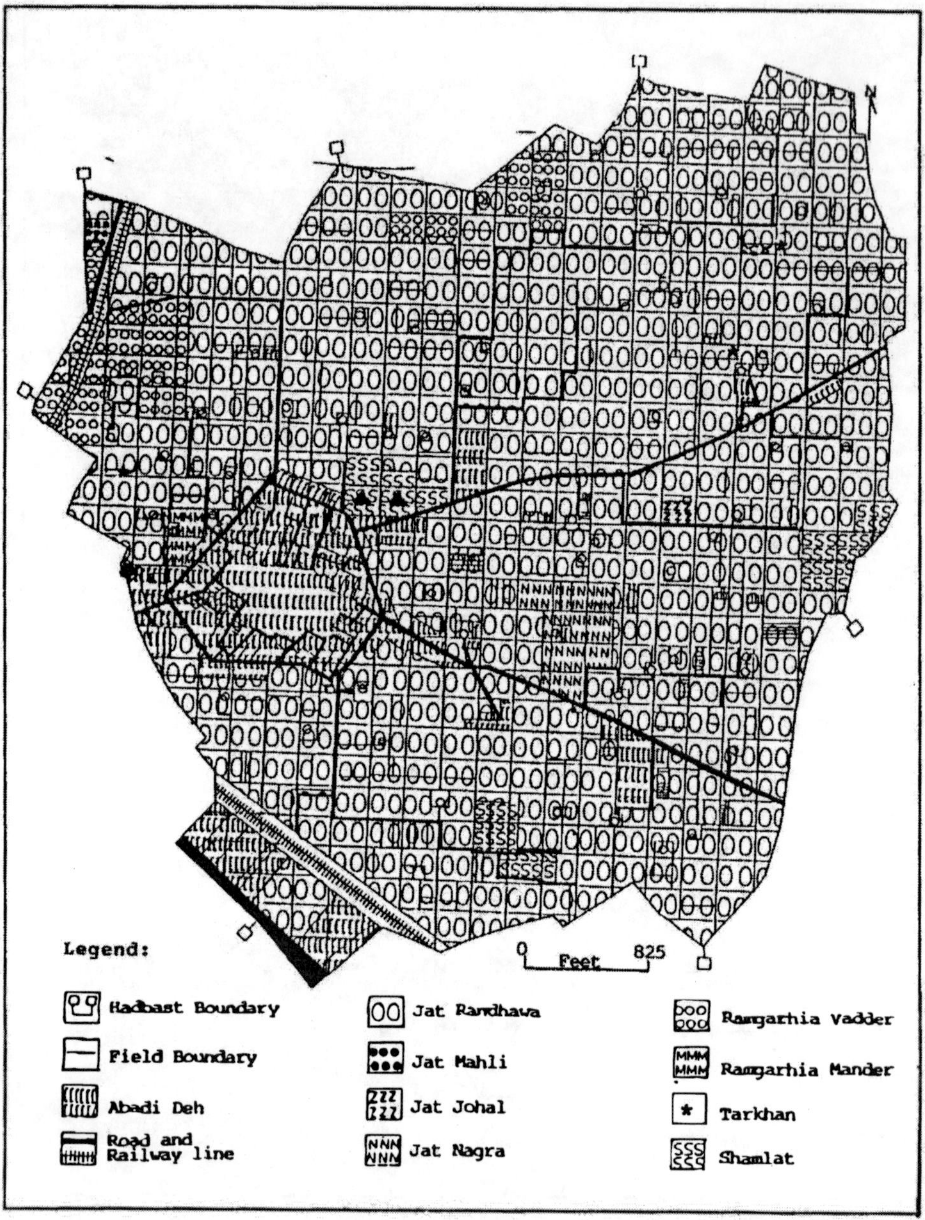

Fig. 9.18 : Village Dhenowali : Distribution of Fields According to Land-Ownings Castes, 1990

not own cultivated land, but they have their own houses in the *abadi deh* and provide labour in farming practices. In brief no remarkable differences in the field pattern owned by the different castes can be discerned as it is a single dominant land-owning caste village. The difference exists only in the amount of land owned by different castes. The economic condition of different land-owning castes seems to have a greater bearing on field patterns. The rich farmers practise mechanised farming. They have big holdings, and large and regular rectangular fields. The holdings and fields of poor farmers are small irrespective of their castes.

Therefore, it is evident from the study of different attributes of field patterns and factors affecting them that the field pattern is not a static phenomenon. The existing field pattern has undergone various evolutionary stages and provides a link and a key in the understanding of the processes of settling in an area of a particular cultural group.

GLOSSARY OF THE VERNACULAR TERMS

Abadi Deh : Settlement.

Arr : A narrow water channel for taking tube well water to the fields.

Banjar Jadid : The land which has been left fallow for atleast 3 years or 3 cropping seasons.

Banjar Kadim : The land which has been left fallow for atleast 8 years or 8 cropping seasons.

Barani : Non-irrigated rain fed land.

Barani Awal : Superior, first ranking *barani* land adjacent to the *abadi deh.*

Barani Daum : Inferior, second ranking *barani* land away from the *abadi deh.*

Barani Saum : The most inferior, third ranking *barani* land near the village territory.

Chahi : Land irrigated by *chah* (well).

Dofasli : Two crops in a year.

Deh : A hamlet

Gairmumkin : Uncultivable.

Hadbast : Boundary.

Jammabandi : A book of property records.

Jat : A yeoman, peasant community of North-west India

Johal : A *gotra* (clan) of the jats.

Kanal : A unit of land measurement, comprising an area of about 500 square yards.

Khasra : Number : Field identification number.

Kande : Thorn

Khatri : A community associated with business.

Killa : A unit of land measurement comprising an area of 8 *kanals* or $198^1 \times 220^+$, equivalent to acre.

Lal Kitab : A record book of property units.

Mahli : A *gotra* (clan) of the Jats.

Mander : A *gotra* (clan) of the Ramgarhia.

Meera Chahi : *Chah* (well) irrigated land generally found away from the *abadi deh*.

Marla : A unit of land measurement comprising an area of about 30 square yards.

Nagra : A *gotra* (clan of the Jats).

Nyayi Chahi : *Chah* (well) irrigated land adjacent to the *abadi deh*.

Phirni : A circular road around *abadi deh*.

Ramgarhia : Craftsman caste.

Randhawa : A *gotra* (clan) of the Jats.

Sekhon : A *gotra* (clan) of the Jats.

Shamlat : Village common land.

Shajra Kishtawar : Village map showing the fields and their boundaries with *Khasra* number, also known as *massavi*.

Shajra Nasab : Genealogical tree of owners of the settlement also know as *Vanshavali*.

Sunnar : Goldsmith caste.

Tarkhan : Carpenter caste.

Teenfasli : Three crops in a year.

Tibba : An uneven sandy track which cannot retain water for long, therefore not good for agriculture.

Vadder : A *gotra* (clan) of the Ramgarhia.

Warr : A fence, hedge or row.

Wat : A low mud-dyke which separates two fields or property units and is used as path for movement of human beings and cattle and as dams to retain water within the field.

References and Notes

1. Mukerji, A.B. (1961). 'Field Patterns in the Upper-Doab Jat Villages', *Journal of Geography*, Vol.3 No. I pp. 1-9.
2. Nitz, H.J. (1966). 'Geographcial Studies in the Field Patterns of Northern India and Germany', *The Geographer*, Vol. 13 pp. 61-74.
3. Grover, Neelam (1985). 'Field, Patterns in the Chandigarh Dun Gujar Village—An Interaction Study of Ecology, Culture and History', in A.B. Mukerji and Aizasudin Ahmad (eds.) *Indian Culture Society and Economy, Geographical Essays in Honour of Prof. Asok Mitra*, New Delhi : Inter-India Publications, p. 46.
4. See ref. 1 p. 4.
5. See ref. 2, p. 61.

6. Lal K.K. (1972), 'Field Patterns After Consolidation: A Land Reform Measure', *The National Geographical Journal of India*, Vol. 18, Parts 3-4, pp. 213-222, ref. p. 213.
7. Randhawa, M.S. (1974). *Green Revolution : A Case Study of Punjab*, Delhi : Vikas Publishing House, p. 40.
8. *See* ref. 7, p. 40.
9. *See* ref. 1, p. 4.
10. *See* ref. 2, p. 61.

10

CULTURAL BASIS OF BHOJA FLURFORM IN THE SIWALIK HILL KANET LANDSCAPE

NEELAM GROVER

Any field landscape is essentially an expression of the community decision through their culture and mode of living. It involves a study of the field pattern and its association with the size, shape, distance from the settlement, land type, and land-use of the fields. These attributes are not static nor is the present day field landscape a spontaneous creation. It has evolved gradually over a period of time, each phase being closely related to, and in fact the product of, the then existing political conditions, land tenure system, technological know-how, the culture group and their *genre de vie*. 'It is therefore obvious that it is not only important to describe the various field patterns, but also, as far as possible to trace the origin and evolution of the patterns in order to find out the socio-economic and political factors in operation at the time of the settlement'.[1] A reconstruction of the field landscape as it existed at different periods of time has been attempted for a single village in order to draw generalizations and be able to identify on the landscape a typical Kanet village. The village selected is Behlon within the Morni Hill tract of the Siwalik Hills in the Naraingarh *tehsil* of Ambala district in Haryana, a classic habitat of the Kanets.

The Morni hills almost wholly consist of 'a chaotic jumble of low hills on stony ravines.' They are 'a portion of the Siwalik range which is here not separated from the Himalayan formation but gradually merges into it'. Bounded to the north by the *Dun*, to the west by the Ghaggar water-gap and by the piedmont to the east and south, the region is divided into two sections, the *Pahar* and *Ghar*, following a well-recognized local distinction. The latter is the low country adjacent to the plains, while the former, as the name suggests, includes the higher mountain ranges of the tract (Fig. 10.1). It is particularly the ***Pahar*** section which is dominated by the Kanet culture area. 'The region is composed of friable sand-stone which has been torn by weather action and perennial streams rising in the southern slopes of the Himalaya'.[2]

Note : Reprinted from *Punjab University Research Bulletin* (Arts), Vol. XXII, No. 2, October, 1991.

Here, 'the land lies in great waves or ridges.... the soil is light and stony, offering no resistance to the rapid drainage, and the whole tract is in consequence intersected with sandy torrents and deep ravines.'[3] In most senses, it is a classic badland topography.

The Kanets, as suggested by several historians and as emphasised by local traditions, records and folklore, are fallen Rajputs, those who accepted and widely practised *karewa* (widow remarriage). Being opposed to one of the four basic precepts which distinguished the traditional Rajputs[4] such people came to be known as *Kuneet* or *niti ke virudh* (those against the law) and later Kanet.

Claiming descent from the Rajputs of Ujjain and Malwa in Rajasthan, the Kanets living in the Morni hills definitely came from the adjacent Lower Himalaya.[5] This suggestion is further supported by the distribution pattern of the Kanets in the larger region. There is almost no Kanet settlement in the piedmont south of Siwalik Hills and there are few, if any, in the Siwalik Hills south of the Dun. It is interesting to note that the Morni hills is a southward continuation of the Sirmur Hills and is not separated from the latter by any stretch of the *Dun*.

Further they entered the Morni Hills at a time when internal social differentiation within the Rajputs (whereby those who became the feudal landlords and the kings of principalities and native states continued to be regarded as Rajputs, while the others who took to farming as their major occupation, became owner cultivators) had been effected and the Kanets had emerged as a separate caste and culture group with an identity of their own, and were accepted as such by the rural communities of the region. The absence of any reference to the Kanets in the Medieval, Mughal, or British historical literature of Malwa suggests that the Rajputs were not differentiated till their entry into the Lower Himalaya. Thus the Kanets belonged to the Rajput stock before social differentiation took place. This is further supported by the presence of Rajput clans such as the Sispawar, among the Kanets as well. Subsequently, their interaction with the Rajputs underwent a change, it became more economic than social and cultural. The former came to be employed by the latter to cultivate their fields. This became specially marked in areas such as the Lower Himalaya, where the Rajputs and the Kanets were interspersed. On the other hand, in the Morni region the Kanets have been dominant both numerically and in terms of the ownership of land. Also, while the Rajputs remained the land owners, though not cultivators, the Kanets emerged as owner and tenant cultivators. The emergence of the Kanets and their continuation as a distinct culture group is reflected in their characteristic settlement landscape.

Fundamental to the understanding of the evolution of the Kanet field landscape is an understanding of the hierarchical territorial structure of the

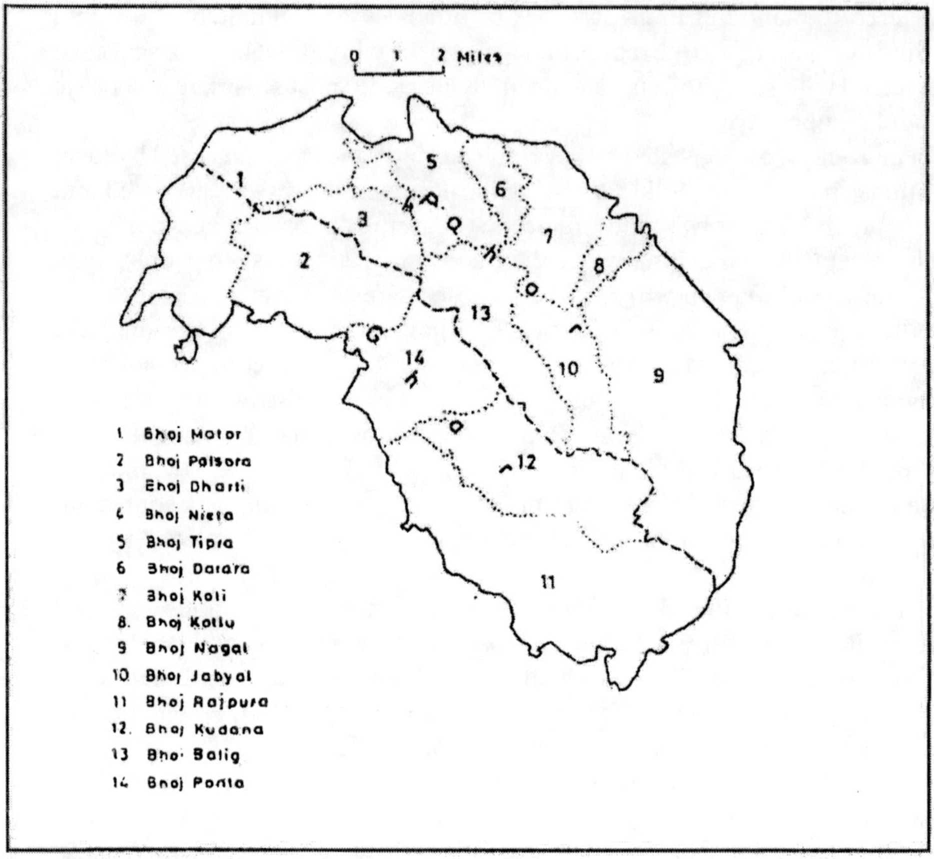

Fig. 10.1 : Morni Hill Tract

Morni hills within which each village territory is a niche carved out of a larger territory.

Covering 93 square miles of hilly country, of which only 11 are cultivated, the region is divided into 14 circles or *bhojes* (synonymous to the *pargana* or *taluka*) comprising 171 separate hamlets. Each of these hamlets is a distinct revenue unit and, though many of them are small uninhabited villages covering a few scattered outlying fields, it has been (and still is) the custom to treat the *bhojes* alone as recognised estates[6] (Fig.10.1). Each *bhoje* contains within it several *mauzas* (village territory, *gemarkung)* which may include one or more than one *abadi* or *bas* (settlement), completing the hierarchical territorial structure (Fig. 10.2 and Fig. 10.3). Any farmer identifies himself with the hamlet, the village territory, and *bhoje* by virtue of its revenue status, unlike the plains where the farmer identifies himself with the village territory alone.

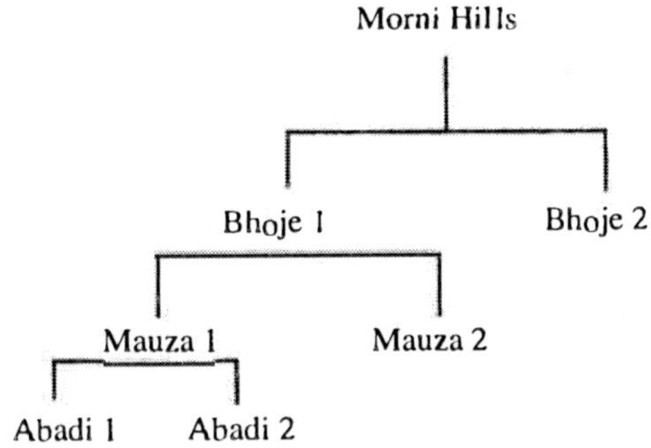

Fig. 10.2 : Morni Hill Tract : Hierarchical Territorial Structure

The names of several *bhojes* clearly suggest their identification with fairly well demarcated physiographic facets having uniform land fertility and productivity. The physiographic concordance is primarily with the ridge crest and the valleys, hence, the *bhojes* are named after them. *Bhoje Dharti*, for example, extends over the ridge crest, the word *Dharti* being derived from *Dhar*, which means ridge crest; and *Bhoje Darrara*, which comes from *darar*, meaning a deeply cut, steep-sided valley , which is indicated by the shape of the valley. Thus, the boundaries of the *bhojes* coincide with the ridge and valleys unlike in the plains where there is no prominent physiographic demarcation.

Within a *bhoje* the villages occur as fragmented production territories (Fig.10.4). Settlements are located on isolated hill tops between which spreads the cultivated land or are tucked into the back slope of the valley. These are interspersed with large forested lands. The pattern of fields and the forests is in itself a crucial indicator of the process and pattern of settling in the Morni Hills. It is indicative of the fact that the villages have been created through the gradual removal of forests in small randomly occurring patches. One of the important processes involved in the creation of cultivated land and village territories has been the practice of *khil* cultivation. This involved the creation of, 'fresh lands, locally known as *khil*, for cultivation' by the destructive process of *jooming*. The 'jungle is burnt recklessly, leaving a great scar up the hillside. The ashes mingle with the soil, and a few poor crops are harvested, after which the land is either abandoned altogether for years, or laboriously worked up into terraced fields, which may eventually yield a fair crop under regular cultivation'.[7] The *jagirdari* status of the Morni *ilaqa* (which was gifted to Mir Jafar Ali Khan of Saharanpur) and absence of any levy, allowed for the widespread practice of *khil* cultivation. This process continued till as late as the end of the nineteenth century when the Morni *ilaqa* (area) was declared a reserved forest in 1892 by the British. 'The fact that the people have been hitherto allowed to clear lands

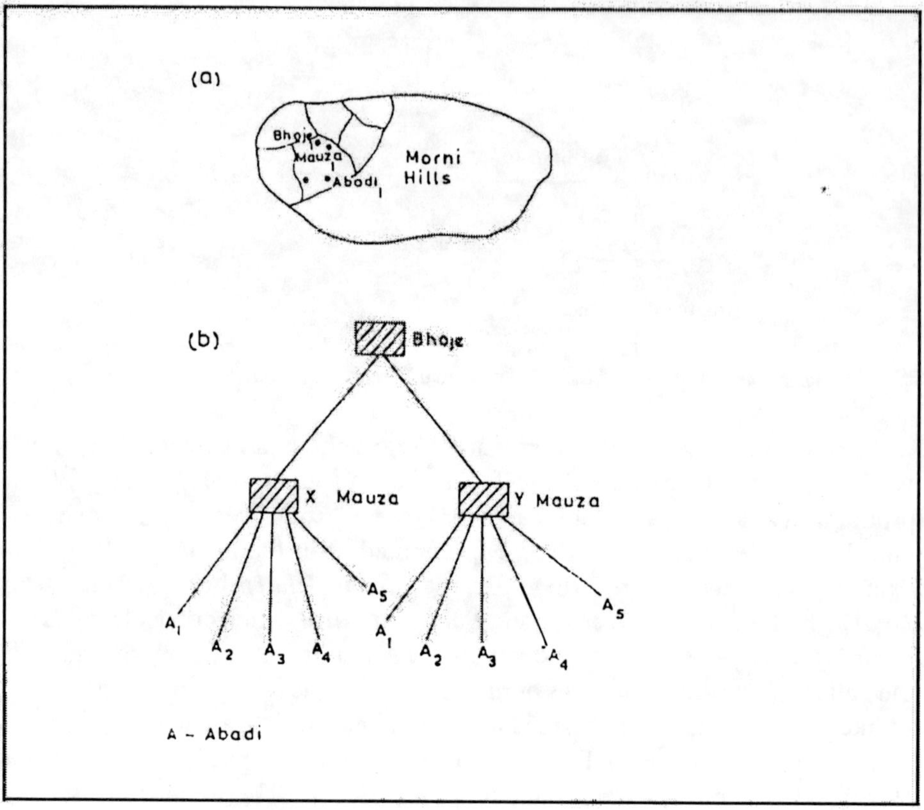

Fig. 10.3.: Morni Hill Tract : Enmeshed Hierachical Territorial Structure

in this way from jungle owned by the Mir may give them rights in the lands already cleared'.[8] Hitherto all new *khils* were to be charged a levy.

The practice of creating *khils* has great antiquity, dating back to the Buddhist and earlier periods, when the term *khilya* was used.[9] Also its prevalence till recent years in the hilly regions of the Outer and Lower Himalayan slopes and the Siwalik Hills is well known.

This process led to the emergence on the landscape of the characteristic dispersed pattern of settlement which came to be organized within the hierarchical territorial structure of the *bhoje* (circle), *mauza* (village territory), and the *abadi* or *bas* (settlement). Creation of new *khils* within the *mauza* led to a gradual proliferation of settlements as a new *bas* (in addition to the main *abadi*) were established at the sites of *khil* cultivation and the latter perhaps emerged as individual villages. Thus, with the imposition of restrictions on the practice of *khil* cultivation, the process of evolution of the settlement fabric has been brought to a halt.

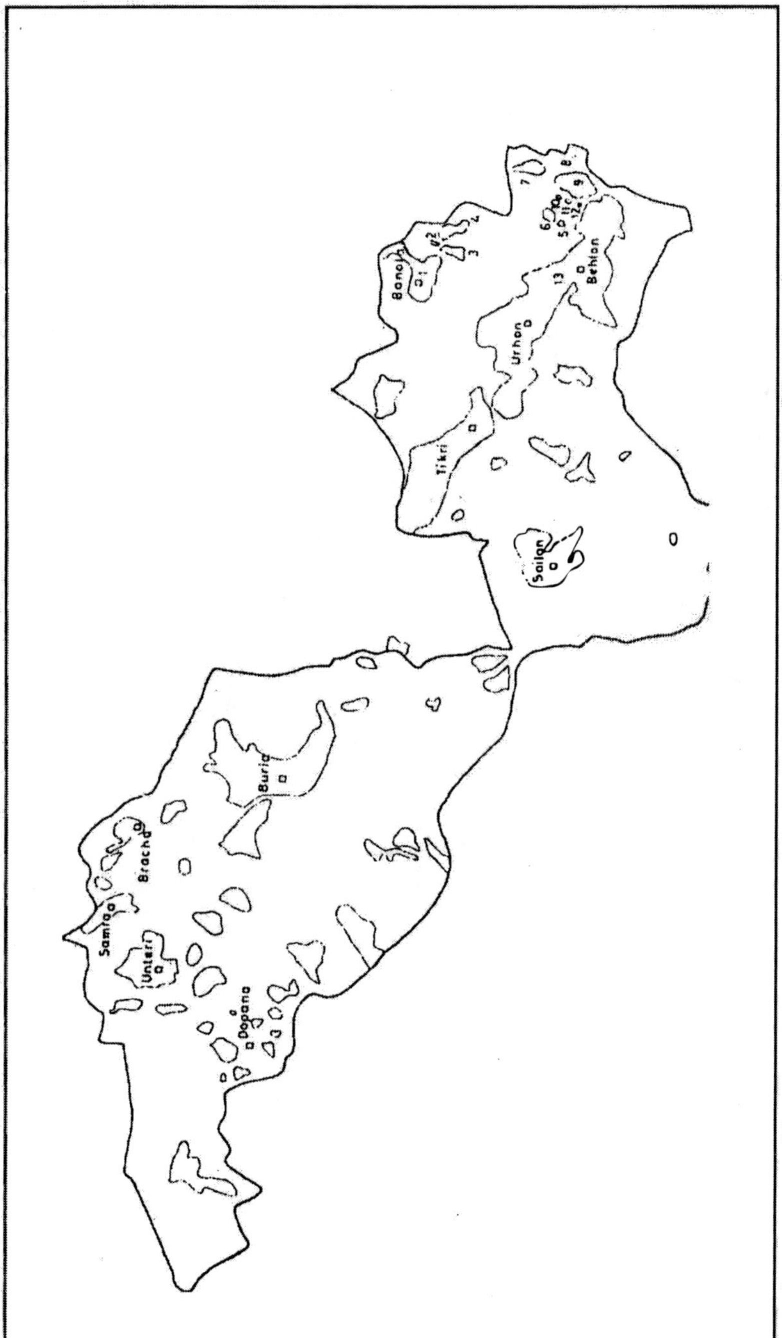

Fig. 10.4 : Bhoj Dharti : Spatial Structure

On a smaller scale, *khil* cultivation has also played an effective role in the evolution of the village territories, which have essentially evolved through the process of accretion. For example, in Behlon, (the uni-clan, uni-caste village of the Sispawar Kanets selected for a fuller understanding of the typical Kanet field landscape) the location of the *khil* land suggests the expansion of the village territory, essentially the cultivated land, outwards from the *abadi* by accretion of fairly widely distributed disjunct patches of cultivated land initially created as *khil*. Behlon, in *Bhoje Dharti*, extends along the ridge top. The settlement occupies an isolated hill top. The village territory is somewhat elongated along the ridge, which provides the only available flat, cultivable land (Fig. 10.5). The isotropic surface as in the plains is absent. Variations in the soil and land types and hence, productivity and fertility are pronounced (Fig. 10.6). Following the well-known local classification, the land is known as either *kulahu, obar* or *khil*, 'the *kulahu* being irrigated from streams, the *obar* covering all level or terraced cultivation, and *khil*, the rough jungle cultivation..'[10] *Obar* land is further sub-divided into *obar awal* and *obar dom*, the former including only the highly manured fields adjacent to the *abadi* and the latter, the outlying fields yielding comparatively poor crops. However, the role of distance from the *abadi* as a control of land productivity, fertility and utilisation is secondary to terrain constraints which mar the development of the classic annular arrangement in concentric circles of the settlement, arable, and commons marks as in the Teutonic Model. Another serious deviation is the striking absence of a *charand*, *charaga* the commons mark, from the village territory. This is primarily attributed to the limited availability of flat land for cultivation, all of which is, therefore, under the plough; the presence of vast stretches of forested lands and scrub jungle between village territories which serve as grazing grounds, and the fact that the people had unlimited rights as the entire *ilaqa* was held as the *jagir* and no levy or cess charged. Generally the cattle are stall-fed.

Along the ridge top or the valleys, the village territories extend in a manner that they do no cut across the ridge or river. Even though in patches, the cultivated land is on the same side. The slopes had to be converted into terraces for cultivation. Thus, individual villages extend up the slope to a line above which the slope is far too steep to be converted into terraces. The weak rock structure of the Siwalik Hills again exerts a control on the height upto which terracing is possible, it being largely confined to the gentler slope. This practice of terrace cultivation clearly suggests that the Kanets moved into this area after developing the technique of terrace building and terrace cultivation from the adjacent Himalaya where terrace cultivation has been widely practised.[11]

Morni Hills and particularly the *bhoje* is one of the few areas south of the *Dun* where field terracing is practised. This is related not only to ecology which makes terracing possible but is also attributable to the introduction of this trait into this area by Kanets who came down from the Himalaya.

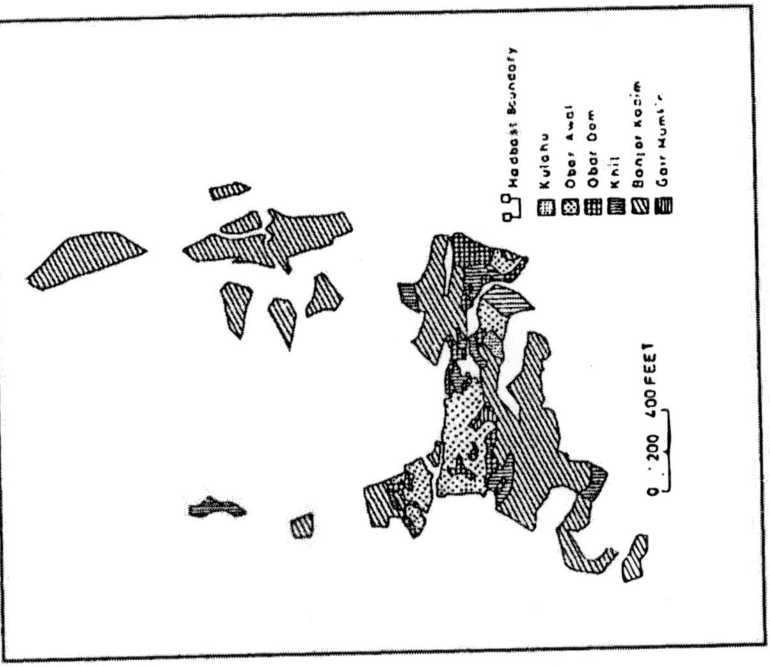

Fig. 10.6 : Behlon : Land Types

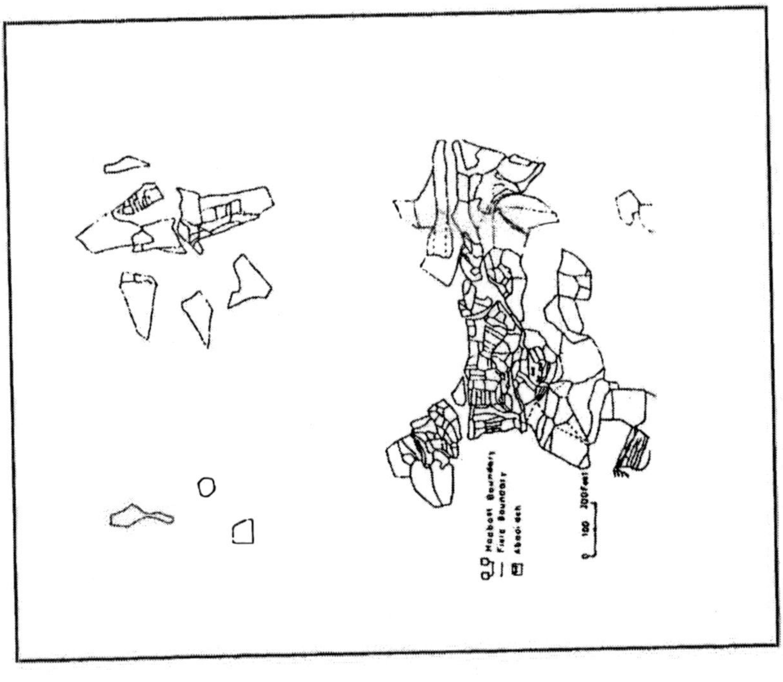

Fig. 10.5 : Behlon : Village Territory

The most striking feature of the field landscape in Behlon, as in other Kanet villages, is the immense irregularity of shape and highly fragmented nature of the cultivated patches. In the first stage itself the cultivated land comprised 18 disjunct patches varying in location, shape, and size (Fig. 10.7). Following the sub-division of the village territory among the three founder owners, 42 fields emerged on the landscape constituting the initial field pattern. Their size, as observed in the map, is highly variable. Variations in shape, size, and location are essentially the result of terrain constraints. Thus, the fields occupying the flat ridge top, north of the *abadi* are the largest and most contiguous. These are blockish but somewhat elongated giving them an almost rectangular shape, primarily the result of the nature of the available flat, cultivable land along the ridge top. The fields to the south of the *abadi*, although highly irregular in shape, are definitely elongated along the contour lines. These are terraces created along the hill slope. All peripheral fields are disjunct and irregular, the result of both the process of expansion of cultivated land by the creation of *khils*, and the terrain constraints, whereby slopes which are too steep cannot be converted into terraces. Since the entire land was held as a *jagir* the process involved in the acquisition of the village territory has primarily been forcible occupation of land through creating cultivated patches by the removal of vegetation which then gave the people proprietary rights. Thus, some part of the village territory continued as Mir's property and was looked upon perhaps as the *bara* land, the potential areas for further expansion of cultivated land. This was so because the Mir was an absentee landlord.

Although the initial division of the village territory is unequal, with Chahia (Fig. 10.8) having the maximum, most continuous stretch of the best land, all members of the community are represented in all parts of the village territory and have a share in all the land types *obar awal, obar dom, banjar kadim* and *khil*. The disjunct nature of the fields suggests the sporadic extension of cultivated land from the *abadi* outwards, down or up the adjacent hill slopes. The presence of *khil* land on the periphery (Fig. 10.6) is strong evidence of this process. Thus, with the exception of the extent of land owned by each founder member, the pan-Indian tradition of equitable distribution in terms of land fertility, productivity, and distance from the *abadi* is maintained.

The evolution of the field pattern has been extremely rapid. Within a single generation the number of fields increased from 42 to 100, the result of intense sub-division among the minimal lineage members (Fig. 10.9). The field landscape reveals intense fragmentation and an intermingling of the properties of all owners as each has a share in every part of the village territory. The blockish shape is completely transformed into small fairly regular rectangular strips running both along and across the contour lines. However, the extent of sub-division is maximum in the only contiguous patch of flat cultivable land

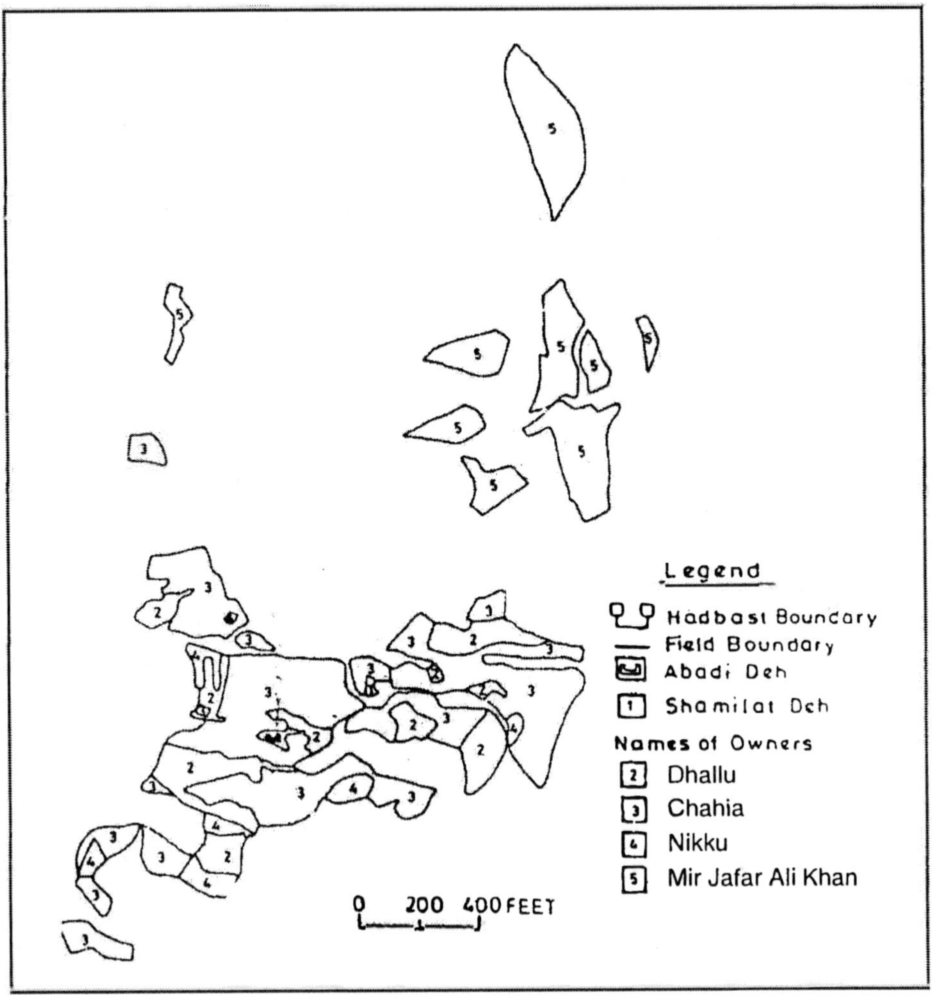

Fig. 10.7 : Behlon Ownership of Cultivated Land Circa 1800

north of the *abadi*. The operation of the principle of equitable distribution in which each has a share in the *obar awal* land most suitable for cultivation, and its limited availability explain the highest incidence of fragmentation here. As a result, the size of the fields north of and adjacent to the *abadi* is drastically reduced. The sub-division of fields to the east and south of the *abadi* is relatively limited on account of the poorer quality of land with low productivity, primarily *banjar kadim* and *khil*. With the proliferation of male heirs down the line of generations, by 1887-88 further sub-division of the fields resulted in the

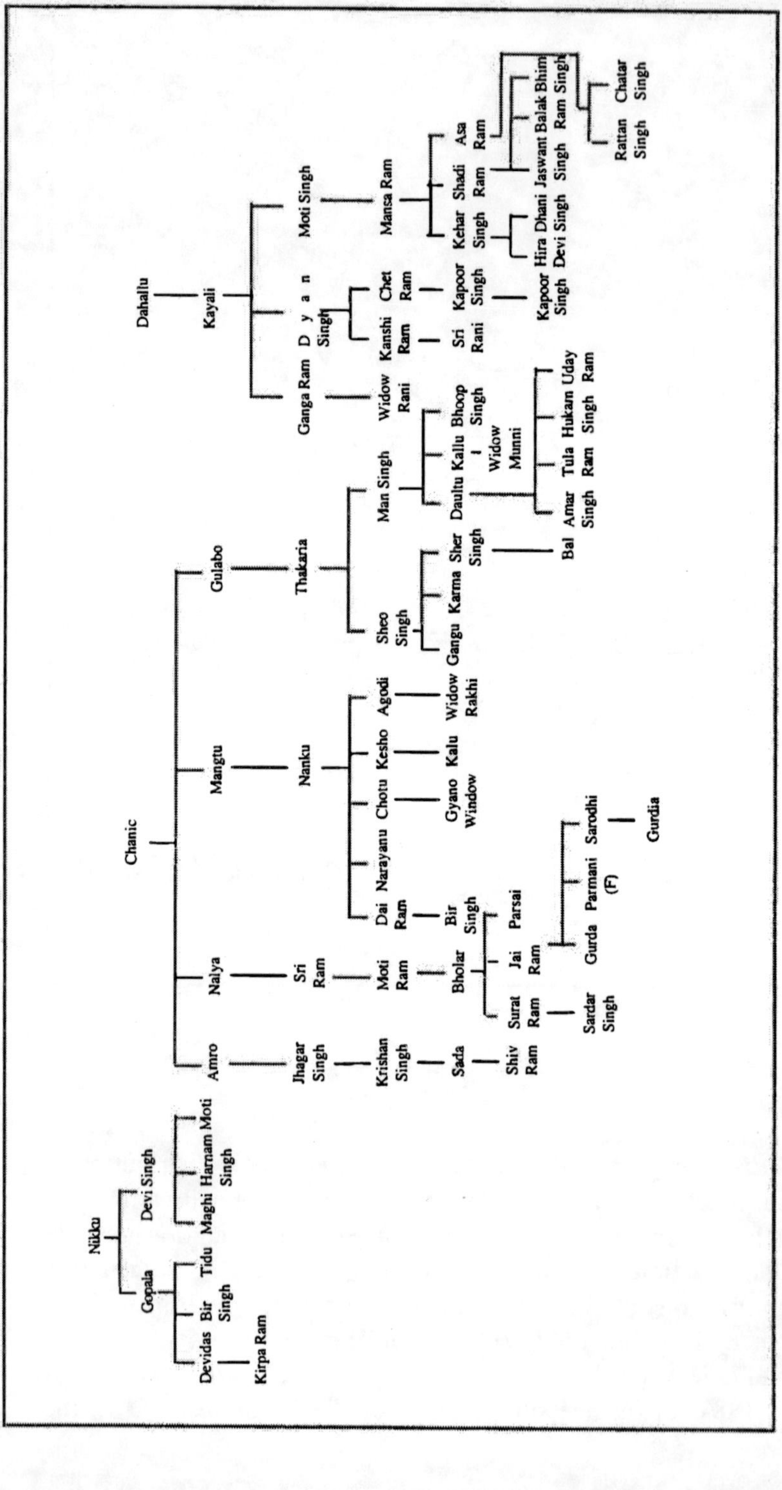

Fig. 10.8 : Genealogical Tree (*Sajra Nasab*) of Behlon

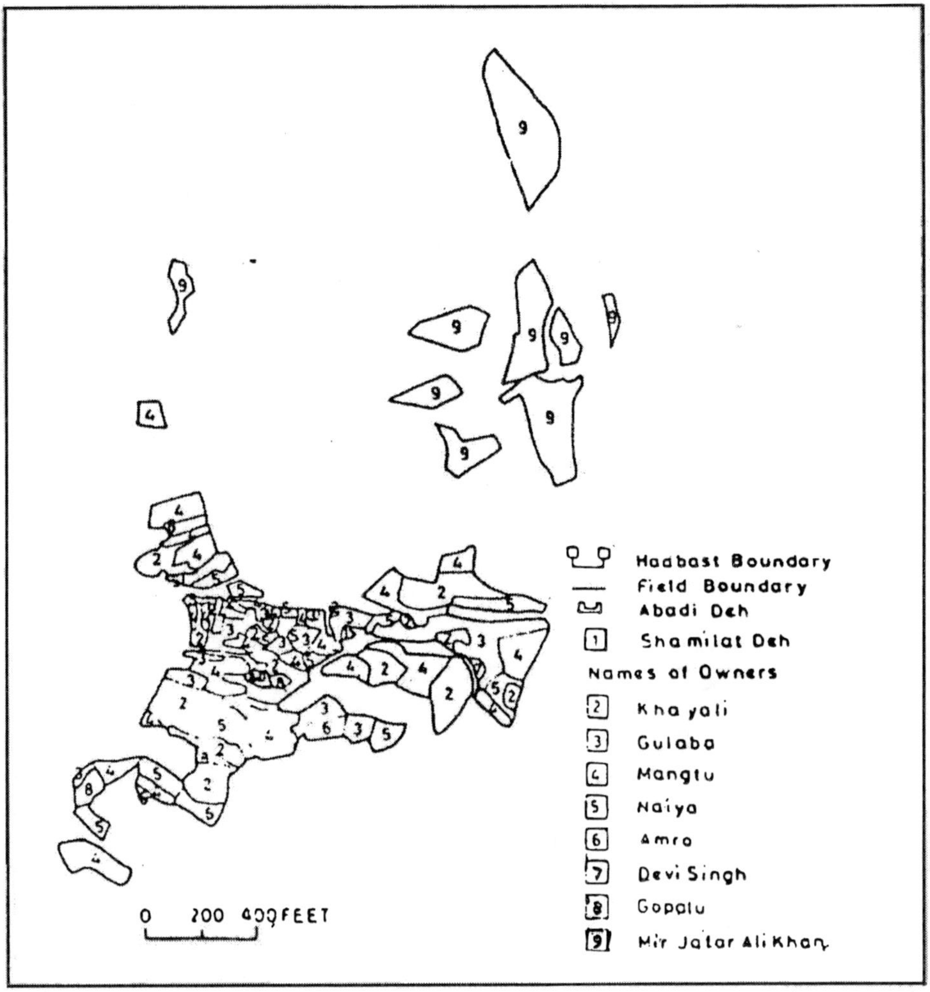

Fig. 10.9 : Behlon : Ownership of Cultivated Land Circa 1840

emergence on the landscape of 208 fields (Fig.10.10). Although the incidence of fragmentation increased, sub-division, leading to the creation of new field boundaries, introduced an element of regularity in the shape of the fields. The cultivated patches, however, continued to be disjunct having no geometrical, regular shape whatever. While in the flat land the shape of the fields varies from rectangular strips to squarish forms, along the hill slopes, the terrace fields are linear or curvi-linear strips running along the contours. In the former there is no definite alignment of the fields with the contours.

On the field landscape there is a marked absence of the contiguous *tak* or holding. Fragmented holdings constituted of single fields spread across the

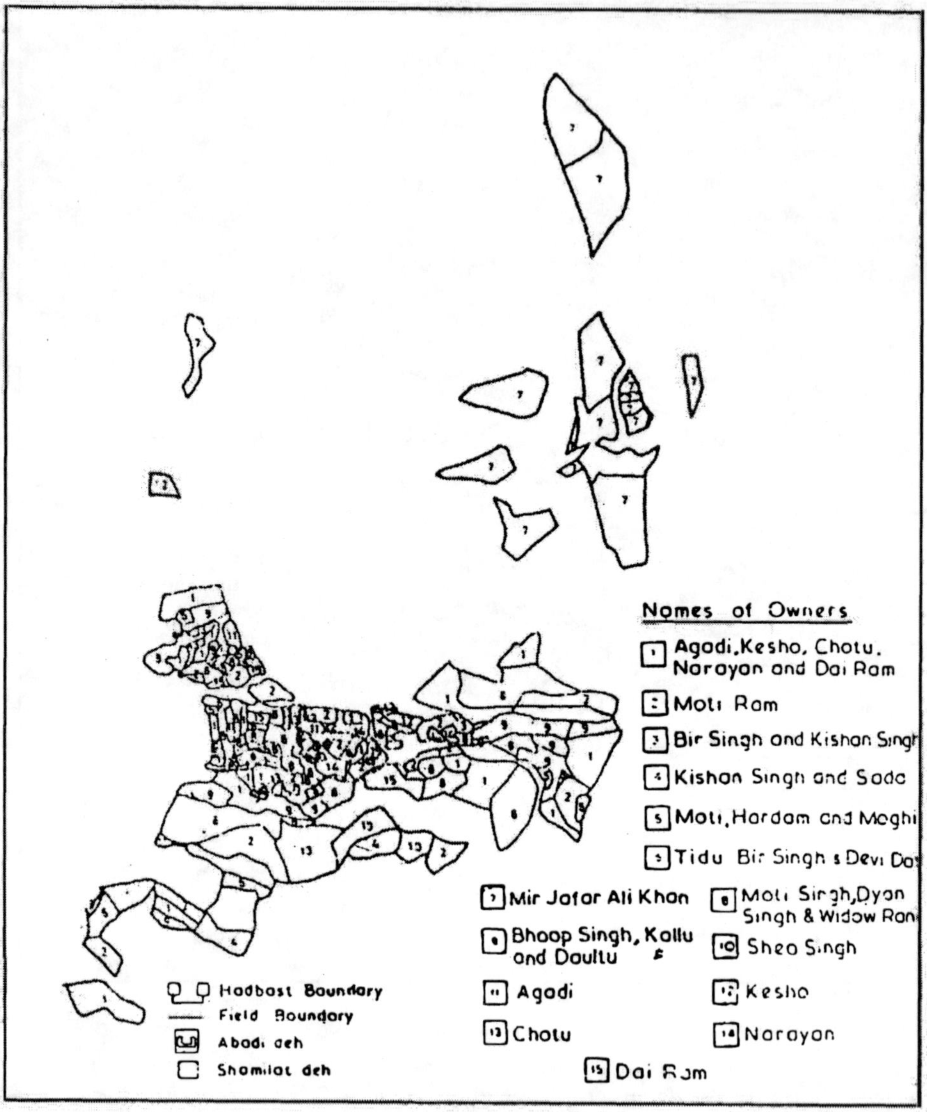

Fig. 10.10 : Behlon : Ownership of Cultivated Land Circa 1887-88

village territory are owned by members of the minimal lineage. This is the result of the limited availability of cultivated land and the desire of each member of the community to share equally the benefits or handicaps resulting from the sites of fields in different parts of the village territory. Thus, the hierarchical order in the spatial structure of the village territor includes the village territory, *khet* (field) and *kyari* (cultivation unit) (Fig. 10.11). No expansion of cultivated land is evident.

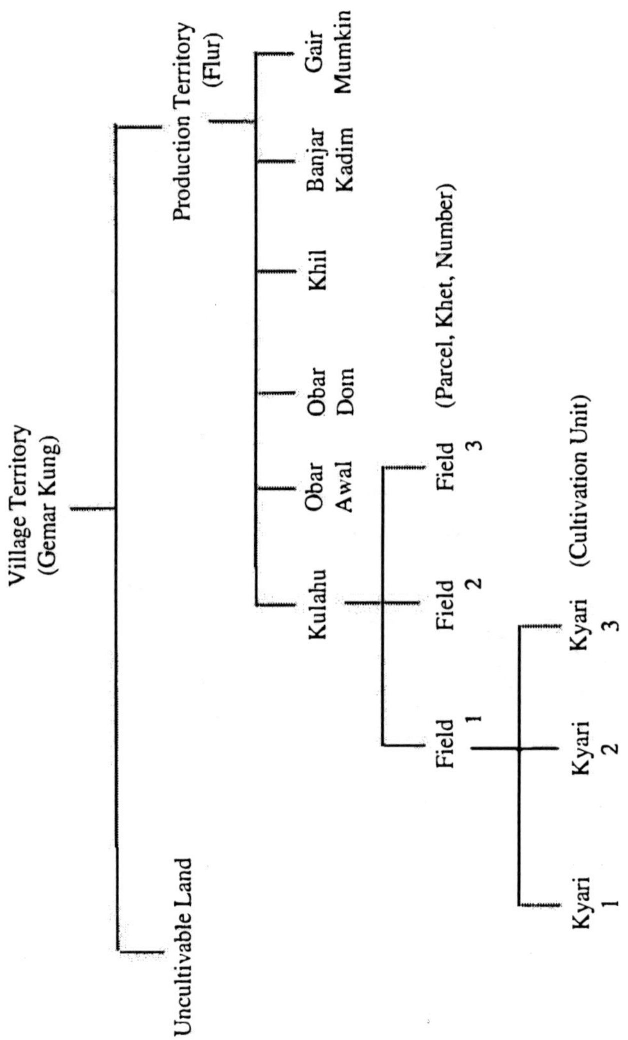

Fig. 10.11 : Model of Spatial Structure of Village Territory

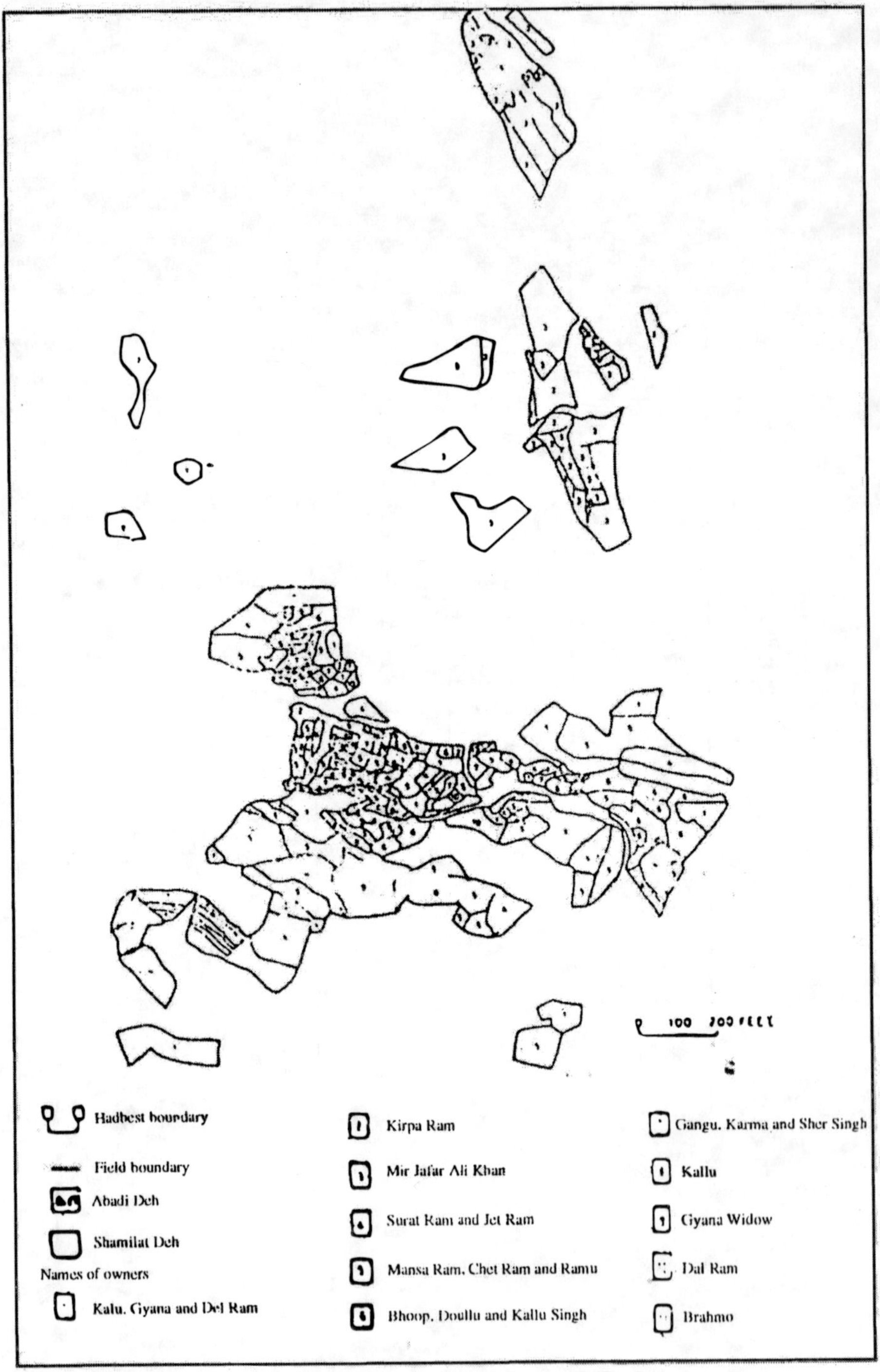

Fig. 10.12 : Behlon Ownership of Cultivated Land Circa, 1917-18

Table 10.1 : Behlon : Number of Fields

Year	*Number of Fields*
1800	42
1840	100
1887-88	208
1917-18	227
1977-78	279

Fields are both individually and jointly owned. A definite relationship of the kind of ownership with the type of land and distance from the *abadi* can be observed. Practically all fields adjacent to the *abadi* to the north of it on the flat ridge top are individually owned. The *obar awal* land here is highly manured, being the best cultivated land it is the most efficiently and intensively utilised. Thus, fields here are individually owned for each member to enjoy the benefits in an area where good, flat, cultivable land is limited. On the other hand, the terraced fields on the hill slope adjacent to the *abadi* on the south and with increasing distance from it in all directions, are jointly owned. This is primarily attributed to the terrain constraints whereby the tasks of initial clearance of the hill slopes of vegetation and their conversion into flat terraces for cultivation warranted joint effort. Moreover, as mentioned earlier, the former process gave them joint proprietary rights. The land here is either *khil* or *banjar kadim*, both of which are poor lands. This joint ownership, being both a multi-generational ownership extending across the third, fourth, and fifth generations (Fig. 10.10 and Fig. 10.8) and an inter-minimal lineage one, wherein members of different minimal lineages of different generations are joint owners, is complex. Ownership, however, is strictly within the maximal lineage.

With the British intervention in the area by the late nineteenth and early twentieth century, a revenue settlement was effected in Behlon as in all the villages of the Morni Hills. This led to some reorganisation of the fields (Fig. 10.12) and the redrawing of the field map. By 1917-18 the number of fields increased to 227 the result of both sub-division of existing fields, particularly those belonging to the Mir, which were perhaps being prepared for cultivation covering only *banjar kadim* land; and an extension of cultivated land down the hill slope (Table 10.1). By now with the imposition of restriction on the creation of *khil* and the declaration of the area as a reserved forest, any further expansion came to a halt.

As the map reveals (Fig. 10.12), the most striking development is the emergence on the field landscape of *taks* or holdings, comprised of fields belonging to the same owner. This introduces an element of regularity in the field pattern and immediately reduces fragmentation. This is primarily the result of a reduction in the number of owners, the sub-division into smaller fields of

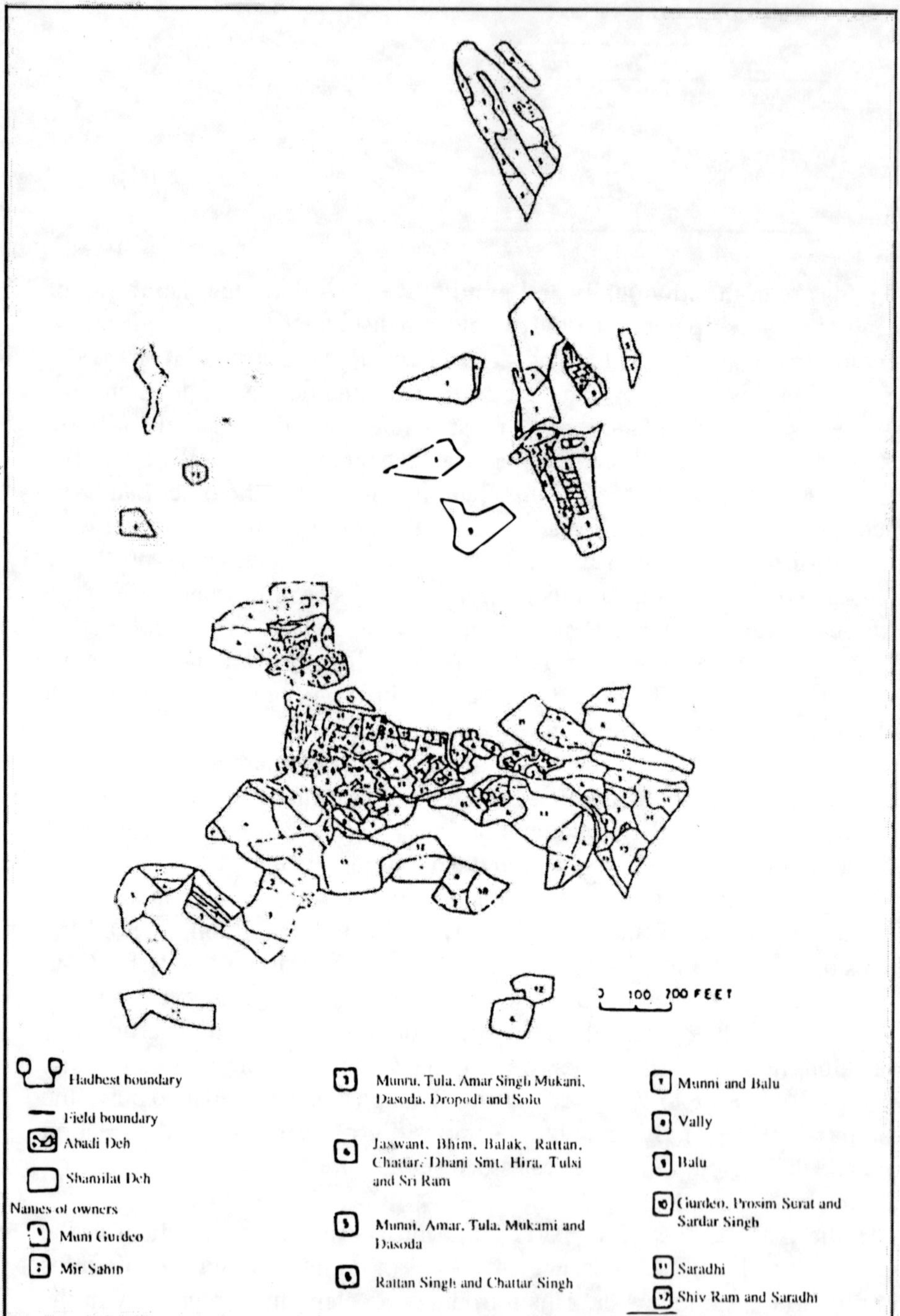

Fig. 10.13 : Behlon : Ownership of Cultivated Land Circa, 1977-78

earlier larger ones, and in some cases on account of the resumption of fields of an adjacent owner after his death or by mutual agreement by members of the same maximal lineage even though belonging to a different minimal lineage. This is apparently an in-built mechanism for retaining the land within the family. Single and joint ownership of fields presists, the pattern of distribution remaining unchanged. Joint ownership is complex involving not merely multi-generational joint ownership but also inter-minimal lineage ownership. For example, a part of the field belonging to Agadi, Kheso, Chotu, Naryana and Dai Ram (Fig. 10.10) is taken over by Daultu, Kallu, and Bhoop Singh (Fig. 10.12) who belong to different minimal lineages but the same maximal lineage (Fig. 10.8). Sub-division everywhere has led to a reduction in the field size, though the shapes continue to remain highly irregular. The incidence of sub-division is now the maximum in outlying, peripheral fields, which are, however, still larger than the ones adjacent to the *abadi*.

Consistent evolution in the field pattern has resulted in a further increase in the number of fields (Fig. 10.13). This is attributed to the sub-division of existing fields. By 1977-78 as the map reveals, (Fig. 10.13) there has been an extension of cultivation into the land formerly held by the Mir probably on account of forcible acquisition of the land by the occupants. Ownership of land is now more joint than individual even in the area to the north of the *abadi*. This change is attributable mainly to the proliferation of male heirs and the limited availability of land for cultivation.

The present field landscape reveals small, irregular, fragmented holdings, occurring in disjunct patches; each comprising several fields. With the exception of the flat land north of the *abadi,* the fields everywhere are highly irregular in shape. The extremely small and narrow rectangular strips adjacent to the *abadi* and occurring at the same level are separated by low mud or stone dykes. The fields contained within the terraces are also similarly divided, while between two adjacent terraces at two levels, it is the terrace scarp which forms the field boundary. All fields are again divided into *kyaris* or cultivation units. While the terrace fields are generally aligned along the contours, no such definite alignment is identifiable for either the fields on the level land or the *kyaris* anywhere.

The implicit linearity of practically all the fields, whether along or across the contours, is perhaps maintained to facilitate the use of the iron-toothed wooden plough. The length then reduces the number of turns to be taken and allows for the more efficient utilization of land.

The relationship of the field pattern to the crop grown is hardly evident as the range of crops grown is restricted to maize and wheat. There are two cropping seasons, the *rabi* (winter crop) dominated by wheat, and *kharif* (summer crop) dominated by maize. All cultivation is *barani* (rain-fed), with the exception of a small patch to the east of the *abadi*, which is *kulahu*, irrigated by water from local *kuls* (streams). Land under *shamilat* is limited and has remained unchanged through the evolution of the field pattern.

References and Notes

1. Nitz, H.J. (1996). 'Geographical Studies in the Field Patterns of Northern India and Germany', *The Geographer*, Vol. XIII, pp. 61-74.
2. Beazley, J.G. (1919). *Assessment Report of the Morni Hill Tract Naraingarh Tehsil of the Ambala District*, Lahore: The Civil and Military Gazette Press, p. 1.
3. Kensington, A. (1893). *Settlement Report of the Ambala District*, Lahore : The Civil and Military Gazette Press, p. 46.
4. Darling Malcolm, (1925). *The Punjab Peasant in Prosperity and Debt*, Geoffrey Cumberlege : Oxford University Press, p. 33.
5. Mukerji, A.B. (1982). 'Cultural Ecology of Rural Settlements of the Chandigarh Dun Region', *India : Cultural Patterns and Processes*, Allen G. Noble and Ashok K. Dutt (Eds.), Boulder : Westview Press, pp. 337-366, ref. pp. 337, 341.
6. *See* ref. 3, p. 39.
7. *See* ref. 3, p. 46.
8. *See* ref. 3, p. 46.
9. Mukerji, A.B. (1961). 'Field Patterns in the Upper Doab Jat Villages,' *Journal of Geography*, Vol. III, No. 1, pp.1-8, ref. p.3.
10. *See* ref. 3, p. 44.
11. *See* ref. 5, pp. 341, 345.

11

THEMES IN COLONIAL URBAN LANDSCAPE IN INDIA

A Study of Bungalow Names in the Hill Stations

K.D. SHARMA

The hill stations in India, as in the other Third World former colonies, represent a major component in the structure of colonial urban influence. The establishment of hill stations in India represents a unique phenomenon of urban growth in which an alien culture group created an environment, physical and social, with the sole objective of providing itself with a sense of being a part of the Metropolitan society 'at Home' and of maintaining the life styles peculiar to that society. The existence and *modus operandi* of the representative colonial cultural institutions related to the functions of residence, religion and recreation, is illustrative of this *leitmotif*. The spatial expression of these factors is manifested in the characteristic colonial urban landscape in the hill stations.

While the religious and recreational functions were manifested on the townscape of the hill stations as structures operating at community level and gave a distinct character to individual hill stations, it were the numerous domestic buildings in which individual attitudes found full expression. In contrast to the uniformity of 'official' and public landscape of 'the plains', the private environment of the hills was consciously varied and utilised for individual expression.[1] These variations in individual expression, however, were confined within the ambit of 'Englishness' of the various landscape elements. For example, a military officer in 1828 stated that at Subathu, 'European shrubs and trees greeted the eye ...'[2]. Similarly, it was reported that 'the climate of Sonawar (near Simla) is thoroughly English . . . .'[3] Such an attitude was not restricted to the hill stations in the Himalayas alone. In 1877 a member of the colonial community stated, 'having seen (Ootacumunde) I affirm it to be a paradise. The afternoon was rainy and the road, muddy but such beautiful English rain, such delicious English mud.'[4]

The townscape of hill stations in India thus represents an attempt to consciously recreate a familiar landscape by transplanting elements from their

Note : Reprinted from *Punjab University Research Bulletin* (Arts). Vo,. XXI, No. 1, April 1990.

cultural hearth. This is best illustrated in the practice of naming buildings, whether monumental or domestic, a trait not common to the Indian life style. The naming of domestic buildings was specially prominent in the hill stations because, as mentioned earlier, these provided a more private social environment to the members of the colonial community.

Objective

This study is an attempt to identify the elements and special themes through an analysis of the names given to domestic structures in the hill stations in India as a part of an understanding of the phenomenon of cultural landscape transference effected by members of the colonial community in India. For this purpose the hill stations of Kasauli, Mussoorie and Nainital have been selected.

Study Towns

Kasauli was established as a cantonment town in 1842. It is located 13 kilometres from Dharampur off the Kalka-Shimla highway. The town is sited on a ridge at a general elevation of 6,000 feet. Mussoorie is also sited on a ridge with comparable elevation. It was established in 1826 as a sanatorium. It is located about 30 kilometres from Dehra Dun. Nainital is a valley town and although the site was known, the regular growth of the town began only after 1842.

Data Base and Methodology

The data base for the present study comprises of the names of individual bungalows in Kasauli, Mussoorie and Nainital. The names of ceremonial and monumental structures, and those of structures related to public and community functions such as hotels, offices, churches, clubs as also the residences of the members of top echelons of administration, have been excluded from the present analysis as these were not considered truly representative of the personal preferences at the individual level.

The bungalow names in Kasauli have been taken from Kasauli and One Hundred Years of Kasauli Club published in June 1980. In addition, intensive filed work at Kasauli helped in collecting additional information in placing these names in a perspective relevant to this study. The bungalow names for Mussoorie and Nainital were collected from old Survey of India guide maps.

These names were intensively analysed to identify the recurrent themes. This analysis is based mainly on the information available in the following sources :

1. Ekwall, Eilert (1960). *The Concise Oxford Dictionary of English Place Names*, Clarendon Press, Oxford.

2. Mitchell, J.B. (1954). *Historical Geography*, English University Press Ltd., London.
3. Money, D.C. (1954). *Introduction to Human Geography*, University Tutorial Press Ltd., London.
4. Dury, G.H. (1957). *Map Interpretation*, Sir Isaac Pitman & Sons Ltd., London.
5. *The Oxford English Dictionary*, Clarendon Press, Oxford.
6. *The Standard Reference Atlas*, George Philip & Sons Ltd., 1956, London.

A classification of the bungalow names on the basis of etymology would be highly desirable but has not been attempted because it would not be relevant in a study in the Indian context and would comprise a deviation from the focus of this study. However, in the bungalow names relevant to different aspects of ecological setting there is a general convergence of etymological meaning of bungalow names and bungalow sites.

General Attributes

Almost half of the total 421 bungalow names analysed for this study are identified as simple names structurally and the other half comprises of compound names. The names in these categories are not confined to any one aspect of the overall landscape of the study towns. Interestingly, a similar division can also be identified in terms of bungalow names containing a generic and those without a generic. Almost half the bungalow names analysed in this study have a generic. The generic component generally occurs as a suffix, and only in a few bungalow names, as a prefix. The generic components identified are cove, combe, common, dun, dale, del, drum, field, feld, ford, glen, tun, ton, wick wood and wald. These generic components refer to either (i) a valley or some attribute of a valley or a similar feature, for example a recess, a hollow or a valley (cove, dun, dale), a small wooded valley (dell), short valley on side of hill (combe), valley with a river (glen), or (ii) to the attribute of a site or specifically to the existence of a town or hamlet. For example, a farm or a village (tun, ton), an enclosure (field, feld), a shallow place for crossing a river by wading (ford), an open upland site containing trees (wood, wald), a pasture town or hamlet (wick). The use of wick as a generic is rare except in place names (Oxford English Dictionary). The practice of giving a place name in Britain to an individual building in India could either be related to these names serving as symbols of identification or to the classical development of the isolated farmstead found only in Europe. There is also an equally strong possibility that the house in India served as a surrogate for the settlement in the perception of the individual members of the colonial community in India.

Major Themes in Bungalow Names

The process of identifying the dwelling places by giving them names is not common to the Indian tradition and was introduced in India on a large scale by the colonial community. The main objective was to give a personal identity to the dwelling places and the desire to maintain a European identity in a place thousands of miles away from there. The bungalow names reveal a strong tendency of borrowing from the cultural wealth which the inhabitants brought with them. The specific name given to a bungalow thus, reflects the taste, the attitude toward the particular aspect of the site, and the perception of the inhabitants of these bungalows.

The major themes and the number of bungalow names related to each theme in the study towns is given in Table 11.1. It must, however, be pointed out here, that the exact number of bungalow names in a particular theme is not as important as the trend represented by these.

An analysis of the 421 names from Kasauli, Mussoorie and Nainital reveal five major themes. These are, ecology, phenomenon of cultural landscape transference, perception, personal names and aesthetic names in the same order of importance (Table 11.1). Of these, ecology, phenomenon of cultural landscape transference, and personal names are comprised of sub-divisions related to the main theme.

The predominance of a particular theme in a particular study town indicates the composite taste and perception of its inhabitants and may not follow the general trend. For example, as against the general trend in which ecology emerges as the major theme (167 out of 421 bungalow names) the bungalow names in Kasauli suggest a cultural landscape transference as the main theme (23 out of 76 bungalow names). Similarly, perception emerges as the second most important theme in the bungalow names in Nainital (25 out of 86 names) against the general trend of cultural landscape transference being the second most important theme (Table 11.1).

A description of the major themes, their sub-divisions and examples of bungalow names in these is as follows.

Ecology

This theme refers to the physical environment of the site of the study towns. It is the most important theme in the naming of bungalows at the aggregate level (167 out of 421 bungalow names) as well as at Mussoorie (124 out of 259 bungalow names) and Nainital (27 out of 86 bungalow names). For a better understanding of naming of bungalows on the basis of ecology, this theme is sub-divided according to different ecological elements. These sub- divisions

Table 11.1 : Distribution of Bungalow Names According to Major Themes

Themes	*Total Number of Bungalow Names*	*Kasauli*	*Mussoorie*	*Nainital*
I. Ecology	167	16	124	27
(a) Geomorphic features	42	3	29	10
(b) Hydrological features	15	-	13	2
(c) Vegetation	85	9	63	13
(d) Birds and Animals	11	2	9	-
(e) Climatic Advantage	14	2	10	2
II. Cultural Transference	90	23	51	16
(a) Literature	9	-	7	2
(b) Place Name	63	14	38	11
(c) Religion	9	5	2	2
(d) Regional Association	9	4	4	1
III. Perception	80	10	45	25
IV. Personal Names	62	16	35	11
(a) Surname	28	8	14	6
(b) Female first name	21	6	13	2
(c) Male first name	13	2	8	3
V. Aesthetic Names	22	11	4	7
Total	421	76	259	86

relate to geomorphic features, hydrological features, type and species of vegetation, birds and animals and climatic advantage of the site of the bungalow. Among these sub-divisions bungalow names based on type and species of vegetation are the most common at the aggregate, as well as the individual hill station level. Such names as 'Poplars', 'The Firs', 'Ivy Lodge', 'Dahlia Cottage', 'Briar wood', 'Fernlee' and 'Oak Cottage' relate to this sub-theme. The examples of names related to other themes are, 'The Mount', 'The Ridge' and 'Dale Cottage' relate to geomorphic features, 'Dave Cottage' 'Glenbrook' 'Abergelalie' relate to hydrological features, 'Ravenswood', 'Antler's cottage' to birds and animals, and 'Sunny Bank', 'Sunny Side', and 'Airyland' to climatic advantage.

Cultural Landscape Transference

This theme refers to the bungalow names derived from some cultural trait or feature which forms a part of Western cultural landscape. In the study of process of cultural landscape transference, this theme has a special importance because the names given to individual bungalows can be definitely traced back to their areas of origin. In other words, the names comprising this theme represent an attempt at a duplication of the original, but in a different setting. Cultural landscape transference is the second most important theme at the aggregate level (90 out of 421 names) but represents a deviation in Kasauli where it is the major theme (23 out of 76 bungalow names) and in Nainital where it occupies the third place (16 out of 86 names).

The names comprising this theme relate to literature, place names, religion and regional associations which represents the sub-divisions of this theme. Among these bungalows, names derived from place names are most common (63 out of 90 bungalow names) at the aggregate,as well as the individual study town level (Table 11.1) Such names as 'Belvedere', 'Balmont', 'Kynaston', 'Kentucky' and 'Bathwell' belong to this cetegory. Such names as Kenilworth', 'Waverly', and 'Westword House', relate to literature. The religious basis is reflected in such bungalow names as 'Abbey', 'Rectory Lodge', and 'Parsonage' while 'Blarney' 'Benmore', 'Monte Cristo', 'Scotch House', 'Isla Cottage', reflect regional associations.

Perception

This theme relates to the manner in which an individual perceived the house to be. It is the third most important theme at the aggregate level, the second in Nainital, and fourth at Kasauli (Table 11.1). There are no sub-divisions of this theme because, generally, although the specific name given to a bungalow may vary, the common perception pertains to viewing the houses as a comfortable, cosy place of rest. Some examples of bungalow names reflecting this theme are 'Beau Sejour', 'Rest House', 'Rest Port', 'Elysium', and 'Harmony'.

Personal Names

Personal names form the fourth important theme in the naming of bungalows at the aggregate level but are as important as ecology as a theme in Kasauli (Table 11.1). This theme includes three subdivisions, surnames, female first names and male first names in the same order of importance except in Nainital (Table 11.1). This theme best represents the attempt at personalisation of the landscape. Generally the bungalow names in this theme contain two components, the personal name and some aspect from any of the other themes, for example, 'Annie Ville', 'Carroll Villa', 'Donald Cottage', 'Eric's Own', 'Gilbert House' and 'Ludlow Castle'.

Aesthetic Names

This theme essentially refers to the naming of bungalows to emphasise some aesthetic aspect of the site. This is the fifth important theme in the naming of the bungalows (22 out of 421 bungalow names). The bungalow names reflecting this theme almost without exception, emphasise the natural view obtained from the site. Considering the importance attached to the view from a bungalow[5] this is not surprising. The small number of bungalow names reflecting this theme could probably be related to the restrictions placed by the site of the hill station. Some examples of bungalow names in this category are 'Fair View', 'Snow View', 'Dale View', 'Pleasure View', 'River View' and 'Spring View'.

Conclusion

The establishment of hill stations in India, as in other former colonies in the Third World, represents a unique phenomenon by which the colonial community created a physical and social environment for the preservation and maintenance of a life style in imitation of the culture traits of the Metropolitan society 'at Home'. The townscape of hill stations, therefore, contains attributes which reflect the attempts of the colonial community to recreate an environment similar to the one they had left behind and conducive to the maintenance of colonial life style. It is suggested that the names given to domestic structures by members of the colonial community represent an important element of the townscape of hill stations and can help in an understanding of the process of recreation of a familiar landscape at the individual level. Such an understanding would also be of considerable help in a study of the phenomenon of transference of cultural landscape by a culture group in an alien environment.

The analysis of bungalow names at Kasauli, Mussoorie and Nainital suggests five major themes which form the basis of these names. The themes relate to various aspects of ecology, transference of names from the cultural landscape 'at Home', perception of the bungalow as a home at the individual level, personal names, and aesthetic aspects at the site of the bungalow in the same order of importance. Interestingly, although these themes recurr in the individual study towns, not many bungalow names are repeated. Structurally, the bungalow names divide themselves into two neat halves comprising simple and compound names. Similarly, about half of the total bungalow names analysed contain a generic component, usually as a suffix.

The themes of bungalow names identified in this study are considered fairly comprehensive but by no means exhaustive. It would require a wider data base and a more comprehensive analysis to fully understand and interpret this important element of the colonial urban landscape in India.

References and Notes

1. King, A.D. (1976). *Colonial Urban Development : Culture, Social Power and Environment*, London : Routledge and Kegan Paul, p.172.
2. Archer, E. (1833). *Tours in Upper India and in Parts of the Himalaya Mountains*, Vol. 1, London : Richard Bentley, p. 207.

3. Royal Commission (1863)., p.4 , quoted in King, A.D. (1976). *Colonial Urban Development: Culture, Social Power and Environment*, London: Routledge and Kegan Paul, p. 165.
4. Brown, H., ed. (1948). *The Sahibs : The Life and Ways of the British in India as Recorded by Themselves*, London : Hodge, p. 27
5. *See* ref. 1, pp. 131-132 and 172.
6. Lowenthal, D. and Prince, H.C. (1965) 'English Landscape Tastes', *Geographical Review*, pp. 186-222.
7. Mitchell Nora (1972). *The Indian Hill Station: Kodaikanal*, Chicago: University of Chicago, Research Paper No. 141.
8. Reaney, P. H. (1960). *Origin of English Place Names*, London: Routledge and Kegan Paul.
9. Brayan, P.W. (1967). 'Geography and Landscape', *Geography*, pp. 1-8.

12

SOCIAL AREAS AS ANALYTICAL CATEGORY IN URBAN MORPHOGENESIS AND TOWNSCAPE STUDIES

Some Examples from North India

K.D. SHARMA

The study of urban morphogenesis and townscape has been one of the earliest major concerns in urban geography and continues to be so in many parts of the world. The studies related to this aspect of urban places generally divide the towns into morphogenetic segments on the basis of the location and chronological emergence of religious and public structures and club together large areas inhabited during a particular historical period. Considering the large extent of the area encompassed and the major historical societal influences, these segments naturally reveal broad similarities in the attributes of the morphological components. The entire emphasis in such an approach remains on the built environment with little consideration of the fact that the built environment is really the spatial expression of the time specific socio-economic and cultural environments and social contexts. The built environment among other things, is an attempt, as well as a consequence of, the efforts of the resident population to provide and create a symbolic representation to the existing social order, and has distinct and lasting spatial manifestations.[1]

In this generally accepted analytical framework many of the fundamental questions raised below remain unanswered : What is the arrangement of space within the identified historical- morphological segments? How have the different social, economic and cultural groups living adjacent to each other within a given segment created the required socio-economic environment to conserve and maintain their identities? How have the inherent conflicts between social status and functional requirements and functional interdependence been resolved locationally? Is there really a spatial order reflecting the social order, and if so, what are its spatial manifestations ? How does an existing socio-spatial order respond to new socio-economic impulses? And how are these responses revealed in the various structural elements of the townscape?

A number of studies, focussing on the influence of traditional social organisation in determining the residential patterns in Indian cities have been made.[2] However, these generally analyse the subject in a purely sociological context and in a non-spatial framework.

It is hypothesised that in the case of the Indian cities, the townscape of each historical segment can be further sub-divided into the constituent *mohallas*, the basic spatial units which have evolved through history and have well recognised, perceived, institutional and legal boundaries, and distinct spatial and social attributes.[3] The various terminological analogues of the term *mohalla* are *para, pol, peth, tola, wara* etc.

It is proposed that a more meaningful understanding of the individual cities can be achieved by using the *mohalla* as an analytical category to formulate generalisations about the internal arrangement and socio-spatial structure of urban space within the broad historical periods in India.[4]

Objective

The old parts of the city, often contained within the city wall, present an intricately woven and complex townscape comprised of the engravings of various cultural and historical processes. As a consequence of its importance in the functional organisation and demographic structure of Indian cities, it is imperative that this segment should be understood in terms of all the implications of its evolution, morphology and functions.

This study is an attempt to analyse the evolution of the present townscape and the socio-spatial structure, particularly that of the walled/old parts of the city by using the various attributes of the *mohalla*.

Study Towns

The examples used in this study pertain to two urban places, Moradabad and Chamba. The first is a Class I city located in Uttar Pradesh, the second a Class IV town in Himachal Pradesh. Moradabab was established in 1625 A.D. by Rustam Khan governor of Katehr. The town was named after Murad Baksh, one of the princes.[5] Its history prior to its establishment has little to offer.

Chamba is one of the oldest former Hindu native state capital towns in the Himalayas. It was established around 920 A.D.[6] Due to its location in a relatively inaccessible region, the town escaped the impact of successive Muslim invasions. The British influence in the town is also insignificant. Thus, Chamba represents an uninterrupted history of urban growth from the pre-Muslim period till today.

Moradabad and Chamba present a fairly wide spectrum of history, the two major cultural influences the Hindu and the Muslim, and the fairly common and widespread process of establishment of towns in a feudal spatial organisation.

The Study of Urban Morphogenesis

The study of urban morphogenesis usually begins by an identification of the limits of different morphogenetic segments in a city. Among the major structural features, inherent in the townscape and used for demarcating the old parts of the city, is the city wall. However, due to the loss of its functional utility since the late medieval period, and the subsequent ecological growth of the city, this townscape element is often lost in the maze of buildings and streets, and is at best identifiable only in parts as a relict feature.

However, two aspects related to the existence of this feature suggest that the present townscape should continue to contain evidence of its former location. These aspects are; one, the wall was a means of functional and social containment; and two, as the only means of physical ingress/egress the wall contained diagnostic physical structures.

The landscape expression of the first is an arrangement of space in the form of *mohallas* according to the social order representing a specific historical period; that of the second is identifiable in the existence of structures which continue to be called gates. In the absence of the physical structure of a gate there is generally a *mohalla* named after the gate.[7] In addition, depending upon the nature and level of performance of urban functions, some supporting evidence is also available in the form of a number of mohallas named after such associated features as an inn (*sarai*) or simply indicating the location of an outpost (*chowki*).

In Moradabad it is possible to identify the limits of the walled city with the help of such (*mohalla*) names and structures as *Mahal Sarai, Chowki Hasan Khan, Sarai Sheikh Mahmood, Amroha Gate, Sidhi Sarai, Sarai Pukhta, Multasil Sarai, Sambhal Gate* and *Sarai Kishen Lal.* These are located adjacent to the city wall as a wreath enclosing the old parts of the city (Fig.12.1).

Although Chamba never had a city wall, there is a gate at the former main entrance to the town.

Socio-Spatial Structure and Urban Morphogenesis

This locational arrangement of *mohallas*, and the names of these *mohallas* indicating the various social and economic groups, are spatial manifestations of the social hierarchical structure and an embodiment of the ritual - purity pollution principles. These can provide significant clues in the identification of the walled/old parts of the city.

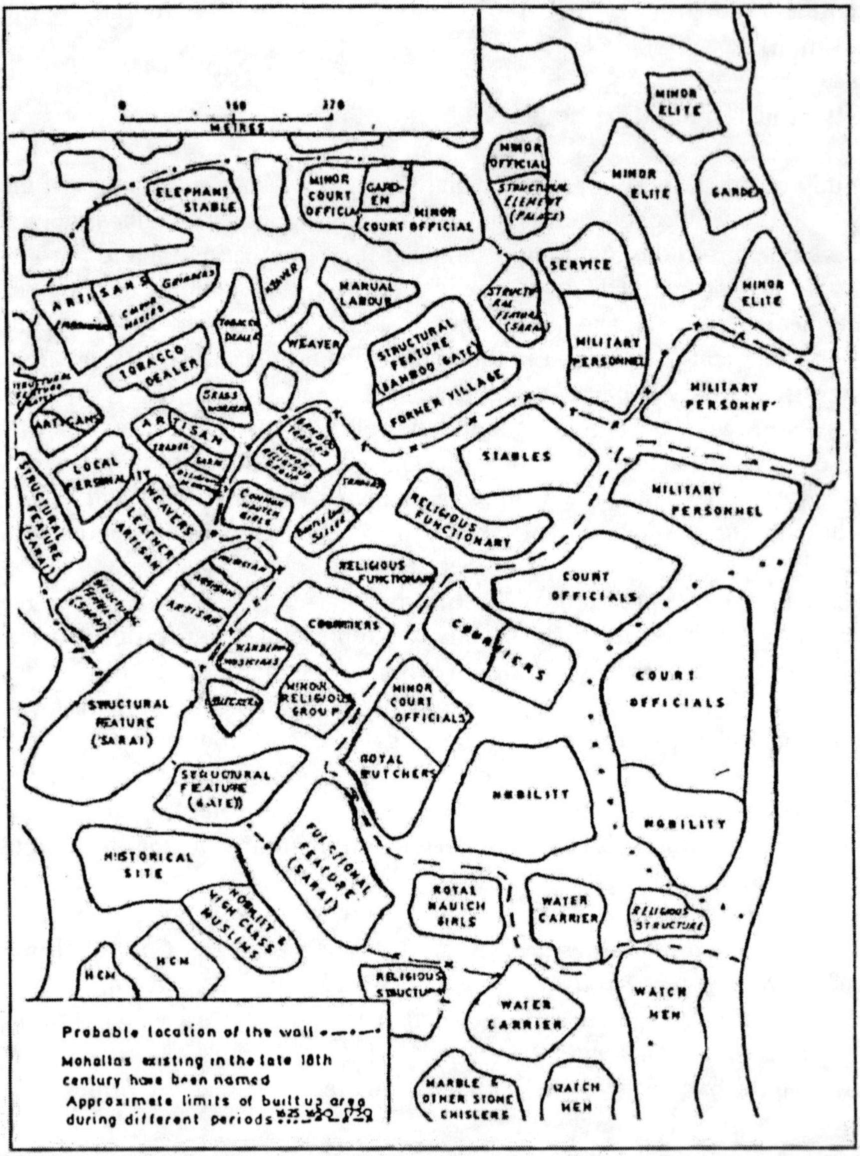

Fig. 12.1 : Moradabad City : Late 18th Century

In Moradabad, the locational arrangement of *mohallas* in the pre-British areas, represents the typical feudal social structure in which distance from the core is an indication of the declining socio- economic status. The only exceptions to this pattern are induced by economic interdependence resulting in an interspersing of coreligious and non-coreligious service caste *mohallas.*[8] The distinctive feature of such a pattern inside the walled city is that it is non-repetitive and non-analogous. The following general scheme based on the data collected from the study towns is suggested for possible use in other towns.

Thus, one complete set would comprise the following major hierarchical socio - economic groups and their appendages: (i) Nobility, (ii) Elite (iii) Artisan and Trading classes, and (iv) Service groups. Each of these board groups comprising a set can be further sub-divided to clearly identify and understand the locational implications of individual *mohallas* (Fig. 12.2). A repetition of any of these categories, revealed through a repetition of the *mohalla* name, will indicate a different phase of growth and accretion and will be an analogue to the original set.

Generally, the first phase in the establishment of a town comprises the settlement of the nobility and the elite classes in quick succession or almost simultaneously. This is followed by the occupation of land by the artisan and trading classes which are followed by service groups.[9]

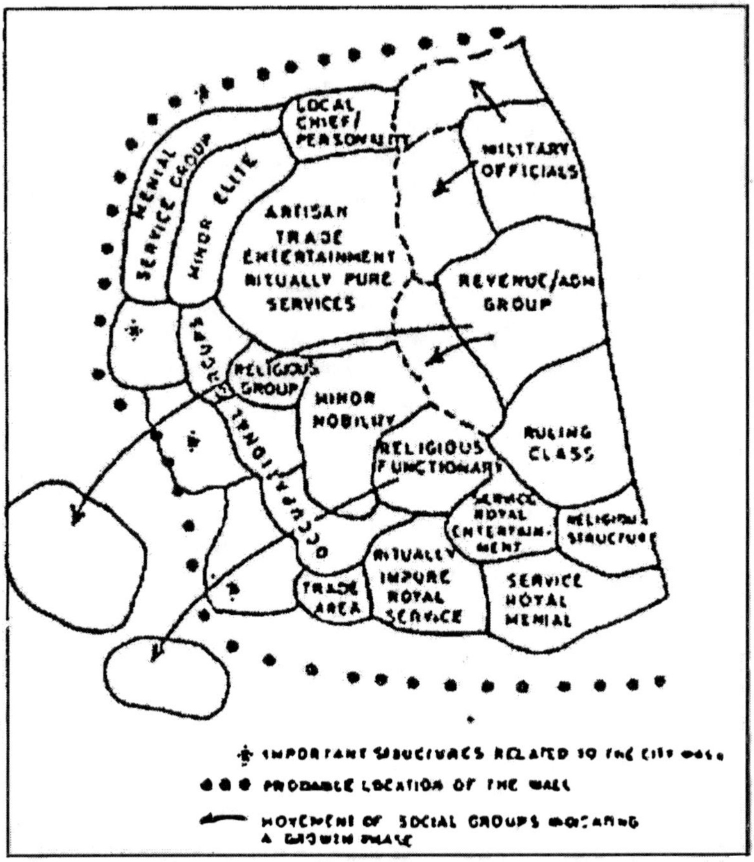

Fig. 12.2 : Generalised Model of Arrangement of *Mohallas* Inside the Walled City

Table 12.1 : Generalised Social Hierarchical Structure

Broad Group	*Sub - Group*		*Specific Group*
1. Nobility	(a) Royalty	(i)	Founding Family
	(b) Minor Nobility	(ii)	Royal Relations
	(c) Local Nobility	(iii)	Landed Gentry, Local Chief etc.
2. Elite	(a) Courtiers	(i)	Civil, Military, Religious
	(b) Minor Elite	(ii)	Court officials, local personalities, money lenders, military/defence officials, court dancers and musicians, astrologers
	(c) Local Elite	(iii)	Traders, landlords, religious and social personalities
3. Artisan and Trading	—	(i)	Metal and stone workers, cloth weavers, flower and perfume workers, minor trading classes
4. Service	(a) Ritually Pure	(i)	Menial - water carriers, household servants, cattle keepers, tank and well diggers.
		(ii)	Others—Watchmen, healers wandering musicians
	(b) Ritually impure	(i)	Menial—Sweepers, grave diggers, cobblers
		(ii)	Others—Butchers etc.

In addition some religious/ritualistic practices with distinct landscape expressions can also be useful. Chamba town, although one of the oldest native state capital towns in Himachal Pradesh, never had a city wall. However, the locational arrangement of *mohallas*, inhabited by different social and occupational groups, reveals a clear demarcation between an inner and an outer segment of the town or between the upper and lower parts of the town as identified by its residents.

The separation between the two has been institutionalised by using a circuit along which an annual religious procession moves. (Fig. 12.3). The circuit has fourteen halting points, one each in a *mohalla* inhabited by either a high socio-economic status or a ritually pure artisan group (Fig.12.3). The location of the halting point is fixed. Its spatial expression is a small open space called *marhi* (Fig.12.3). The various *marhis* are visited in a fixed sequence starting from the one located near the former royal palace. The sequence itself is indicative of the different levels of social hierarchy in the town. In Chamba town therefore, the demarcation between the inner/ritually pure and the outer parts can be based upon areas enclosed by the circuit.[10] Interestingly the circuit includes such ritually pure groups as blacksmiths, and other metal workers and transporters etc. and excludes the ritually impure groups such as leather tanners, butchers and sweepers.

Components of the Townscape

The old parts of most of the cities in India contain the landscape expression of at least few hundred years in the areas that have emerged and evolved in different historical contexts. The townscape of old parts should be analysed in terms of their constituent building blocks representing different time periods and phases in the expansion and in-filling of these parts. It is not necessary that the various growth phases must comprise concentric belts or annules representing a uniform gradient of social and economic status with reference to the centre. The ground reality is shaped by the actual topographical attributes of the site and how these were perceived during different historical periods by different socio-economic groups. The distribution and location of the building blocks reveals how the topographical attributes were perceived and used in different socio-economic contexts and during different historical periods. The *mohalla*, as a spatial unit, is an aggregate of such blocks. The townscape components in these *mohallas* representing different cultural and historical influences should, therefore, reveal diagnostic features.

The three principle components of townscape, the streets, the buildings and building blocks, contain in their various attributes, the evidences of cultural and societal influences specific to different historical periods. Homogeneity in cultural perceptions and norms produces a homogeneity in attitudes towards building forms and functions. This is particularly true of the domestic buildings. In addition there are evidences to suggest that these are the extensions of a vernacular building style and by virtue of this have a tendency to persist on the urban landscape in their almost original forms for a considerable period of time. In Chamba town, the traditional houses still contain such diagnostic features as the *oan*, the *obri*, the *bhor*, and the *prol*. These features are also characteristic of the houses in the rural areas of Chamba region.[11]

These facts when considered along with the evolved, socially homogenous and perceptual character of the *mohalla* establish a strong case for the interpretation and analysis of townscape elements as spatial manifestations of different socio-ecological influences characterising different *mohallas* in a town.

It must be pointed out here that such an analysis must go beyond a rational interpretation of the ground plan as an indicator of a building only as a functional unit. This is because in addition to the functional aspects there also exist elements of culture which may appear to be non-rational when viewed outside their specific cultural-historical contexts. Such elements, therefore, can be understood only through an understanding of the buildings in terms of cultural historical contexts.[12]

It is axiomatic that a great majority of houses in the old parts of the city would reveal the impact of urban influences. However, these influences operate mainly within the context of the urban life style values and generally relate to the use of better building material and addition/alteration in some of the sub-units of a house. It is proposed that although the exterior of a house in the old

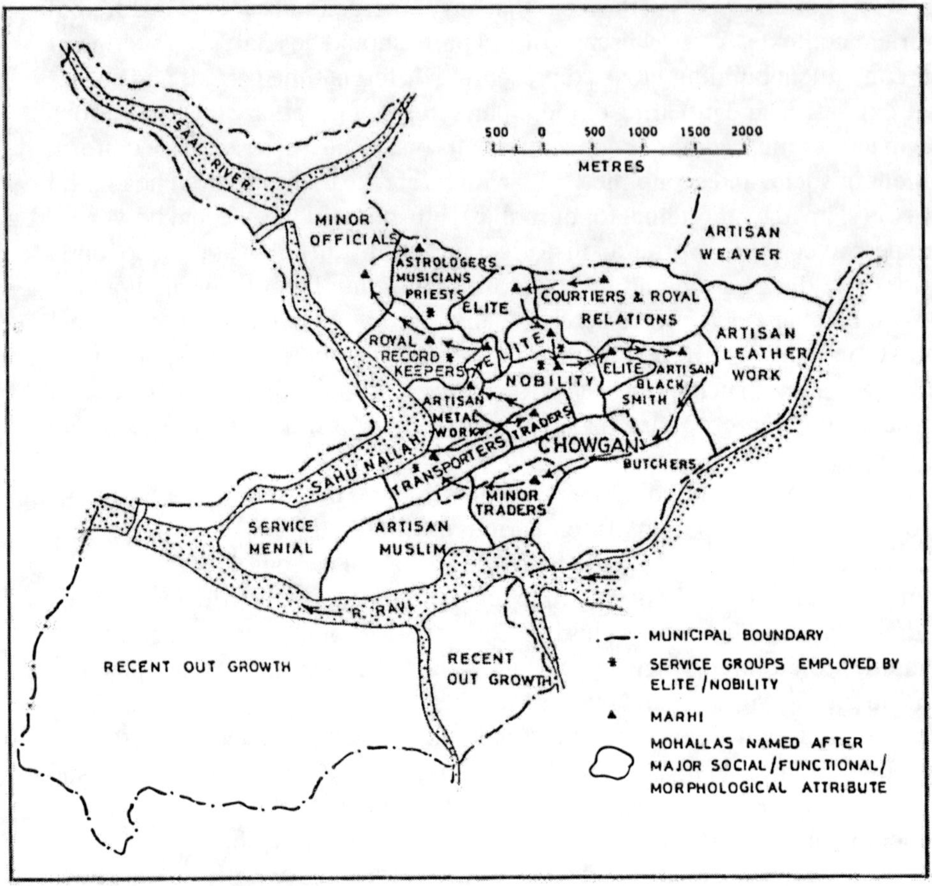

Fig. 12.3 : Chamba Town

parts of the city may appear remarkably modern, the basic character of the domestic building remains traditional and contains sufficient elements of its original form.

House types as an element of cultural landscape in India have received due attention. However, almost all the studies remain confined to the rural areas.[13] It is proposed that since the cultural and socio-economic influences operate in a regional framework it should be possible to identify traditional urban house type also.[14] Admittedly, this would involve considerable effort in terms of empirical testing before regional and cultural models can be proposed, but it would also help in fulfilling a vast lacuna existing in our present understanding of the urban phenomena.

Conclusion

An exclusive concern with the built environment *per se* in urban morphogenesis and townscape studies leaves a vital gap in our understanding regarding the influence of cultural and societal factors. In Indian cities and towns the basic spatial unit in which these influences are manifested in the form of buildings, domestic as well as monumental, is the *mohalla*. In addition, the various *mohallas* through their locational arrangement also represent the various phases and processes of growth in the evolution of an urban place. As an area inhabited by socially, culturally and economically homogenous groups, the various *mohallas* through their relative locations to each other, also help in understanding the impact of societal influences in the organisation of urban space. These attributes of the *mohalla* provide a strong base for its use as an analytical category in urban morphogenesis and townscape studies in India.

The use of *mohalla* in this study of Moradabad and Chamba, suggests that this basic spatial unit of Indian urban places has a structural, social, historic, economic and also symbolic identity which can be used at the individual and the aggregate level for a reconstruction of the phases of urban growth as well as for the identification of morphogenetic areas in townscape studies.

References and Notes

1. Conzen, Michael P. (1977). 'Analytical Approaches to the Urban Landscape', in K.W. Butzer (Ed.). *Dimensions of Human Geography : Essays on Some Familiar and Neglected Themes*. University of Chicago Research Paper No. 186, pp. 136-37.
2. Some of the studies are, Bopegamage, A. (1957). *Delhi : A Study in Urban Sociology*, Bombay: Bombay University Press; Mukerjee, R. and Singh, B. (1961). *Social Profiles of a Metropolis : Social and Economic Structure of Lucknow, Capital of Uttar Pradesh, 1954-56*, New York : Asia Publishing House; Gadgil, D.R. (1965). *Sholapur City*, Poona : Gokhale Institute of Politics and Economics; Bose N. K. (1968). *Calcutta : A Social Survey*, Bombay : Lalvani Publishing House; Fox, R.G. (1969). *From Zamindar to Ballot Box : Community Change in a Northern Indian Market Town*, Ithaca: Cornell University Press; Gore, M.S. (1970). *Immigrants and Neighbourhoods*, Bombay: Tata Institute of Social Services; Mitra, A. (1970). *Delhi : Capital City*, New Delhi: Thomson Press; and Vatuk, S. (1972). *Kinship and Urbanization : White Collar Migrants in Northern India*, Berkeley: University of California Press.
3. From the etymology of the word it would appear that the term came into general use during the Muslim Period. The word , corrupted from the Persian *Mahalla, Muhalla* refers to a district, division, quarter (of a city or town) , ward, parish, a camp , a review or muster (of troops). See Platts, John T. (1977). *A Dictionary of Urdu, Classic Hindi and English*, Reprint, New Delhi: Oriental Books, p. 1010. The term also denotes streets, squares, courts, wards, places, houses, districts, departments (in India). See Steingass F. (1977). *A Comprehensive Persian - English Dictionary*, Reprint, New Delhi : Cosmo Publications, p. 1182. That the term basically referred to a spatial unit inhabited by a homogenous community is evidenced from the practice of recognising a *Mahall - Dar*, the headman of a *mohalla*.
4. The fact that the *raison d'etre* of the *mohalla* is social and cultural affinity of the residents is well recorded. See, Brush, John E. (1962). 'The Morphology of Indian Cities' in Turner Roy (Ed.), *India's Urban Future*, Bombay: Oxford University Press, p. 60; Breeze, Gerald (1963). 'Urban Development Problems in India', *Annals of the Association of American Geographers*, Sept. No.3, p. 261; and Smailes, Arthur E. (1969). 'The Indian City : A Descriptive Model', *Geographische Zeitschrift*, Sept., pp. 181-182. Interestingly Smailes refers to the *mohalla* only as Muslim quarters, *ibid*, p.188.

5. *Imperial Gazetteer of India*, Vol XVII, London : Clarendon Press, p. 422.
6. *See* ref, 5, Vol. X, p.130.
7. Two types of gates can be generally identified ; (i) those located where the major streets met the city wall. These gates were generally named after the name of the city towards which he gate opened and indicated the political, military and economic linkages of the city; and (ii) those located inside the city to enclose specific areas in times of disturbances. These were generally named after the resident community of the area to be enclosed by the gate.
8. The same social class but at different levels of social stratification tends to locate at convenient proximity of concentration of the relevant elite group. Economic interaction as well as economic interdependence leads to the placing of *mohallas* in proximity to each other but in an interspersing pattern of co-relgious and non-coreligious groups. The arrangement of *mohallas* is therefore, determined by the simultaneous functioning of two principles; (i) ritual- social space; and (ii) economic social space.
9. Sharma, K.D. (198 l) 'Evidence of Mohallas in the Reconstruction of Urban Morphogenesis', *Geographical Aspects*, August, p. 44.
10. The symbolic preservation of ritual and social purity through the creation or institution of a zone is not uncommon. The existence of sacred zones has also been identified in Varanasi, although 'most pilgrims are unaware of these zones of the sacred city'. See Eck, Diana L. (1983). *Banaras : City of Light* ,London: Routledge and Kegan Paul, pp. 351 - 357.
11. Another fine example, located hundreds of kilometres away and in a different cultural ecological region is the *wara* house in Poona city as well as the rural areas of the region.
12. Mukerji A.B. (1967). 'Musalman House Type in Meerut Rural Area', *Folklore*, No. 4 p.117.
13. Spate, O.H. K. and Learmonth, A.T.A. (1967). *India and Pakistan*, London : Methuen & Co. p. 201.
14. For an interesting study in this field see Hirt, Howard F. (1982). 'Caste and Urban House Type in South India : Brahmin House in Tamilnadu and Karnataka', in Allen G. Noble and Ashok K. Dutt (Eds.), *India : Cultural Patterns and Processes*, Colorado : Westview Press, pp. 125-145.

13

THE INDIAN CITY

A Conglomeration of Culture

(Case of Ahmedabad)

ANJANA DESAI

Introduction

Cities, in general, are agglomerations of communities of various cultures. Heterogeneity is reflected not only in the composition of population but also in the landuse pattern and living styles of people. Different groups of people with diverse backgrounds of language, religion, caste and occupation have agglomerated in Indian cities. The degree of cohesion can be observed through the clustering of communities over space. Social spaces characterized by ethnic groups can be recognised in the landuse through the location of the regional school, temple, library or shops belonging to specific communities. These communities make an effort to meet at religious festivals, as also social and cultural functions. The general life-style, food and dress, tradition, values and beliefs reflect the diversity of Indian cities. A 'social space' created by the concentration of an ethnic group can be easily recognized by the distinct aroma of regional food and regional style of dress. Such diversity has made the Indian urban environment rather complex.

Irrespective of this complexity a distinct order prevails. An individual identifies himself with a group for socio-economic and cultural interaction. The main means of communication is language and hence, the basic rationale for communities is to form linguistic groups. A finer mosaic of divisions is further implanted on the basis of religion, caste, occupation or socio-economic status of individuals in a community. Within a group, communities need to interact not only to speak in their own language but also for the social, cultural, psychological and financial support the community offers. In fact, it is this support from their own community that enables the survival in Indian cities of a large number of people from rural background staying without their families. The community, thus, becomes a pivotal factor although even in this 'community-living' a hierarchical order prevails on the basis of socio-economic status. People of higher socio-economic status may not form clusters for staying close to each other as they can support themselves financially and psychologically, but for people of a lower socio-economic status the community support system is vital. However, broadly speaking the community does provide all groups the required base to satisfy their cultural needs by participating in regional functions and communicating within their group.

Hence, an Indian city is not just an agglomeration of communities, but a "Conglomeration" where the identity of diverse communities is visible and yet a fusion of culture has taken place— an Indian diversity which is cemented on the basis of traditional unity.

This study mainly focuses on the behavioural aspects among linguistic groups of interstate migrants in an urban environment of an Indian city, namely Ahmedabad. It is important to know the living order of social space—how ethnic conglomerations have taken place, having an intermediate social system where neither 'segregation' nor 'melting pot' are in vogue and what is the retionale behind 'distancing' among communities, which allows them to live, work and interact within and between groups of people. In this study interactions among migrants and local residents have not been looked into. The issue of social harmony among groups is not only important in India but also at a global level, as it brings out human problems which often divide space and people.

Seven linguistic interstate migrant groups to the city of Ahmedabad in 1981 have been analysed to examine empirically the extent to which migrants conglomerate at their destination where the local language, in this case Gujarati, is quite unfamiliar to them.

Methods and Data Aspects

The migrant groups selected include the Sindhi and Punjabi from the north, Bengali from the east, and Tamil, Telugu, Kannadiga and Malayalee from the south. Sampling has been done on a random, stratified basis by age, sex, socio-economic status and by geographical locations within the Ahmedabad Standard Urban Area. In the absence of information regarding location of migrants in Ahmedabad, first socio-cultural organisations have been contacted for addresses of migrants, and then more addresses have been acquired from the migrants themselves. In addition,other sources of information such as the telephone directory, textile mills, government and private offices, have also been consulted. In every migrant group, some old members have been interviewed (without a structured field questionnaire) for an in-depth understanding of their experience and exposure regarding the urban process of assimilation, adaptibility, likes and dislikes of the place and of the local people. Nearly 1,500 migrants and 300 local residents have been interviewed during 1988-89. Secondary data sources include census reports and booklets published by regional organisations. Techniques used are tabular, graphical, cartographic and statistical testing.

The Study Area

Ahmedabad a typical pre-industrial city, set up in 1411 A.D., and covering an area of 190.5 square kilometres with a population of 3.3 million in 1991, exudes a dual character of traditional and modern structure and function. Pluralistic with many communities, its spatial organisation is based on socio-cultural parameters, rather than on economic factors.

The concentration of the cotton textile industry here has exerted a strong economic pull factor attracting, from both far and near, a large number of migrants since 1950s. These include both skilled and unskilled workers from rural as well as urban background. However, increasing unemployment following closure of textile mills, a declining economy and repeated social disturbances during 1980-89 brought to the surface the 'sons of soil' *versus* migrant workers feelings. During 1971-81, interstate migrants to Ahmedabad reduced from 31 per cent to 28 per cent possibly due to economic setback and social unrest. In 1981 the relative importance of Ahmedabad was further reduced for migrants, who found Surat and Vadodara more attractive for jobs The proportion of interstate migrants (by cities) was highest in Surat (35 per cent) compared to that in Ahmedabad (31per cent) in 1981. However, in absolute terms, Ahmedabad still attracts the largest number of migrants in Gujarat.

The states located close to Gujarat comprise the major senders of migrants to Ahmedabad *e.g.* Rajasthan accounts for 31.81per cent of the migrants, while Uttar Pradesh 26.24 per cent, Maharashtra 20.76 per cent and Madhya Pradesh (6.22 per cent) (as in 1981). Thus, nearly 85 per cent of the total interstate migrants (2,78,269 in 1981) are drawn from these four states. Nearly 8 per cent of the remaining come from the four southern states (Karnataka, Andhra Pradesh, Tamilnadu, and Kerala) and only 2 per cent from the eastern states. The composition of migrants, in general, follows the 'distance decay' concept in that, the number of migrants decreases with the increase in distance from the point of attraction *i.e.* Ahmedabad. On the basis of the mother tongue, 70 per cent of Ahmedabad's population is Gujarati speaking, while Urdu, Hindi and other languages are spoken by only 25 per cent of the population. Sindhi, Punjabi, Tamil, Telugu, Malayalam, Kannada and Bengali are spoken by the remaining 5 per cent population including the migrant groups. Of these groups, the Sindhi community in Ahmedabad constitutes the largest group followed by Punjabi, Malayalam, Tamil, Telugu, Kannada and Bengali (as in 1981).

Regarding the quality of human resources migrating from the above mentioned states to Ahmedabad, in terms of literacy, in 1971, 52 per cent of migrants coming from Andhra Pradesh were illiterate, followed by Karnataka (45per cent), Tamil Nadu (43 per cent), Punjab (28 per cent), and West Bengal (24 per cent). The lowest percentage of illiterate migrants were among those who come from Kerala. The percentage of migrants having graduate and post-graduate degrees was higher among those coming from West Bengal and Kerala. Economic imperatives having been the main reason of migrants coming to Ahmedabad, the interstate migrants are mainly engaged in manufacturing, processing, servicing, repair and household industry (51per cent), followed by trade and commerce (17 per cent), and 75 per cent of them belong to the working community. Nearly 45 per cent of all migrants are in the age group of 30-59 years (1981) and 57 per cent of them have lived here for more than ten years.

Process of Adaptation

As a typical Indian urban centre, Ahmedabad's pluralistic nature finds expression in terms of cultural conglomerations, both in location and in the system of interaction among communities. For reasons which may be economic, social, cultural or religious, focal points are created for living or for meeting. The consequent psychological feeling of oneness through language, caste, religion and life-style has brought about a social mosaic which is dynamic and vibrant in nature.

Several factors contribute to the adaptation and assimilation of migrants in the new place of destination. In the initial stage, a sense of 'distancing' is strong as migrants bring along with them their original cultural roots. The process of assimilation varies by age, sex, socio-economic status, period of stay, and geographical location. Connections with friends and relatives of own community often are helpful in over-coming the initial problems of migrants. In the second stage, filtering takes place through exposure to other cultures over a period of time and willingness to interact with other communities. In the next stage, an image is built in the mind of the migrants regarding the place and the people. Moreover, this image also gets affected by the degree of social and economic satisfaction of the migrants. Hence, their attitude and behaviour are in response to the development of this mental image. In this entire process of integration, it is not necessary that each group of migrants will be following a particular style. There are many factors, as discussed here, that control the process of adaptation and assimilation.

On the basis of empirical evidence it has been found that geographical location of communities is an important factor in the formation of a conglomeration (Fig.13.1). In Ahmedabad, the initial concentration of Sindhis is found in Sardarnagar and Kubernagar areas located near the airport. Later, people with higher economic status have shifted to Vadaj or to Maninagar. In

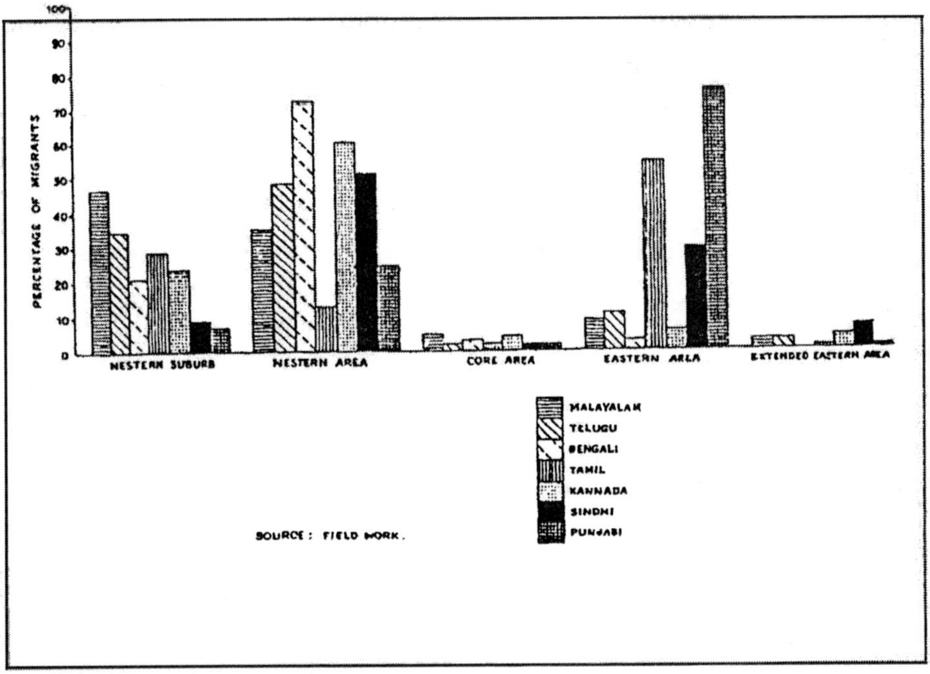

Fig. 13.1 : Location of Migrants of Specific Sampled Communities in Ahmedabad

the core area they have a large concentration in Revdi Bazaar. Similar to the Sindhis, the Punjabi and Sikh communities too are mainly business oriented. As such, they are largely concentrated in the eastern industrial sectors of Ahmedabad, especially in Asarva, Saraspur, Rakhial, Khokra and Maninagar. In the western part of the city, concentrations are seen in Vadaj, Naranpura, Navrangpura, Thaltej and Sabarmati areas. The Bengali community initially settled in the core area of the city in Khanpur, Ratanpole, Manekchawk, and Kalupur. Later, however, most of them scattered in different parts of the city, such as Navrangpura, Maninagar, Kankaria, and Sabarmati. In the early 1940s, a large number of Tamil workers with low socio-economic status came to work in the textile mills of Ahmedabad and settled in the industrial area in the east, in Khokra Memdabad, Kankaria, Behrampura and also in Ranip and Sabarmati in the northern part. Tamils with higher socio-economic status have settled on the western side of the city. Both elite and vernacular groups of Tamils are mainly service oriented. The Telugu community came to Ahmedabad to seek jobs in textile mills in early 1940s, especially in the weaving sections of the mills, and as such are concentrated in Rakhial and Bapunagar. The Telugu community having higher socio-economic status came in the late sixties and settled in the western part of the city in Navrangpura, Naranpura, Jodhpur and Vastrapur. A similar pattern of location is also seen in case of the Kannada community. Among the groups from the south, the Malayalees are the largest in number in Ahmedabad. They are generally scattered, with few concentrations at Khokra, Behrampura, Ishanpur, Naroda, Ghatlodia, Jodhpur, Vastrapur, Thaltej and in Vasna.

In areas where people of lower socio-economic status concentrate, regional schools, shops, temples, libraries or halls of associations where people interact, can be observed. These then become the focal centres of regional interaction (Fig. 13.2), such as religious festivals and cultural activities. Beyond the linguistic domain, finer levels of interactions exist among various groups of migrants on the basis of caste and religion.

Cultural roots of migrants, their individual socio-economic characteristics and connection with own community, contribute much towards the process of conglomeration. Migrants initially experience a sense of rootlessness and prefer to keep close contacts with people back home through frequent visits, by sending money, inviting friends and relatives and attending marriages and festivals at home (Fig. 13.3). Such rootlessness is felt most by people of the 30-39 age group and also by males. Young people however, adjust well, while the elderly have enough exposure to local culture at Ahmedabad over time. Those migrants who stay close to their community or have established good communication, either with their own community or with local residents, have little problem of adjustment. If migrants were given a choice, nearly 58 per cent would have stayed near their own community in order to share social functions and similar life-style and communication in their own language. (Fig. 13.4). To emphasise their cultural identity, different communities enjoy regional dance, music, drama, food and dress. Nearly 65 per cent of migrants feel the need of regional associations to cater to their cultural and social needs. More than 80 per cent are members of their respective cultural or religions organisations (Fig. 13.5).

Eventually, the connection with root culture gradually gets fainter with time in response to benefits that migrants achieve, exposure they get and their acceptance by local residents. However, interaction among people speaking the same language and of the same caste and religion is frequent. About 57 per cent have their best friends from within their community. Among the Tamils and Sindhis the percentage of best friends speaking the same language is higher, that is 77 per cent and 82 per cent respectively. In addition, frequently visited neighbours by migrants, also comprise members of the community speaking the same language. This is particularly characteristic of the Tamils. Migrants were initially helped by relatives and friends speaking the same language, in getting houses, jobs and also in getting admission for children in schools. They

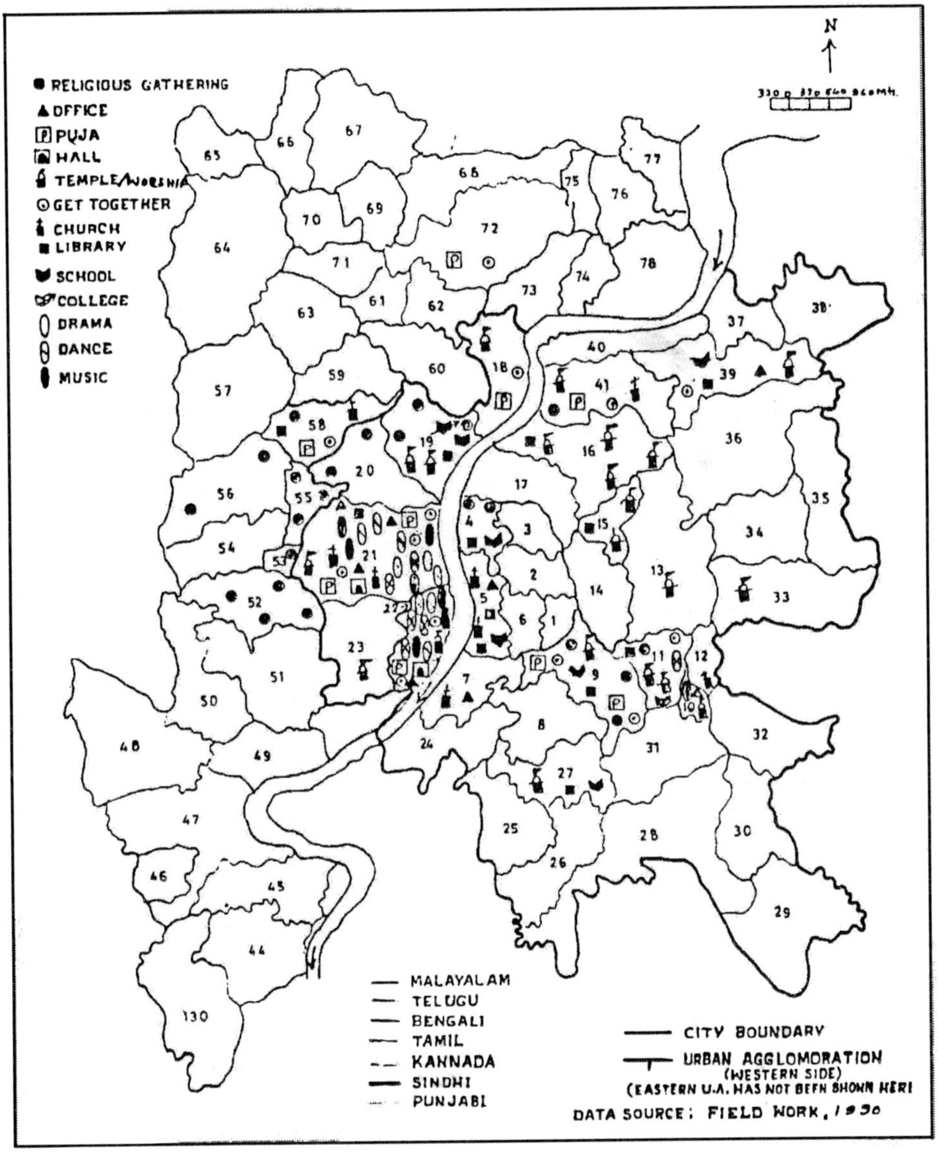

Fig. 13.2 : Socio-Cultural Institutions and Meeting Places of Migrants in Ahmedabad

depend on each other as members of a social system to tide over various problems at new destinations.

In general, migrants have acted in their own channelised groups to form conglomerations. Their individual decisions for location and interactions are often influenced by group decisions and behaviour which is biased with emotion, attachment, belief and incomplete information. Consequently, bias is seen in

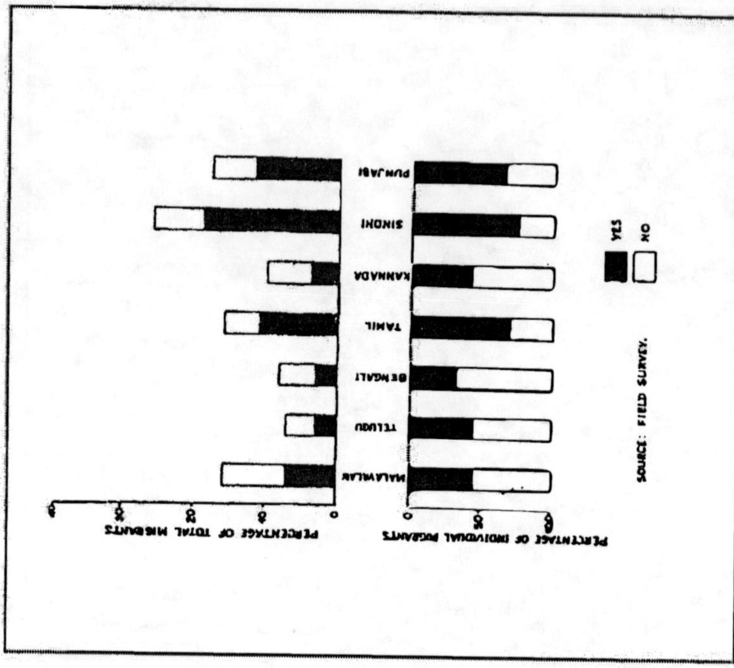

Fig. 13.4 : Preference to Stay Near Own Community by Language of Migrants

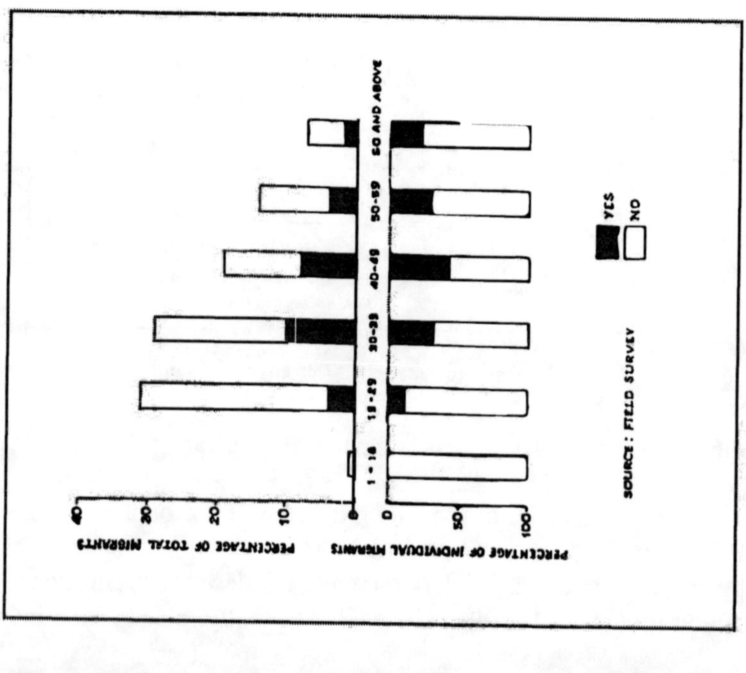

Fig. 13.3 : Feelings of Rootlessness by Age-Group of Migrants

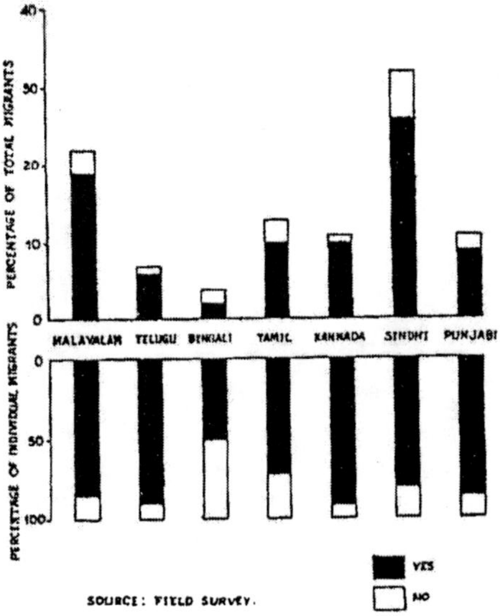

Fig. 13.5 : Socio-Cultural Interaction of Migrants (Within their Own Communities)

human locations and interactions differently in different communities. The process of adjustment is not the same for migrants belonging to various groups. However, each migrant has worked out a way with the help of own community and based on own socio-economic background. It is very important for a migrant to achieve a balance in having an identity of ethinicity on one hand, and trying to participate in local base on the other hand, in the form of a 'Conglomeration' where each mineral can be identified in a cemented rock.

References and Notes

1. Agnew J., Mercer J. and Sopher, David (ed.) (1984). *The City in Cultural Context*, London : Allen and Unwin.
2. Baldassare, M. (ed.) (1983). *Cities and Urban Living*, New York : Columbia University Press.
3. Banerjee, B.G. (1986). *Socialization: An Ethnic Study*, New Delhi: Inter-India Publications.
4. Desai Anjana (1985). *Environmental Perception: The Human Factor in Urban Planning*, New Delhi: Ashish Publishing House.

14

KIPLING'S CALCUTTA

Literary Glimpses

RAMESH C. DHUSSA AND THEODORE I. ERSKI

Introduction

The perceptual dimension of places is one in which geographers are inherently interested. Literary geography provides an understanding of the feelings, values and attitudes of people which are unavailable in 'scientific' modes of analysis. Objective descriptions, coupled with feelings, provide geographers with a complete picture of a place. In the field of literary geography, Rudyard Kipling has remained largely unexplored. Employing his writings in *City of the Dreadful Night*[1] insights into the city of Calcutta, India, have been gained, as it existed approximately one hundred years ago.

According to Salter and Lloyd, 'As geographers, we ought to benefit by capturing this power of literature and directing it toward a deeper understanding of the humanized, cultural landscapes of the earth'.[2] In this manner imaginative literature is not so much a scientific report, but a powerful medium of expressing and conveying subtle images. In Noble and Dhussa's review of literary geography, literature is presented as 'data banks' containing perceptual information of various writers.[3] This study incorporates this idea as Calcutta is investigated through the works of Rudyard Kipling.

Rudyard Kipling

Born in Bombay, India, in 1865, Rudyard Kipling was the son of a professor of architectural sculpture at a local art school. Kipling's early years were plagued with the death of his younger sister and numerous trips to and from England.[4] His early writings consisted of various literary works in newspapers, journals, and magazines. Poetry was his main focus during this period of his life. Books by Kipling began to appear in 1890 with *The Light That Failed*. Throughout his life Kipling travelled the globe and made a first hand written account of all his adventures. In 1898 he travelled to Calcutta, India, the subsequent account of that journey, *City of the Dreadful Night*[5] is used in this study.

Calcutta

From its humble beginnings as a Portugese trading post in the mid 1600s, Calcutta grew to a population of well over one million people by the late 1800s. Calcutta owes much of its growth in size and status to England, which used Calcutta as a base of operations for British imperialism in India until 1911. Dutt observes, 'There is, however, no doubt that it was the erection of the British encampment that began the impetus of growth that eventually turned the trading centre into a gigantic metropolis'[6] (Fig.14.1). Until 1912, the capital was located in Calcutta, which resulted in the city's growth and fame towards the end of the 19th century.

A contemporary author provides a picture of Calcutta in the following words, 'Thanks to its harbour and its numerous industries, its metal foundaries and chemical and pharmaceutical works, its flour mills and its lines, jute [burlap], and cotton factories, Calcutta boasted the third highest average wages per inhabitant of any Indian city, immediately after Delhi and Bombay'.[7]

Weather

Kipling emphasises the negative effect of the heat on the British. Being from Great Britain, the British were of course ill suited for such high temperatures, for example, '... if he stepped out of his brougham [buggy] and walked about in the sunshine for ten minutes he would get a bad headache. In half-an-hour he would probably catch sunstroke'.[8] To protect themselves from such ailments the British constantly wore hats, much to the amusement of the natives.

Kipling also briefly mentions that economic activity in Calcutta went up during the hot months (in northern India May and June are the hottest and driest months of the year). The traffic and people of Calcutta virtually went into a frenzy in an effort to stay afloat financially. Kipling observes, 'The hot months are the busiest in the year and money's the tightest'.[9] During these two months no farming took place as the temperatures were too high and irrigation was not yet widely practiced. As a result, the leisure time of May and June developed into the 'marriage season', and is still the traditional time when marriages take place in Bengal in particular and in northern India in general. As a result of so many marriages, rural people flocked to Calcutta to purchase items necessary for marriage, such as clothing and ornaments.

Of course it was not always hot in Calcutta, although it may not have seemed so to most Europeans. Kipling portrays this point with: 'Even in February, the weather would, [in the] up-country, be called muggy and stifling, but Calcutta is convinced that it is her cold season'.[10]

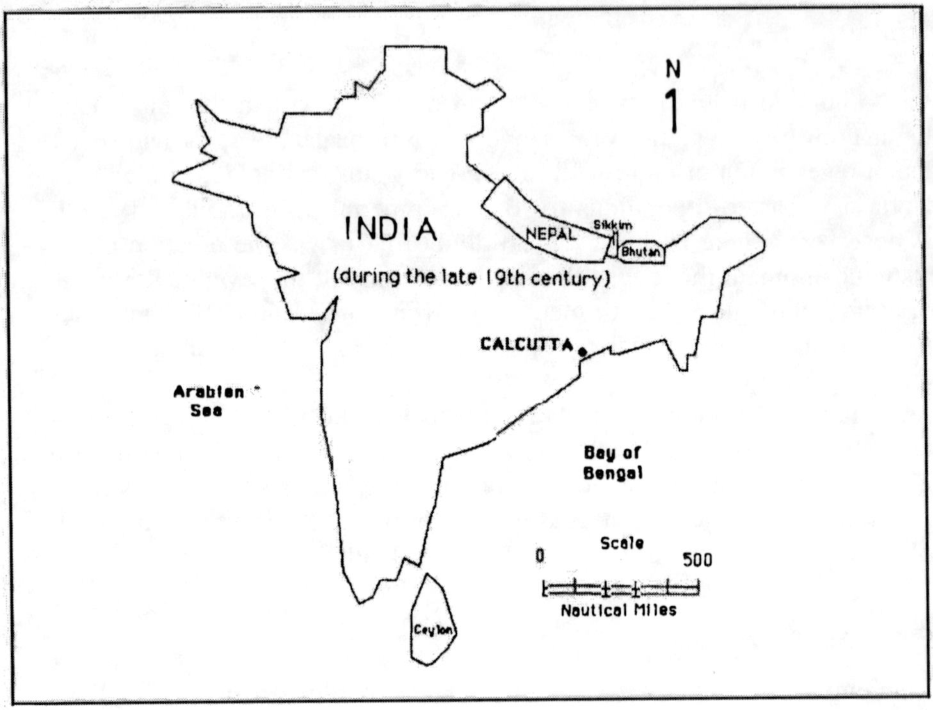

Fig. 14.1 : Location of Study Area

Smellscape

One of the first things Kipling notices about Calcutta is the ever present smoke hanging in the air. A juxtaposition of smoke, constant traffic problems, jumbled telephone lines, crime and poverty, filth, noise, and a pervading stench in the air, creates a vivid image of what Kipling means when he says, 'This is a city'.[11]

Indeed, the stench of the city was not a trivial matter, as various references to it are made throughout the text. For example, 'It resembles the essence of corruption that has rotted for the second time—the clammy odour of blue slime'.[12] Later, 'The air is heavy with a faint, sour stench the essence of long-neglected abominations...'.[13] A native of Calcutta would of course not notice any stench, but for Kipling all such characteristics of Calcutta stood out. When he asked a native about the stench the reply was, 'Wait till the wind blows off the Salt Lakes where all the sewage goes, and *then* you'll smell something'.[14]

Fire Brigade

Kipling also noticed a well organized fire brigade established and run by the British; able to be on the street, racing towards a fire in as little as 1 minute 40 seconds. Not that it did much good, however, as the maze of streets and bewildering lights would frequently confuse them, and the entire episode would often times end up in a futile effort to locate the fire. Thus, it can be seen that the chaotic elements of the city could frequently foil even the best organized public service efforts. As a result many houses in Calcutta undoubtedly burned down.

House Types

The most common type of house in the city was three stories tall with a small courtyard surrounded by a balcony. Kipling gives readers a picturesque account of the typical house in Calcutta with the following passage: 'The houses with their breakneck staircases, their hundred corners, low roofs, hidden courtyards and winding passages...'[15]. In addition to the structural characteristics of the edifices, one also learns that houses in some sections of the city were of higher value than those in less economically fortunate areas. From the headquarters of the Calcutta Police, located at No.22 Lal Bazar area (in the heart of the city), Kipling writes of the various socio-economic areas found in Calcutta, ranging from stinky to posh living conditions. In one instance a character in the narrative says to Kipling, 'Let us see handsome houses where the wealthy nobles dwell'.[16]

Kipling's Spatial Perceptions

Kipling had the elements of an astute geographer who is blessed with a mental map of the cardinal directions. To the north lay the 'great human jungle of the native city'.[17] Three miles of city, then three more miles of suburb comprise the area in which, 'The lamplit dusk on this side is full of noises and shouts and smells'.[18]

The southern part of the city, called *Maidaan* (which Kipling spells Maiden) and Chouringhi, was 'where the respectabilities live and the Police have very little to do'.[19] It is here where high government officials lived and the governor's residence was located. The Victorian landscape of the *maidaan* and the surrounding area is why it is referred to by Kipling as 'The City of Palaces'. Here we also find the imposing 'Victoria Memorial', built entirely of marble. From Kipling's account, *Maidaan* (a Hindi word literally meaning a vast open space for recreation, and used as a place of *rendezvous* for people, especially in the evenings) was a huge public area where cows and oxen feed freely amongst impressive memorial statues built by the British. Such is evident from the following passage, 'The level (referring to *Maidaan*) is studded with brazen statues of eminent gentlemen riding fretful horses on diabolically severe curbs.

The expanse dwarfs the statues, dwarfs everything except the frontage of the faraway Chouringhi Road. It is big—it is impressive'.[20]

Kipling's depiction confirms that the central market place of Calcutta was located close to the bank of the river Hoogly which marked the western limit of the city. He states, 'From the east goes up to the sky the clamour of Sealdah (the main railroad station of central Calcutta), the rumble of the trains and the voices of all Bow Bazar chaffering and making merry. Westward are the business quarters, hushed now, the lamps of the shipping on the river, and the twinkling lights on the Howrah side'.[21] Thus, Kipling's directional perceptions reflect an innate ability to create mental maps for his readers.

Neighbourhoods

Kipling also refers to various bazars (market places) and neighbourhoods in the city. For example, the *Machua Bazar* (literally meaning the market place of fishermen—an area comprised primarily of ocean going vessel docks and poor fishermen) was a good place to find prostitutes, and was where, 'natives sleep and lie all about all over the place'.[22] This suggests how unkept and seedy an area this neighbourhood was. Other shady areas of Calcutta included the *Burra Bazar* (*Burra* meaning big, as in a large market place filled with many small shops and large crowds) and *Jora Bagan* (Jora literally meaning pair, *Bagan* meaning garden—thus a place with two gardens), and *Colootollah.*

Colootollah's actual spelling should be *Kolh Tola, Kolhu* (a Hindi word for a mustard seed pressing machine) referring to those lower caste people employed in the mustard and linseed oil pressing business, and *Tola* referring to a neighbourhood.[23] This area (*Colootollah*) of Calcutta was where the police were continually aggravated by 'petty cases, that keep men up all night and make 'em swear'.[24] The Lal Bazar and Bow Bazar were the rowdiest areas in the city, while *Bamun Bustee* was the quietest (*Bamun* meaning Brahmin, *Bustee* meaning neighbourhood, thus a 'Brahmin neighbourhood', comprised of upper class, priviledged and educated people).

Kipling also writes of the heavy gambling and crime prevalent in the Chinese quarter (a 'Chinatown' type of neighbourhood existing in Calcutta), where the police had little control. This lack of control could have stemmed from the fact that the local police might have been bribed by the Chinese not to interfere with their lives. The Chinese had the resources to do this as many of them came to be lucratively employed in either dentistry or shoe making after they emigrated to India during the British colonial times. Lastly, Kipling speaks of *Dhurrumtollah* (*Dhurrum*, a derivation from the Hindi word *Dharma*, meaning virtues and upstanding morals, *Tollah*, a derivation from *Tola*, meaning neighbourhood, suggesting a place of high moral nature), where a heavy Eurasian population lived.[25] (Fig. 14.2).

People

Kipling notes that though there were many 'natives', it was impossible to simply lump them all together in one culturally homogeneous group. Indians living in Calcutta from various parts of the country were considerably different from each other, carrying with them a language and sub-culture different from that of Bengali (Calcutta) residents. This idea is underscored in the following passage, 'Do not bring a north-country servant into Calcutta. He is sure to get into trouble, because he does not understand the customs of the city'.[26]

In the late 1800s, natives, British administrators, and a considerable number of Chinese immigrants lived in Calcutta. According to Kipling the total population was about one million. Of this million, one hundred thousand were of a transient, or 'floating' nature.[27] Kipling provides the reader with a sense of the complex nature of many different people living in one city. At one point in his writing he describes such complexity as 'an ethnological museum where all the specimens are playing comedies and tragedies'.[28]

The entire city was a microcosm of the cultural mosaic of India, but a unique variety of people could be found by the docks of the sea going vessels. An understanding of the amount of shipping that Calcutta handled could be deduced simply by observing the raw number of sailors found in the city, as evidenced in the following observation, 'Two thousand sailors of fifty nationalities are adrift in Calcutta every Sunday, and of these perhaps two hundred are distinctly worse from liquor'.[29] Kipling refers to sailors from Germany, Italy, America, and numerous, other places. All these sailors docked in Calcutta, quickly spending their wages on prostitutes and alcohol. Even one hundred years ago the life of a typical sailor was similar to today's as understood from the following: 'They'll quit a good ship for the sake of a spree...as soon as the money's gone they'll ship, but not before'.[30]

Government

All governing for the city of Calcutta was conducted by the Bengal Legislative Council. The business of this council, ranging from street care to capital punishment, was housed in the colonial built octagonal shaped structure called the Writers' Building. Inside there were plush accomodations for the officials, and a throne like chair for Sir Stuart Bayley, the ruler of Bengal of that period.

Kipling makes it clear that the Government of India, housed in Calcutta, was similar to the British Parliament. In one instance he mentions people talking about the current debate over an amendment to a bill, in another, a brief mention about the 'Ayes' and 'Noes' and there being a plurality of votes. These two clues and the additional filibustering by a native speaker, all hint at a democratic form of government similar to that of the British.

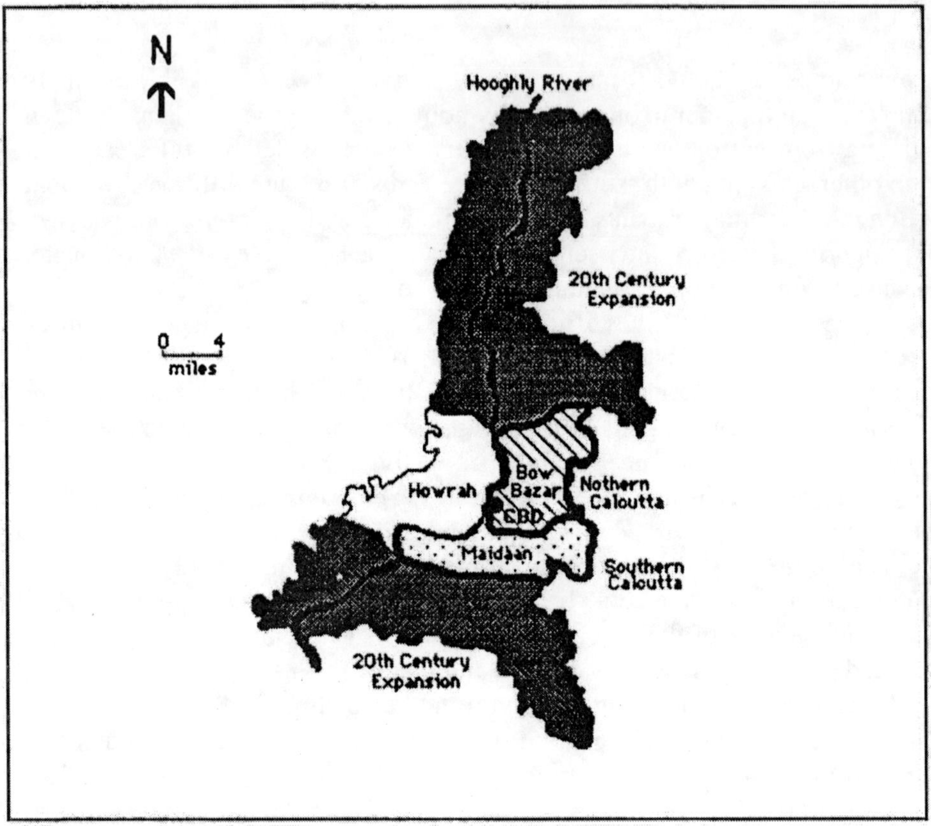

Fig. 14.2 : Kipling's Area Perceptions of Calcutta

In fact, the attitude of the typical British council member was that the system worked too well for the natives, who frequently abused it. When all of India was finally under the British crown by 1876, the British hardly expected to be sitting in session listening to the endless rambling of unimportant native citizens, yet that is exactly the scene Kipling describes. In the following passage British officials grudgingly comment on a native filibuster, 'We taught him what was constitutional and what was unconstitutional in the days when Calcutta smelt. Calcutta smells still, but we must listen to all that he has to say about the plurality of votes and the threshing of wind and the weaving of ropes of sand. It is our fault completely'.[31]

The laws of Calcutta were upheld by a local police force, composed of three thousand natives and one hundred Englishmen, both on foot and horseback. It was a tightly run organization as can be seen from the following passage, 'Very military men are the Calcutta European Police in their set up, and he who knows their composition knows some startling stories of gentlemen-rankers and the like'.[32] In order to perform their duties the police couldn't be anything but military like, as keeping order in even a small part of Calcutta was extremely difficult. This image is eluded to in the following lines, 'From Howarh and Bally (Bally Gunj) and the other suburbs, at least a hundred thousand people come into Calcutta for the day and leave at night. Also Chandernagore (a small town located north of Calcutta) is handy for the fugitive law-breaker, who can enter in the evening and get away before the noon of the next day, having marked his house and broken into it'.[33]

Transportation

Three different modes of transportation were available to the people of Kipling's Calcutta. Of the three, the importance of the railroad in India has especially been emphasised; Kipling goes as far as to devote over three chapters to the disucssion of the rail system in and around Calcutta. In addition to railroad transportation, goods and people were also moved about by steamers going to and from the sea on the Hooghly river. Kipling also mentions people using trams (similar to trolly cars) to get around the city.

The railroad system in India was one of the most important contributions of British occupation. The first rail line was laid down in the 1850s by the 'East Indian Railway Company', which according to Kipling was the richest company in India at that time. In just 40 years, by the time Kipling wrote of India, rail lines stretched fifteen thousand miles.[34] It is evident from his writing that the railway system was fully available to the people of Calcutta even one hundred years ago.

A second important mode of transportation was the steam ship. These ships played a vital role in trade between Calcutta and the rest of the world. All shipping on the Hooghly river, to and from the sea, was directed by the 'Port Office', an enormously wealthy and tightly run organisation. It was through this office that 'two million tons of sea-going shippage yearly find there way up and down the river'.[35] In addition to directing traffic up and down the Hooghly, the 'Port Office' also maintained river banks, expanded jetties and constructed docks for the steamers. With all these tasks performed so expertly, it is easy to see why, when writing about the Port Office, Kipling says, 'This is a place to enter more reverently than the Bengal Legislative Council'.[36]

The people of the city had access to a railroad car transportation (tram) line too. The system, even though it didn't stretch across the entire city, was one of the most convenient ways to get around. There were two drawbacks to the tram system which can be gathered from Kipling's observations. The first concerns the negative attitude of the tram conductors towards the passengers, as seen in the following quote, 'Tram conductors are not polite. Someday one of them will be hurt'.[37] The second drawback was the fact that the trams did not run 24 hours a day as illustrated by, 'All good Calcutta has gone to bed, the last tram has passed, and the peace of the night is upon the world'.[38]

Kipling provides the reader with an understanding of the significance of the railway system in India. The other major mode of transportation in Calcutta was the steamship. With his lengthy account of the job of the 'Port Office', one can see how vital steamer travel was going up and down the Hoogly river. The account of the tram in Calcutta gives a reader both a perspective of the use of trams during the 1890s and an understanding of the extent of its use in Calcutta.

Conclusion

As seen in this study, *City of the Dreadful Night*[39] is particularly helpful in understanding the city of Calcutta as it existed during the 1890s. Another famous work by Kipling, *Kim*[40] might be useful in understanding the life and culture of the people in the north-western section of the outer Himalayan region. Kipling's vast collection of works, both in prose and poetry, are simply gold mines for researchers in literary geography on Indian themes. The humanistic dimension of the geographic environment of India as evoked from the writings of Rudyard Kipling will make one's understanding of India richer and more meaningul.

References and Notes

1. Kipling, Rudyard (1912). *City of the Dreadful Night*, New York : P.F. Collier & Son.
2. Salter C. and Lloyd, F.J. (1977). *Landscapes in Literature*, Washington : Association of American Geographers, p.1.
3. Noble, Allen G. and Dhussa, Ramesh G. (1990). 'A Review of Literary Georgaphy', *Journal of Cultural Geography*, p. 50.
4. Pinney, Thomas (ed). (1990). Rudyard Kipling: *Something of Myself and Other Autobiographical Writings*, Cambridge : Cambridge University Press.
5. *See* ref. 1.
6. Dutt, Ashok K. and Geib, Marget M. (1987). *Fully Annotated Atlas of South Asia*, Boulder : Westview Press, p. 132.
7. Lapierre, Dominique (1985). *The City of Joy*, New York : Doubleday & Company Inc., p. 26.
8. *See* ref. 1, p. 173.
9. *See* ref. 1, p. 208.
10. *See* ref. 1, p. 208.
11. *See* ref. 1, p. 163.
12. *See* ref. 1, p. 165.
13. *See* ref. 1, p. 216.
14. *See* ref. 1, p. 165.
15. *See* ref. 1, p. 233.
16. *See* ref. 1, p. 174.
17. *See* ref. 1, p. 174.
18. *See* ref. 1, p. 207.

19. *See* ref. 1, p. 207.
20. *See* ref. 1, p. 175.
21. *See* ref. 1, pp. 207-208.
22. *See* ref. 1, p. 212.
23. Noble, Allen G. and Dhussa, Ramesh C. (1982). 'Ethnic Neighbourhoods in Indian Cities : The Paras of Dumka and the Bengali Tola of Bhagalpur', in Allen G. Noble and Ashok K. Dutt (eds.), *India : Cultural Patterns and Processes*, Boulder : Westview Press, pp. 81-103 ref. p. 83.
24. *See* ref. 1, p. 212.
25. Gist, Noel P. and Wright, Roy D. (1973). *Marginality and Identity : Anglo Indians as a Racially Mixed Minority in India*, Leiden : E.J. Brill, p. 115.
26. *See* ref. 1, p. 168.
27. *See* ref. 1, p. 207.
28. *See* ref. 1, p. 201.
29. *See* ref. 1, p. 209.
30. *See* ref. 1, p. 204.
31. *See* ref. 1, p. 187.
32. *See* ref. 1, p. 209.
33. *See* ref. 1, p. 212.
34. *See* ref. 1, p. 279.
35. *See* ref. 1, p. 197.
36. *See* ref. 1, p. 196.
37. *See* ref. 1, p. 239.
38. *See* ref. 1, p. 237.
39. *See* ref. 1.
40. Kipling, Rudyard (1916). *Kim*, New York : Charles Scribner's Sons.

15

COSMIC ORDER AND CITYSCAPE OF VARANASI (KASHI)

Sun Images and Cultural Astronomy

RANA P.B. SINGH AND JOHN MCKIM MALVILLE

Introduction

The city plan of Varanasi was developed according to a cosmic order. In testing this hypothesis it was observed that the temples and shrines related to Sun (Aditya) are placed in a meaningful spatial pattern corresponding to the spatial geometry and movement of the Sun, the association of cosmic north and Kashi-North, and the celebration of seasonal festivities in a sequential order referring to solstices and equinoxes. Probably, this pattern had developed in the pre-Brahmanical tradition, and later on was superseded by the Shaiva tradition. The nomenclature and iconographic features further indicate the mythology and the inherent scientific meanings.

Worshiping the Sun by making offerings of Ganga water amidst chanting of sacred verses, circling oil lamps of *aarti* and bells after the ritual bathing in the Ganga is an integral part of the Varanasi morning scene along the Ganga. In fact, Sun worship dates back to the Vedic period. Known as Aditya, the child of Aditi and Prajapati, the Sun in early Vedic mythology is related to seven mainfestive forms corresponding perhaps to the seven days of the week or the seven planets visible to the naked eye which circle the geocentric earth. However, by the 5th century, at the time of the *Brahmanas* and the *Puranas* the number of manifestive forms of Aditya increased to twelve, associated with the number of months of the solar year. A 13th-14th century text, the *Kashi Khanda* of the *Skanda Purana* which initially mentions twelve Adityas, towards the later chapters mentions two more Aditya shrines (Sumantva and Karna) establishing therefore, fourteen Adityas of Varanasi.[1]

Although the present religious landscape of Varanasi is not exactly dominated by Sun shrines, in Puranic literature the city is described as the 'City of Light'—*Kashi* or 'City of the Sun'—*Alarkapuri*. According to Eliade,

Note : Reprinted from *National Geographical Journal of India*, Vol. 41 (1) March, 1995.

'the process of sacralization is greatly assisted by the benevolent and in many ways passive nature of this divinity'.[2] According to the *Purusha Sukta* of the *Rig Veda*, the Sun was born of the eye of the cosmic giant, *Purusha* (the primordial man), and, 'at death, when man's body and soul become one more part of that cosmic giant, his eye will go back into the Sun'.[3] The spatial manifestation of this homology is narrated in the *Kashi Khand* ; according to it *Lolarka* and *Keshava* are the eyes of the '*Kashi-Goddess*'. The city is perceived as 'goddess' whose territory is defined by the two tributaries - the Varana in the north and the Asi in the south, symbolized as the arms of the goddess.

Historical evidences suggest that during the Gupta period (4th-5th century) Sun worship was common in north India and by the 12th century it had been established as a major part of Hindu rituals.[4] The *Puranic* treatises of this period praise the Sun God in Varanasi, *e.g.* the *Matsya*, the *Kurma*, and the *Vamana Puranas*. After some passage of time and changes brought about by religious-cultural influences, Shiva worship superseded the ritual arena and sun shrines became part of or ancillary to the Shiva shrines. Since the Puranic period the practice of worshipping multiple gods together is accepted as the common norm in Varanasi. These gods included Shiva, Parvati (goddess), Vishnu, Ganesha, and Surya (Sun). That the Vedic king Divodasa as per the *Kashi Khand,* removed all the divinities from Kashi, except his family deity, Surya is indicative of the supremacy of sun worship at Kashi.

Varanasi possessed 14 Adityas (Fig. 15.1), each of which was associated with a shrine or temple. In the late 12th century and the following years of Mughal invasion and occupation of the city (12th-18th centuries) these structures were practically razed to the ground. However, neither the Sun nor the spirit of a devout people vanquished, and the locations of the Adityas remained in the communal memory of the city. Presently the 14 Adityas of Varanasi are marked by a variety of typically small and unpretentious features: Sun discs, lotus-form stones, or images of Surya alone or riding his chariot, set into walls or installed in small shrines on the precincts of temples (Figs. 15.4 and 15.5). The popular pilgrimage texts describe the sacred journey to these 14 Sun shrines. However, popularity of this has waned, and now many of the sites are not easy to locate.

On the basis of detailed field study, the location of all the 14 Aditya has been marked and mapped. During February-March 1994 two 'Garmin GPS-75 Receivers' were also used to map out the differential Global Positioning System measurements and locations were identified (Table 15.1, Figs.15.1 and 15.2). The corresponde-nce of locations and historical descriptions throw fresh light on a new dimension to understand the cosmic order and cityscape of Varanasi. A form of 'non-destructive archaeo-astronomy' is apparently discernible from the mapping of the sites of the original Aditya temples which reveal something

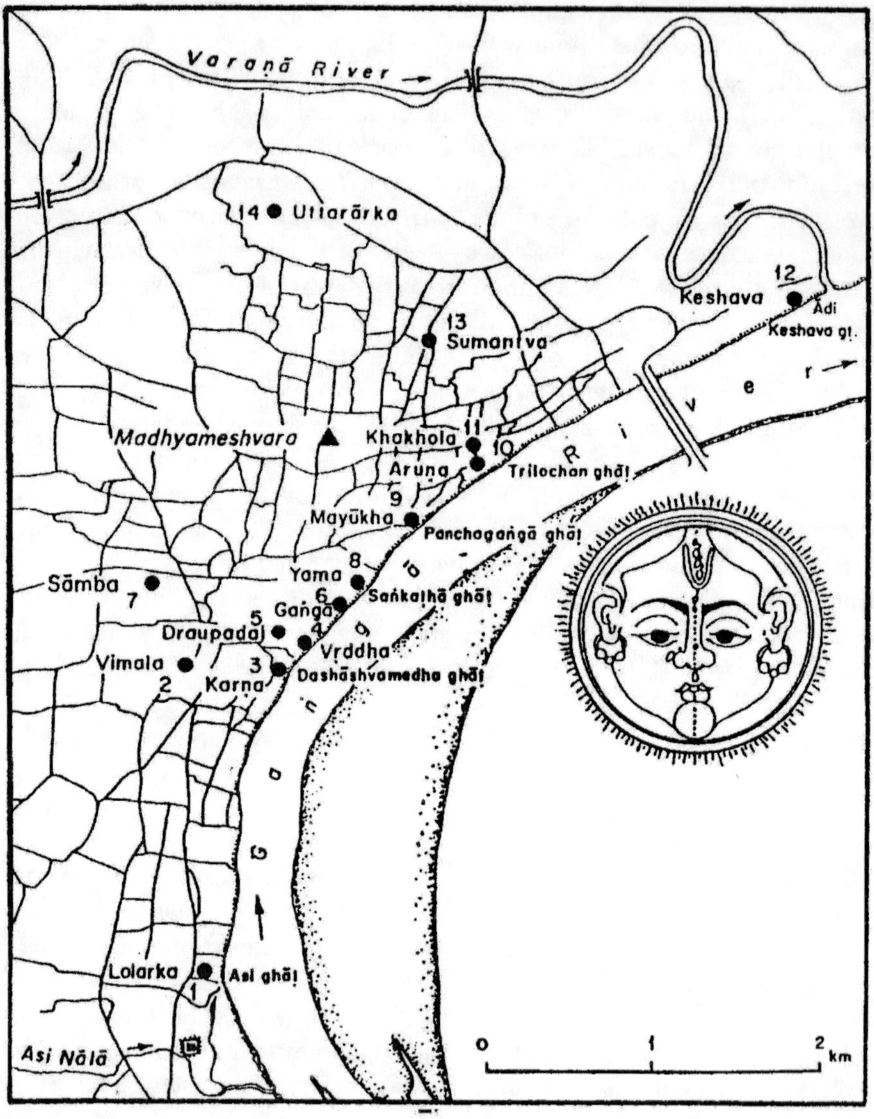

Fig. 15.1 : Vārāṇasī : Sun/Āditya Shrines

of the general nature of city planning as also the human dimension, preserved through the rituals in memory of the shrines and their mythological associations. In other words, this analysis also reflects upon cultural astronomy. The perception and practices of phenomena related to Sun shrines —their locations and movement of Sun—mostly depend upon the meaning and significance that Hindu culture ascribes to its constructed universe. However, what is considered

Table 15.1 : Varanasi : Location of Sun Shrines

Sl. No.	*Aditya/ Sun*	*Location in the City (Area House No)*	*Latitude, N*			*Latitude, E*		
			0°	*0'*	*0"*	*0°*	*0'*	*0"*
1.	Lolarka	Lolarka Kunda, Asi	25	17	29.1	83	00	21.4
2.	Vimala	Jangambari -- Khari Kuwan : D35/273	25	18	25.1	83	00	16.8
3.	Karna	Ramamandir, Dashashvamedha Ghat : D1/111	25	18	24.1	83	00	36.7
4.	Vriddha	Mir Ghat: D3/15	25	18	34.2	83	00	43.
5.	Draupada	near Vishvanath temple : CK 35/21	25	18	38.1	83	00	37.0
6.	Ganga	Lalita Ghat : K1/68	25	18	34.9	83	00	48.1
7.	Samba	Suraj Kunda : D51/90	25	18	43.9	83	00	10.5
8.	Yama	Sankatha Ghat : K7/164	25	18	43.7	83	00	58.4
9.	Mayukha	Manglagauri temple, in pillar: K24/34	25	18	53.1	83	01	02.8
10.	Aruna	Trilochan Temple : A2/80	25	19	09.7	83	01	24.0
11.	Khakhola	near Kameshvara Temple backside :A2/9K	25	19	12.8	83	01	21.4
12.	Keshava	Adi Keshava Temple : A37/51	25	19	44.7	83	02	34.8
13.	Sumantva	Hanuman Phataka : A31/91	25	19	37.3	83	01	12.2
14.	Uttararka	Bakaria Kund	25	20	04.4	83	00	41.2
	Madhyameshvara	Dara Nagar : K53/63	25	19	16.1	83	00	51.2

here as empirical reality may not be relevant to other cultures and even if it be so, the meaning is likely to be very different.

Cosmic Order and Cityscape

It has been observed that 10 of the 14 Adityas lie approximately along the sides of the triangle formed by the Uttararka, Karna and Khakhola. The longest side is formed by the north-south line connecting Karna with the 'northern Sun' Uttararka, covering a distance of 3.1 km. Each of the two axis in the northern and southern sides of the triangle have a length of 2001 m and 1997 m, respectively, equal to within 0.2 per cent and hence suggestive of intentional design (Fig.15.2).

With only negligible variation of 1.14° E, the line connecting Karna and Uttararka approximately represents the cosmic north . Considering Samba as the centre of the cosmic north in Varanasi, the azimuths for rest of the shrines are calculated from east to south ranging from 18° 36' at Uttararka to 172°30' at Lolarka. Yama falls exactly at 90°—symbolizing the east, facing the Ganga river (Table 15.2).

Table 15.2 : Varanasi : Azimuths of the Sun Shrines

Sl. No.	*Aditya/ Sun image*	*Azimuth from the Cosmic North : Samba as centre*			*Azimuth : S, N : Samba - Yama as the eastern axis base*		
		0°	*0'*	*0.0°*	*0°*	*0'*	*0.0°*
1.	Lolarka	72	30	172.5	82	30	82.5
2.	Vimala	163	54	163.9	73	54	73.9
3.	Karna	128	30	128.5	38	30	38.5
4.	Vriddha	104	00	104.0	14	00	14.0
5.	Draupada	104	00	104.0	14	00	14.0
6.	Ganga	97	12	97.2	7	12	7.3
7.	Samba	—Samba as centre—			—	—	—
					Samba – Yama, the eastern axis		
8.	Yama	90	00	90.0			
					Towards North (+)		
9.	Mayukha	79	48	79.8	10	12	10.2
10.	Aruna	70	24	70.4	19	36	19.6
11.	Khakhola	66	30	66.5	23	30	23.5
12.	Keshava	66	30	66.5	23	30	23.5
13	Sumantya	48	06	48.1	41	54	41.9
14.	Uttararka	18	36	18.6	71	24	71.4

Further, the line connecting Samba and Yama symbolizing the Winter Solstice, *Makra Samkranti* (14th January in ancient Hindu astronomy, and 21st December as per Western astronomy) divides the holy territory and the Aditya shrines into two equal portions. Six of the Adityas fall on either side of this line with an unequal degree of azimuthal variation. The axis connecting Samba,

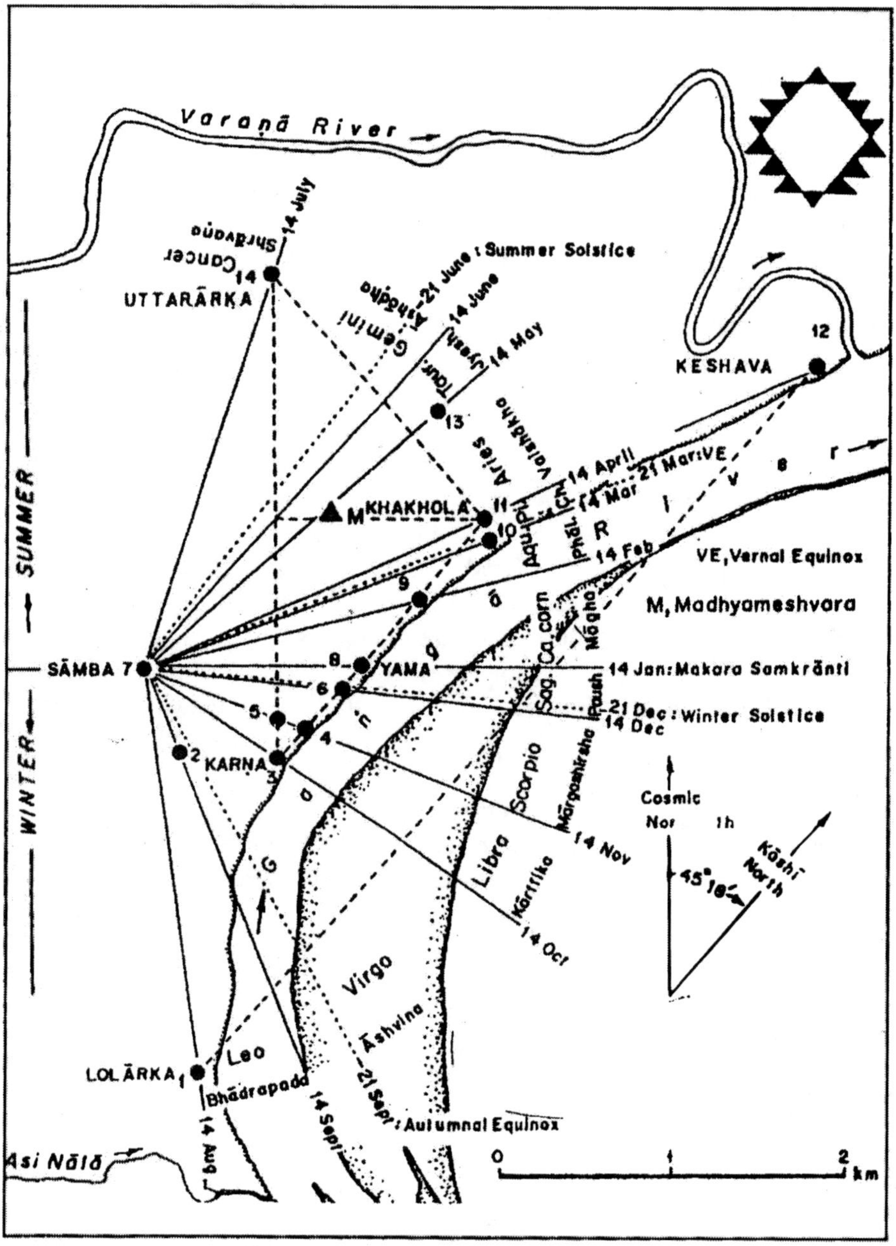

Fig. 15.2 : Vārāṇasī : Sun/Āditya Shrines : Cosmic Order and Time

Khakhola and Keshava is at 23° 30' north from the east coinciding with the position of the Sun on reaching the Tropic of Cancer referred to as Vernal equinox or *Mesha Samkranti* (14th April as per ancient Hindu astronomy, and

21st March as per Western astronomy). This point refers to the movement of the Sun towards the North Pole, reaching its peak at 70° 24', Uttararka, coinciding with the time of Summer Solstice, *Karka Samkranti* (14th July in ancient Hindu astronomy and 21st June in Western astronomy).

Southwards from the eastern tangent (Samba-Yama) there are six sun shrines, including Karna which lies at 38° 30' close to the position of the Sun at the Autumnal equinox or *Tula Samkranti* (14th October in ancient Hindu astronomy, and 21st September in Western astronomy). The position of Lolarka (82° 30') marks the movement of the Sun from the zodiac sign Cancer to Leo (period of 14th August to 14th September). The movement of the Sun and its influence spatially correspond to the sun shrines from the south to the north in terms of Hindu months (Table 15.3 and Fig. 15.2).

Parallel to the Karna-Khakhola tangent lies the axis connecting Lolarka and Keshava, covering most part of the main channel of the Ganga and its *ghats* (steps to the water). This angular variation is 44° 42' north from the Samba Yama axis, *i.e.* 45° 18' east from the Cosmic north—this is perceived as the Kashi-North, popularly symbolized by the northerly flow of the Ganga. The geometry of the Adityas thus combines the northern directionality of both the macrocosm (Sun) and the mesocosm (the Ganga river). The importance of these two directions within the city is not surprising. The pole of the heaven establishes the order of the cosmos, and the direction of the flow of the Ganga establishes the source of the world and the great axis of death and re-birth.

Samba at Surya Kunda appears to be the focus for lines radiating outward to the dawn sun at various days throughout the year and marks the cycle of solar rhythm, starting at Lolarka referring to the Hindu month of *Bhadrapada* (14th August-14th September, zodiac Leo) and completed at Uttararka referring to the Hindu month of *Shravana* (14th July-14th August zodiac Cancer). On the basis of the months and auspicious occasions prescribed for the different Adityas, the twelve zodiacs and their corresponding months can be compared (Table 15.3 and Fig. 15.2).

Only between Sumantva and Uttararka, there appears a gap in the zodiac Gemini and the Hindu month of *Ashadha*. However, by theoretical analogy this can be measured; if a line connecting the mid point on the axis joining Karna and Uttararka and Samba be drawn, this problem can easily be solved. Ultimately the total pattern of spatial geometry of Sun shrines and the corresponding zodiacs and months can be projected.

The line connecting Karna, Draupada and Uttararka Adityas represents the Cosmic north, while the tangent connecting Karna, Vriddha, Ganga, Yama, Mayukha and Khakhola moves parallel to the flow of the Ganga. The triangle made by connecting Karna, Khakhola and Uttararka (Fig.15.2) shows the common core of the Sun shrines. The shrine of Madhyameshvara (a form of

Table 15.3 : Varanasi : Sun Shrines and Cultural – Astronomical Connotation

Sl. No.	*Aditya/Sun Image*	*Hindu month of worship*	*Corresponding Zodiac & Period*			*Ref in The KKh: verse*	*Religious merit*
			Hindu	*Latin*	*Period 14 – 14*		
		1	*2*	*3*	*4* *a b*	*5*	*6*
1.	Lolarka	Bhadrapada	Simla	Leo	Aug-Sep	46	Liberation
2.	Vimala	Ashvaina	Kanya	Virgo	Sep – Oct	51.83-100	relief from leprosy
3.	Karna	Karttika	Tula	Libra	Oct – Nov	85.45	blessings of the Sungod
4.	Vriddha	Margashirsha	Vrishchika	Scorpio	Nov-Dec	51.27.42	long life
5.	Draupada	Margashirsha	Vrishchika	Scorpio	Nov-Dec	49.1.24	relief from hunger
6.	Ganga	Pausha	Dhanu	Sangittarius	Dec-Jan	51.101-115	merit of bathing in the Ganga
7.	Samba	Whole year/ Ashadha	Mithuna	Gemini	Jun-Jul	48	relief from skin deseases
8.	Yama	Magha	Makara	Capricorn	Jan-Feb	51.105-115	relief from fear to death
9.	Mayukha	Phalguna	Kumbha	Aquarius	Feb-Mar	49.25-95	poorlessness
10.	Aruna	Chaitra	Mina	Piscus	Mar-Apr	51.21-26	sorrowlessness
11.	Khakhola	Vaishakha	Mesha	Aries	Apr-May	50	burning sins and impurity
12.	Keshava	Vaishakha	Mesha	Aries	Apr-May	51.44-82	get wisdom, sinlessness
13.	Sumantva	Jyeshtha	Vrishabha	Taurus	May-Jun	65.6	relief from leprosy
14.	Uttararka	Shravana	Karka	Cancer	Jul-Aug	47	peace, relief from poverty

(KKh, The *Kashi Khanda* of the *Skanda Purana*. Period : 14 – 14, 14th of month *a* to 14th of month *b*)

Shiva, *i.e.* 'Lord of the centre of cosmos') lies inside the triangle close to the perpendicular bisector of its longest side. Madhyameshvara lies 45 metres north of the precise midpoint of the 2452 metres north-south side of the triangle. The location of this historically important temple has been unique in the spatial pattern of Sun shrines; as in earlier times it was surrounded and protected by the triangle of sun temples. Presently, however, this site is not commonly visited by pilgrims.

The Festivities, Cultural Astronomy and Lifeworld

The Sun symbolized the divine form of 'fire' in the sky, described in Vedic mythology as the chief controller of the cycle of cosmic rhythm. The Sun also symbolized the male (awakened, *jagrata*) and water the female (sleep, *sushupta*) modes of existence. That is how the special time of movement of Sun from one zodiac to another marks an auspicious time for bathing—the most commonly performed ritual in Varanasi. This movement is known as *samkranti*, named after the movement of the Sun from one zodiac to another; there being 12 such instances.[5] The meeting of male and female modes of existence at this time generates a special energy purifying the humans. This represents the meeting of macrocosmos (Sun), mesocosmos (water and earth) and microcosmos (the human beings). The efficacy of the water spots for bathing is often eulogized to be equal to, or many times greater than one of the standard Vedic sacrifices. According to Puranic mythology, sacred bathing on the occasion of *Samkranti* gives all the religious merit and removes all types of sins. Sacred bathing on the occasions of solstices, equinoxes, eclipses and on full-moon / new-moon days, followed by visitation to sun shrines is highly acredited. This is still a common practice in Varanasi.

According to Puranic sources worship of the Sun god thrice in *Magha* is equivalent to six months of worship. And, worship of the Sun after a holy bath in the Ganga on all the Sundays throughout the year provides all sort of religious merit. These mythical eulogies are accepted as traditions and still followed by many worshippers. Further, if the solar eclipse falls on Sunday there exists a special astrological and auspicious moment called *chudamani yoga* during which a sacred bath in the Ganga followed by worship of the Sun images are prescribed as most merit-giving and mystical. The Sun images of Varanasi and their association with festive offerings and the body symbolism are given in Table 15.4.

One of the aims of cultural astronomy is to interpret the relationships between a society and the perceived terrestrial environment. The meanings imposed and the traditional practices in this context are situtation-dependent and may vary from one culture to another and even within the sub-groups of

Table 15.4 : Varanasi : Sun Shrines and their Festive and Symbolic Association

Sl. No.	*Aditya/ Sun*	*Corresponding Puranic name* 1	*Water ritual with* 2	*Offering to Sun* 3	*Primordial Man's body Symbol* 4	*Associated Planet* 5
1.	Lorarka	Yama	pumpkin	golden lion	stomach	Sun/Saturn
2.	Vimala	Hiranyareta	raw sugar	grains	waist	Ketu
3.	Karna	Divakara	banana	cloths	hips	venus
4.	Vriddha	Mitra	coconut	house	sexual glands	Mars
5.	Draupada	Mitra	coconut	house	sexual glands	Mars
6.	Ganga	Vishnu	lemon	cloths & vehicle	thighs	Jupiter
7.	Samba	Ravi	nutmeg	cloths & food	heart	Mercury
8.	Yama	Varuna	sesame	firewood	knees	Rahu
9.	Mayukha	Surya	Citron	cow & grass	claves	Saturn
10.	Aruna	Bhanu	pomgranate	buds & flowers	feet	Jupiter
11.	Khakhola	Tapana	grapes	ram	head	Mars
12.	Keshava	Tapana	grapes	ram	head	Mars
13.	Sumantva	Indra	mango	cow	face	Venus
14.	Uttararka	Gabhasti	cucumber	ghee/clarified butter	chest	Moon

Source : 1 and 2—*Bhavisya Purana* a 10th century CE text, *Varaha Purana*; 3—Kamalakara Bhatta's *Nirnayasindu*, an early 17th century CE text, 4—Varahamihira's *Brihatsamhita*, a 6th century CE Vedic text, *Taittariya Brahmana*; 5—Astrological traditions in Varanasi.

the same culture. In cultural astronomy the system of this relationship is studied within the sphere of this cultural system as a whole.[6] Phenomenologically the symbolic organization of the astrological counselling process and its interpretation in Hindu lifeworld can be investigated. In Hindu society it is commonly believed that the planetary changes in the context of cosmic rhythm of time have direct influence on the human sphere, especially with reference to sickness, accidents and good or bad happenings in one's life. Of course one can say that these phenomena are more the subject of psychological mechanisms. Since the Sun is recognised as the Lord of all the Planets, its role in astrological counselling becomes important in the Hindu psyche.

Based on detailed investigation of astrological counselling in Varanasi, according to Pugh, 'Astrology schematises four mutually coincident aspects of the person: a physical, or body (*sharirik*) aspect, a psychological (*mansik*) aspect, and two social aspects of familial (*parivarik*) and community (*samajik*) relationship'.[7]

According to astrological correspondences, all the Sun images of Varanasi are associated with respective zodiacs and further with the associative planets and parts of the body. These relationships are also linked with the tissues of the body which constitute the person's physiological processes. Astrologers while interpreting the horoscopes explain the role and impact of various planets with respect to zodiacs and the Sun. This is a complicated process of narration and acceptance, and more a subject of the complex system of faith healing. In specific cases with the help of an expert priest, people perform varieties of rituals for relief from the bad effects of the Sun and associated planets. However, more commonly people take a holy dip in the Ganga river and visit the Sun images and worship them while reciting the holiest *Gayatri mantra*:'Let us obtain adorable splendor of the Sun; may He arouse our minds'.

Detailed ritual processes now remain only as a part of past mythology, although the roots and faith continues, only the degree varies. For experiencing and understanding the cosmic spirit we need to comprehend the psychological ecology underlying physical ecology by performing some of the rituals on specific occasions such as solstices, equinoxes and solar eclipse. This would help to promote sustainable-reverential development.

The Surya Yantra

Similar to other forms of *Yantra*, representing the complex network of interconnectedness and wholeness, the Surya Yantra also summarizes at a glance the whole sense of the Hindu world of myth and symbol. The structural elements in the design of Surya Yantra includes : (i) square outer frame, composed of two straight lines turned according to a regular pattern where exist 16 spots associated with eight directional deities and the associated planets; (ii) a series of 8 stylized lotus petals, and (iii) a hexagram design of creative energy whose root lies at the centre. The two latter parts are associated with 12 zodiacs, 2 eclipses (solar and lunar), 14 images of Sun (in Varanasi) and finally the root symbol of primordial seed (*bija*)—a point at the centre expressed as *Om*. By this mechanism there appear 45 spots. According to Puranic sources the Sun God gives highest merit if worshipped with the support of *Surya Yantra* (Fig. 15.3). This *Yantra* also refers to the asymmetrical proportional ratio of 7 : 5 : 3 from where any straight-side the total comes to 15, and thus finally 45 (Fig.15.3). Both of these mystic designs (cosmograms) show an image of the harmony of the universe.

Sun Images : The Iconographic View and Historical Context

Verifiying the locations, iconography and some popular mytho-logical tales with inherent meaning and context on a scientific scale the description of the Adityas as contained in the *Kashi Khanda* has been reviewed. Their references and merits as given in the *Kashi Khanda*, are presented in Table 15.3.

Lolarka : The 'agitated' sun. This image marks the southern most limit of the city, and provides liberation from the dangers and curses of Yama ('Lord of death'). This site is considered to be the oldest among the holy spots in Varanasi, and is referred to in an 8th century BCE text, the *Shatapatha Brahmana*. A tiny Sun disk lies at the inner steps close to the inner sanctum pond (*kunda*). A *Gahadavala* inscription of 1151 CE refers that queen Goshala Devi performed rituals and gave donations at this site. Lolarka represents the fertility aspect of the Sun, demonstrated by the festival of *Lolarka Shasthi* which occurs on the sixth day of the waxing moon of *Bhadrapada* (Aug-Sept).

Vimala : Vimala's sun. According to *Puranic* mythologies a Brahmin named Vimala installed the Sun disc on a raised platform of marble for relief from leprosy and mental tension. This disc is still known for possessing that power. Presently the Sun disc is located in a congested room owned by a milkman (*gvala*). The forty lotus petals around the Sun face represent the rays of the Sun.(Fig. 15.4)

Karna : Karna's sun. Karna, the son of Surya and Kunti, one of the major figures of the Mahabharata, was distinguished by unlimited generosity. This image symbolises that quality of the Sun which provides without surcease the energy needed by life on the earth. The original shrine had already been destroyed at the time of writing of the *Kashi Khanda* in the 13th century; but the site was there, and visitors to the site were advised that they could still obtain merit. The present icon showing the Sun riding on a chariot drawn by seven horses is probably a replica of the old one. Now in ruins it is set badly in a corner within the the compound of the Rama temple.

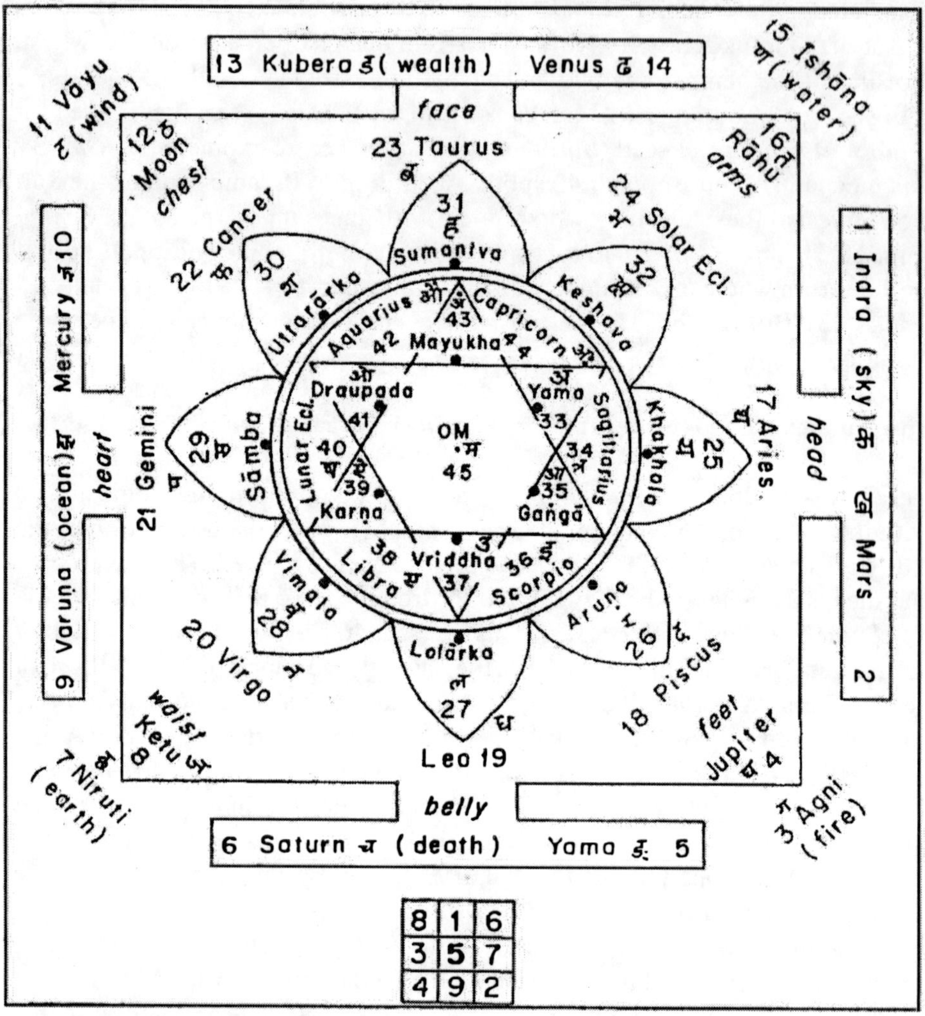

Fig. 15.3 : Vārāṇasī : The Surya Yantra

Vriddha : The Old Man's Sun. This image is associated with an old Brahmin, Harita, who by performing austerities overcame the sufferings of old age and death. Location of this shrine falls within the cyclic rhythm of the Sun from the autumnal equinox to the winter solstice, associated thus to the forthcoming shortest day of the year. This is symbolised as the 'old and tired' condition of the Sun. This image has two parts : a disc set into the wall close to the step, and a small statue in the nearby room set into the wall. The image shows a four-armed Sun sitting in a meditative pose.The disc has two layers of lotus petals around the Sun figure, and in both the number of lotus petals is twelve (Fig. 15.4).

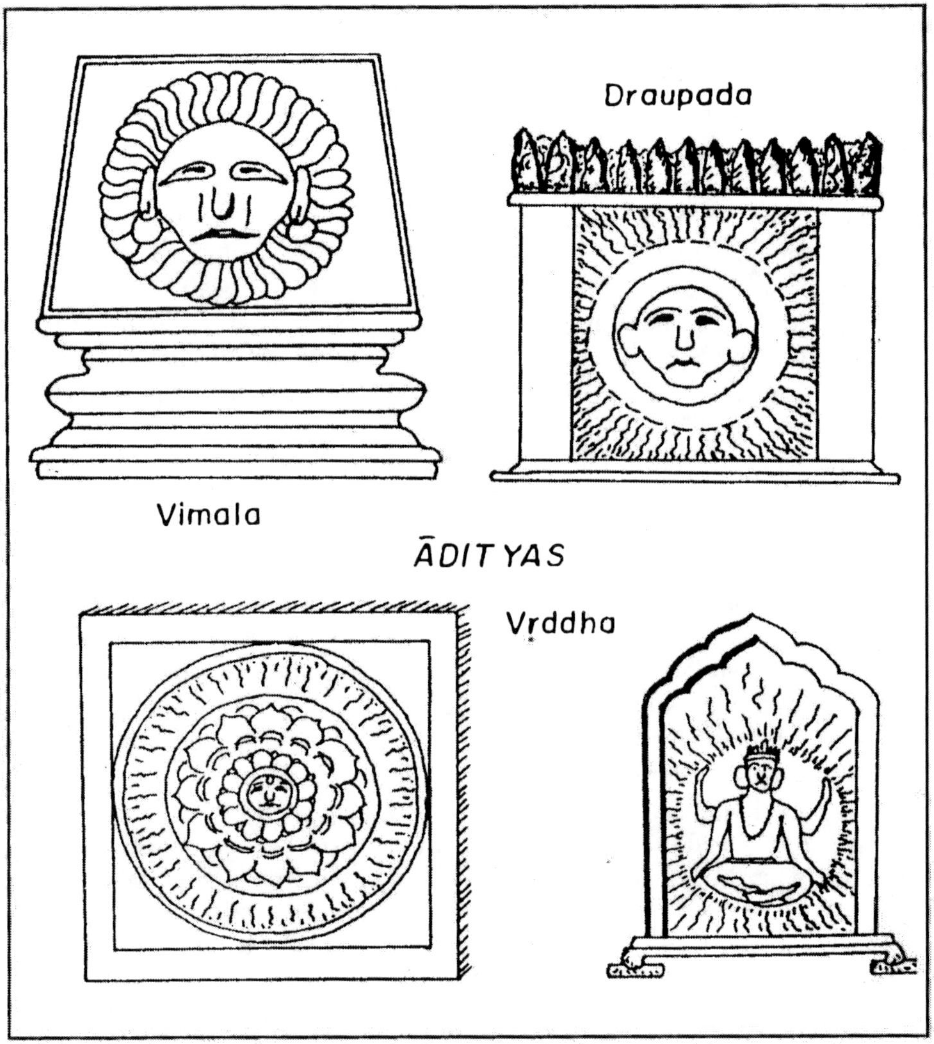

Fig. 15.4 : Ādityas

Draupada : Draupadi's Sun. During the period of the Pandava's exile Draupadi had difficulty finding food. Upon reaching Kashi, Draupadi performed special rituals and established a Sun image and as blessing the Sun God gave her a magical pot and a ladle which provided unlimited food. The Sun had promised that those who worshiped this image shall never go hungry. This image is carved on a stone slab with twelve elongated stone marks on its upper side. Many south Indian pilgrims venerate these as a form of the goddess (Fig. 15.4).

Ganga : Ganga's Sun. This image represents the merging power of the

Sun and water, and gives the merit of Ganga worship and purification. During 1960s the original statue was destroyed, and a replica made of marble was installed. Attached to it lies a small statue of Bhagiratha, a mythical king who brought the Ganga from the heaven to the earth.

Samba : Samba's Sun. This image was established by Krishna's son Samba to get relief from leprosy. The *Linga Purana*, a 5th century text, describes the site and its glory. Sherring[8] believes that there were 12 wells dedicated to the Sun at this site, but they have all been lost. The small temple having Samba Aditya consists of a large flat stone with a carving of four layers of 12 lotus petals, representing 12 months of a year, around the face of the image. This was constructed by Surjana Hada, the King of Rajasthan, around 1580 CE (Fig.15.5). Today, worshipping the Samba Aditya relieves one of all forms of skin diseases. In the vicinity there are some Sun discs too .

Yama : Yama's Sun. This site, lying between Lolarka and Uttararka combines the merit earned from worshipping one's ancestors in the south at Rameshvaram and worshipping at Manikaranika in the north. It is commonly believed that this image which shows a human-like figure with sign of rays on the head was installed by Yama himself. It lies on the steps to Sankatha Ghat and faces the south—the direction of Yama, 'Lord of Death'. (Fig. 15.6).

Mayukha : The Sun of Rays. According to *Puranic* description the Sun god fully disappeared while worshipping Shiva, therefore Shiva himself later manifested him in the form of Mayukha (rays). This indicates the narrative of a total solar eclipse and the solar carona which remained in the sky after the Sun disappeared from view. This happened on 10th May 1054. The eclipse occurred at the ascending node of the Moon. *i.e.* when the Moon was moving from south to north, and therefore, was associated with the demon Rahu. This happened before the writing of the *Puranic* story narrated in the *Kashi Khanda* and probabaly later on narrated in the mythology. A full solar eclipse again took place this century on 18th April 1931. The icon of Mayukha is contained in the Mangla Gauri temple and consists of a small disc set in a pillar containing the Sun's face surrounded by closed rays (Fig. 15.6).

Aruna : Aruna's Sun. Aruna was the son of Vinita and the eldest brother of Garuda. The image of Aruna refers to the first rays of the Sun, and also his charioteer driving seven horses across the sky. The number seven is an ubiquitous sacred number, symbolizing in this situation the seven realms of the heaven as well as the seven colours of light. The period of worshipping Aruna refers to the Hindu month of *Chaitra*, the first month in the cosmic rhythm. Aruna's image, lying in the wall at the lower side in the circumambulatory path of the Trilochana temple, consists of a 20 cm disc surrounded by rays and sitting on a lotus with seven petals. (Fig. 15.6).

Khakhola, or Vinita : Vinita's Sun. The *Puranic* story refers to Vinita,

Fig. 15.5 : Sambha Aditya

mother of Aruna, who for fear of the bad influence of the hot and disturbed Sun, stuttered' producing a sound like *khakhola* : *kha* meaning sky; and hence, the name. The fascinating and complex story associated with this Aditya contains hints of a number of astronomical phenomena that may have been witnessed by the inhabitants of the city. Between CE 1080 and 1275 the Sun had a major outbreak of sunspots that were visible to the unaided eye; at no time before or since then so many spots have been detected on the Sun by the naked eye.[9] This incidence may be described with mythological connotation in the *Kashi Khanda*. The present status of Khakhola, of course deserted, clearly shows the sky, sunspots, and their reflection in a tiny-model pond (Fig.15.6).

Keshava : Keshava / Vishnu's Sun. There are five Sun discs in and around Adi Keshava temple, each facing east across the Ganga river. The oldest disc is set in the wall showing the Sun's face surrounded by forty-eight rays in petal form, while other discs show only twenty-four rays. Khakhola and Keshava lie in the line linking Samba, and also are closest to the axis of the vernal equinox.

Sumantva : Sumantva's Sun. Manifested by the sage Sumantva after being cured of leprosy, this is a well-preseverd and sharp image, approximately 40 cm across, of *Surya* holding a sword in the right hand and a lotus in his left

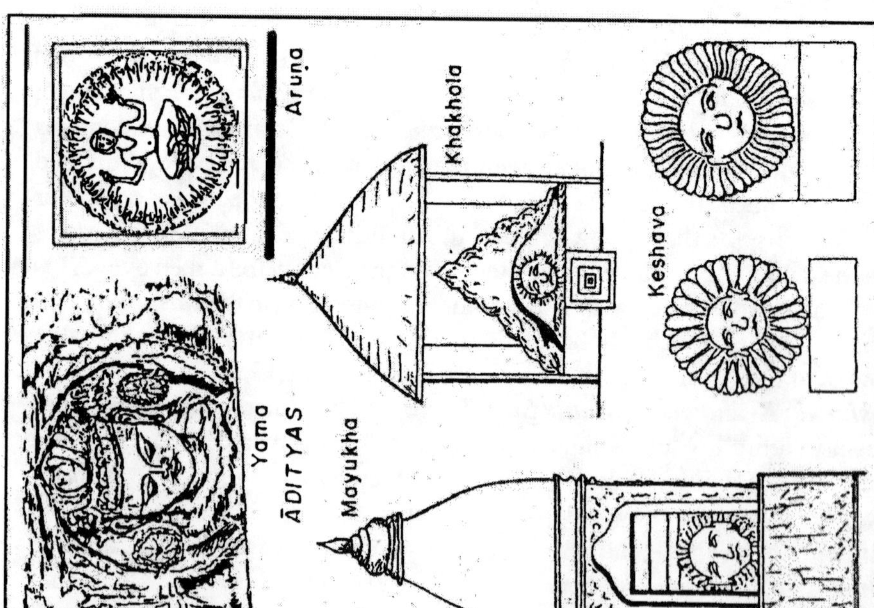

Fig. 15.6 : Adityas

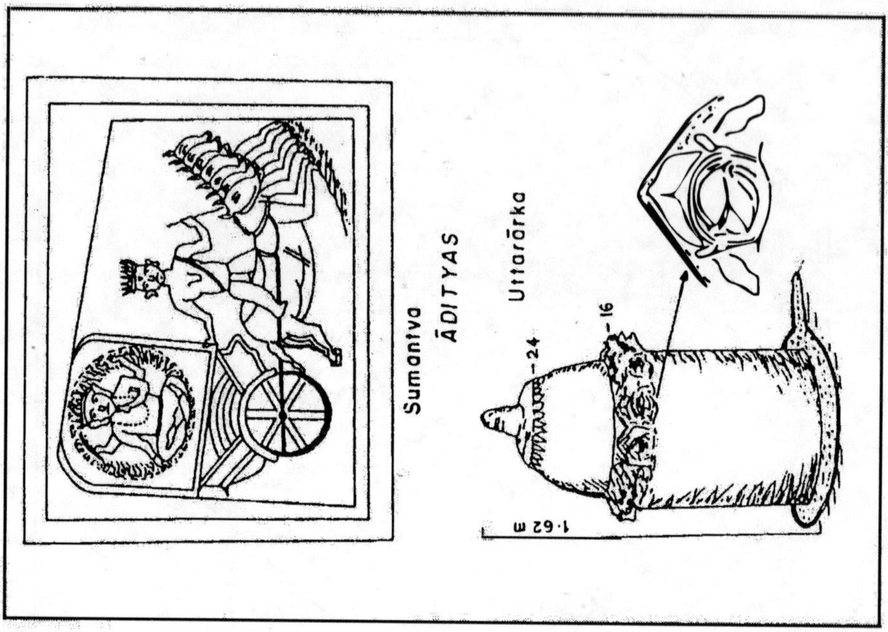

Fig. 15.7 : Sumantva, Adiryas, Uttararka

sitting in his chariot in meditative pose, carried forward by Aruna and seven horses. The chariot wheel consists of 12 axes, symbolizing 12 months. His four arms refer to his control over four directions. This image is set in the wall on a big stone slab (Fig. 15.7).

Uttararka : The Northern Sun. Representing the northern most point of Varanasi, Uttararka protects Kashi from the northern side by the Sun's strongest rays, referring to the summer solstice. Although the original temple and structure were demolished during Mughal rule, however the main image is presently preserved in a well-protected compound. The image consists of a 1.62 metre tall *yupa* or *lingam* like form with three parts the base, now modified in the form of *yoni* (vulva), a series of 16 double lotus buds 2/3 up its side, and a series of 24 petals near its top (Fig 15.7). Sherring's statement that it is a part of a Buddhist pillar is not acceptable;[10] however his description of the iconographic feature is praiseworthy. Nearby, the once famous sacred pond, *Arka Kunda* (now called Bakaria Kund) is now almost completely lost to encroachments and is filled with garbage and waste.

Conclusion

The complex network and structure of the spatial pattern of Sun shrines and their association with the movement of the Sun throw light on the cosmological sense of 'city planning' in the ancient period. Probably this pattern was older than the Brahmanical tradition, although in time it has been superseded by the Brahmanical (in fact Shaiva) tradition. Eliade[11] characterizes the sky deities (specially the Sun) as typically aloof and distant, possessing little involvement with daily human concerns. In Varanasi, however, the Sun is directly involved in the making of cosmic ordering, and in the life of the city, daily as the dawn Sun, but also traditionally as an agent for Lord Shiva—the patron deity of this city, and also as one among the 'five ancillary divinities' (*panchadevas*) worshipped together. These solar attributes must have emerged from local folk tradition and further elaborated in mythology and epic literature. Cosmological order and cityscape of Varanasi can further be explained with the study of spatial patterning of other deities and series of sacred journeys.[12]

References and Notes

1. Singh, Rana P. B. (1987). 'The Pilgrimage Mandala of Varanasi (Kashi): A Study in Sacred Geography', *National Geographical Journal of India*, 33 (4), pp. 493-524; Singh, Rana P.B. (1994). 'Varanasi : Cosmic Order and Cityscape, Sun Images and Shrines', *Architecture+Design*, XI (6), Nov-Dec. pp. 75-79.
2. Eliade, Mircea (1958). *Patterns in Comparative Religion*, London: Sheed and Ward, London.
3. *See* ref. 2, p. 144.

4. Mishra, V.B. (1973). *Religious Beliefs and Practices of North India During the Early Medieval Period*, Leiden: E.J. Brill.
5. Singh, Rana P.B. (1994) 'Water Symbolism and Sacred Landscape in Hinduism : A Study of Benares (Varanasi)', *Erdkunde,* Bonn, Germany, Band 48, No. 3, September, pp. 210-227.
6. Ruggles, C.L.N. and Saunders, N.J. (eds.) (1993). *Astronomies and Cultures*, University Press of Colorado, Niwot, CO.
7. Pugh, Judy F. (1983). 'Astrological Counselling in Contemporary India' *Culture, Medicine and Psychiatry*, 7, pp. 1-21.1984. Pugh Judy F. (1984). 'Concepts of Person and Situation in North Indian Counselling', *Contributions to Asian Studies*, 18, pp. 85-105.
8. Sherring, M.A. (1868). *Benares : The Sacred City of the Hindus*, Delhi: Cheap Publications, reprinted in 1990, p. 127.
9. Malville, John M. (1985). 'Sun Worship in Contemporary India', *Man in India*, 63 (3), pp. 207-233.
10. *See* ref. 8, p. 281.
11. *See* ref. 2, pp. 124-125.
12. Singh, Rana P.B. (ed.) (1993). *Banaras (Varanasi) : Cosmic Order, Sacred City, Hindu Traditions*, Varanasi: Tara Book Agency; Singh, Rana P.B. (1994). 'The Sacred Geometry of India's Holy City, Varanasi : Kashi as Cosmogram', *The Spirit and Power of Place*, Rana P.B. Singh, (ed.) National Geographical Society of India, Pub. 41; also in *National Geographical Journal of India*, 40, pp. 189-216.

16

HINDU COSMOLOGY AND THE ORIENTATION AND SEGREGATION OF SOCIAL GROUPS IN VILLAGES IN NORTHWESTERN INDIA

JAI PAL SINGH AND MUMTAZ KHAN

Introduction

This essay presents a cultural analysis of the orientation and segregation of social groups in villages in a part of the norhtwestern plains of India (Fig. 16.1). It pertains to Cultural Geography and in particular to Settlement Geography. It specifically interprets the cultụrally created landscape in terms of culture theory and explores the social order in the built environment in terms of a cultural schema.

The phenomena of directional orientation and segregation of social groups in villages and towns in the Indic culture relam are very old. The prehistoric city of Mohenjodaro oriented its streets and rectangular blocks towards the points of the compass.[1] Orientation and segregation appear to have been based on religious beliefs, though excavations in Mohenjodaro could not link it to any temple. But Kosambi asserts that 'the 'citadel' mound undoubtedly corresponds to the temple-Zikkurat structures in Mesopotamia' and the '... 'Great Bath' at Mohenjodaro... must have been a ritual tank'.[2] But later Sanskrit texts on town planning make it abundantly clear that the orientation of streets and segregation of caste groups in settlements were ideally linked to the eight gods (*dik-pālas*) of the cardinal and intermediate points.[3]

Hindu religion ordains a caste hierarchy in society. It also recognises socially high and low segments in the occupied space of a settlement. The occupied space, so to say, is hierarchically divided into a number of quarters. The position of each quarter in the hierarchical order depends on the status of the guardian deity controlling the quarter. From a ritual perspective the space of a village thus slopes down from the high quarter towards the low quarter. This ideal gradient may well be termed a social gradient of the settlement. To avoid confusion, it is necessary to point out that the social gradient is independent of

Note : Reprinted from *Geografiska Annaler*, Vol. 81, Ser. B, 1999.

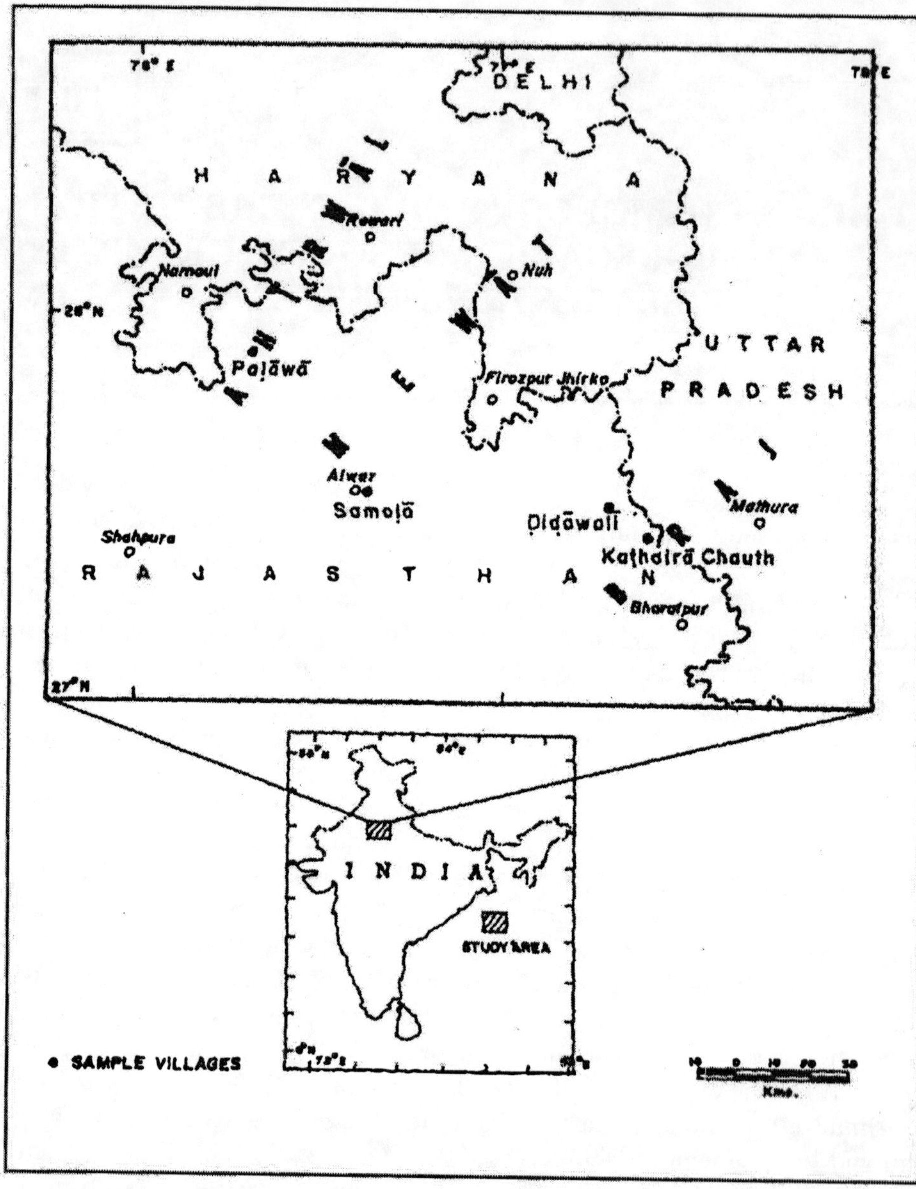

Fig. 16.1 : Location of Sample Villages

the actual physical gradient of the site.[4] It is therefore reasonable to suppose that town planners in ancient India tried to harmonize the caste hierarchy with the hierarchical order of the different quarters of the site, consecrated for settlement.

There is a definite socio-religious order in the rural landscape in India. The *mohallās* (wards, also called *paṭṭī, pānā, wānā, tolā, bās,* etc. depending on the region) of various social groups in a village are not only segregated, but their segregation has a definite directional orientation. A number of social scientists have reported the segregation of castes in Indian settlements[5] but none has treated exhaustively the directional orientation of these caste *mohallās*.

The orientation is also manifested in a number of other features of the landscape. Folk houses do not face southwards. The various types of haystacks (*bāgaṅa or paila*) and dungcake-stacks (*biṭoṅā*) have their long axis east-west and their short axis north-south. Parenthetically, it may be pointed out that dungcakes are used for cremation along with wooden logs. The funeral pyre, on the other hand, has a longer north-south axis and a shorter east-west axis. A dungcake-stack will never have a longer north-south axis because it might then look like a funeral pyre and invite bad luck to the family. Ideally, fields of *kharīf* (wet season) crops are sown in furrows running east-west, while furrows for *rabī* (dry season) crops run north-south. However, *sarson* (mustard) linings of fields of *rabī* foodcrops run east-west. Thus orientation is writ large on the Indian rural landscape and brings order to it. Such eminent geographers as Spencer and Spate[6] missed out the religiously ordained order in the Indian landscape, though they do point out the segregation of social groups in Indian villages.

Indian civilization is ancient and has had a continuous existence for nearly five thousand years. Its cultural traditions are deeply rooted in history.[7] Therefore, settlements in India cannot be comprehended without an understanding of cultural traditions, including that of town planning and localized folk traditions. Villagers follow these traditions in allocating house sites to various caste groups at the time of founding a village or a town.

In line with the interpretive methodologies of Cultural Geography[8] the paper is grounded in cultural theory to interpret the structure of a built environment. The Hindu cosmological schema is a language that communicates the social order of the built environment. It provides a divine model for structuring settlements. In India, traditional environments were built in the image of the cosmos. *Silpaśāstras* (texts giving rules for architecture) invariably supplanted cosmological schema on the consecrated site for building environments. This study interprets built environments in terms of Hindu cosmological schema. It relates the location of caste *mohallās* in built environments with the segmented space of the cosmology. Specifically, it attempts to find out how far the traditional built environments conform to the cosmological schema. Therefore, study of cosmological shcemata is essential for understanding the morphology of villages.

This essay is broadly organised into two sections. The first deals with the Hindu concepts that have some bearing on the layout and structure of built environments and discusses the hierarchical composition of Hindu society, the oriented space of divine cosmology and the integration of social order in the spatial order. Also discussed briefly are the Mohammad Habib-Buddha Prakash debate and the theory of twin settlements, which are relevant to our central theme. The second part analyses regional particularities and ecological influences on village morphology. It presents case studies of four villages to adduce empirical evidence about the impress of cosmology on built environments.

Composition of Society

Hindu society is a complex whole, encompassing a great variety of social groups, both ethnic and occupational.[9] Its structure is not so simple as to be explained by pairs of opposites as some structural sociologists have tried to do. Indian civlization has a unique way of integrating different social groups into its fold. It integrates them into one whole by bringing them under the all-encompassing umbrella of Hindu culture and religion; yet it allows each group to retain its identity. Caste endogamy and hierarchy and the differentiated social customs and traditions of individual castes sustain and reinforce these diverse identities. Society is thus loosely integrated. Superficially, it is a single whole; but one need not probe deeply to recognise its division into a large number of social groups (especially castes). This division is as old as the coming of Āryans to this land, probably older.[10] The Āryan division of society into Āryans and Dāsas or Dasyūs produced a fundamental dichotomy in the Indian social order. This dichotomy subsequently took the form of *Āryas* and *Śūdras* and finally converted into 'Caste Hindus' and untouchables, or, as the latter are presently designated, *Harijans* (in popular usage) or scheduled castes (in official parlance).

The *Puruṣa-Sūkta* (Hymn of Man) of the *Ṛg Veda* (the most ancient of all Indian scriptures) definitely gave a broad hierarchical order to various *varnas* by tracing their origin to the 'horizontally and hierarchically' divided body of primeval man. 'The hierarchical division of the body serves [as] a suitable metaphor for the hierarchical division of society'.[11] In the post-Vedic period, the lawgivers fitted the various occupational and ethnic groups then extant into the hierarchical continuum of the four *varnas*. The system was favourable to high castes and unfavourable to low. Yet it contributed to unity and harmony in the hierarchically divided society by assigning certain roles to each caste and linking castes with one another by the all-pervasive *jajmānī* system.[12] Thus it is a 'coherent social system based on the principle of inequality'.[13]

The role of ideology of caste hierarchy is central to any study of segregation of castes in a Hindu settlement. The ideology mediates the relationship between social groups fundamentally unequal in terms of power. In other words, it negotiates the relations of dominance and subordination in built environments.

Hindu ideology of caste hierarchy claimed to represent the interest of upper castes as the interest of the whole society. The hegemonic influence of upper castes was great enough to persuade low castes to accept the ideology of caste hierarchy as *ṛta* (moral law). Thus caste hierarchy became the ruling ideology accepted by all the sections of society. Hegemonic upper castes persuaded low castes to occupy 'low' segments of social space in built environments. Low castes were convinced that it was their lot to live in a world where they were asked to live in a separate, peripheral *mohallā*. The complex social relations between upper and lower castes are embedded in the built environment. They transform it into a mosaic of social areas with sub-cultural identities associated with low castes.

The caste hierarchy is pan-Indian, but the placement of various castes on the continuum of caste hierarchy differs from one region to another. In the plains of northwestern India, where most peasants own their land, the caste hierarchy is different from that of other regions. Here the Brāhmans as usual are at the top; then come Rājpūts and Baṇiyās on the second rung of the ladder. In the middle are placed Jāts, Ahīrs, Gūjars, Meos and others, who, but for the Meos, are widely distributed peasant proprietors. These are the dominant landowning castes.[14] The status of a peasant caste is determined by the ownership of land. Landowners are placed high on the hierarchy as compared to the landless. Then come the clean service castes such as *Khātī* (carpenter), *Luhār* (blacksmith), *Kumhār* (potter), *Nāi* (barber) and others. On the lowest rungs are placed *Dhānak* (pig keeper), *Julāhā* (weaver), *Chamār* (leather worker), *Bhangi* (scavenger) and other groups regarded as untouchables.

In the religious sphere the Brāhmaṇs occupy the ritually highest position. In the temporal world, however, the caste which owns the most land dominates in the daily affairs of village society.[15] The *jajmāns* (drawn mainly from the dominant land-owning castes) used to make payments in kind to the various service castes, including the Brāhmaṇs, providing fixed customary shares of their produce for services rendered under the *jajmānī* system prior to its gradual erosion in the post-independence period. Therefore, the welfare of all other castes depended on the landowning caste. At the time of founding of a village, the dominant caste allotted to itself the ritually highest site for house building. The other castes who were invited to settle were assigned house sites along the social gradient of the *kheṅā* (settlement site) in conformity with their status on the caste hierarchy.

Indian villages are microcosms of Hindu society[16] whose hierarchical divisions are reflected in the hierarchically differentiated segments of occupied village space. It is therefore relevant to discuss the concept of space and place and the tradition of settlement planning in ancient India.

The Concept of Space and Place in Hindu Society

Dating back to *Ṛg Veda*, there is a very old tradition of 'adoration of the deities of *vāstu* (site) and *dik* (quarters)'[17] in India. It was probably during the post-Vedic period, when Hinduism (weakened for a time by the advance of Buddhism and other new religions reasserted itself, that many of the Āryan gods who occupied high positions during the Vedic age 'were relegated to the minor position of the guardians of quarters (*dikpálas*)'.[18] Assigning deities to *diks* enhanced the significance of cardinal and intermediate directions and offered new vistas for ancient Indian thinkers to conceptualise about cosmology.

Hinduism has its own way of transforming space into place. The ritual of site and *dik* worship consecrates a place and makes it fit for human habitation, thereby transforming space into place.[19]

Ancient Hindus conceptualised space as bounded,[20] so as to provide security to man and relieve him of agoraphobia. The concept of boundedness finds its expression in a host of spatial features. A house, a temple, a fort, a village or a town are all bounded, so much so that, even if their physical limits are not evident, they are conceived to be bounded ritually.[21]

To ancient Hindus space took some geometric form. In the *Vāstusāstras* it has been recognized that 'the form of the *Vāstupuruśamaṇḍala* is square which is an essential form and which can be converted into a triangle, a hexagon, an octagon and a circle of equal area retaining its symbolism'.[22] Thus the site plan of a house, a temple, a fort, a village or a town may be shaped either as a square or as any other geometric variant of it.

Hindu Cosmology

Hindu cosmology is a physical schema for building environments. It provides a conceptual framework for planners and villagers for the layout of settlements.[23] Moreover, it is imbued with sacred meaning which communicates the hierarchical social order of the built environment. Hence, the use of Hindu cosmology as a schema for interpreting built environments is in order for the task at hand. Hindus have drawn a schema of oriented place defined by cardinal and intermediate points in order to provide orientation and security to man in the world.[24] Though Man is not represented at the centre of the cosmology, he is in a sense assumed to be there, because it may be seen as designed to provide protection for him. It also helps to produce social and religious harmony in a multicaste society in a place and a sense of harmony with the cosmos,[25] 'to energize each other'.[26] Hindus ritually classify the consecrated places into various components and assign values to them. The eight points of the compass have guardian deities, *dik-pālas*, who not only provide protection to the quarters they rule but also convey specific values and properties to them.[27] To reinforce the security and sanctity of oriented places there are eight *dik-gajas* (elephant guardians of quarters) corresponding to the eight *dik-pālas* (Fig. 16.2).

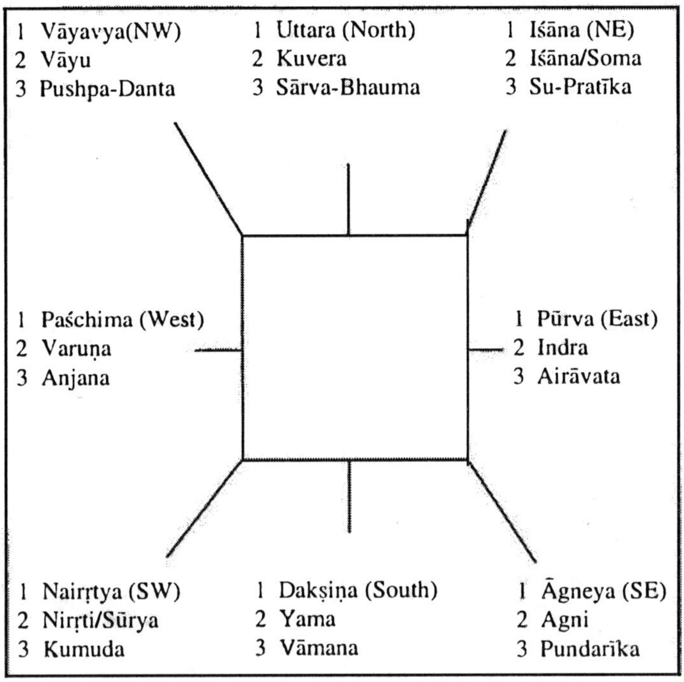

Index : 1 Direction; 2 Dik-Pāla; 3 Dik-Gaja

Fig. 16.2 : Hindu Cosmology

The eight guardian deities are : Indra, god of atmosphere and bestower of rain, reigning in the east; Agni, god of fire, guardian of the southeast quarter; Yama, god of death, ruler of the south; Nirṛti, goddess of decay and goblins, guardian of the south-west; Varuna, god of the seas, rivers and night, protector of the west; Vāyu, god of wind, the *dik-pāla* of the northwest; Kuvera, god of wealth, reigning over the north; and Isāna, the upward-looking face of lord Śiva, ruler of the northeast. Some texts substitute Sūrya (the Sun) for Nirṛti and Soma, the ritual elixir or the Moon, the lord of herbs, for Iśāna.

The Brāhmaṇic cosmology in pan-Indian. But to derive a naturalistic explanation of the Brāhmaṇic cosmology, it will be useful to speculate about the locale within India where the gods of the Āryan pantheon were originally established as the guardians of the eight quarters. In our view the locale where the cosmology was given its theocratic character might have been somewhere on the plains of northwestern India.[28] Our explanation follows.

To begin with, assigning the eastern quarter to Indra is significant in that he brings rainfall from the east and fights *Vritra*, the demon of drought, with his thunderbolt. In fact, easterly monsoon winds bring rainfall to the drought-prone north-western region of India. The assumed northwestern genesis of the cosmology would also mean that in the Sun's daily march it will always remain to the south of the zenith. Consequently the three southern quarters will be hot and are appropriately alloted to Agni (fire), Yama (son of the Sun) and in some formulations, Śurya (Sun).. Sunset symbolises the fading of glory. The south-western quarter is therefore alternatively allotted to Nirṛti, the godess of decay. Moreover, the southern frontier of the first Āryan homeland in India impinged upon a region known as *Marusthalī* (desert, etymologically the land of death), a hot, arid and inhospitable land.

In their advance towards eastern India, the ancient Āryans left behind them the *Sapta-Sindhava* (land of seven rivers). Possibly for this reason Varuna was alloted the western quarter. The most prevalent wind in this part of India is the northwesterly. This fact helps to explain the Āryan belief that Vāyu (wind) reigns over the northwestern quarter.

The northern plains of India are rimmed by the Himalayas, which the Āryans regarded as the *ratna-garbhā* (treasure house (womb) of gold and jewels). This would explain the allocation of this quarter to Kuvera, the god of riches. Allotment of the northeastern quarter to Soma or Isāna may possibly be due to the location of the ecological niche of the *Soma* plant (*Asclepias acida*), so important in ancient Vedic rituals, or the situation of Mount Kailāsa, the seat of Siva (Isāna), to the northeast of the postulated origin locale.

It is pertinent to our enquiry to seek cultural explanations of Hindu cosmology in addition to the physical arguments adduced above. The auspiciousness of the northeastern quarter, for example, relates to its assignment to Isāna, the upward-looking face of the god Siva. The upward-looking 'face of Isāna can be thought to somehow involve movement between heaven and earth'.[29] For the same reason, village temples are ideally located in the northeast quarter. Opposite, the southwest, is the Nirṛti's quarter of decay.

The east-west axis in the cosmology is very important. It is the path traversed by celestial bodies in their daily march. Two prominent Āryan deities define this axis: Indra on the east and Varuna on the west. Varuna, a deity of high moral stature and upholder of the law, is placed in opposition to the libertine Indra. Furthermore, the despised Mlecchas were in the east, while Āryans came from the west. Probably for this reason, the west is thought to be high and the east low.

The regent of the north, Kuvera, the god of wealth, is placed in opposition to Yama, the guardian of south and god of death. Descending bracing *Vāyu* (winds) from the mountainous northwest are placed in oppostion to hot, upward-moving *Agni* (fire) of the southeast. The former belongs to the sky and the latter to the earth. These might lend weight to Moore's suggestion that there is

' a north-south cycling of human lives'.[30] Therefore, the major axis of transport within a settlement runs east-west along the path of celestial bodies, while the minor axis would run north-south along the path that human beings travel in their life cycles. This sort of axiality is very well developed in the planned walled city of Jaipur and partially in the nearby town of Alwar. The pattern is also a common characteristic of the regularly laid out Brāhmaṇic villages in the east coastal plains from Orissa to Tamilnadu[31] and in the grid pattern temple towns of South India.[32] Most villages, however, for want of any central controlling authority, usually do not have axiality of streets; nor do most cities.

To sum up, the northern half of the cosmic schema is auspicious, ruled as it is by benevolent gods. The western quarter is the best quarter. The southern half is relatively inauspicious because it is governed by the gods of death, decay and hot *Agni*.

The Integration of the Social Order into the Spatial Order

Settlements are created by the conscious efforts of humans. They are thought before they are built.[33] Hindu cosmology provided the fundamental guiding principles for village and town planning in ancient India. Town planners as well as ordinary folk followed rules for the orientation and segregation of various castes within a settlement according to the above-mentioned qualities of guardian deities. These rules still function for integrating the fragmented social order into a segmented spatial order. The values and properties of the various sectors of a *kheṅā* can be comprehended in accordance with the ruling deity of each sector. The elaborate schema of place defined by cardinal points and protected by guardian deities was conceptualised to harmonize the social with the cosmic order, assigning values and properties to various quarters in accordance with their guardian deities and allotting house sites for various castes consonant with their hierarchical status. Thus the southern sector of a settlement, governed by Yama, the god of death, will contain the houses of *Chaṇḍālas* (a name applied to untouchables) and the cremation ground, some distance away from the main settlement. The south-east, governed by Agni, the god of fire, is regarded as hot and should properly be allotted to blacksmiths. East, belonging to Indra, the king of heaven and god of war, is where military commanders should reside. The northeast, the domain of Isāna, the divine physician, or to Soma, lord of herbs and plants, should be allotted to vegetable, fruit and flower-sellers. The north, identified with Kuvera, the god of wealth, provides the fittest location for the habitations of the wealthy. The northwest, belonging to Vāyu, the god of wind, is the most suitable locale for the houses of weavers and wine-distillers. The west belongs to Varuna, originally king of gods and men and moral governor; therefore, the palace of the king should be located in this quarter. The southwest belonging to Nirṛti, the goddess of decay and degradation, is regarded as fit for the *āsramas* (temporary lodges) of mendicants, particularly *sanyāsīns* (world renouncers).

The planning schemata in the *Silpāsastras* and *Purāṇas* are, however, sometimes contradictory. This may be due to the fact that they were written during different times at different places and hence accommodated on the cosmologic grid the specific hierarchies of castes peculiar to those periods and regions. But one thing common to all of them is that they allot house sites to various professions and castes according to the deities of the cardinal points or *padavinyāsa* and thus have directional orientation.[34]

Segregation of Castes

The texts of ancient town planners prescribed segregation of various castes into different caste *mohallās* for the cohesive and harmonious social life of each group. Each caste had, and still has, its own way of life and social traditions. In observing social rituals, especially those connected with the life cycle, the members of a given caste interact to a great extent among themselves. Each caste has its own gods or revered ancestors, whom others may ignore. Therefore, the social space of interaction of the members of a caste is mainly confined to its own discrete caste *mohallā*.[35] Orientation of castes to the points of the compass, will, even in the absence of other pressures, itself lead to segregation. Thus segregation is an inherent characteristic of orientation. Caste segregation, especially the exclusion of untouchables from the main body of settlement, has been a topic of debate among historians. It is therefore in order to consider this debate.

The Mohammad Habib-Buddha Prakash Debate

Regrettably, Indian settlement geographers have paid little attention to the views of historians regarding the structure and composition of Indian settlements. Two prominent Indian historians, Mohammad Habib and Buddha Prakash, have debated whether traditional Indian rulers allowed all castes, low as well as high, to live within the four walls of a city.[36] Habib, while propounding the 'theory of urban revolution' during the Sultanate period (1206-1526), holds that 'The so called Ghorian (Ghurid) conquest of India (in the late twelfth century) was really a revolution of Indian city labour led by the Ghorian Turks'. He asserted that previously artisans, workers and craftsmen were not allowed to live within the outer walls of cities and towns. Inside the fortified cities were only the '*seths* (big traders), *banyas* (merchants), brokers, clerks, *jotshis* (astrologers), teachers of all kinds, *vaids* (medical practitioners) temple priests and all other non-fighting elements'.[37] Buddha Prakash, on the other hand, asserts that 'all the castes, classes and crafts were located inside the cities. Brāhmaṇas, Kṣatriyas, Vaiśyas, Śūdras and Mlecchas (impure groups) lived in different quarters and sectors of the cities. Artisans and craftsmen had their separate streets. Brewers, potters, fishermen, goldsmiths, ironsmiths and artisans, including weavers, carpenters, barbers, launderers and cobblers, carried on their trades in separate localities inside the cities'.[38]

Ignoring the Sanskritic sources for obtaining descriptions of pre-Ghurid Indian cities, Habib painted an erroneous picture of their social composition. Buddha Prakash, on the other hand, appears to assert that all castes were accommodated within the city walls. But such was not the social composition of the cities. Numerous scholars have pointed out that during the ancient and medieval periods only clean service castes were allowed to live along with caste Hindu within the cities, while a large number of unclean and low service castes, collectively termed as *bahyas,* (people living outside towns) lived outside the city walls.[39] Moreover, a number of scholars have also noted this pattern in Indian settlements even in the contemporary period.[40] To sum up, pre-Ghurid Indian cities had a dichotomous character.

The Theory of Twin Settlements

The division of Hindu society into 'caste Hindus' and untouchables (now the scheduled castes) on the principle of ritual purity/pollution is reflected in the twin character of rural settlements. Many students of Indian village communities have discussed the segregation of untouchables some distance away from the main settlement. Among geographers, K.N. Singh deserves credit for putting forward the concept of twin settlements.[41] According to this concept, the main part of such settlements is inhabited by caste Hindus and the subsidiary part by untouchables, the two generally being separated by an open space of varying width.

Singh propounds a religio-ritual explanation of the origin of twin settlements. His model postulates the need for ritual distance between the castes based on notions of purity and pollution.However, he believes that the model has certain limitations and adds to it the concept of secular-dominance. Landowning castes, he avers, acquire secular power and govern the settling and socio-economic patterning of village society.[42]

But K.N. Singh fails to recognise that hierarchy permeates the social space of a settlement. It has a diversity of both high and low components. Therefore, the allocation of house sites to various castes is done according to rituals that place castes harmoniously on the social gradient of the *kheṅā*, according to their position in the caste hierarchy. Twin settlements are thus the result of socio-religious, not secular forces. In fact, it is the religious mould that gives form and coherance to rural settlements in India.

Another shortcoming of Singh's model is that it does not take into consideration the orientation of caste *mohallās* in a village. In ignoring orientation, K.N. Singh failed to grasp the basic logic behind the patterning of social groups. The *mohallās* of scheduled castes are generally located to the south of the village cores. Yet the very maps of the two villages on the basis of which he seeks to substantitate his theory have Chamar *mohallās* in their southern sectors.

Regional Particularities

Before discussing regional particularities it may be pointed out that India is a vast country exhibiting marked regional variations in the social morphology of settlements. First of all, there is a broad difference between north and south Indian villages owing to the influence of *Nāgara* (North Indian) and *Dravina* (South Indian) traditions of town planning. Further, there are smaller regions where orientation of caste *mohallās* in settlements is quite different from that of the villages in our study area. For example, Sharma reported that height is one of the determining factors in the social morphology of Himalayan towns.[43] Lewandowski found that spatial arrangement of caste *mohallās* in the temple town of Madurai is related to the Minakshi temple;[44] Gowdra averred that ritual circles condition the morphology of a Mysore village.[45] Moreover, social morphology of villages in vast tribal areas in central and northeastern India is dependent on factors other than Hindu cosmology.

The regional focus of our study is southern Harayna, north-eastern Rajasthan and an adjoining part of western Uttar Pradesh, areas with which the authors have had a lifelong association and in which they have extensively travelled for the study of cultural landscape especially of the folk regions of Mewāt and Ahīrwāl. Schwartzberg has delineated Mewāt and its adjoining regions on the basis of folk perceptions.[46]

As noted above, in the plains of northwestern India, the western quarter of a settlement is considered the highest social sector. From it, social space is believed to slope down towards the east. This hierarchical conception is so pervasive and dominant in this region that it has led peasant castes to call west high and east low. The accordance of high social status to the western quarter of a settlement is attributable to a number of factors: (i) Varuna, the guardian deity of the western quarter, was originally considered as the king of gods and men, and the upholder of *ṛta*, the supreme deity to whom grandiose cosmic functions were ascribed. Despite his subsequent decline elsewhere, the folk of this part of India probably long continued to assign him a high position and therefore regarded the western sector to be the highest segment of the social space.[47] Although they have now forgotten Varuna's erstwhile role as the pre-eminent deity, the tradition of assigning the highest social value to the western quarter continues. (ii) In the *Gṛhya Sūtras* (domestic manuals), which laid down the rules for domestic rites and ceremonies, the western side is associated with the benevolent aspects of the ancestors, the material prosperity of the household and the fertility of women.[48] This might have influenced the folk to place the western side of a *kheṅā* on a higher social plane. (iii) Until recently, rural houses characteristically faced inward, towards the centre of a village. Therefore, the houses located in the western quarter will face eastwards. Eastward-facing houses are considered auspicious, 'for the direction of the rising Sun holds the greatest cosmological significance for the Hindus'.[49] (iv) The *Silpaśāstras* extol eastward declivities and condemn westward declivities in settlement sites.[50] (v) Some *Śāstras* prescribe that the ruler's palace be located in the western quarter of a settlement so that the ruler may see the rising Sun. This injunction might have influenced the dominant castes, who have a strong tendency to imitate royal behaviour and to house themselves likewise in the western quarter of a village. (vi) Last but not the least, the privileging of the west may be related to the Harappan tradition of housing the elite and the ruling aristocracy in the raised western citadel and the common people in the lower eastern *mohallā*.

Assigning low status to Indra's eastern quarter may seem rather anomalous. However, despite Indra's initial high place among the Āryan gods, he later fell from that position. The *Purāṇas* depict him as being a libertine. The pastoral people of Braj (a region mainly in western Uttar Pradesh, but overlapping into Rajasthan) used to worship Indra. But the belief that Lord Kṛśna persuaded the cowherds of Braj not to worship Indra[51] may be responsible for lowering the status of the eastern quarter.

Our earlier discussion of Hindu cosmology makes clear why the folk consider the southern sector, presided over by Yama, the god of death, to be the socially lowest segment of the settlement. The south is also considered inauspicious because Lankā, the kingdom of the mythological demon King Rāvaṇa, is located in that direction. Therefore, this sector is allotted to untouchables. There is normally some intervening distance between the main village and the southern sub-settlement of the untouchables, and folk houses in the main portion of the village usually do not face south.

Thus, the sectoral properties of village social space are complex. People conceive not only a gradient from west to east but are also convinced that the south is ritually the lowest. In the ritually ordained twin character of a village, the main part is occupied by caste Hindus and a small subsidiary part by untouchables, the occupants of Yama's quarter. If this part of the village is treated as a separate sub-settlement, then the main part of the village has a general high western to low eastern gradient. The west is ideally occupied by the landowning (dominant) castes. To their east are their latecoming relatives and further east the clean service castes according to their position on the caste hierarchy, the lowest along the eastern margin. The subsidiary part of the settlement has its own scheme of segregation, the lowest among the untouchables occupying the southernmost fringe.

Ecological Underpinnings

The orientation of the castes in rural settlements is also partially linked to the regional ecology. The westerly wind is the most prevalent in the study region for all but a couple of summer monsoon months. Therefore, the western part of a village receives unpolluted fresh air, while the east receives polluted air. Among the least frequent winds are those from the south. Therefore, the dominant castes occupy the western quarter and the untouchables, following such unclean occupations as removing and skinning dead animals and tanning their hides, which produce foul smells, are housed in the southern sector. They formerly used animal fats to light lamps, which also produced bad odours. As they are housed in the sector from which the winds blow least, there are thus minimum chances of polluted air affecting the main part of the village. For the same reason the untouchables and the cremation ground are also located to the south of the settlement.

Empirical Evidence

We have selected four villages from northeastern Rajasthan to find out exactly how far their social morphology conforms to the cosmological schema discussed above. All four selected villages are typical multi-caste settlements, which may be referred to either by their proper names, as given in Fig.16.1, or, as is customary in this part of India, by the names of their dominant castes. The selected villages with their caste labels are Paḷāwā (Ahir̃), Kaṭhairā Chauth (Jāt), Ḍiḍāwalī (Gujar) and Samolá (Meo). We have chosen these four castes to provide a diversified base of empirical evidence and because these are the castes that constitute most of the land-holding population of the region.

The study area has three well-marked caste regions: Ahir̃wal in the west, Mewāt in the centre and Jāt dominated Braj in the east.[52] We have selected Palāwā from Ahīrwāl, Samoḷā from Mewat and Kaṭhaira Chauth from Braj. However, Gūjar villages are scattered over a wide area. The selection of Ḍiḍawali was therefore more or less arbitrary.

The land revenue department maintains cadastral maps of the villages, but these lack details of the built-up area. We had to survey the selected villages on a very large scale, with a plane table, to provide a house-by-house map of their castewise residential patterns.

Of the four villages, Paḷāwa has the most complex spatial arrangement of various castes (Fig.16.3). Although most of the village area is taken up by the dominant Ahir̄ caste, there is a paradoxical presence in the centre of the village of several households of scheduled castes, namely Chamārs and Bhangīs. Paḷāwā was therefore intensively investigated so as to solve the riddle of this anomalous situation. Land revenue settlement records pertaining to the village were consulted. We also spent several days interviewing village elders about Paḷāwā's history. We asked the elders to identify on our map the houses in which their forefathers lived in order to find out the stages of expansion of the built-up area of the village. Thereby we identified three stages in its evolution.

According to oral tradition the village was founded about three hundred years ago. The first stage of its morphological evolution lasted for approximately a century. The village then had two clusters of houses, aligned on an east-west axis. The western cluster was inhabited by Ahir̄s and Khātīs (carpenters), the latter to the east of the Ahir̄s. The eastern cluster was inhabited by Chamārs, Dhaṇaks and Bhangīs, all traditionally untouchable castes. In the eastern cluster Chamārs occupied the western part, the Dhāṇaks the eastern and the Bhangīs the southern part. The two clusters were separated from each other by a thirty-metre-wide open space (Fig. 16.3). The village thus had a twin character at this stage.

The second stage of morphogenesis extended from approximately 1800A.D. to 1900 A.D. During this stage the built-up area of the Ahir̄ *mohallā* expanded considerably in all directions. Eastward it coalesced with the cluster of houses of the scheduled castes. Thus the twins were fused into one, though the Ahir̄ houses had their backs towards the cluster of scheduled castes to maintain a semblance of separateness. Moreover, four new castes-Nāis (barbers), Kumhārs (potters), Swāmīs (religious mendicants) and Baṇiyās (merchants) were added to the village. The Kumhārs were housed in the southern quarter. They could not be housed in the Agni's south-eastern quarter because of a physical depression there. The Nāis, whose profession of hair cutting may be symbolically linked with the goddess of decay, were accommodated on the southwestern fringe in the Nirṛti's quarter. The Baṇiyās and Swāmis were housed in the Kuvera's northern quarter, the latter being appropriately located adjacent to the village temple. Also added were a few families of Ahir̄s belonging to the same *gotra* (lineage) as the original settlers. They were housed in the east on communal land some distance away from the *mohallā* of scheduled castes, as there was no longer any space available to house them on the *kheṇā* proper. Their addition and the later expansion of their *mohallā* placed the original area of scheduled castes in the central part of the village.

The third stage of the morphogenesis extends from around the turn of the century to the present day. In this period the built-up area of the Ahir̄ *mohallās*

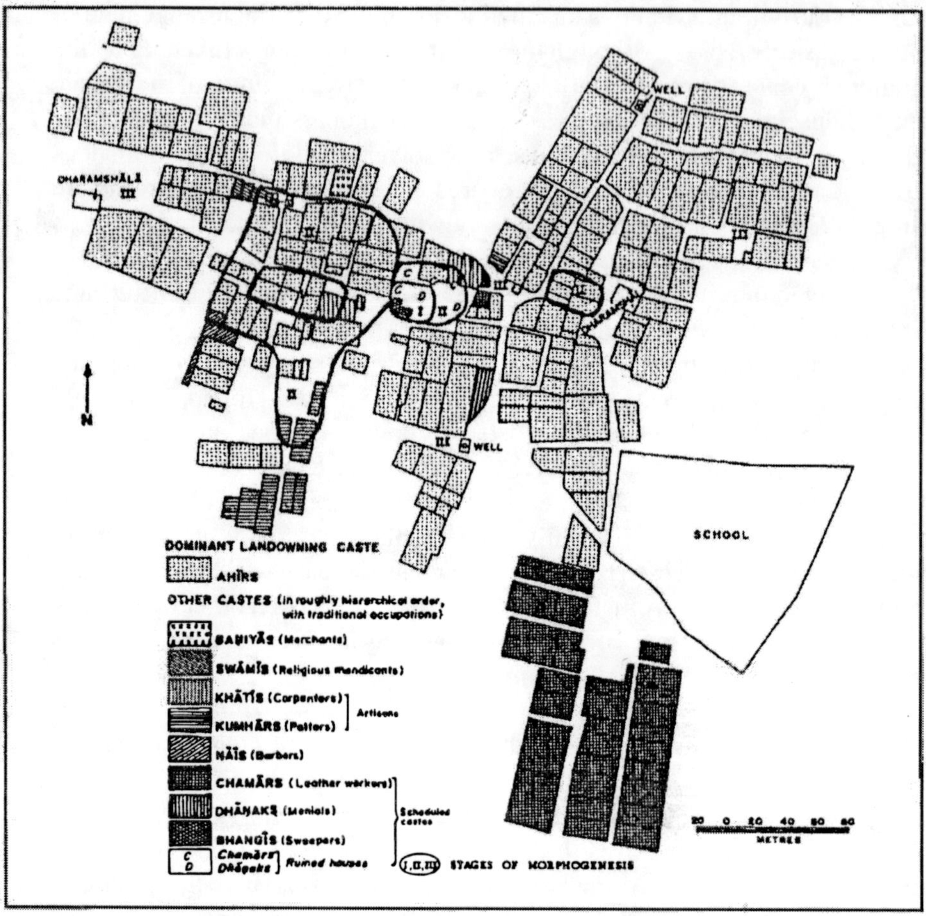

Fig. 16.3 : Palawa : Castewise residential pattern

expanded so much that it completely engulfed the early cluster of scheduled caste households. The Chamārs, Bhangīs and Dhāṇaks were not allowed to expand beyond their original *mohallā*. They thus remained huddled together in the central part of the village. However, during the late 1960s the Rajasthan government allotted land to scheduled castes for house building at the south eastern fringe of the village. Thus, a large new Chamār *mohallā* came into existence there.

As in Paḷāwā, in each of the remaining three villages we interviewed village elders about the oral history of the village, the location and orientation of caste *mohallās* and the social gradient of the village site. We also enquired about the identity of the original *mohallās* of the village founders.

Kaṭhairā Chauth (Fig. 16.4) is a village in which the dominant caste of Játs own all the agricultural land. Their houses in the western, northern and eastern quarters occupy about two-thirds of the inhabited village space. Chamārs and Bhangīs are housed in a separate *mohallā* in Yama's southern quarter. Their *mohallā* exists as a more or less distinct sub-unit separated from the main village by a relatively wide lane. The houses in Chamār *mohallā* face inwards onto two intersecting narrow, inner lanes. Two houses of Bhangīs lie on the eastern fringe of the Chamār *mohallā*.

The settlement history of the village reveals that it originally had three clusters of houses: Jāṭs in the west, Gadariyās (shepherds) in the east, and Chamārs in the south. Later, a more powerful group of Jāṭs of the same clan came to settle in the village and somehow obtained possession of 60 per cent of the agricultural land. This is exceptional; generally descendants of the original settlers possess the largest share of village land. The newcomers settled in the northern part of the village. Gradually, the Jāṭ houses of both *paṭṭīs* (wards) expanded to cover most of the western, northern and eastern parts of village. Meanwhile, the Brāhmaṇs and Banīyas were accommodated between the houses of the two groups of the Jāṭs. A Naī household was also provided with a house site just southwest of the Brāhmaṇs. The late coming Khātīs and Kumhārs were housed in the northeast. The cremation ground in Kaṭhairā Chauth is located in Yama's southern quarter and the village temple in Isāna's northeastern quarter.

Gūjars, the dominant landowning caste in Diḍawālī (Fig. 16.5), occupy about two-thirds of the inhabited space of the village in its western, northern and eastern sectors. The Āgarīs (who used to manufacture salt) and Nāīs are located in the south-eastern quarter between the Gūjar and Chamār households. The Chamārs and Bhangīs are appropriately housed in Yama's southern quarter, but their *mohallā* is not separated from the main village. Although some Chamār houses are juxtaposed against those of Gūjars and Āgarīs, to maintain a semblance of separateness the houses of both the latter groups have their backs towards the former. Moreover, the Chamār *mohallā* has its own internal lanes on to which its houses open.

The village elders told us that at the time of founding of the village, Gūjars occupied the western quarter, while Āgarīs were housed to their southeast. The Chamārs occupied a separate *mohallā* in the south. With the passage of time, descendants of the Gūjar founding fathers and their later arriving clansmen expanded to occupy the northern, western and eastern sectors. Two other castes, Nāis and Bhangīs, who came much later, were given house sites according to their status in the caste hierarchy, the former between the houses of Gūjars and Āgarīs, the latter to the south adjacent to the Chamār *mohallā*. The cremation ground is located in the south-eastern quarter, while the village temple, as in Kaṭhairā Chauth, lies in Isañ's northeastern quarter.

In the village of Samolā (Fig. 16.6) Meos are the dominant landowning caste. The Meos are a peasant caste who converted to Islam in the late medieval

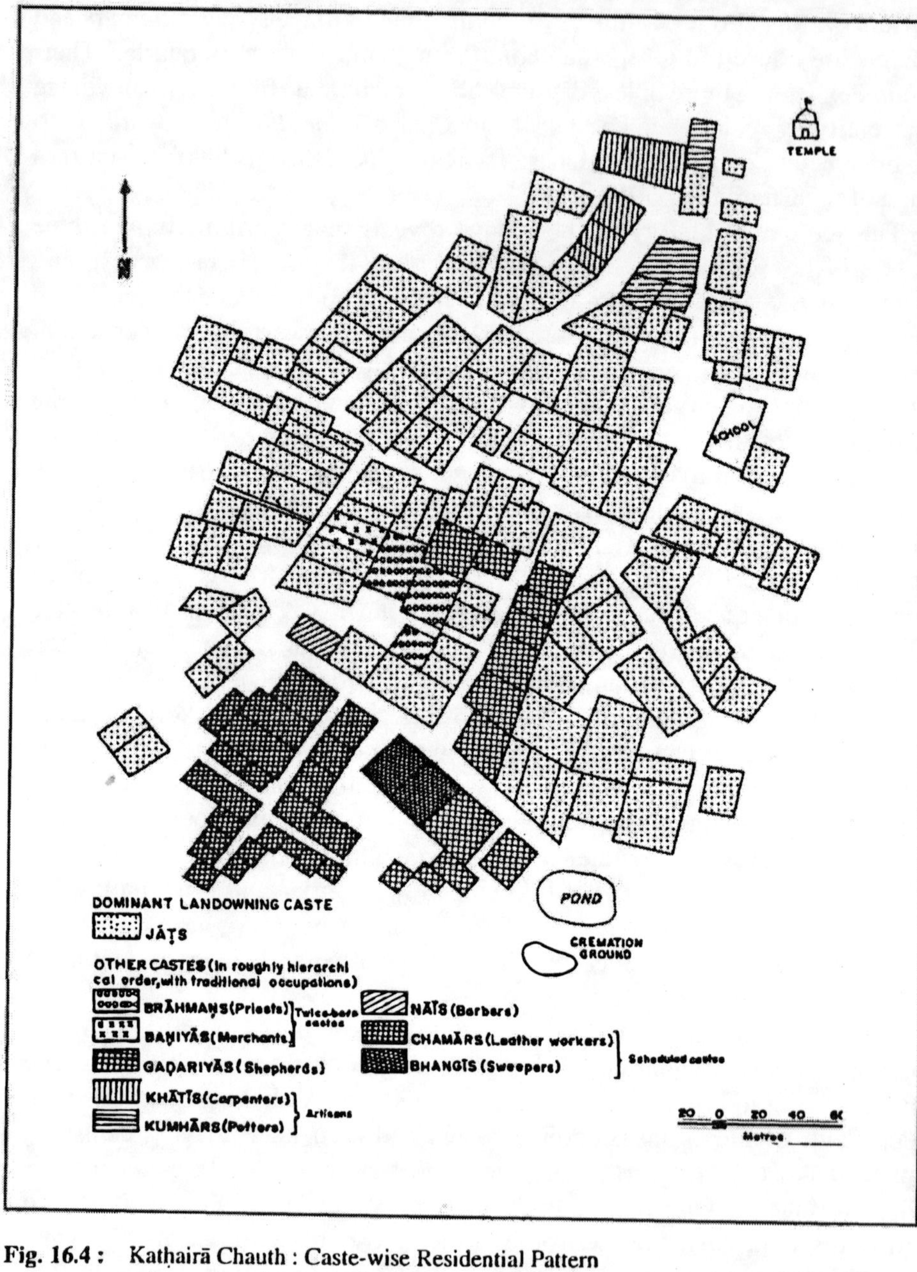

Fig. 16.4 : Kathairā Chauth : Caste-wise Residential Pattern

period (thirteenth to fifteenth century). Despite this conversion, they have not been able to free themselves from certain forms of Brāhmaṇic caste-based behaviour. They still believe in a caste hierarchy. The social morphology of the Muslim Meo villages therefore resembles that of Hindu villages. In addition to

Meos, Samolā has half a dozen Chamār, two Jāṭ, a Brāhmaṇ, an Ahir̄ and a Mirāsī (bards) household.

Samolā is located adjacent to the city of Alwar. The impact of the city's recent expansion is discernible in the morphology of the village. Some non-Meo castes came to settle in the village only during the last decade. To reap the advantage of proximity to Alwar city, all the newcomers have built their houses along the Alwar-Itarana road. However, some Chamār households have left the village. Moreover, the Meos relocated the houses of Chamārs into the northeastern sector during the post-independence era. These changes have produced the present deviant social morphology of the village. It would therefore be worthwhile to probe the history of the village. The village elders told us that during the early part of this century Samolā was a small village, located on a slightly raised mound to the south of the village pond. It had only three castes; Meos, Mirāsīs and Chamārs. The Meos occupied the raised mound entirely. The Mirāsīs had their houses on the eastern fringe of the village and the Chamārs were housed in a small separate *mohallā* to the south of the Meos. The burial ground was located to the east of the village. Around the pond a large patch of village common land was reserved for pasture. At this stage the village had a twin character.

Gradually, the built-up area of the village expanded considerably. The village common around the pond was occupied by the Meos. The Meos shifted the Chamārs and the Mirāsīs to the northeastern fringe of the village. While the Mirāsī household adjoins Meos, most of the Chamār houses are separated by a narrow lane. Recently a Chamár householder constructed his house on a plot in the southern part of the village along the main road, as did newcomers of other castes. They all valued the roadside location more than the ritually high quarter.

Discussion

Originally, each of the four villages had a dichotomous character. The main part was occupied by caste Hindus and a separate subsidiary part by scheduled castes, demonstrating that the hierarchic division of Hindu society into high and low castes permeated the built environment (Fig. 16.7). However, in only three villages were the scheduled caste *mohallās* appropriately located in the southern sector. In the fourth village (Paḷāwā) the scheduled caste *mohallā* had a deviant location in the eastern sector, though it was in conformity with the folk belief that west is high and east is low.

Regional ecology also played its role in moulding the social morphology of villages at the initial stage. The influence of prevailing west winds is visible in the location of dominant castes in the western sector. The scheduled castes with unclean professions are housed in the low-wind-frequency southern sector.

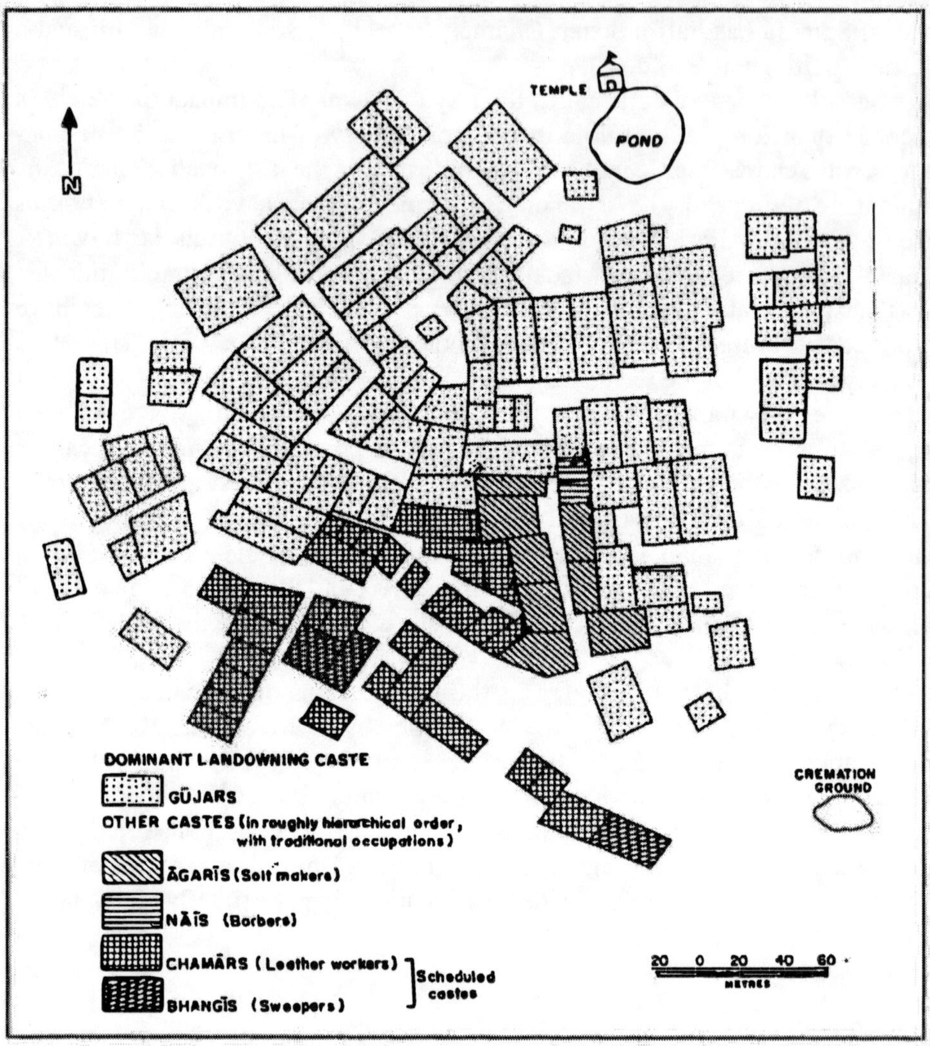

Fig. 16.5 : Diḍawālī : Caste-wise Residential Pattern

With the expansion of the built-up area and the later addition of various castes, the twins coalesced into compact morphological units in all four villages. In the villages of Kaṭhairā Chauth and Ḍiḍāwalī the scheduled caste *mohallā* retained its southern location, while in Paḷāwā it was virtually engulfed by a successive expansion of caste Hindu houses. However, the scheduled caste *mohallā* was relocated at a later date, both in Paḷāwá towards the southeast and in Samolā towards the northeast.

Another common characteristic of the four villages is that at the initial stage the ritually clean service castes were housed east of the houses of the

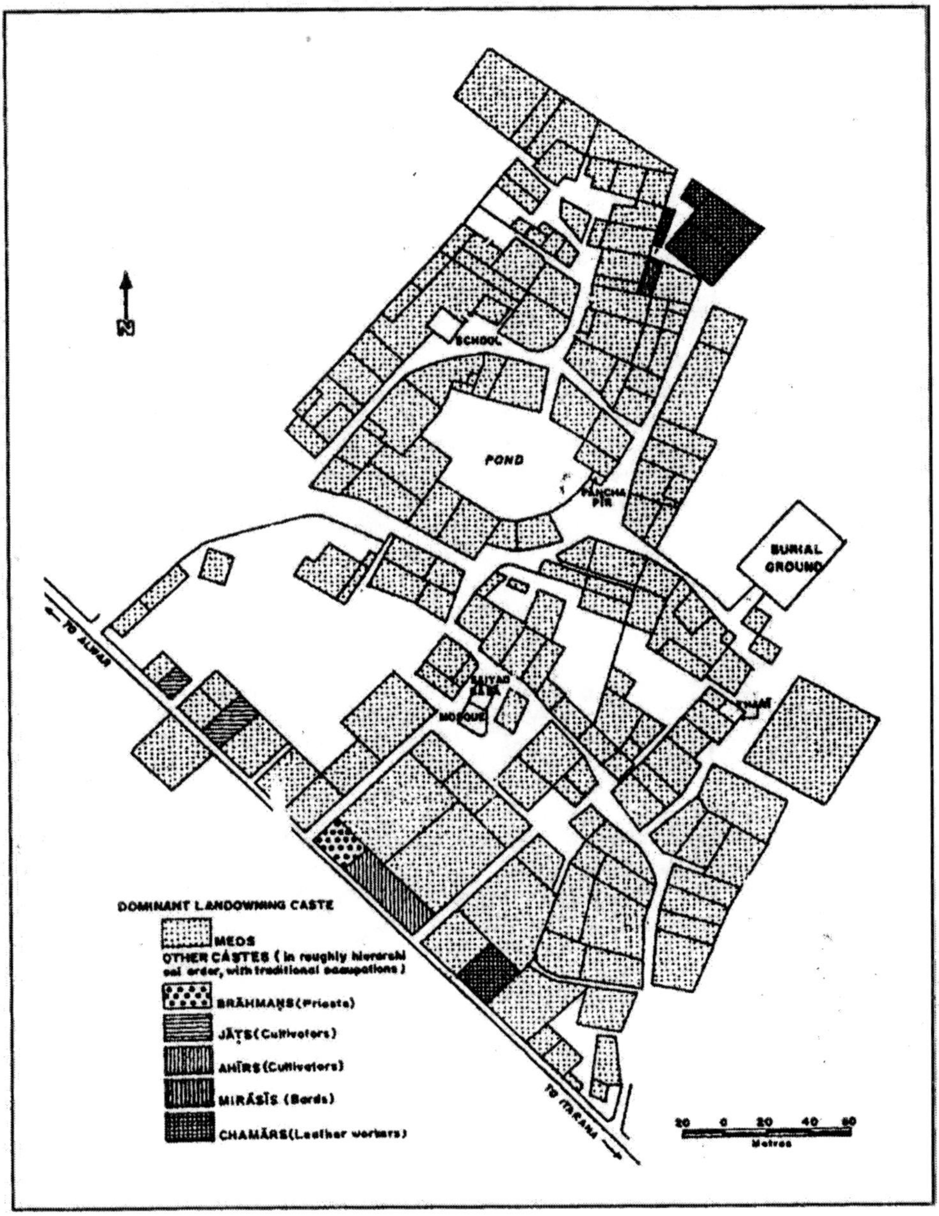

Fig. 16.6 : Samolā : Caste-wise Residential Pattern

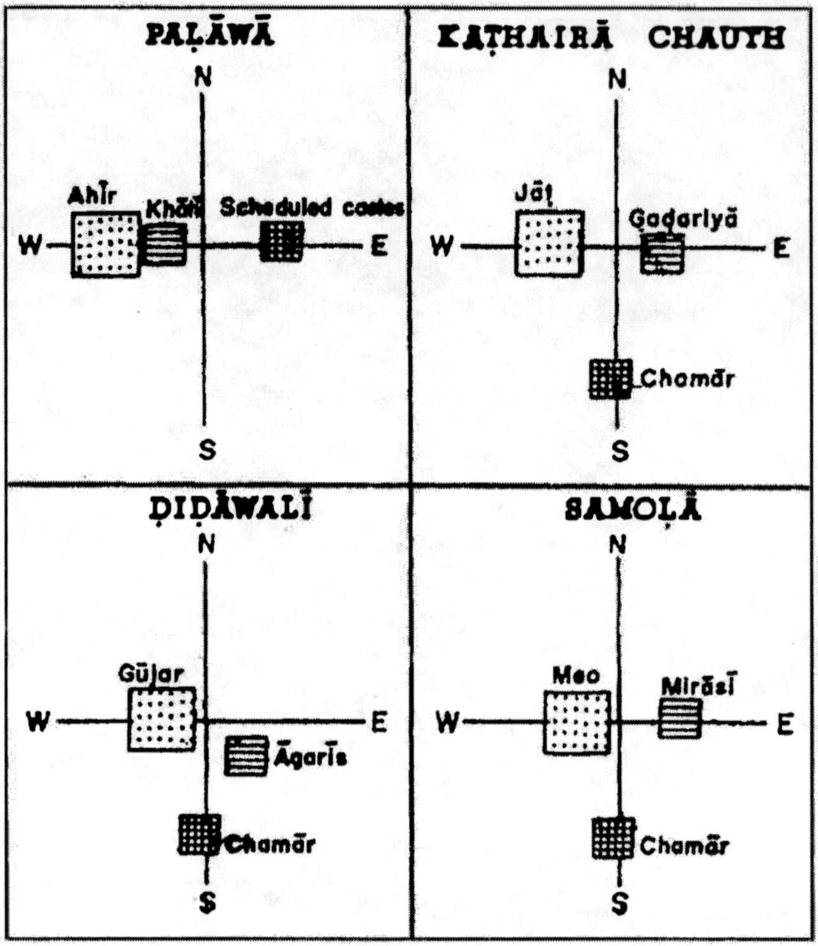

Fig. 16.7 : Schematic Diagram : Sample Villages at the Time of their Founding

descendants of founders in conformity with the regional belief in a west-east gradient in the occupied space. Later additions of new castes do not conform fully to the schema of Hindu cosmology, the extreme deviant case being that of Samolā where the Jāṭ and Ahīr households preferred southern roadside location to the 'high' western quarter.

To sum up, in the early stages of morphogenesis, the precepts of cosmology were generally adhered to, while in the later stages the influence of cosmology gradually waned.

Conclusion

The phenomenon of orientation of castes to the sacred directions is a system evolved by Indian civilization to integrate the fragmented social order into a segmented spatial order. The orientation is based on a Hindu cosmological schema which integrates the cardinal directions (quarters) into an organised whole. The cosmology ascribes sanctity to the quarters by assigning a deity to each of them and impresses cardinality on settlements. It conveys the message that the castes inhabiting a village are living in harmony with the cosmic order, each occupying a quarter in consonance with its hierarchic status and traditional occupation. The cosmology thus produces a sense of functional oneness among the segregated social groups in a settlement.

Hierarchy permeates all things in Hindu culture. The built environment tends to replicate the hierarchic divisions of society. The high castes, occupied the ritually high quarters as a symbol of high status. They prevented the lower castes from enjoying that privilege and relegated them to lower quarters. More importantly, the basic dichotomous division of Hindu society into caste Hindus and former untouchables is reflected in the twin character of many rural settlements. The caste Hindus occupy the main part of the settlements, while the scheduled castes are segregated into a separate cluster of houses normally located in the southern quarter. As they are regarded ritually beyond the pale of Hindu society, their *mohallās* appropriately lie on the periphery of the cosmological schema. In many cases, owing to the expansion of built-up areas and other factors that have little to do with tradition, the twin character of settlements becomes blurred as the two components coalesce. In such cases, examination of the past is necessary in order to understand the present social morphology of the villages.

Orientation of castes to the sacred directions automatically leads to their segregation into caste *mohallās*. Segregation is thus an inherent outcome of the phenomenon of orientation.

India is a vast country exhibiting great regional variations in its cultural traditions. Folk of different regions have modified the cosmological schema to conform to their local traditions. The empirical evidence we have examined leads us to the conclusion that in the plains of northwestern India, the phenomenon of orientation manifests itself in a particularistic form and the impress of Hindu cosmology on rural settlements is only partial. While the west-east axiality in the ordering of castes seems to be somewhat more prominent than the north-south, there also tends to be a dichotomous division of social space along the latter axis. Folk do recognise that the western quarter is a high social space. For this reason, it is occupied by the descendants of the founding fathers of many, if not most villages. To the east of them are housed the clean service castes and the late coming fellow clansmen. Folk explicitly assert that the social space of the *kheṇā* slopes down from west to east. Barring a few exceptions, as in Paḷāwā, the scheduled castes are or were originally housed in the southern quarter. Generally, this quarter also contains the cremation ground. Villagers believe that the southern quarter is the lowest segment of the village. Moreover, in many contexts they do not consider the *mohallā* of the scheduled castes to be an integral part of the village. For caste Hindus and even Meos it is often regarded as a separate sub-unit. To sum up, in folk reckoning, the social space of a village is divided into three quarters: the western, the eastern and the southern.

The Hindu cosmological schema applied well in the initial stages of rural morphogenesis when village morphology was simple and the grip of the *varṇa* system strong. Later additions of new castes and the expansion of built-up areas have often created deviant situations in which the applicability of Hindu cosmology as a critical interpretive schema of landscape is partial.[53] However, it would be interesting to interpret the built environment of traditional cities and individual Hindu houses in north India as Beck and Moore have done for parts of the south.[54]

It is necessary to point out that the cardinality of the folk is proximate, based as it is on perceived directions. There are often significant discrepancies between the perceived and actual directions. Whether a scheduled caste *mohallā* is located slightly to the east of due south or a bit to the west of it, for the folk it is still regarded as the south. Some margin of imprecision must therefore be allowed when interpreting the orientation of caste *mohallās* in a village.

The phenomenon of orientation of castes in the villages is currently undergoing significant changes. These are related to modernity, the new outlook of the younger generation, the expansion of built-up areas, and the loosening of the hold of dominant castes over village society. The younger generation is becoming oblivious to the social gradient of the *kheṇā*. We have therefore tried to provide a snapshot of the phenomenon before its disappearance from the village scene.

References and Notes

1. Stanislawski, D. (1946). 'The Origin and Spread of the Grid Pattern Town', *Geographical Review*, 36, pp. 105-20.
2. Kosambi, D.D. (1956). *An Introduction to the Study of Indian History*, Bombay: Popular Prakashan, pp.63 and 64.
3. Transliteration of local words (excepting the standardized ones) follows a uniform pattern in this paper, but where other scholars have been quoted we have followed their spellings of the words and method of transliteration.

4. During the fieldwork in the region we witnessed an interesting incident. A person was standing at the gate of his house (situated at the lowest part of the village) and helplessly looking at the rainwater coming down the village lanes, accumulating in front of his house and making the area muddy. Another villager passing by taunted him because of the lowly site of his dwelling, but the houseowner proudly replied that his house occupied the highest spot of the village. The fellow villager, who understood the social context of his reply, could not dispute it and nodded in agreement because in the framework of social space his house was indeed sited on the highest spot of the village, *i.e.* its western most edge.
5. Spate, O.H.K. (1952). 'The Indian Village', *Geography*, 37, pp. 142-52; Spate, O.H.K. (1954). *India and Pakistan: Land, People and Economy*, London: Methuen; Karve, I. (1958). 'The Structure of an Indian Village', *Deccan College Bulletin*, reprinted in A.R. Desai, (ed) (1969): *Rural Sociology in India*, Bombay: Popular Prakashan, pp. 186-92; Bailey, F.G. (1960). 'An Oriya Hill Village: I', in M.N. Srinivas, (ed.) *India's Villages*, London: J.K. Publishers, pp. 122-30; Gough, K. (1960). 'The Social Structure of a Tanjore Village', in M.N. Srinivas (ed.), *India's Villages*, London: J.K. Publishers, pp. 90-102; Rosser. C. (1960). 'A "Hermit" Village in Kulu', in M.N. Srinivas (ed.) *India's Villages*, London: J.K. Publishers, pp. 77-89; Karve, I. and Damle, Y.B. (1963). 'Group Relations in Village Community', *Deccan College Monograph Series 24*, Poona; Chauhan, B.R. (1964). 'Residential Concentrations and Caste Groupings in the Rural Setting', *Man in India*, 44: 175-81; Ray, A. (1964). *Villages, Towns and Secular Buildings in Ancient India*, Calcutta: Firma K.L. Mukhopadhyaya; Mandelbaum, D.G. (1970). *Society in India*, Bombay: Popular Prakashan; Singh, K.N. (1972). 'An Approach to the Study of the Morphology of the Indian Village', in R.L. Singh, *et al.* (ed.), *Rural Settlements in Monsoon Asia*, Varanasi: National Geographical Society of India, pp. 203-14; Hasan, M. (1976). 'The Social Organisation of Residence in Urban India', *Discussion Paper Series No. 17*, Deparment of Geography, Syracuse University, Syracuse; Bichsel, U. and Kunz, R. (1985). 'Segregation as a Structural Feature of North Indian Settlements', in A.B. Mukerji, and A. Ahmad, (ed.), *India: Culture, Society and Economy*, New Delhi: Inter India Publications, pp. 1-25.
6. Spencer, J.E. (1973). *Oriental Asia: Themes Toward a Geography*, Englewood Cliffs, New Jersey: Prentice Hall; Spate, O.H.K. (1952): 'The Indian Village', *Geography*, 37, pp. 142-52; Spate, O.H.K. (1954). *India and Pakistan: Land, People and Economy*, London: Methuen.
7. Kosambi, D.D. (1956). *An Introduction to the Study of Indian History*, Bombay: Popular Prakashan; Kosambi, D.D. (1981). *The Culture and Civilization of Ancient India in Historical Outline*, New Delhi: Vikas Publishing House; Sharma, R.S. (1983a). *Material Culture and Social Formations in Ancient India*, Delhi: Macmillan India; Sharma, R.S. (1983b). *Perspectives in Social and Economic History of Early India*, New Delhi: Munshiram Manoharlal Publishers.
8. Cosgrove, D.E. (1984). *Social Formation and Symbolic Landscape*, London: Croom Helm; Cosgrove, D. (1989). 'A Terrain of Metaphor: Cultural Geography 1988-1989', *Progress in Human Geography*, 13, pp. 566-75; Cosgrove, D. (1990). '... Then We Take Berlin: Cultural Geography 1989-90', *Progress in Human Geography*, 14, pp. 560-8; Cosgrove, D. (1992). 'Orders and a New World: Cultural Geography 1990-91', *Progress in Human Geography*, 16, pp. 272-80; Cosgrove, D. and Daniels, S. (eds.) (1988). *The Iconography of Landscape*, Cambridge University Press, Cambridge; Cosgrove, D. and Jackson, P. (1987). 'New Directions in Cultural Geography', *Area*, 19, pp. 95- 101; Duncan, J.S. (1990). *The City as Text: The Politics of Landscape Interpretation in the Kandyan Kingdom*, Cambridge University Press, Cambridge; Duncan, J.S. (1993). 'Landscapes of the Self/Landscapes of the Other(s): Cultural Geography 1991-92', *Progress in Human Geography*, 17, pp. 367-77; Jackson, P. (1989). *Maps of Meaning: An Introduction to Cultural Geography*, London: Unwin Hyman; King, A.D. (ed.) (1980). *Buildings and Society: Essays on the Social Development of the Built Environment*, London: Routledge & Kegan Paul; Knox, P.L. (1987). 'The Social Production of the Built Environment', *Progress in Human Geography*, 11, pp. 354-77; Rapoport, A. (1982). *The Meaning of the Built Environment*, Beverly Hills: Sage Publications;

Relph, E. (1975). *Place and Placelessness*, London: Pion Ltd; Rowntree, L. (1987). 'Cultural/Humanistic Geography', *Progress in Human Geography*, 11, pp. 558-64; Rowntree, L.B. (1988). 'Orthodoxy and New Directions: Cultural/Humanistic Geography', *Progress in Human Geography*, 12, pp. 575-86; Rowntree, L.B. and Conkey, M.W. (1980). 'Symbolism and the Cultural Landscape', *Annals of the Association of American Geographers*, 70, pp. 459-74.

9. Berreman, G.D. (1979). *Caste and Other Inequities*, Folklore Institute, Meerut; Bose, N.K. (1975). *The Structure of Hindu Society*, New Delhi: Orient Longman; Desai, A.R. (1969). *Rural Sociology in India*, Bombay: Popular Prakashan; Karve, I. (1961). *Hindu Society—An Interpretation*, Poona: Deshmukh Prakashan; Mandelbaum, D.G. (1970). *Society in India*, Bombay: Popular Prakashan; Marriott, M. (ed.) (1961). *Village India: Studies in the Little Community*, Bombay: Asia Publishing House; Panikkar, K.M. (1956). *Hindu Society at the Cross Roads*, Bombay: Asia Publishing House; Schwartzberg, J.E. (1993). 'Caste and the Study of Regional Development', in A. Ahmad (ed.), *Social Structure and Regional Development: A Social Geography Perspective*, Jaipur: Rawat Publications, pp. 21-49.
10. Karve, I. (1961). *Hindu Society - An Interpretation*, Poona: Deshmukh Prakashan, pp. 4, 45, 54, 55, 65.
11. Das, V. (1977). *Structure and Cognition: Aspects of Hindu Caste and Ritual*, Delhi: Oxford University Press, p.126; Das, V. (1976). 'The Uses of Liminality: Society and Cosmos in Hinduism', *Contributions to Indian Sociology*, (NS)10, pp. 245-63.
12. Wiser, W.H. (1936). *The Hindu Jajmani System*, Lucknow: Lucknow Publishing House; Kushner, G. (1967). 'A Synthetic Model of the Hindu Jajmani System', *Man in India*, 47, pp. 35-60.
13. Fruzzetti, L. and Ostor, A. (1984). *Kinship and Ritual in Bengal: Anthropological Essays*, New Delhi: South Asian Publishers, p. 27.
14. The concept of 'dominant caste', initially introduced by Srinivas (1955) and later (1959) elaborated by him has evinced keen interest among social anthropologists. Srinivas holds that 'A caste may be said to be 'dominant' when it preponderates numerically over the other castes, and when it also wields preponderant economic and political power. A large and powerful caste group can be more easily dominant if its position in the local caste hierarchy is not too low', Srinivas M.N. (1955). 'The Social System of a Mysore Village', in M. Marriott(ed.), *Village India*, Chicago: Chicago University Press, pp. 1-36, ref. p.18; Srinivas, M.N. (1959). 'The Dominant Caste in Rampura', *American Anthropologist*, 61 (i), pp. 1-16. Though it has been pointed out that 'its usefulness as an analytical tool seems now to have been greatly diminished' Oomen, T.K. (1970). 'The Concept of Dominant Caste: Some Queries', *Contributions to Indian Sociology*, NS (4), pp. 75-83; it remains a good descriptive model for understanding rural Indian society.
15. The concept of dominant caste extends the Hindu theory of kingship to a lower level. According to this theory, the higher status and secular power of both a king and of a dominant caste is legitimized by the Brāhmaṇs, who serve both in religious matters The well-documented case of the Maratha community who were raised from the position of Sudras to that of Kastriyas testifies to this. Karve, I. (1961): *Hindu Society—An Interpretation*, Poona: Deshmukh Prakashan, pp. 41-2.
16. Lewandowski, S. (1980). 'The Hindu Temple in South India', in A.D. King, (ed), *Buildings and Society: Essays on the Social Development of the Built Environment*, London: Routledge & Kegan Paul, pp. 123-50, ref. p. 132.
17. Sastri, K.A.N. (ed.) (1957). *A Comprehensive History of India. Vol. II*, Calcutta: Orient Longmans, p. 81.
18. *See* ref. 17, p. 379.
19. Singh, J.P. (1983). 'The Nature of Space and Place in Indian Culture', *Presidential Address* read at the Fifteenth Rajasthan Geography Conference, Pratapgarh (Rajasthan); 'The Sastras have recommended an elaborate ritual for consecrating a site', Dutt, B.B. (1977). Town Planning in Ancient India, Delhi: New Asian Publishers, pp. 60-62; A number of rites and

rituals connected with the selection of sites, laying foundations, entering a house, and so forth 'are common to tribal, Brāhmaṇical sacredotal and hieratic and Brāhmaṇical current and folk practices in the villages', Ray, A. (1964). *Villages, Towns and Secular Buildings in Ancient India*, Calcutta: Firma K.L. Mukhopadhyaya, p.122; These rituals were presumably 'adopted from relatively primitive societies and were eventually integrated into the Brāhmaṇical socio-religious pattern, with later additions and accretions', Ray, A. (1964). *Villages, Towns and Secular Buildings in Ancient India*, Calcutta: Firma K.L. Mukhopadhyaya, p. 110.

20. Dowson, J. (1957). *Classical Dictionary of Hindu Mythology and Religion, Geography, History and Literature*, London: Routledge & Kegan Paul, p. 180.
21. The existence of rigid and traditionally demarcated ritual circles, both in the settled part of the village and the entire village territory, has been identified in a Mysore village, Gowdra, G.K. (1971). 'Ritual Circles in a Mysore Village', *Sociological Bulletin*, 20, pp. 24-38.
22. Shukla, L.K. (1972). *A Study of Hindu Art and Architecture*, Varanasi: Chowkhamba Publishers.
23. Rapoport, A. (1982). *The Meaning of the Built Environment*, Beverly Hills: Sage Publications, p.134.
24. A good introduction to ancient cosmographical schema, both anthropocentric and non-anthropocentric, is provided by Blacker and Loewe, see Blacker, C. and Loewe, M. (ed.) (1975). *Ancient Cosmologies*, London: George Allen & Unwin; especially the chapter on Indian cosmographies by Gombrich (pp. 110-42). A perceptive exposition of Indian cosmological schema, with numerous illustrations, may be found in Schwartzberg's study, Schwartzberg, J.E. (1992). 'Cosmological Mapping', in J.B. Harley and D. Wooward (eds.), *The History of Cartography Vol. 2, Book 1, Caligraphy in the Traditional Islamic and South Asian Societies*, Chicago: The University of Chicago Press, pp. 332-85; Tuan notes that anthropocentric views are common in both Old and New World schemata, Tuan, Y.F. (1977). *Space and Place: The Perspective of Experience*, London: Edward Arnold, p. 91.
25. The existence of terrestrial-celestial models reflecting the supernatural affinity between human dwellings and the cosmic order is clearly evident in the archaeological remains of ancient civilizations stretching from the Nile to the Indus valley and which later on diffused to a much wider area. Noble, W.A. (1985). 'Terrestrial-Celestial Models and the Renaissance of Monumental Architecture in South Asia', in A. B. Mukerji, and A. Ahmad, (eds.), *India: Culture, Society and Economy*, New Delhi: Inter India Publications, pp. 117-59; In China and Japan the imperial cities were 'oriented to the cardinal points, and there are clear indications that the whole spatial geometry rested on a cosmic-ritual system, on a model of the cosmos (the heaven) with its replica on earth'. Nitz, H.J. (1987). 'Order in Land Organisation: Historical Spatial Planning in Rural Areas in the Medieval Kingdoms of South India', in V.S. Datye, *et al.* (eds.), *Explorations in the Tropics*, Prof. K.R. Dikshit Felicitation Volume Committee, Pune, pp. 258-79, ref. p. 259; The system of divination-based canons in the form Rossides labels as 'ecomancy' has a tremendous influence, even on modern Japanese society and its living space, Rossides, N. (1982). 'The Influence of Ecomancy on Japanese Living Space', *Ekistics*, 49, pp. 278-84; It is also pertinent to point out here that these schemes have also been utilised for political purposes. Bohle has successfully demonstrated how religion and 'ritual politics' were systematically functionalized by the feudal states for their own stability and defence. Bohle, H.G. (1987). 'Spatial Planning and Ritual Politics: The Evolution of Temple Towns and Urban Systems in Medieval South India', in V.S. Datye, *et al.* (eds.), *Explorations in the Tropics*, Prof. K.R. Dikhsit Felicitation Volume Committee, Pune, pp. 280-9.
26. Wheatley, P. (1991). 'Review' of Duncan, J.S., *The City as Text: The Politics of Landscape Interpretation in the Kandyan Kingdom*, Cambridge: Cambridge University Press, *American Anthropologist*, 93, pp. 762-3, ref. p. 762.
27. Singh, J.P. (1982). 'Social Morphology of Indian Villages', *Annals of the Association of Rajasthan Geographers* , 2, pp. 1-5; Singh, J.P. (1983). 'The Nature of Space and Place in

Indian Culture', *Presidential Address* read at the Fifteenth Rajasthan Geography Conference, Pratapgarh (Rajasthan); Singh, J.P. and Khan, M. (1994). *Cosmology in Hindu Culture*, Unpublished Manuscript.

28. Singh, J.P. and Khan, M. (1994). *Cosmology in Hindu Culture*, Unpublished Manuscript.
29. Beck, B.E.F. (1976). The Symbolic Mergerx of Body, Space and Cosmos in Hindu Tamil Nadu', *Contributions to Hindu Sociology*, (NS) 10, pp. 213-43.
30. Moore, M.A. (1989). 'The Kerala House as a Hindu cosmos'. *Contributions to Indian Sociology*, (NS)23, pp. 169-202, ref. p. 183.
31. Nitz, H.J. (1987). 'Order in Land Organisation: Historical Spatial Planning in Rural Areas in the Medieval Kingdoms of South India', in V.S. Datye, *et al.* (eds.), *Explorations in the Tropics*, Prof. K.R. Dikshit Felicitation Volume Committee, Pune, pp. 258-79.
32. Bohle, H.G. (1987). 'Spatial Planning and Ritual Politics: The Evolution of Temple Towns and Urban Systems in Medieval South India', in V.S. Datye, *et al.* (eds.), *Explorations in the Tropics*, Prof. K.R. Dikhsit Felicitation Volume Committee, Pune, pp. 280-9.
33. Rapoport, A. (1980). 'Vernacular Architecture and the Cultural Determinants of Form', in A.D. King (ed.), *Buildings and Society: Essays on the Social Development of the Built Environment*, London: Routledge & Kegan Paul, pp. 283-305, ref. p. 284.
34. Though both the *Dravina* (south Indian) and *Nagara* (north Indian) schools of town planning follow the orientation and segregation of social groups, they differ fundamentally in that the former allocates sites based on *padavinyasa* (division of a site into squares), whereas the latter do so with reference to the king's palace or the streets.
35. Caste segregation is a prominent attribute of Indian towns. It is widely held that it has been losing its strength in modern times. However, Hasan questions the assertion that urbanization and industrialization are changing the social organization and residential segregation of Indian towns. By examining the residential pattern and social interaction of the urban population he has shown that socio-spatial boundaries between the members of social groups still exist. Hasan, M. (1976). 'The Social Organisation of Residence in Urban India', *Discussion Paper Series No. 17*, Deparment of Geography, Syracuse University, Syracuse. Similarly, D'souza has shown that with increasing industrialization and urbanization the degree of segregation of the scheduled caste population in the towns of Amritsar district has tended to increase. Moreoever, such groups also tend to be increasingly concentrated in peripheral localities. D'souza, V.S. (1978). 'Does Urbanism Desegregate Scheduled Castes? Evidence From a District in Punjab', in S. Saberwal, (ed), *Process and Institution in Urban India - Sociological Studies*, New Delhi: Vikas Publishing House, pp. 219-39.
36. Singh, J.P. and Khan, M. (1998). 'Dichotomy in the Built Environment: the Mohammad Habib-Buddha Prakash Debate', forthcoming in *Indian Historical Review* 25.
37. Habib, M. (1952). 'Introduction', in Elliot, H.M. and Dowson, J. *History of India as Told by Its Own Historians*, Vol. II, Aligarh: Cosmopolitan Publishers, pp. 1- 102, ref. pp. 54 and 52.
38. Prakash, B. (1965). *Aspects of Indian History and Civilization*, Agra: Shiv Lal Agarwala, p. 244.
39. Ray, A. (1964). *Villages, Towns and Secular Buildings in Ancient India*, Calcutta: Firma K.L. Mukhopadhyaya, pp. 25, 33, 59, 71; Sharma, R.S. (1958). *Sudras in Ancient India*, Delhi: Motilala Banarsidas, p. 227; Sharma, P.L. (1981). *Morphology of Rural Settlements in Deeg Tehsil*, Unpublished MA Dissertation, Department of Geography, Government Arts College, Alwar (Rajasthan), p. 25; Ghoshal, U.N. (1965). *Studies in Indian History and Culture*, Calcutta: Orient Longmans, p. 324; Mukerji, A.B. (1969). 'Rural Settlements in the Jatakas', *Geographical Review of India*, 31, pp. 17-32., ref. p. 27.
40. Spate, O.H.K. (1952): 'The Indian Village', Geography, 37, pp. 142-52, ref. p. 148; Karve, I. (1958). 'The Structure of an Indian Village', *Deccan College Bulletin*, Reprinted in A.R. Desai, (ed.) (1969). *Rural Sociology in India*, Bombay: Popular Prakashan, pp. 186-92, ref. p. 191; Bohle, H.G. (1987). 'Spatial Planning and Ritual Politics: The Evolution of Temple Towns and Urban Systems in Medieval South India', in V.S. Datye *et al.* (eds.), *Explorations in the Tropics*. Prof. K.R. Dikhsit Felicitation Volume Committee, Pune, p. 282; Singh, J.P.

and Khan, M. (1992). 'The Cultural Landscape of Mewat', in A. Ahmad, (ed.). *Social Structure and Regional Development: A Social Geography Perspective*, Jaipur: Rawat Publications, pp. 160-2; Sharma, K.D. (1993), 'Social Morphology of Himalayan towns, in A. Ahmad, (ed.), *Social Structure and Regional Development: A Social Geography Perspective*, Jaipur: Rawat Publications, pp. 129-49.

41. Singh, K.N. (1972). 'An Approach to the Study of the Morphology of the Indian Village', in R.L. Singh, *et al.* (ed.). *Rural Settlements in Monsoon Asia*, Varanasi: National Geographical Society of India, pp. 203-14.
42. *See* ref.41, pp. 207-212.
43. Sharma, K.D. (1993). 'Social Morphology of Himalayan Towns, in A. Ahmad, (ed.), *Social Structure and Regional Development: A Social Geography Perspective*, Jaipur: Rawat Publications, pp. 129-49.
44. *See* ref. 16.
45. Gowdra, G.K. (1971). 'Ritual Circles in a Mysore Village', *Sociological Bulletin*, 20, pp. 24-38.
46. Schwartzberg, J.E. (1985). 'Folk Regions in Northwestern India', in A. B.Mukerji, and A. Ahmad, (eds.), *India: Culture, Society and Economy*, New Delhi: Inter India Publications, pp. 205-35.
47. Chattopadhyaya avers that, 'The decline of Varuna along with his *rita* also marked the beginnings of the distinctly class-divided society and the moral order codes that followed were invariably tainted by the outlook and interests of the ruling class. As Varuna was superseded by Indra, the ancient law of *rita* had to give place to the law of *karma*. And this was how, as early as in the days of the *Chandogya Upanisad*, this law of *karma* was expounded to justify caste-rule and caste-exploitation'. Chattopadhyaya, D.P. (1985). *Lokayata: A Study in Ancient Indian Materialism*, New Delhi: People's Publishing House, 1985, p. 642.
48. See ref. 11, p. 107.
49. See ref. 16, p. 132.
50. Dutt, B.B. (1977). *Town Planning in Ancient India*, Delhi: New Asian Publishers, pp. 52-3.
51. Wilson, H.H. (1961). (Reprint of 1840 original), *The Vishnu Purana: A System of Hindu Mythology and Tradition*, Calcutta: Punthi Pustak, pp. 417-21.
52. Schwartzberg, J.E. (1985): 'Folk Regions in Northwestern India', in A.B. Mukerji, and A. Ahmad, (eds.), *India: Culture Society and Economy*, New Delhi: Inter India Publications, pp. 205-35; Singh, J.P. and Khan, M. (1992). 'The Cultural Landscape of Mewat', in A. Ahmad, (ed), *Social Structure and Regional Development: A Social Geography Perspective*, Jaipur: Rawat Publications, pp. 160-2.
53. Pred, A. (1991). 'Review' of Duncan, J.S. *The City as Text: The Politics of Landscape Interpretation in the Kandyan Kingdom*, Cambridge: Cambridge University Press, Journal of Historical Geography, 17, pp. 115-7, ref. p. 117; Beck, B.E.F. (1976), The Symbolic Merger of Body, Space and Cosmos in Hindu Tamil Nadu', *Contributions to Hindu Sociology*, (NS) 10, pp. 213-43.
54. Moore, M.A. (1989), 'The Kerala House as a Hindu Cosmos', *Contributions to Indian Sociology*, (NS)23, pp. 169-202.

SECTION-III

CULTURE PROCESS AND CHANGE

17

INTERACTION, EXCHANGE AND THE DYNAMICS OF LINGUISTIC ACCULTURATION

Portuguese Loan-Words in Bengali and Marathi Languages

SWAPNA BANERJEE GUHA AND SUMIT GUHATHAKURTA

Introduction

The cultural and social history of India unfolds a chequered process of interaction among distant and contiguous language communities arising out of the process of trade, domination and colonisation. The early interactions that took place between the indigenous populations and migrant-nomads from Western Asia and Central Europe led to large scale cultural exchange and assimilation. It was followed by Islamic and later European invasions taking the form of commercial exchange, political domination and colonial rule. In migration of people from different cultures and economies and their residence in various parts of the country that took place due to the above processes created a basis for acculturation and exchange between the migrant and native communities. Cross cultural diversity in elements such as language, music and architecture bears marks of such processes.

Various regions of India with their diverse physiographic, economic and socio-cultural characteristics did not experience the above process uniformly and there existed distinct locations of choice of various foreign groups/powers. Interactive systems of these areas differed considerably from the situations of other areas of lesser attraction. In general, the coastal tracts were far more attractive for the trading European powers than the interior areas because of their prospects towards commerce and trade. Languages of these tracts hence, show marks of such interactions more than that of the interior ones. Essentially, the above socio-spatial interaction patterns can be considered as a part of the general process of language development and growth in different regions of India marked by frequent borrowing of loan-words. Due to the differentials in technological or political status of European and native communities, it was found that the borrowing was heavier in the latter languages. This study attempts to trace the loan-words borrowed from Portuguese by two coastal languages of India, *viz.* Bengali from the east and Marathi from the west, and on the basis of these words, analyse the nature of social interaction prevailing between the said foreign community and the respective indigenous populations in these two distant locations.

A Brief Account of the Portuguese Advent in Bengal and the West Coast

The official impact of Portuguese presence in India is essentially said to be '*opera bouffe*'—strong on cultural aspects (albeit in selective regions) but of less significance in wider national affairs. In the regions of their presence, customs and practices (especially in some communities) still bear the Portuguese influence. The impact, however, is seen more importantly in the usage of Portuguese loan-words by the regional languages, especially their dialects, indicating a wider interaction between Portuguese and the native population at large.

The foremost advent of the Portuguese in India was to the southern west coast in 1500 A.D, through the establishment of factories in Calicut, Cochin and Cannanore. Actually the maritime trade in the west coast came in the hands of the Portuguese by securing a foothold in Calicut and Cochin following which they could dislodge the Arab traders, who had till then monopolised the control of the littoral trade of India. The advent of the Portuguese in the south western coast was followed by their entry in the land locked island of Goa. In the port of Goa, the Portuguese started operating a supplement trading centre *vis-a-vis* Calicut, which had been the only port operating between the Gulf of Cambay and Cape Comorin till then. Subsequently, Portuguese fortresses were erected at Cochin and cities at Mylapore and Nagapatam were founded near the Gulf of Cambay. The cities of Chaul, Bassein and Damaon are the Portuguese landmarks of this region.[1]

During the same period the Portuguese started visiting Bengal in the eastern part of India, off the coast of Bay of Bengal. Similar to the west coast, trade and socio-cultural relations became significant here. Actually, much before the late 16th century, private traders had begun to call at Chittagong (referred to by the Portuguese as Porto Grande) in eastern Bengal and Satgaon in western Bengal. Sultan Mahmud Shah of Bengal had permitted these Portuguese traders to set up their factories and custom houses in some settlements. Founded in 1537 Hughli, in western Bengal, became a major centre of the Portuguese in the east. Following the decline of Satgaon caused by the silting of river Saraswati, Hughli, (especially since 1580) grew into a substantial town with a population of 5000 by 1603. Religious services were provided by members of various orders who chose to live in the area to form a large Augustinian monastery at Bandel, founded in the late sixteenth century a few miles upstream from Hughli. Portuguese presence also continued to remain substantial in Chittagong and Arakan.[2]

In most of these places in Bengal, as well as on the west coast, the Portuguese did not have any official presence. It helped the private traders to mingle with the indigenous society and develop a spectrum of interaction and symbiosis. Besides the well known big ports, Portuguese set up numerous small ports in Bengal and along the western coast. Geographical names given to these places bear the imprint , *e.g.* Dom Banik islands, Firanghi Bazar, Bandel, etc. Mixed Portuguese (with words from both Protuguese and the local language) was then a very popular 'lingua franca' used by the indigenous people in these areas.[3]

During the mid-17th century (1632), the Portuguese port at Hughli was attacked by the then Mughal Emperor, Shah Jahan. The city was taken over and pillaged. Bassein fort fell in 1739 followed by the fall of Bassein city. The Portuguese thereafter ruled only Goa, Daman and Diu.[4]

The history of the Portuguese presence in India was brief but largely informal and therefore, interactive. From the early 16th century till the mid-eighteenth century when their influence declined with the rise of the British power, a variagated interactive framework was created by the Portuguese dominion in Cochin, Calicut, Bombay, Thane, Bassein, Salsette and Chaul, through commercial connections in Surat, Bijapur and Bengal and by the colonisation of Goa, Daman and Diu. An array of divergent forces and factors were at work in this system *e.g.* direct dominion, commercial connection, political influence, religious propoganda and missionary activity. Due to these divergent interactive patterns, the impact and influence of Portuguese remained even after their rule or physical presence was over. The presence of loan-words from Portuguese in Bengali and Marathi bear imprints of this. The loan-words simultaneously speak of co-association of many vernaculars in some of the above mentioned cities, influence of other languages such as Persian or Arabic in either and each of these three languages, (indicating a similar interactive system operating between and among other communities) and last but not least, co-existence of Indo-Portuguese and Anglo-Indian vocabulary.[5]

The following analysis of Portuguese loan-words in Bengali and Marathi exhibits various strains of the above interactive system.

Loan-words and Their Characteristics: Interaction Between Societies

A language community, at peace or in belligerence, with a contiguous or distant language community of similar or different stock or brought in contact with the other in the wake of migration or colonisation, borrows words to formalise diversities and interaction and in the process enriches its own vocabulary. In normal circumstances of durable commercial and cultural exchange, a two-way adoption takes place with the technologically and culturally weaker community generally adopting more words from the language of the dominant partner. An extreme case in point is the massive borrowing of Chinese words and related ideographic writing system by Japan from China, having a much higher civilisation under the Hans. Presence of words with Dravidian, Arabic or Persian roots in Indian languages of Indo-European origin, exemplified a more general situation. Borrowing of words and their integration in the host language by adoption and adaptation thus, can be said to be the result of colonisation, political domination, trade, commerce or enbloc migration. Loan-words, therefore, present interesting examples of geographical diffusion in language contact situations. In contrast to bilingualism, the other product of language contact, which is more individual or people oriented, loan-words talk of a deeper societal interaction which is more language oriented, atuned to language growth and development. Hagerstrand's migration model[6] can be applied in such studies concerning migration of words from one language to the other and their subsequent integration.[7]

The presence of Portuguese loan-words in Bengali and Marathi indicate a divergent and variegated type of interaction caused by factors, at times, operating at variance. Such words, also prove that language is not a closed system but an open one with multifarious interactive levels and that, assimilation of words by adoption and adaptation is a universal feature of language evolution. Words of Portuguese origin such as *anaras*, *nilam*, *alkatra* in Bengali or *kaju*, *pagar*, *sabu* in Marathi, signifying different aspects of life are hence, used by the speakers of these respective languages without any inhibition, naturally and spontaneously, as scores of words of Norman-French origin are used with as much ease and spontaneity by the English speaking people, or the Turko-Arabic words by the Rumanians. Although morphologically and phonetically they are indistinguishable from other words of the host language, they have a history of their own, experiencing a convergence of political, economic and cultural interaction.

Loan-words from Portuguese in Bengali and Marathi : Nature and Degree of Interaction

The number and nature of loan-words from Portuguese in Bengali and Marathi vary. The tables in the present paper do not claim to be exhaustive, but they are representative enough to indicate the trends in the interactive system.

In Bengali there are 90 Portuguese loan-words. Fifty of these are found only in Bengali, while the remaining 40 are also identified in Marathi, although their form, pronunciation and orthography may vary occasionally. In Marathi, there are 56 Portuguese loan-words of which 16 are found solely in Marathi, while 40 are also identified in Bengali as mentioned above. In total 110 Portuguese loan-words have been found to be incorporated in these two Indian languages.

Depending on their nature and utility, the loan-words have been classified into four major groups, *viz.* (i) flora and fauna, (ii) interaction (iii) commerce and technology, and (iv)house-hold and daily use. Words denoting interaction have been further sub-classified into 3 sub-groups, *viz.* (a) social, (b) political, and (c) religious; indicating the nature of interaction.

Not all the loan-words included are purely of European origin. Some are Asiatic-Portuguese, some Portuguese adaptations of Amerindian words and some Indo-Portuguese. Most of the names are of birds or plants or items which the Portuguese had come across in their colonising campaign in South America, the Asiatic region, and Africa, and had included them in their vocabulary, signifying an earlier interactive system existing between them and other host communities in which the Portuguese were the receptors. The words denoting plants, birds, utility items, technology, etc., the Portuguese carried with them wherever they settled for trade and commercial gains. Hence, the associated names followed. In such cases the Portuguese essentially acted as agents of diffusion of resources, technology and way of life, which later became an integral part of the daily life of the Indian host society.

The above interactive process is also demonstrated by replacement of existing words for objects known to Bengali and Marathi language communities, by Portuguese loan-words and the survival of the latter. Many of these words were widely used in the littoral pockets of Portuguese influence in the 17th century and found their way into the inland dialects. Some of these words have

Table 17.1A : Portuguese Loan-words in Bengali Relating to Flora and Fauna

Sr. No.	*Portuguese word*	*Bengali word*	*Meaning in English*	*Marathi counterpart*
1.	Ata	Ātá	Custard apple	Sitāphál
2.	Anona	Noná	- do-	Rāmphál
3.	Pato	Pátihaṅs	A kind of duck	Badák
4.	Peru	Perú (now replaced by tharki)	A bird-turkey	Tarkíy
5.	Zapota	Sabéda	Fruit	Chikú

Complied by the Authors.

been retained till now by the respective languages despite the strong influence of the English language.

Loan-words, denoting religious interaction, have been restricted to those which have wider social significance. Words relating to rites and practices of the Roman Catholic faith and confined to a very small Luso-Indian community, have been excluded. Similarly, a number of loan-words evidenced in middle Bengali and middle Marathi literature have also been left out for being archaic and obsolete. Some of the loan-words included in the tables had entered the respective languages replacing already existing native words representing articles of daily use, for reasons of novelty and social prestige. They have almost become obsolete now with the changing social attitude of the speakers, *e.g. tolo* (shallow plate) in Bengali.

The following table consists of Portuguese loan-words relating to flora and fauna borrowed by Bengali and Marathi.[8] Table 17.1A consists of the loan-words borrowed only by Bengali while Table 17.1-B presents Portuguese loan-words borrowed only by Marathi. Table 17.1-C consists of the common Portuguese loan-words borrowed simultaneously by Bengali and Marathi.

The plants and birds mentioned in the following tables once considered as exotic in India, were introduced by the Portuguese. Except '*Couve*', having

Table 17.1B : Portuguese Loan-words in Marathi Relating to Flora and Fauna

Sr. No.	*Portuguese word*	*Marathi word*	*Meaning in English*	*Bengali counterpart*
1.	Batata	Batātáh	Potato	Āloo

Compiled by the authors.

Table 17.1C : Portuguese Loan-words in Bengali and Marathi Relating to Flora and Fauna

Sr. No.	*Portuguese word*	*Bengali word*	*Marathi word*	*Meaning in English*
1.	Ananás	Anāas	Anānás	Fruit pineapple
2.	Couve	Kopí	Kobī	Vegetable : cauliflower
3.	Papaia	Peńpe	Papayá Papai	Fruit/Veg. : papaya
4.	Caju	Kājú	Kaju	Fruit : cashew
5.	Pera	Peyará	Perú	Fruit : guava
6.	Cintra	Santára (not in use)	Santrá	Fruit : orange
7.	Afonso	Alfońso	Hapús	A variety of mango
8.	Peres	Pairí	Pairí	A variety of mango
9.	Tabaco	Tamāk	Tambákhu	Eatable leaf tobacco
10.	Cacatú	Kakatuwa	Kakakúwa	A bird-Cacatu
11.	Cha	Chá	Chahá	Tea

Compiled by the authors.

European origin, the others were as much unknown to the Portuguese before they started explorations in different continents. Through interaction with the native populations they gathered knowledge about these plants and birds of South American, Malayan, African, Indian or Chinese origin and adapted their names to suit their tongue. These Portuguese names were further modified morphologically and phonetically in Bengali and Marathi to constitute an essential part of their flora and fauna terms. The following group of tables consists of the words denoting interaction of social, political and religious nature between the migrant Portuguese society in India, and Bengali and Marathi speaking population residing in the eastern and western parts of the country respectively. As mentioned earlier, these words, dënoting interaction, have been further sub-classified depending on the specific nature of interaction as exemplified in the related tables.

Table 17.2.1A : Portuguese Loan-words in Bengali Denoting Socio-Economic Interaction

Sr. No.	*Portuguese word*	*Bengali word*	*Meaning in English*	*Marathi Counterpart*
1.	Quarenta	Koreńta (not in use)	Forty	Chalís
2.	Trenta	Tereńta (not in use)	Thirty	Tees
3.	Primeira	Premāra	A card game	-
4.	Vinte	Bintí	A card game	-
5.	Viole	Baila/Behalá	Violin	Vyaóleen
6.	Entaro	Entār	Enough	Bharpúr
7.	Acabar	Kābár	Over	Akhẽr
8.	Maria	Mairí	Swear word in the name of the mother	Dewashapath
9.	Resto	Résto	Wherewithal, money available for expenditure (usually in journey)	Watkhúrcha
10.	Bombardeiro	Bombéte	Pirate (in Portuguese: Bombardier)	Chãnchá

Compiled by the authors.

Table 17.2.1.A. shows considerable evidence of social interaction exemplified by names of card games (*korenta*, *terenta*, *premara*, *binti*), high frequency use words (*entar*, *kabar*, *resto*), musical instrument name (*behala*), swear word (*mairi*), etc. Use of loan-words '*pasar*' and '*pagar*', in common Marathi parlance, is an indication of a high degree of social interaction with considerable seepage.

Table 17.2.1.B : Portuguese Loan-words in Marathi Denoting Socio-Economic Interaction

Sr. No.	*Portuguese word*	*Bengali word*	*Meaning in English*	*Bengali Counterpart*
1.	Passar	Pasar	Death	Biǵato/Mríto
2.	Pagar	Pagár	Salary/Pay	Betón

Compiled by the authors.

Table 17.2.1.C : Common Portuguese Loan-words in Bengali and Marathi Denoting Socio-Economic Interaction

Sr. No.	*Portuguese word*	*Marathi word*	*Bengali word*	*Meaning in English*
1.	Sorte	Sorát	Surtí	Lottery
2.	Armada	Aarmad	Harmād	Pirate (in Portuguese : warship)

Compiled by the authors.

Table 17.2.2 A : Portuguese Loan-words in Bengali Denoting Political Interaction

Sr. No.	*Portuguese word*	*Bengali word*	*Meaning in English*	*Marathi Counterpart*
1.	Cafre	Kaphrí	Negro	Negro
2.	Dinamarquês	Dinemár	Dane	Deních

Compiled by the authors.

Table 17.2.2 B : Portuguese Loan-words in Marathi Denoting Political Interaction

Sr. No.	*Portuguese word*	*Marathi word*	*Meaning in English*	*Bengali Counterpart*
1.	Tronco	Turáng	Jail	Karagār/Jel

Compiled by the authors.

The political loan-words are commonly shared by Bengali and Marathi language communities. It seems both had come to know about western arms and ammunition and European nationalities considerably through the Portuguese.

With the Portuguese, the penchant for colonisation went hand in hand with their zeal for proselytisation. Religious conversion in course of time thus gave rise to an Indo-Portuguese community in the pockets of Portuguese dominance. This accounted for large-scale infiltration of Portuguese loan-words into Bengali and Marathi, most of which have been left out of the present tables for reasons stated earlier. The tables include only the religious loan-words of wider social acceptance.

Table 17.2.2 C *: Portuguese Loan-words in Bengali and Marathi Denoting Political Interaction*

Sr. No.	*Portuguese word*	*Bengali word*	*Marathi word*	*Meaning in English*
1.	Cartucha	Kārtuj	Kartús	Cartridge
2.	Pistola	Pistól	Pistool	Pistol
3.	Bomba	Bōmá	Bōmá	Bomb
4.	Ingles	Ińgrej	Angréj	English
5.	Francez	Farashi	Frasisí	French

Compiled by the authors.

Table 17.2.3 A : Portuguese Loan-words in Bengali Denoting Religious Interaction

Sr. No.	*Portuguese word*	*Bengali word*	*Meaning in English*	*Marathi Counterpart*
1.	Igreja	Girjá	Church	Church

Compiled by the authors.

Table 17.2.3 B : Portuguese Loan-words in Marathi Denoting Religious Interaction

Sr. No.	*Portuguese word*	*Marathi word*	*Meaning in English*	*Bengali Counterpart*
1.	Natal	Natál	Christmas	Kristmas Barodin

Compiled by the authors.

Table 17.2.3.C : Portuguese Loan-words in Bengali and Marathi Denoting Religious Interaction

Sr. No.	*Portuguese word*	*Bengali word*	*Marathi word*	*Meaning in English*
1.	Cristão	Khristán	Kiristān	Christian
2.	Cruz	Krush	Krush	Cross
3.	Padre	Pādrī́	Pādrī́	Father/ Bishop

Compiled by the authors.

The following group of tables indicate the interaction between the migrant Portuguese society and the concerned Indian societies in terms of commerce and technology.

Table 17.3 A : Portuguese Loan-words in Bengali Relating to Commerce and Technology

Sr. No.	*Portuguese word*	*Bengali word*	*Marathi word*	*Meaning in English*
1.	Grade	Garaad	Gaj	Window frame
2.	Janela	Jānāla	Khirkī́	Window
3.	Escrevente	Kerānī́	Karkoon	Clerk
4.	Toca	Túka/Tóka	Nakkál	Copying
5.	Marca	Markā	Síkka	Seal
6.	Calafaté	Kālāpātī́	Lambī́	Caulking
7.	Mastio	Mastúl	Dolkathi	Mast
8.	Alfinete	Alpín	Tachnī́	Pin
9.	Armario	Ālmārī́	Kapāt	Almirah
10.	Espada	Ispát	Polād	Steel

(Contd)

Table 17.3A(Contd.)

11.	Cris	Kirĭch	Koytá	Knife
12.	Prego	Perék	Khilá/Molá	Nail
13.	Funil	Fundĭl	Nálke/ Narsal	Funnel
14.	Salsa	Sálshā	Raktashódha	Blood purifier
15.	Arsenico	Séṅko	Somál	Arsenic
16.	Carana	Katán (not much in use)	Súra/Súri	Knife
17.	Alcatrão	Ālkatrá	Dambār	Pitch
18.	Cana	Khāná (not much in use)	Khádde	Ditch

Compiled by the authors.

Table 17.3 B : Portuguese Loan-words in Marathi Relating to Commerce and Technology

Sr. No.	*Portuguese word*	*Marathi word*	*Bengali Counterpart*	*Meaning in English*
1.	Visagra	Vijāgar/ Vijagari	Kabjá	Vise
2.	Prancha	Paranchĭ	Bhāra	Scaffold
3.	Ripa	Reep	Aarkath	A side pannel
4.	Tanque	Tankĭ	Tyank	Water tank
5.	Contrato	Contrat	Chuktĭ	Contract
6.	Louvado	Lawád	Shalĭsh	Agent/Go-between
7.	Arratel	Rattál	Oján	Weight
8.	Umbreira	Umbártha/ Umbára	Chowkáth	Threshold
9.	Granade	Garnār	Grenēd	Grenade

Compiled by the authors.

Table 17.3 C : Common Portuguese Loan-words in Bengali and Marathi Relating to Commerce and Technology

Sr. No.	*Portuguese word*	*Bengali word*	*Marathi Counterpart*	*Meaning in English*
1.	Verga	Bārga	Wárga	Wooden rod in the ceiling
2.	Gudão	Gudaṁ	Gudám	Godown
3.	Leilão	Neelám	Leeláo	Auction
4.	Mestre	Mistrí̆	Mestrí̆	Technician
5.	Capitão	Kaptán	Kaptán	Captain
6.	Boia	Boyá	Boia	Buoy
7.	Bomba	Bōmá (not much in use)	Bomb	Waterpipe

Compiled by the authors.

The inadequacy of the material culture and the desire for novelty provoke the receptor language community towards adopting loan-words of the above type. Words that signify a new tool, an innovative method or an unfamiliar commercial system readily become choices for adoption by the host language community. Such items are exemplified by the above tables.

The following group of tables consist of Portuguese loan-words relating to household and daily use, thereby indicating a close interaction of day to day level between the migrant Portuguese and the respective Indian communities.

Initially socio-cultural traditions associated with household activities may give rise to reluctance and poor receptiveness to alien domestic utilities. But gradually new utility words are borrowed for pragmatic reasons, while old ones are at times even replaced for the sake of novelty and social prestige, if the source language community commands a higher status. A sizeable scale of Portuguese loan-words for household articles and daily use included in the above tables illustrates such a process.

Table 17.4 A : Portuguese Loan-words in Bengali Relating to Household and Daily Use

Sr. No.	*Portuguese word*	*Bengali word*	*Meaning in English*	*Marathi Counterpart*
1.	Cadeira	Kedārá	Armchair	Khurchi
2.	Tigela	Tijél (not in use)	Plate	Patéle
3.	Toalha	Toālé	Towel	Towél
4.	Talha	Toló	A kind of plate	Gadgé
5.	Boião	Boyám	Small jar	Baráni
6.	Biscoito	Biscoot	Biscuit	Biskīt
7.	Vasilha	Besālī́	Vessel	Bhānda
8.	Botelha	Botál	Bottle	Batlī́
9.	Camara	Kāmrá	Room	Kholī́
10.	Touca	Toká	A kind of hat	Topī́
11.	Botão	Botām	Button	Battán
12.	Loba+Abade	Lobáde	Gown	Jhagá
13.	Saia	Shaía	Petticoat	Parkár
14.	Pires	Pirích	Saucer	Bashī́

Compiled by the authors.

Table 17.4 B : Common Portuguese Loan-words in Marathi Relating to Household and Daily Use

Sr. No.	*Portuguese word*	*Marathi word*	*Meaning in English*	*Bengali Counterpart*
1.	Bucha	Booch	Cork	Chhipī́
2.	Bacia	Bashī́	Small plate	Plēt

Compiled by the authors.

Table 17.4 C : Common Portuguese Loan-words in Bengali and Marathi Relating to Household and Daily Use

Sr. No.	*Portuguese word*	*Bengali word*	*Marathi word*	*Meaning in English*
1.	Achar	Aachar	Aachar (less used)	Pickle
2.	Estirar	Istirí/Istrí	Istirí	Iron
3.	Camisa	Kameez	Khameez	Shirt
4.	Gamela	Gamlā	Ghaméla	Shallow bucket
5.	Chave	Chabĭ́	Chawĭ́	Key
6.	Prato	Parát	Parath	Big plate
7.	Pão	Paṅrúti	Pao	Bread
8.	Pipa	Pipē	Pip	Wooden jar
9.	Fita	Fitā	Fith (less used)	Tape
10.	Balde	Baltĭ́	Baldí/Badlí	Bucket
11.	Sabão	Sabaan	Sabú/Saban	Soap
12.	Jalape	Joláp	Juláb	Purgative

Compiled by the authors.

Conclusion

Language is a repository of the total interactive system of a community. The multi-dimensional experience that a community gathers in the course of its journey on earth finds a congruent expression in words, a good part of which are the relics of the past, and very often, a product of confrontation or interaction with other communities. Bengali and Marathi are no exception to this basic postulation. For common historical and political reasons, during Muslim rule, these two languages had borrowed heavily from Arabic and Persian, as in the recent past from English legions of words have been and are still being borrowed, because of its abiding influence in public life. Between these two periods there was a European interregnum from the early 16th century to the late 18th century. The Portuguese, among other powers like the French, Dutch and Danish, competing for colonial gains, managed to establish pockets of informal interaction through culture and trade in Bengal and the west coast and created considerable political, econcmic and socio-cultural impact on the local people. The loan-words which they provided to the languages of these two respective regions were thus, mainly derived through verbal communication unlike Arabic, Persian or English, in case of which literary and official sources dominate. Hence, Luso-Indians and Portuguese settlers were largely responsible for diffusion of Portuguese loan-words in this process. The presence of Portuguese loan-words in Bengali and Marathi also offers a contrasting example of multifarious informal impact to the official impact of Portuguese rule in Goa and the resultant inclusion of innumerable Portuguese loan-words in Konkani, the local language.

The above analysis although deals with one aspect of acculturation in India, namely linguistic, it does throw light on the general problems of acculturation. It highlights the fact that on most ocassions of borrowing, language has complemented and reflected non-linguistic acculturation, which implies that language and non-linguistic aspects of culture seldom act differentially.[9]

References and Notes

1. De Souza, T.R. (1985) (ed). *Indo-Portuguese History*, New Delhi : Concept Publishing Company.
2. Campos, J.J.A. (1919). *History of the Portuguese in Bengal*, Calcutta : Butterworth & Company.
3. Chatterjee, S.K. (1976). *The Origin and Development of the Bengali Language*, London: George Allen & Unwin Ltd.
4. Danvers, R.C. (1988). *Portuguese in India*, London : W.H. Allen.
5. Dalgado, M.S.R. and Soares, A.X. (1936). *Portuguese Vocables in Asiatic Languages*, First AES Reprint 1988, New Delhi: J. Jetley for Asian Educational Services.
6. Hagerstrand, T. (1952). 'The Propagation of Innovation Waves', *Lund Studies in Geography*, Series B, Human Geography, No.4, Lund, Gleerup.
7. Afendras, E. (1969). 'Sociolinguistic History, Sociolinguistic Geography and Bilingualism', *International Days of Linguistics*, Rome : Luigi Sturzo. Institute, pp. 663-82.
8. The lists of Portuguese loan-words in Bengali and Marathi have been prepared by the authors with the help of the following dictionaries and glossaries. Dalgado, M.S.R. and Soares A.X. (1936). *Portuguese Vocables in Asiatic Languages*, First AES Reprint, 1988, New Delhi: J. Jetley for Asian Educational Services; Bandyopadhyay, H. (1988). *Bangiya Shabdakosh, Vol. I & Vol. II*, New Delhi : Sahitya Academy ; Chatterjee, S.K. (1976). *The Origin and the Development of the Bengali Language*, London : George Allen & Unwin Ltd; Das, P.P. (ed) (1988). *Banglaye Atithi ShabderAbidhan*, Calcutta : Annapurna Prakashani ; Kulkarni, K.P. (1993). *Marathi Vyutpatti Kosh, 2nd Edition*, Pune : Shuvada Saraswati Prakashan; Joshi P.N. (ed) (1970). *Adarsh Marathi Shabdakosh*, Pune : Vidarbha Marathwada Book & Co.
9. Dozier, E.P. (1968). 'Two Examples of Linguistic Acculturation : The Yaqui of Sonora and Arizona and the Tewa of New Mexico', *Language*, Vol. 32 (1). pp. 146-157.

18

SEQUENTIAL CHANGES IN OCCUPATIONAL STRUCTURE AND RESOURCE AVAILABILITY AMONG THE KATKARI OF MAHARASHTRA

RENUKA PATNAIK

Introduction

The Katkaris of Maharashtra are a fairy large cultural group concentrated mainly in the two northern Konkan districts of Thane and Raigad and, distributed on the Sahyadris and the adjacent plateau. Traditionally involved in occupations associated with either the forest or river such as fishing, hunting, food gathering and especially the preparation of *katha* from acacia trees and charcoal from a variety of trees, the Katkaris have like many other tribes in India, undergone a marked change in their occupational structure . These changes are related to changes in the natural resource base and the larger infrastructural fabric in which their particular mode of living existed and functioned. Ethnographic literature on the Indian tribes frequently provides information and insight to the gradual shrinking of the natural resource base of forest caused by external factors which also abridge their rights and limit their access to them. This forces them either to move out of their traditional habitat temporarily or device survival strategies by changing over to completely different occupations and modes of living for neither of which their traditional skill or traditional cultural traits are efficient or even suitable. We have, thus, a telling example of the concept of maladaptation, the condition in which at least some of the original traditional culture traits when applied to the changed habitat lineaments, proved to be totally non-functional. The Katkaris too reveal a history of sequential changes in the mode of living associated with the process of maladaptation, mainly caused by the partial snapping of the umbilical linkage with their traditional resource base of forest, and conditions of very low literacy, low sex-ratio, and very small family size, that do not permit a revelation of the process.

ETHNOGRAPHIC OUTLINE

Origin and Past Movements of the Katkaris

According to Campbell,[1] the word Katkari, the name of the tribe, is derived from *Kath* which in Sanskrit means *Kvath*, something boiled. One of the essential features in the making of *katha*, one of the traditional occupations of the Katkaris is that of boiling the *Acacia catechu* bark, and then allowing the thick liquid syrup to solidify. This relationship of the name of the tribe ultimately with the Sanskrit word, thus, suggests that the Katkaris are a very old tribe. Also the logic of relating the name Katkari to the occupation is convincing. It is tempting to suggest another relationship of the name Katkari with *katha* in that it was and still remains an essential ingredient in the preparation of *paan* (betel leaf) for chewing, a practice that was widely prevalent among the Hindus in ancient times and continues till today. These evidences do suggest that Katkaris are indeed an ancient tribe.

However, there is no concensus regarding their place of origin and past movements (Fig. 18.1). On the basis of the existence of a social division, *athavas* among the Katkaris, and *athavas* being the original name of Surat district in Gujarat state, Enthoven identifies the present-day district of Surat in Gujarat as the original area of concentration of the Katkaris from where they moved into Maharashtra.[2] According to him, the Katkaris, who were perhaps at one time distributed over large areas in Gujarat, where they are supposed to be derived from the Bhil tribe as indicated by a marked similarity in the dialects and customs of the two tribes, moved to their present location in Maharashtra from Gujarat during an early period.[3] Campbell, on the other hand, is of the opinion that the Katkaris moved into their present Sahyadri habitat from the Konkan.[4] Campbell further describes the Katkaris as migratory people in Ratnagiri district, often travelling northwards to the Khandesh region.[5] As a result of their wandering, at one point of time, they were also reported from northern Satara and to its west along the foot of the Sahyadris.[6] According to Vyas, however, the Katkaris were taken by the Bohra contractors from the Sahyadri habitat to the Udaipur region of Rajasthan where they worked as contract labourers in the preparation of *katha* from *Acacia catechu* trees which formed extensive forest in that region.[7] Through this process most of the Katkaris eventually settled in those areas over a fairly long period. But once the forests were heavily depleted they lost their traditional occupation and gradually moved back to the Sahyadri habitat. Some of those who did not return were eventually employed as agricultural and other kinds of labourers.

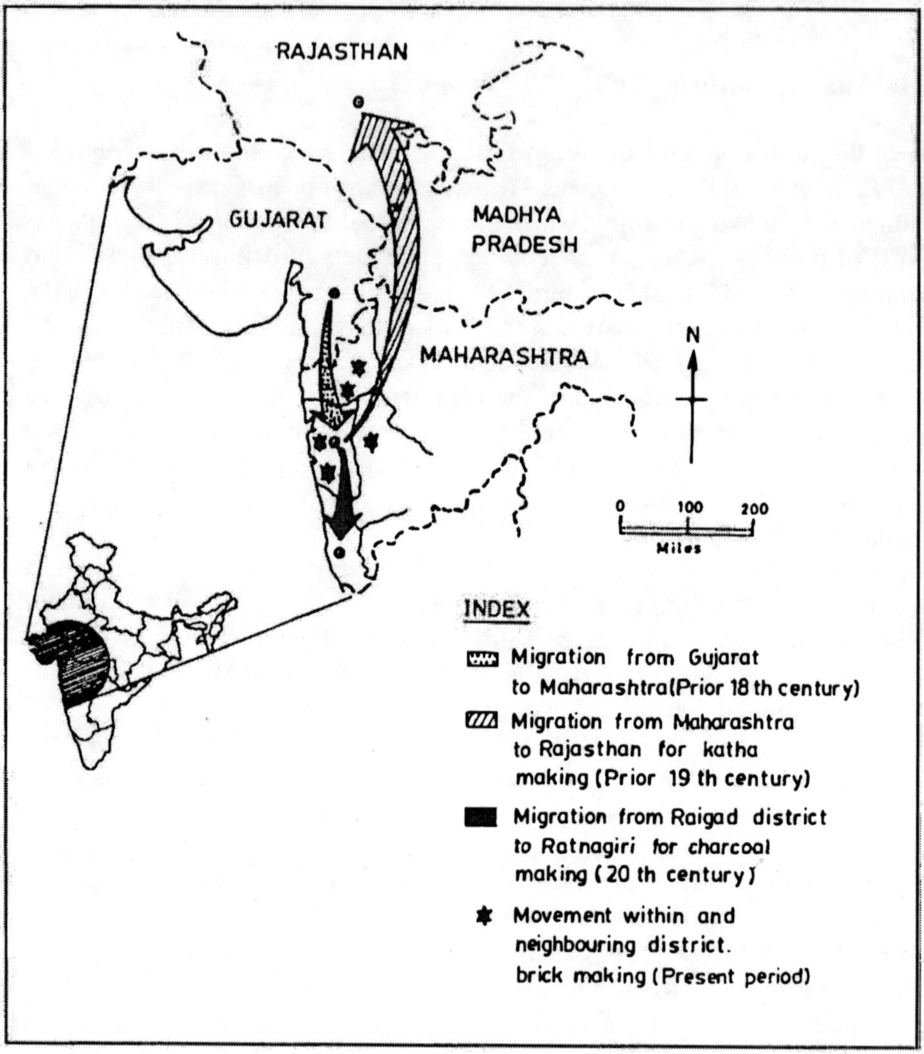

Fig. 18.1 : Migratory Movement of the Katkari Tribe (A Spatial—Temporal Pattern)

Social Divisions

There are two main social divisons among the Katkaris, the Son and the Dhors. These two groups are endogamous in nature and each of them is divided into a number of exogamous classes. The two first order divisions are based on their food habits, thus relating their formation to the available resource use patterns. The former do not eat the flesh of the cow and the brown-faced monkey.[8] The

Sons are also known as Marathas. It may be suggested that the Sons have been Hinduised to such a degree that they do not eat either beef or the flesh of the brown-faced monkey. Ethnographers suggest that both the Sons and the Dhors were eating practically all the animals of the forest in the past.[9] The habit of eating of a large variety of small and large animals was so widespread and so deeply rooted that the long array of the lower order social groups have their surnames derived from these animals as also from the trees that constitute the forest in which the animals reside. These interesting etymological features have been noted by many writers. The surnames actually have been identified as clan names.[10] Although the exogamous clan names have no significance for resource utilisation, they still suggest their distribution over large distances which has hindered mutual interaction thereby preventing them from evolving into a special community.

Ecological Setting of Katkari Habitat

The case study Katkari villages of Kashele, Malegaon, Raspe, Mograj, Tembre, Singdole, Nandgaon and Kothimbe, fall in the Raigad district which forms part of the Konkan coastal lowland of Maharashtra. It is surrounded by district Thane to the north, district Pune to the east and Ratnagiri to the south. The Arabian sea forms its western boundary, for about 250 kms (Fig. 18.2).

Raigarh district comprises highly eroded and broken escarpment (Fig. 18.3), the topography being undulating due to the presence of many spurs projecting from the Western Ghats. The angle of slope although generally three degrees, at some points reaches even seven degrees. The middle and lower slopes are equally intensely eroded. Thus the amount of land topographically suitable even for the simpler mode of field cultivation is very limited. The escarpment slopes and the coastal plateau comprise the Katkari habitat from where the resources of the upland as well as the escarpment can be utilised.

The area receives more than nintey-five percent of its annual rainfall during the monsoon months of June to September. In areas of high rainfall, highly porous soil causes the water to percolate into the lower horizons leading to acute shortage of water in the summer season. Moreover, the preponderantly westward sloping terrain promotes the rapid flow of the regional river waters towards the sea. The shortage of water and the infertile lateritic soils discourage the Katkaris from agriculture in the study villages.

The diversity of the soil type here is impressive. While the hill slopes are covered with the reddish and acidic soil characteristic of the Ghats and in the Konkan used for fodder crops, the sands near the coast are suitable for plantation of coconuts. The reddish brown and coffee brown soils are excellent for paddy cultivation.

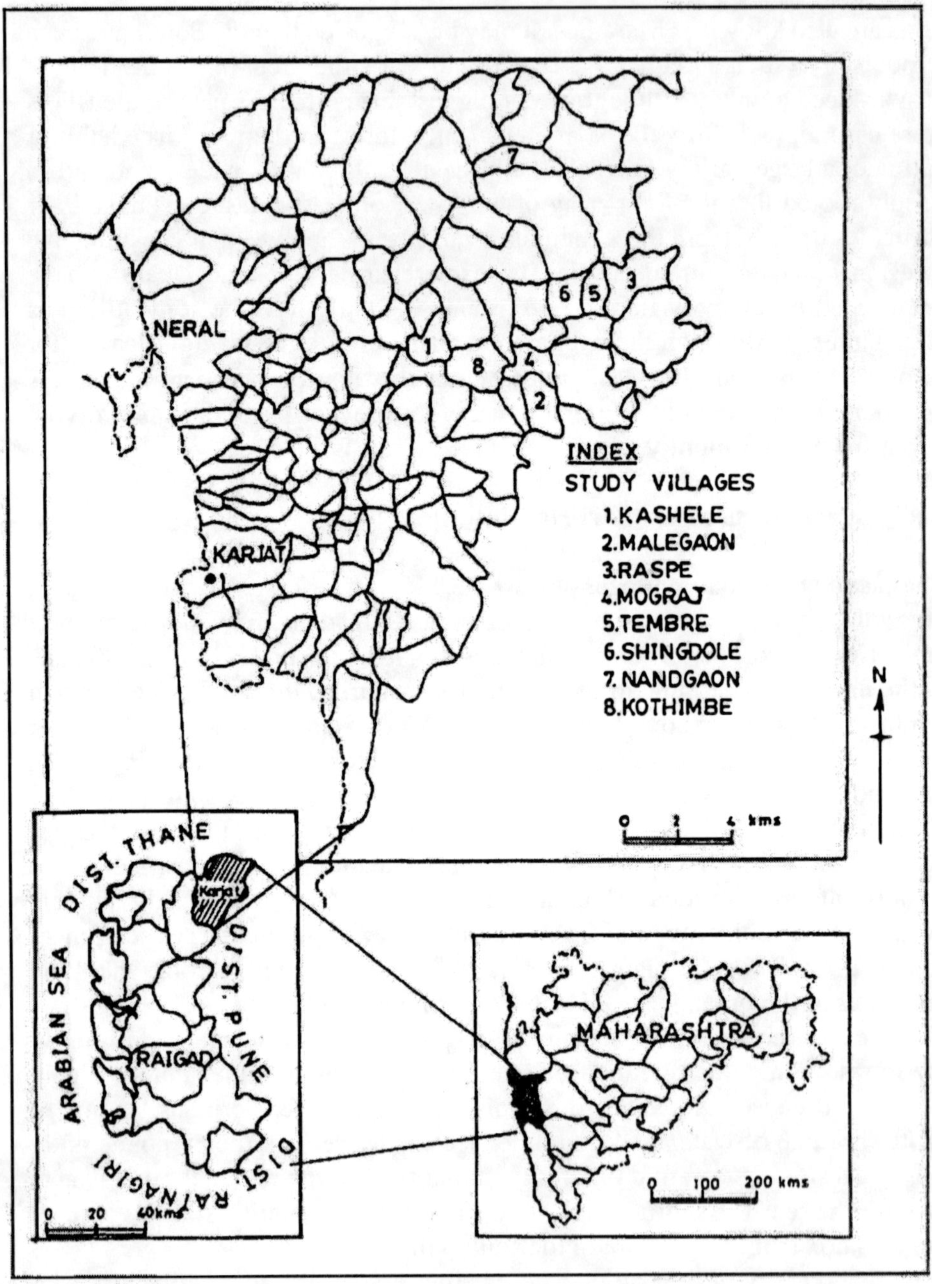

Fig. 18.2 : Karjat Taluka Showing Boundaries of Villages Selected for Case Study

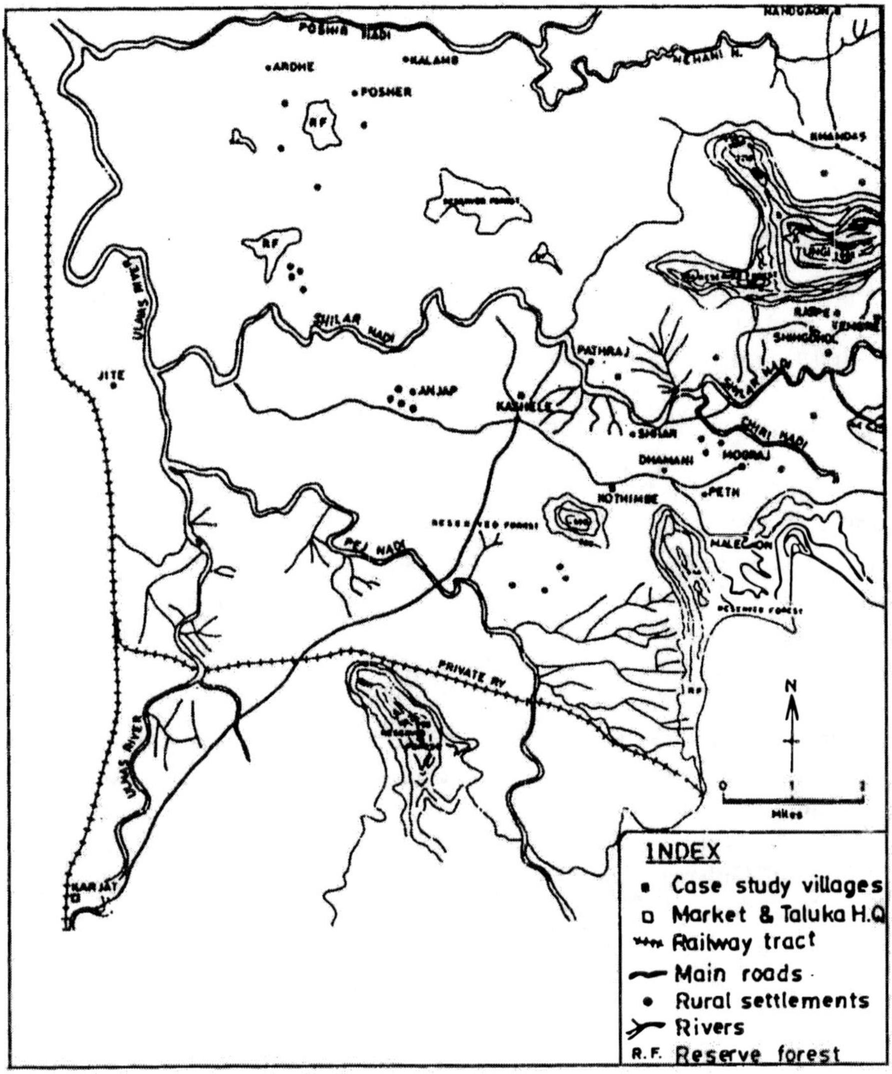

Fig. 18.3 : Physiography of Katkari Habitat (Showing Location of Case Study Villages) based on SOI Sheet No. 47 F/5

The area, at the conjuction of the Western Ghats scarp and the lower slope of the scarp, is traversed by numerous short rivulets such as the Ulhas, Amba, Savitri, Vasistha, Patal Ganga and Kundalika originating in the Sahyadris and merging with the Arabian Sea. Most of the streams are short and carry little discharge. Most of the springs however, are perennial and are used by the Katkaris for fishing. The higher slopes of the hilly areas in Raigad district experiencing heavy rainfall are covered by extensive forests dominated by teak. The Katkari settlements known as Katkari '*vadi*' are usually located inside the forest, on the escarpment, along the foot hills or in plain areas and along roads. The Katkaris of the plains mostly practice agriculture.

The Katkari settlements are usually away from the main village. While the agglomerate settlements are mainly seen inside the forest and on the escarpment, the Katkari settlements along the road and which are generally permanent are linear in form. Katkari families who move frequently usually stay in temporary huts which are square in form and made of mud, wood, grass and leaves. However, roofs of houses along the road and of permanent settlements are often tiled following government assistance. In the absence of a complex marketing system, as with subsistence tribal economies, in this case too, Karjat is the only nearest market town for Katkari villages to bring their surplus produce for exchange or monetary transactions.

The Katkaris generally walk from one place to the other, within a radius of 15 to 20 kms or may use bullock cart. However, their poverty prevents them from utilising the fairly efficient bus transportation available, especially from Kashele to Karjat except for carrying their collection to Karjat market.

Distribution and Demographic Profile of Katkari

The extent of the traditional habitat of the Katkari was much larger than what it is at present and in the beginning of the 19th century included the districts of Surat, Thane and Kolaba and Janjira state, all to the north. In the east the habitat includes some districts of Maharashtra, and towards the south the talukas of Khed, Chiplun, Guhagad, and Madangad of Ratnagiri district.

In Maharashtra the Katkaris are predominantly found in the Raigad and Thane district of the Sahyadri region, followed by Ratnagiri, Pune and Nashik. In the rest of the districts the Katkari population is negligible, less than 50 to 100 persons and in some districts like Buldhana, Bid, Akola, Solapur, it is even below 50 persons (Table I). Thus, it is clear that the Katkari population is more concentrated in the Sahyadri region in the western and coastal parts of Maharashtra. The concentration slowly decreases towards the eastern part of Maharashtra, which is largely a part of the Gondawana region of central India (Fig. 18.4).

Table 18.1 : District wise Distribution of Katkari

Sr. No.	*Name of the district*	*Total Katkari Population*	*Percentage of Katkari population*
1.	Raigad	75,680	43.34
2.	Thane	57,456	32.91
3.	Ratnagiri	21,265	12.12
4.	Pune	8,885	5.09
5.	Nashik	4,444	2.55
6.	Others	6,872	3.99
	Total	1,74,602	100.00

Regardless of location, all villages have a family size of either one to three or four to six persons. The villages located within the forests tend to have a preponderarce of small families. In Mograj for example, 71per cent of the total families classify as small. In villages close to roads and streams small size families become still more marked. The characteristically small size of the family among the Katkaris is related to the fact that most of them are nuclear in nature. There are very few, if any, multi-generational families, for immediately after marriage the young couple separates from the parents and starts living in a new hut.

The sex ratio among the Katkari population reveals a greater balance in the male and female population and thus represents a distinct departure from the general population. For the entire population of 751 persons, the percentage of males and females stands at 52 per cent and 48 per cent respectively. In the eight villages under study the sex ratio among the adult population is evenly balanced although some variations do exist. However, the sex ratio among the minor population (below 18 years age), is very low. This is primarily the result of lack of care and higher death rate among female minors than male minors in the rough and tumble of tribal life. This is so despite the absence of any discrimination against girls and women and the fact that both equally share the burden of earning livelihood.

The Katkaris are predominantly illiterate. The literacy level in all villages is dismally low. In Nandgaon, of the effective population of 63 persons, only one person is literate and in village Tembre the level of literacy is zero. Thus there is a pervasive condition of abyssmal illiteracy. For the entire Katkari population only 17 per cent are literate. The village-wise variation in the level of illiteracy reveals a stark picture of hopelessness about social development. Obviously the efforts of the state, no matter how adequate and sincere, have been brought to a nought as a consequence of the continuous tendency of the Katkaris with their families to move from one place to another; one, in response to the movement of contractors who supply labour for brick making and two, in the process of collecting minor forest products.

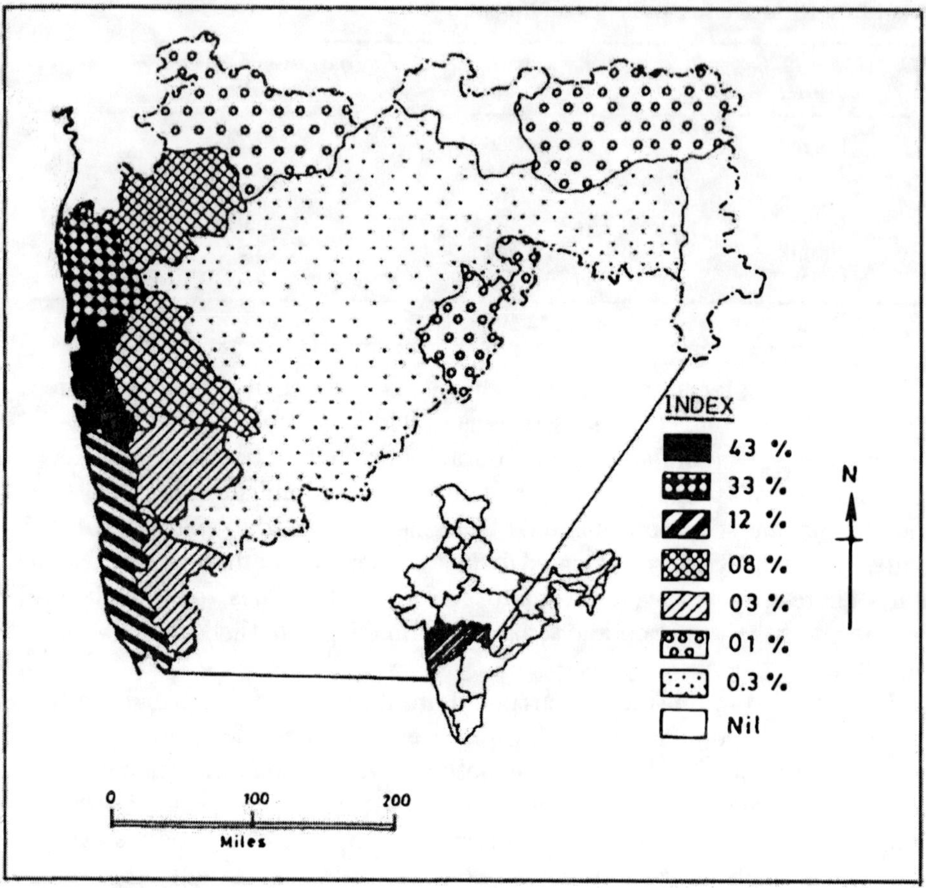

Fig. 18.4 : Concentration of the Katkari Population in Maharashtra

Occupational Pattern

The Katkaris, like most other tribes of Maharashtra and indeed of India have undergone distinct changes in their occupational pattern. Many of the changes have been caused by the felling of trees in the forests including the common species, *Acacia catechu* (*Khair*),which provided the very basic of their main occupation the making of *katha* and the making of charcoal. The emergence of commercial townships requiring a large volume of cargo labour, and the local demand for cheap agricultural labour employed on daily wages have made the Katkaris wage earners. Two interesting features of the pattern of the occupational changes of the Katkaris, although not obvious or apparent, are (a) the conditions which have induced changes in the occupational pattern are all external to the community, and (b) the changes in the occupational pattern have occurred while the community has continued to live within the traditional habitat or, returned to it to live permanently, after staying outside for a short period of time. As a result of the changes that were effected in the Katkari mode of living the occupational pattern has evolved through several stages. Each stage is characterised by a specific major occupation, while the minor occupations were common to all these stages. At least five distinct historical stages through which the occupational pattern has evolved have been identified.

Past Occupation (Major Occupation)

(i) *Katha Making:* About the middle of the 19th century and prior to the deterioration of forests and restriction of forest resource utilisation, the primitive Katkari tribes migrated to the forests of western Udaipur in Rajasthan for distilling *Acacia catcheu* for making of *Katha* from *Khair* trees.[11] According to Vyas, Katkaris were taken to Rajasthan for making of *katha* by Bohra forest contractors. 'The Bohra contractors persuaded some 300 Kathodi families to migrate from their native place to seek fresh employment in the forests of western Udaipur'.[12] In the beginning only males migrated to Udaipur to work in the *Khair* forest and returned to their native place when the *katha* season was over, a highly inconvenient arrangement. However, following strong assurances from contractors many Katkari families migrated and were employed in the forest areas of Udaipur district. However, they received rations which were not enough although they continued to live hoping that the conditions would improve, but as forests in the area depleted, the Katkaris were worst affected and had no option but to return to their native place in Maharashtra, where they gave up making *katha* altogether. *Katha* making which was carried on from November to April, was the main source of income. For the rest they were engaged in hunting, fishing, and collecting roots and fruits to supplement their income.

(ii) Charcoal Making: On their return to Maharashtra, the Katkaris switched over to manufacturing charcoal as an alternative principal source of livelihood and moved to Ratnagiri district. For this the Katkari households received some advance during the rainy season without any written agreement as this was the most difficult period to tide over since most food supplies were exhausted. The understanding was that the Katkari families were to work in the kilns operated by the party that provided them with the advance. All accounts were generally settled by the end of the season in May when work was terminated just before the beginning of the rains in June. The Katkari moved seasonally from the uplands to the Konkan plains to make charcoal from Diwali onwards till the beginning of the rainy season, *i.e.* from November to the end of May.

However, with the new Forest Act, the Government's policy has banned the making of charcoal as it requires felling of trees which leads to deforestation and enviornmental degradation, disturbing the fragile ecosystem. This in addition to the harassment by the kiln operators now has discouraged the Katkaris for moving for the making of charcoal.

Evaluation of Occupational Pattern

Thus, we find that the Katkari economy was based mainly on forest labour and forest produce. Their primary occupation was to prepare *katha,* while gathering and selling of forest produce provided them subsidiary food as well as a subsidiary source of income. Regular hunting and gathering which were the main economic activities formerly, have considerably declined with the decline of forest resources. Further, changes in forest policies have restricted their free handling of forests and their produce. As a result, this landless primitive tribe, the Katkari, has displayed distinct changes in their original occupation (Table 18.2).

Present Occupation

Faced with problems created by changes in forest laws, Government policies, deterioration of forests and also due to exploitation by contractors, the Katkaris gave up their traditional occupation of making *katha*. Further, they had to abandon the preparation of charcoal due to restriction on felling trees. Hence they earn their livelihood mainly as agricultural labourers and at the same time they engage in other activities such as daily wage labour in different seasons among which brick laying is a predominant activity.

(i) *Brick Making:* Brick laying leads to seasonal migration to different nearby places. This movement usually begins from November and lasts till the end of May. They are hired on a daily wage basis and the area of work is not specific or predetermined. Since payment is per family the entire family is engaged in the task of brick making. While men prepare the paste, women fill the moulds and the children shift them under the sun.

(ii) *Agricultural Labour*: Returning to their native place in the rainy season, all Katkari families work as agricultural labour, the men engaged in ploughing and levelling the fields and women in planting, transplanting, cutting, threshing and winnowing the grains on the fields of the landlords.

(iii) *Cargo Labour*: In the absence of a permanent resource base or income, the Katkaris are constantly engaged in different occupations, depending on the availability of resources in space and time. With no resources available to them they move to neighbouring urban places 20 to 30 kms. away from their native place to work as cargo labour, which involves the loading and unloading of trucks and related works.

(iv) *Cultivation*: Since most of the Katkari are landless, the proportion of cultivators, cultivating their own land are very few, varying between four per

Table 18.2 : Model Showing Five Distinct Stages of Socio-Economic Evolution Among the Katkaris

Stage of economic period	*Time period*	*Occupation*		*Resources utilised*	*Resources obtained*	*Habit*	*State of the environment*	*Spatial sphere of interaction*
development		*Primary*	*Secondary*					
I	Up to late 19th century	Catechu making	Hunting gathering & fishing	Animal, Birds, Trees & Khair (Acacia catechu)	Forests & streams	Purely Nomadic	Forested	Within forested areas
II	Early Up to mid 20th century	Charcoal making	Hunting gathering & fishing	Animals, Birds, Fishes,	Forests & streams	Purely Nomadic	Forested	Within forested areas
III	Mid 20th century upto Indepen-dence (1950)	Brick Making, Constru-ction and cargo labourer	Gathering Fishing	Fruits, Fishes, land water	Forests, streams soil mining	Semi Nomadic	Depleting forest resources	Within forest & in adjoining rural areas
IV	1960-80	Agri. & daily wage labourer, Brick making	Gathering fishing	Fruits, fishes Agri. land water	Forests streams & soil mining	Permanent settlement with seasonal migration	Degraded environ-ment	Minimum movement within forest areas but major movement is in rural and urban areas
V	1980 onwards to present situation	Agri. daily wage labourer and brick making in rural areas working in industry in urban area	Gathering fishing	Fruits, fishes Agri, land water	Forests, streams, soil mining urban area	Permanent settlement with seasonal and daily commuting movement	Depleted forest resources	Minimum movement within forest areas but major movement is in rural and urban areas.

cent in case of the village Kothimbe to twenty-two per cent in the village of Raspe. Of the 233 families surveyed only 20 families *i.e* nine per cent own land, that too dry lands with no irrigation facilities. As they do not possess cattle or any other agricultural implements, cattle are hired and seeds purchased from other tribals and non-tribals for agricultural operations.

Minor Occupations

It is well known that for long the tribes of India did not have a monetised economy and hence occupations, as understood by and in the non-tribal society, do not exist among the tribes. Undoubtedly they were engaged in an array of major and minor activities, mostly related to the generation of an assured supply of food for the household. In such a situation the term occupation acquires an altogether different meaning. In our discussion, occupation that constitutes the tribal economy, will be primarily concerned with household activities, with the ultimate aim of sustenance or survival. This particular connotation of occupation has been kept in mind in the discussion of major and minor occupations of the Katkaris.

(i) Hunting: Hunting has been a significant activity of the Katkari who used only the bow and arrow to kill rabbits, hares and monkeys. Though the Katkari is not a hunter by profession they supplement their livelihood by hunting animals. 'In the Ratnagiri district the Katkari is known as '*Vandermarya*' a hunter of monkeys, where his services are engaged for hunting or scaring away herds of monkeys that devestate gardens'.[13] The Katkaris are experts at hunting field rats using a stick or an iron rod and then in collection of their grain. Seasonally the Katkari catches frogs for consumption. However, now following the deterioration of forests and diminishing numbers of animals, only rats and frogs are available and that too with difficulty.

(ii) Gathering: Gathering too is a seasonal occupation for the Katkaris, particularly for those living inside the forest. The gathering and selling of forest produce contributes to their income as well as to their subsistence. Though the collection of fire wood is carried out throughout the year, the principal season for forest collection starts from March to the end of June. Both the Katkari couples and children engage themselves in different activities for the collection of *Jammun* (*Jambolana indica*), different types of berries (*Zyzyphus nummularia*), and mango. The adult members climb up the trees and pick the fruits while women and children collect and put them in baskets. They also collect several types of edible roots and leaves for their daily food.

(iii) Fishing: Fishing is one of the more profitable and advantageous occupations followed by the Katkari. In proximity to streams this activity has been specialized so successfully that it has developed into a reliable means and often a principal source of subsistence for the people during winter and early

Table 18.3 : Occupational Pattern of Katkari Families

Sr. No.	*Villages*	*No. of families inter-viewed*	*Relative location of the study villages*	*Brick making*	*Catch and sell fish*	*Forest coll-ection*	*Agri. labo-urer*	*Daily wage labo-urer*	*Agri. own land*
1.	Kashele	37	Road side	5 (14)*	37 (100)	—	37 (100)	37 (100)	—
2.	Malegaon	25	Near stream and forest	—	25 (100)	25 (100)	25 (100)	25 (100)	—
3.	Raspe	23	Inside forest	—	23 (100)	23 (100)	23 (100)	23 (100)	5 (22)
4.	Mograj	28	Inside forest	28 (100)	05 (18)	28 (100)	28 (100)	28 (100)	4 (14)
5.	Tembre	20	Near stream and inside forest on escarpment	20 (100)	20 (100)	20 (100)	20 (100)	20 (100)	—
6.	Shingdole	25	Near road & stream	10 (40)	20 (80)	25 (100)	25 (100)	25 (100)	2 (7)
7.	Nandgaon	30	Near road & stream	25 (03)	30 (100)	—	30 (100)	30 (100)	7 (23)
8.	Kothimbe	45	Road side	40 (89)	—	—	42 (93)	42 (93)	2 (4)
	Total	233		128 (55)	160 (69)	121 (52)	230 (99)	230 (99)	20 (9)

* = Figures in parantheses indicate percentages calculated out of total families interviewed.

summer. Since the poor soil of the hilly parts of the Konkan makes dependence on agricultural labour for the whole year impossible, the Katkaris engage in fishing from rivers, streams, or ponds, an activity which provides both money and food. The Katkari fish primarily by constructing dams or weirs, or by trapping or poisoning. The most popular method of fishing is by trapping, using a conical apparatus made up of thin bamboo chips.

Diversification of their occupations has been essentially the result of the rapidly diminishing forest with which in the earlier times they had very intimate relationship and on which they depended very largely for their livelihood. This diversification has removed them from the forests and into altogether new sources of livelihood and emerging opportunities.

Temporal Cycle of Work

A simplified account of the occupational pattern at the time of the year when it

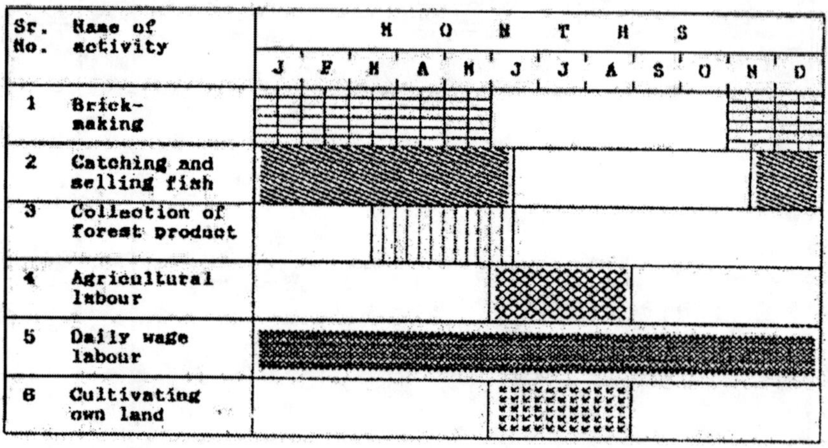

Note 1 : The period of one month represents a proportion of a little more than 8 per cent of the total time of year.

Note 2 : The sum of the percentages of time devoted to particular work does not come to 100 because some of the activity time-slots overlap and so the diagram provides a generalised picture of time-slots for different activities.

Fig. 18.5 : Temporal Cycle of Work Among the Katkaris of Village Mograj and Shingdole

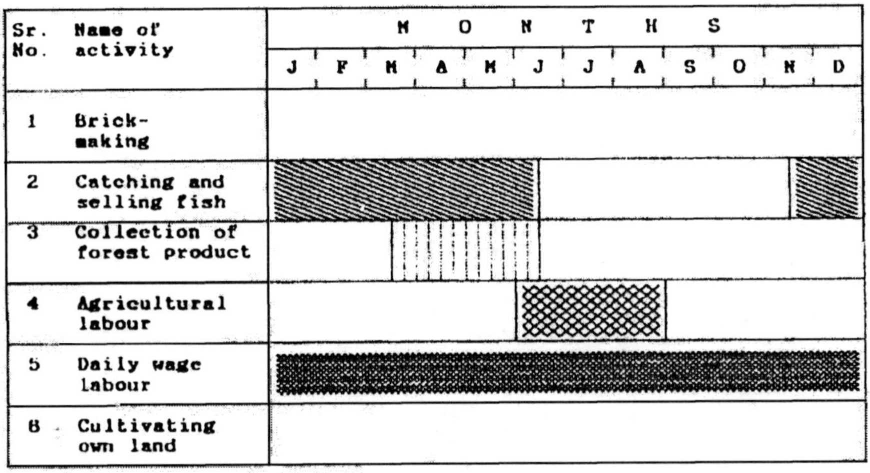

Note 1 : The period of one month represents a proportion of a little more than 8 per cent of the total time of year.

Note 2 : The sum of the percentages of time devoted to particular work does not come to 100 because some of the activity time-slots overlap and so the diagram provides a generalised picture of time-slots for different activities.

Fig. 18.6 : Temporal Cycle of Work Among the Katkaris of Village Malegaon

is performed, and the proportion of households engaged in it is provided in Figs. 18.5 to 18.11. They also present a generalised pattern over a number of years wherein there are changes in the proportion of families engaged in a particular occupation extending over particular months. Of the six occupations being still practised, three are concerned with the sale of labour, one with the exploitation of forests, another with streams and still another with the very limited land.

It needs to be emphasised that the individual Katkari families do not follow only one occupation throughout the year rather it is usual for them to follow different occupations during different months or periods of months in a year. These figures for temporal cycle of work of the different Katkari villages are similar to ergographs but there is an element of generalisation. (Fig. 18.8 Note 2) For each of the eight villages, separate figures have been prepared, which when superimposed reveal some very interesting features about time allotment to particular activities.

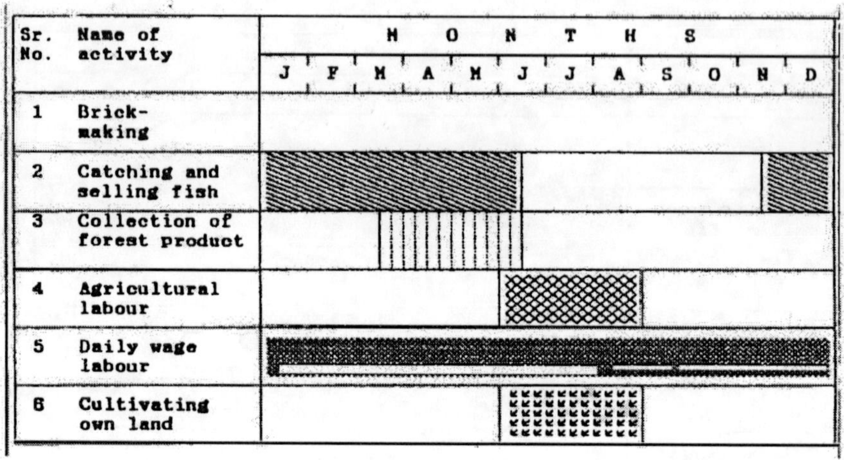

Note 1 : The period of one month represents a proportion of a little more than 8 per cent of the total time of year.

Note 2 : The sum of the percentages of time devoted to particular work does not come to 100 because some of the activity time-slots overlap and so the diagram provides a generalised picture of time-slots for different activities.

Fig. 18.7 : Temporal Cycle of Work Among the Katkaris of Village Raspe

1. Universally, the Katkari families are engaged as daily wage labourers primarily in the neighbouring market towns, throughout the year. There is no dearth of work relating to loading and unloading of a large range of goods and Katkari labour comes handy for contractors and wholesalers. Moreover it provides every Katkari family with some cash throughout the year even though, the money they earn does not get accumulated or converted into one or the other form of capital for which in any case they do not seem to have any dire or urgent need. Accumulation of capital is something that is still unknown to them and despite the efforts of the state to inculcate the habit of saving the Katkaris spend practically their entire daily earning on the day they earn, mainly on the purchase of liquor. Thus, although money is earned, capital is not accumulated.

2. With the exception of Malegaon and Raspe villages (Fig. 18.6 and 18.7) located within the forest, where there are no brick kilns, making of bricks is a common occupation. This activity extends from November to May end, through winter and the early summer months when little rainfall, constant sunshine,

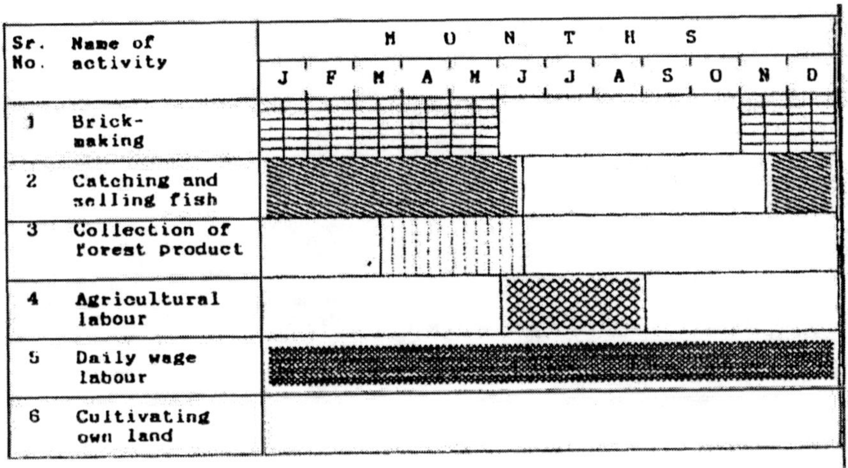

Note 1 : The period of one month represents a proportion of a little more than 8 per cent of the total time of year.

Note 2 : The sum of the percentages of time devoted to particular work does not come to 100 because some of the activity time-slots overlap and so the diagram provides a generalised picture of time-slots for different activities.

Fig. 18.8 : Temporal Cycle of Work Among the Katkaris of Village Tembre

high temperature and hot winds blow over the region, conditions highly favourable for brick making, both the dry and kiln burnt varieties. Also the reddish *murrum* soil derived from weathering of basalt, gives high strength and cohesiveness when made into bricks.

3. Since daily wage labour is carried on throughout the year, there is temporal overlapping of it with all other occupations. Four of the six occupations, catching and selling of fish, collection of forest produce, agricultural labour and cultivation of own land, extends to only three to four months and in terms of money earned are definitely secondary in importance. In terms of time slot too all occupations are secondary to the daily wage labour. With the exception of brick making which spreads over seven months of a year, minor occupations are limited to three to four months. The months during which the Katkaris are

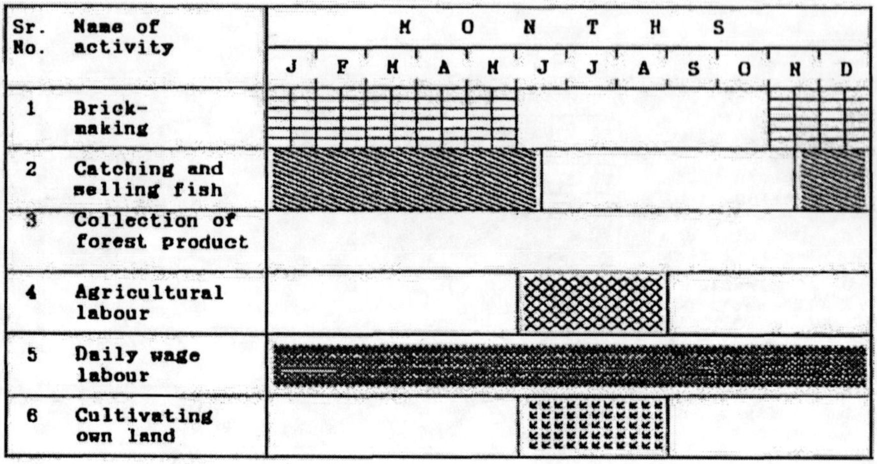

Note 1 : The period of one month represents a proportion of a little more than 8 per cent of the total time of year.

Note 2 : The sum of the percentages of time devoted to particular work does not come to 100 because some of the activity time-slots overlap and so the diagram provides a generalised picture of time-slots for different activities.

Fig. 18.9 : Temporal Cycle of Work Among the Katkaris of Village Nandgaon

engaged in a particular occupation, and the continuation of this engagement is largely dependent on the season and the fluctuations of weather. The Katkari catches and sells fish in the market between May and July when rivers have water and fish breeds in the numerous small pools of water. When the rivers are in flood from August to October they stop fishing. Thereafter fishing is interrupted due to small volume of water in the semi-dry rivers. Similarly, the Katkari are engaged in the collection of minor produce, particularly wild mango, *Jammun* (*Jambolana indica*), and a range of berries, the commonest being those growing on thorny bushes of *Zyzyphus nummularia*, from the beginning of March to the end of May, when the forests are full of these fruits. Since these fruits are seasonal, their collection extends over a short period. The amount of money earned by a Katkari family by selling fish or fruits is difficult to compute because they sell different amounts of these articles on different days within the season and at different prices. Occupation as agricultural labour and

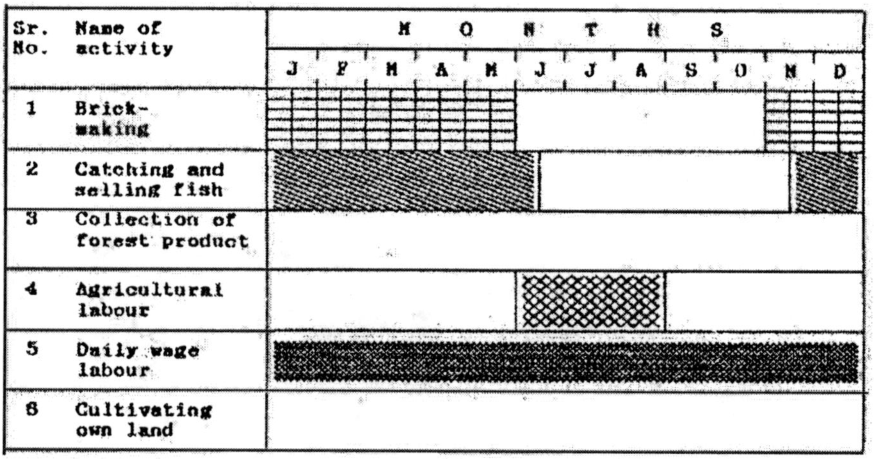

Note 1 : The period of one month represents a proportion of a little more than 8 per cent of the total time of year.

Note 2 : The sum of the percentages of time devoted to particular work does not come to 100 because some of the activity time-slots overlap and so the diagram provides a generalised picture of time-slots for different activities.

Fig. 18.10 : Temporal Cycle of Work Among the Katkaris of Village Kashele

cultivators is limited to only the rainy months of June, July and August, as all cultivation is invariably rainfed.

4. A pattern of multiple occupations during the same period is also evident from these diagrams. While marked seasons and changes in weather conditions are responsible for this, there is also the fact that the period in which a Katkari family is engaged in a particular activity does not hold it through everyday even within the period and therefore, shifts into other occupations are not by design but by chance. As a result several of the occupations overlap each other producing a rather confusing picture.

5. Another noteworthy fact is that all the occupations do not engage the Katkari families in each of the study village; on the other hand the individual periods of the occupations also vary from one village to another. For example in three villages, Tembre, Malegaon and Kashele; the Katkaris do not practice any cultivation while collection of forest produce is not practised in Kothimbe, Nandgaon and in Kashele, fishing is not practised in Kothimbe and brick making is not practised in Raspe and Malegaon. Why a particular occupation does not exist in a village or does not employ the Katkaris in a village is not clear, but

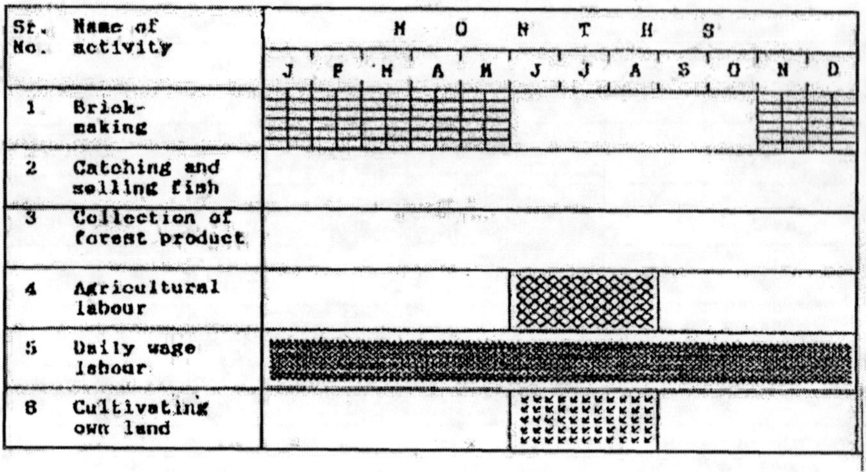

Note 1 : The period of one month represents a proportion of a little more than 8 per cent of the total time of year.

Note 2 : The sum of the percentages of time devoted to particular work does not come to 100 because some of the activity time-slots overlap and so the diagram provides a generalised picture of time-slots for different activities.

Fig. 18.11 : Temporal Cycle of Work Among the Katkaris of Village Kothimbe

some tentative explanations are obvious. Kothimbe for example, has a road-side location distant from both the forest and river. Again, cultivation is not practised in Kashele, Malegaon and Tembre simply because the families do not have lands of their own, so they work on others' lands as agricultural labour.

Family Income Pattern

The Katkari too have experienced monetisation of their economy and wages are increasingly being earned in terms of money for the work they perform rather than in kind. This was convincingly revealed in the field work and interviews. It was found that there was a preponderance of low income families. Except for one village (Mograj) in all the villages, more than 70 pre cent of the families earned between rupees 2,001 to 4,000 per year; which simply means that their monthly income ranged between 167 and 333 rupees, a situation indicative of stark poverty and of families trying precariously to survive on a daily income of six to twelve rupees. Even today therefore, the Katkaris to a very large extent depend on the game they hunt and on minor produce they collect from the forests. They accept payment in kind only some times, and catch fish whenever available in the streams.

The inter-village variations in the annual income is not very striking but if the average income of a family instead of the number of families belonging to different income groups is taken into consideration including all the families of all the villages, marked inter village variations come to the fore.

Computation of Income

Illiterate and unable to keep account of their earnings, the Katkari relate themselves more to the unit of labour they are selling rather than the amount of money they earn. Computation of income is difficult because their earnings are either in the form of daily wages or per product wage. Moreover, the extreme uncertainty of their employment; a family employed in the making of bricks in a particular year may not go for the same job the following year and in another year may switch over to a totally different occupation or may not go at all for a wage earning occupation, also makes computation of the variations in the monthly and seasonal income of the individual families extremely difficult.

A strong correlation, however, exists between family size and income, although income variation is also related to several other factors, such as location of villages; if for instance a village is located along the road, accessibility by bus is better as is the job potential as in villages like Kashele and Kothimbe, while villages such as Raspe and Mograj, located inside the forest depend mainly on collection of forest produce. Also the family structure that is the proportion of the old and children to adult working population shows greater association with income size. Availability of exploitable resources made significant difference to their potential earnings. The general trend observed in the case of the study villages reveals that, the population residing in the forest villages shows a greater propensity to move as their major occupation was related to collection of forest produce and fishing, while those in the plains have moved to the second stage of the occupational pattern of change, namely, agricultural labour. In the overall pattern one observes that there is little difference in their economic situation, except that livelihood strategies differ with differences in geographical location. Those within the reach of urban areas depend mainly on daily wage either brick making or in loading and unloading of trucks and road construction work.

Gender and Occupation

Unlike the surrounding areas where gender based specialsation of occupation is very marked, among the Katkaris both men and women are engaged in practically all major and minor occupations with the exception of employment as cargo labour wherein no women are involved. Within individual activities that constitute an occupation, specialisation of tasks to be done by men or women is clear. A Katkari family is both a social and production unit and the former is reflected in the latter. The entire family, except the old and children, work in a particular occupation and within this occupation the activity slot is performed by men, women, boys and girls separately. In fishing, both men and women using common technique and common equipment catch fish from the streams. Women carry fish on their head in the baskets and sell the catch in the market. Similarly while collecting wood, the male members climb the trees, cut the branches and throw them on the ground and the women collect the pieces of wood, tie them into bundles and carry them as headloads to their huts and occassionally to sell to the neighbouring non-tribals. A similar division of labour is observed when the men climb the trees to pick *jammun* (*Jambolan indica*), berries (*Zyzyphus nummularia*), and wild mango and the women carefully collect each of the items and put them in a basket and carry it on their heads to the market. The fruits collected, are sorted into various sizes and variety which involves enormous labour. An occupation that is increasingly becoming common for Katkari men only is that of cargo labour which involves loading and unloading of cargo from trucks in the neighbouring market towns. In brick-making, in contrast to cargo labour, both men and women are employed. In fact, as mentioned earlier, the entire family participates in this occupation, with the men, women and boys and girls performing their individual chores.

Thus, it is observed, that mainstream development has sucked the Katkari out of their traditional habitat as they become increasingly employed in construction activities of roads, dams, bridges, and buildings. The women too are engaged in the related task of breaking of stone, while the men carry the metal pieces to the site of construction.

In agriculture too while the men are engaged mainly in ploughing the fields and levelling them, it is the women who plant, transplant, thresh, winnow and clean the grain.

Spatial Entrapment

In their search for daily wage labour employment, the Katkaris have a strong tendency to restrict their movement within their traditional habitat, thus depriving themselves of better opportunities. The Katkaris move only over a radius of 2 to 5 kilometres for fishing, about 8 to 10 kilometres to work as agricultural labourers, and 30 to 40 kilometres to work as brick layers and cargo labourers. Thus this situation reveals a condition of spatial entrapment in which they suffer low wage daily employment. Their inability to venture into new areas for jobs and their limited skills and illiteracy are serious constraints. Money which is necessary for travelling is also difficult to come by. Moreover stark differences in the conditions prevailing in the new areas of potential opportunities could be so great that even when markets for labour are large and the wages attractive, the very thought of leaving their habitat for a distant area and the consequent weakening of the bond with the homeland acts as a powerful impediment to such possible movements.

Conclusion

The Katkaris, almost entirely concentrated in the Sahyadri region of Maharashtra, have been undergoing through long historical period, marked changes in their traditional occupational patterns as a response to radical changes in the natural resource base originally available in their ancient habitat. With a continual and major loss of forest resources on which their traditional occupation of making *katha* from *Khair* trees was based, the Katkaris have adopted at least three new and non-traditional occupations, brick making, cargo labour, and agricultural labour, that clearly comprise an adaptive survival strategy. As a result of this they are experiencing what can best be described as spatial entrapment in which the movements of the Katkaris are restricted in the general region in which they are distributed. For the first time in their ethnographic history they are now engaged as wage workers to earn money in occupations in which there is a clear separation of male work from female work. Easily the most interesting feature of the Katkari way of living has been the transformation in the occupational pattern in which the two major occupations of *katha* making and of charcoal making have completely disappeared with either the shrinking of forest or with the abridgement of the Katkari customary rights on the use of forest resources. As a response to the natural resources still available in their habitat, their temporal cycle of work reveals an overlapping in time periods of some residual traditional activities with some of the newly adopted occupations.

References and Notes

1. Campbell, S.J. (1884). *District Gazetteer, Thane, Vol. XIII*, G.C.P, pp. 158-165.
2. Enthoven, R.E. (1922). *The Tribes and Castes of Bombay*, Vol. II, New Delhi: Asian Educational Services, p. 170.
3. *See* ref. 2, p. 170.
4. Campbell. S.J. (1883). *District Gazetteer, Nashik, Vol. XVI*, G.C.P., p. 65.
5. Campbell. S.J. (1880). *District Gazetteer, Ratnagiri, Vol. X*. G.C.P., p. 130.
6. *See* ref. 1 p. 158.
7. Vyas, N.N., (1980). 'Kathodi A Tribe in Transition', *Indian Tribes in Transition*, (eds.) N.N. Vyas and R. S. Mann, Jaipur: Rawat Publications, p. 153.
8. Enthoven, R.E. (1922). *The Tribes and Castes of Bombay*, Vol. II, New Delhi: Asian

Educational Services, p. 182; Weling, A.N. (1934). *The Katharis : A Sociological Study of an Aborginal Tribe of the Bombay Presidency*, Bombay: The Bombay Book Depot, p. 51.
9. See ref. 2. p. 182.
10. Weling, A.N. (1934). The Katharis : A Sociological Study of an Aborginal Tribe of the Bombay Presidency, Bombay: The Bombay Book Depot, p. 52.
11. *See* ref. 7, p. 152.
12. *See* ref. 7, p. 152.
13. *See* ref. 8, p. 43.
14. *See* ref. 13, p. 52.
15. *See* ref. 13, p. 43.

19

WEST INDIAN EMIGRATION TO CANADA—1968-87

An Analysis

MAURICE CARNEY AND ASHOK K. DUTT

Introduction

The movement of people from one country to another has been prevalent since early times. People migrate from one area to another in search of various things. The migration of people from Europe to North America was fueled first by a need for religious freedom and later by a desire for economic betterment. Within the last three hundred years North America has probably absorbed more immigrants than any other area in the world.

Shifts in the Canadian Migration Policy

In Canada, the French were the first Europeans to arrive, followed by the British. The original population of the Dominion of Canada was composed of these two groups. In the 19th and 20th centuries, Canada attracted various European migrants. The immigrant composition of the Canadian population one hundred years ago resulted mainly from Canadian immigration policies. These policies were geared towards attracting Europeans who were willing to settle vast stretches of agricultural land in Canada. Though the Canadian government encouraged immigration, it was against allowing non-white peoples to immigrate to Canada. Kalbach states that in 1815 the Nova Scotia Assembly was appalled at the influx of Negroes from Bermuda, and near the end of the 19th century there was a growing concern over the number of Asians entering the country.[1] The concern over the number of Asians entering Canada continued from its Confederation days into the 1960s. During the 1960s Canada modified its immigration policies which resulted in an influx of non-white immigrants. Among these non-white immigrants who took advantage of the changed immigration policies were the West Indians. Before the 1960s there were West Indians in Canada, but their numbers were infinitesimal. During the 18th and 19th centuries there were West Indians who were brought to the shores of

Atlantic Canada as slaves. Ramcharan notes that non-whites such as Blacks from the West Indies and United States, as well as Chinese, Japanese, and East Indians from Asia only trickled into Canada prior to 1962.[2] Potter contends that entry of Caribbean peoples into Canada was recorded as early as 1796, when a band of runaway slaves was shipped from Jamaica to Nova Scotia.[3]

The West Indians who immigrated to Canada during the 20th century, prior to the 1960s, were for the most part domestic workers. During the period following World War II, Canada like most industrialized nations, witnessed an accelerated economic growth. The economic growth created a demand for workers to fill the menial jobs that the Europeans were vacating. The Canadian government filled this vacuum by recruiting domestics from Third World nations. Ramcharan states that in 1955 the Canadian government established a domestic service scheme with the British Caribbean governments allowing Caribbean women to work in Canadian households in order to fill the demand for household workers.[4]

The class and number of West Indians living in Canada changed drastically in the years following a 1967 immigration law that allowed the entry of non-whites into the country on a point system; over 2,25,000 West Indians have migrated to Canada since 1960s. This point system took into consideration the education, age, skill and employment possibility of the immigrant as the main criteria for entry. It has been argued by many that the increased migration to developed nations has removed the best skilled and professional workers of the Caribbean. As a result of the 1967 immigration act, the West Indians were coming from various countries throughout the Caribbean region. The fact that the Caribbean is relatively close to Canada and for the most part, English-speaking, also heightened their incentive to migrate. They ventured to Canada in the hope of achieving a better standard of living. They settled mainly in the large metropolitan areas throughout Canada where employment opportunities were maximum. The large influx of West Indians into Canada continued into the mid-1970s when stricter immigration policies were implemented by the Canadian government. The stricter policies curbed the influx of immigrants entering Canada in the late 1970s and the 1980s. This study investigates the intended settlement patterns of West Indian immigrants entering Canada from 1968 to 1987.

Factors Influencing West Indian Emigration to Canada, 1968 to 1987

Migration as a way of life has been occurring in the Caribbean for centuries. It is an integral part of the culture and tradition of the Caribbean people. Before World War II most of the migration took place within the Caribbean region. In the post-World War II era, Caribbeans began migrating to industrialized nations mainly because of the greater employment opportunities and the relaxed immigration policies in those industrialized nations. During the period following Second World War until the 1960s, Great Britain received most of the West Indian migrants. The countries sending most of the migrants were Jamaica and Trinidad and Tobago. Though the industrialized countries to which the West Indians have been migrating have varied, not much change has taken place in the countries from which the immigrants are coming.

This study concentrates only on the movement of West Indians to Canada, the countries sending West Indian immigrants to Canada, and the political, economic, social and demographic characteristics of such countries that might have acted as push factors. An attempt has also been made to point out any evidence of a 'brain drain' (the loss of professional and skilled workers of a nation through migration) in the Caribbean. The countries of Jamaica, Trinidad and Tobago, Haiti, and Barbados, which account for 88 percent of the total Caribbean migration in Canada from 1968 to 1987, have been analyzed along with other Caribbean nations (Fig. 19.1). Although these four countries represent 88 per cent of the total Caribbean emigration to Canada, they account for only 32 per cent of the total population in the Caribbean region.

Jamaican Emigration

Of all the source nations in the Caribbean, Jamaica has witnessed the largest number of immigrants leaving for Canada. From 1968 to 1987, some 97, 329 Jamaicans migrated to Canada. Jamaicans leaving for Canada accounted for 43 per cent of all Caribbeans emigrating to Canada during the twenty year period. However, Jamaica's 2.2 million inhabitants only account for 7.3 per cent of the total population of the Caribbean.

Emigration from Jamaica to Canada was minimal during the first period (1968-1972), accounting for only 24 percent of the total migration of Jamaicans during the twenty year period. However, this does not mean that Jamaicans were not leaving the country. Numerous Jamaicans had emigrated to Great Britain prior to this period. Some Jamaicans also left for the United States.

The second period (1973-1977) witnessed the largest flow of immigrants from Jamaica to Canada; some 42,938 Jamaicans left for Canada during this period, which accounted for 43 per cent of the total migration of Jamaicans to Canada during the twenty year period. Apart from economic and legislative pull factors in Canada, many have suggested that the changing political climate in Jamaica added to the large wave of emigration. Anderson states that during the 1970s, migration was propelled by the determination to escape the social upheaval and political violence that followed Prime Minister Manley's decision to follow a socialist direction.[5] The Manley government which took power in

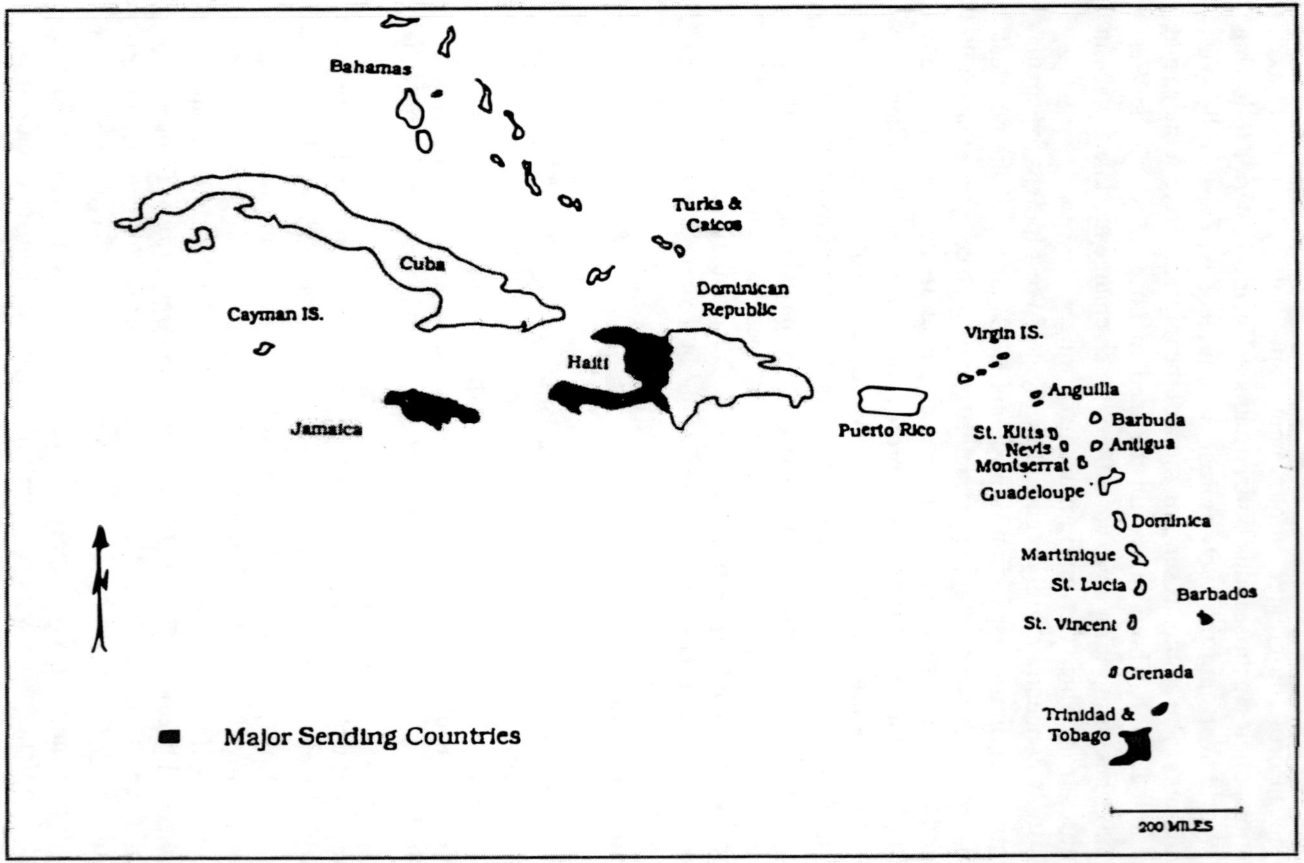

Fig. 19.1 : West Indies : Major Sending Countries
Source : Employment and Immigration, Canada, 1990

1973, voiced its desire to build a socialist society which caused many Jamaicans to consider leaving the country. When Manley developed close ties with the Cuban communist leader Fidel Castro, many Jamaicans may have believed that total socialist reform was imminent, thus influencing more Jamaicans to leave. It is believed that the Jamaicans who left the country were mostly skilled workers. Michael Manley's desire to build a socialist state did not simply engender adverse emigration, it also brought on a 'brain drain' and loss of a large number of skilled workers.

During the 1973 to 1977 period, most of the Jamaicans who left for Canada had intended to be employed in clerical, service and fabrication and assembly repair jobs. However, evidence of the 'brain drain' was not strong during this period. Some skilled workers were leaving, but the professional workers were few.

The peak year of Jamaican migration to Canada was 1974 when 11,284 Jamaicans emigrated. That was the year after Manley made his declaration that his party sought to implement a policy of Democratic Socialism. However, the announcement came at the end of the year and could not have affected the years' migration totals; therefore, one must be cautious in drawing inferences from this situation.[6]

The third period of this study (1978 - 1982) witnessed a decline in Jamaican migration to Canada; only 15,819 Jamaicans emigrated to Canada during this period which accounted for 16 per cent of the total migration from 1968 to 1987. For the first part of this period (1978-1980) Manley was still in power. However, there was a marked decline in the number of people emigrating from Jamaica to Canada. Most of the migrants leaving Jamaica had intended to be employed in clerical, service and fabrication and assembly repair jobs as was the case in the previous period. However, a rise did occur with those people who intended to be employed in professional fields such as medicine, administration and the sciences. Due to the stronger representation of workers in the professional fields, the 'brain drain' theory was more evident during this period.

The final period from 1983-1987 witnessed a continual decline of Jamaicans emigrating to Canada. Jamaicans who migrated to Canada in this period accounted for 17 percent of the total from 1968 to 1987. They were mostly people who had intended to be employed in service and fabrication and assembly repair jobs. A large increase occurred in the percentage of Jamaicans choosing the service area as an intended employment. The percentage of Jamaicans intending to be employed in service jobs increased almost threefold. This may have been a result of the demand for labour in the expanding tertiary sector in Canada during this period. Those Jamaicans looking for professional jobs were negligible. As was the case in the second period (1973 to 1977), the jobs that the Jamaicans sought did not indicate much of a 'brain drain'. From 1973 to 1987 the Jamaicans chose the service, fabrication and assembly repair and clerical fields as the primary areas of intended employment. The professional fields were not well represented which makes an argument for the 'brain drain' more difficult. However, some evidence, though scant, does exist of a 'brain drain' (Fig. 19.2).

In the Jamaican case, the major push factors appear to have been political dissatisfaction, grouped with social and economic duress. Anderson states that some people believed that their personal and social survival depended on their escape from the island.[7]

Trinidadian Emigration

Trinidad and Tobago provided the second largest number of West Indians who migrated to Canada from 1968 to 1987. During this period, 46,348 Trinidadians emigrated to Canada, which accounted for 21 per cent of Caribbean immigration to Canada. As is the case with the Jamaicans, the Trinidadians make up a substantial per centage of West Indian immigrants to Canada, but they account for a small per cent of the total Caribbean population. Trinidadians only comprise 3.9 per cent of the total Caribbean population. Most Trinidadians who migrated to Canada did so during the first two periods (1968 to 1972 and 1973 to 1977). Forty-two per cent left in the first period and 38 per cent in the second. The two periods accounted for 80 per cent of the total Trinidadian migration to Canada from 1968 to 1987.

The percentage of Trinidadians leaving during the first period (1968 to 1972) coincides with the troubled economic situation of the country. During this period the country experienced a sizeable rise in unemployment which might have induced people to emigrate.[8] However, regarding Trinidadian migration to Canada, the high percentage of migration during the second period does not correspond to the economic situation. About Trinidad, Diaz-Briquets stated that 'following the 1973 Organization of Petroleum Exporting Countries (OPEC) oil price increases the national economy boomed'.[9] However, data from the International Financial Report indicates that, although Trinidad's Gross National Product (GNP) increased from 1973 to 1977, it did not rise sharply; neither did Trinidad's production of crude oil. The 'boom' had not begun to be realised until the late 1970s. As was the case with the Jamaicans, the Trinidadians intended to be employed mainly in clerical, service and fabrication and assembly repair jobs during this period (1973 to 1977). A loss of skilled workers and some professional workers did exist but the 'brain drain' was not substantial.

During the period 1978-1982, the emigration rate dropped drastically. Only 10 per cent of those Trinidadians who left for Canada from 1968 to 1987 migrated

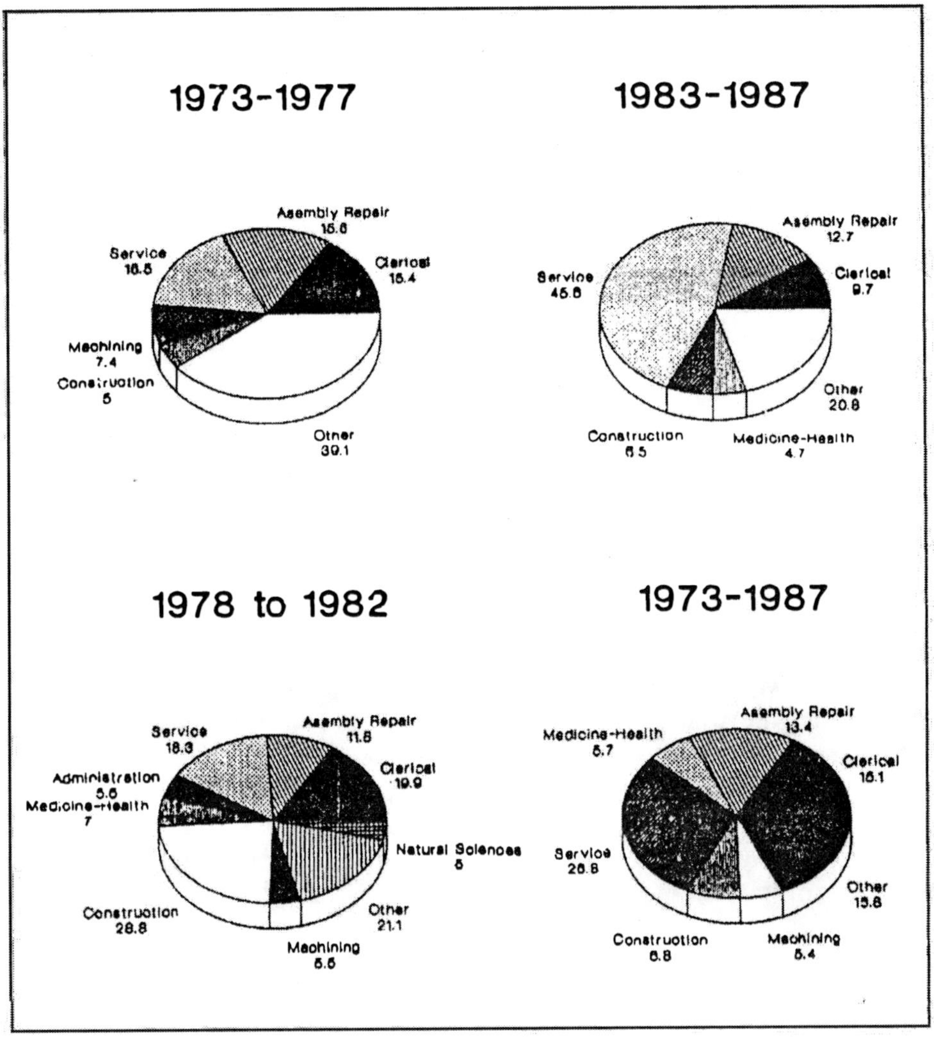

Source : Tabulated from Employment and Immigration Canada, 1990

Fig. 19.2 : Intended Occupations of Jamaican Immigrants to Canada : Precentage to Total for Selected Periods

during this period. The economic growth experienced by Trinidad at this time could have acted as a deterrent for emigrants (Table 19.1). The Trinidadian government used revenues from its petroleum export to invest in the infrastructure of the nation, thus improving the social and economic structure of the nation. Johnson and Henry state that investment in the economy led to return migration and the importation of skilled and professional workers.[10] Although the Trinidadian economy offered skilled and professional opportunities, those emigrants leaving for Canada sought employment in the professional fields more so than the previous period. In fact the 'brain drain' was more evident in this period than the 1973 to 1977 period. This is much like the Jamaicans who showed the strongest evidence of a 'brain drain' during the 1977 to 1982 period. However, unlike the Jamaicans, the Trinidadians seeking work in the professional and skilled fields did continue into the following period.The fourth period (1983 to 1987) was similar to the previous period in that a large number of Trinidadians stayed home. Only 4,524 Trinidadians migrated to Canada during this period, most of them leaving in 1986 and 1987. These Trinidadians accounted for 10 per cent of the total migration from 1968 to 1987. Until 1984, the economy was doing well, therefore the incentive to leave may have been minimized. However, after 1984 Trinidadians began to experience higher unemployment rates. Diaz Briquettes states that high unemployment rates returned when the national economy began to cool off in 1984.[11] This may have acted as a push factor near the end of this period in 1986 and 1987, when most Trinidadians who had migrated during this period (1983 to 1987) did so. The composition of the emigrants leaving during this period was much the same as the previous two periods. They aspired to be employed mainly in service, clerical and fabrication and assembly repair jobs (Fig. 19.3). A 'brain drain' was evident but not to the degree of that of the second period.

Throughout the twenty-year period Trinidad did not appear to lose many of its skilled or professional workers. There was no substantial 'brain drain' except for may be in the third period (1978 to 1982). Most workers who left for Canada had intended to be employed in clerical, service and fabrication and assembly repair jobs. Trinidad showed the strongest argument for the 'brain drain' phenomenon of the four major source countries.

Haitian Emigration

The third largest number of immigrants to Canada from 1968 to 1987 came from Haiti. The 42,393 Haitian immigrants represented 19 per cent of the Caribbeans who immigrated to Canada during the twenty year period. Unlike Jamaica and Trinidad and Tobago, the percentage of Haitians who emigrated to Canada closely relates to the percentage of Haitians in the Caribbean region. Haitians make up 20 per cent of the total Caribbean population.

Unlike the Trinidadians most Haitians did not migrate from 1968 to 1972. Only eleven per cent of the Haitians migrated to Canada from 1968 to 1972. This was the lowest percentage of migration throughout the twenty year period for the Haitians. Economic, social, political and demographic push factors were most probably at work; however, they did not seem to have any substantial impact on emigration.

Table 19.1 : A Comparison of Trinidad's Economic Growth to Emigration to Canada, 1968 to 1987

Year	*Gross National Product (Millions of Trinidadian Dollars)*	*Crude Petroleum Production: Index Numbers (1985=100): Period Averages*	*Trinidadian Emigration to Canada*
1968	1,205	101.1	2,499
1969	1,381	104.1	5,513
1970	1,415	89.3	4,731
1971	1,514	79.4	3,972
1972	1,639	73.7	2,818
1973	1,946	79.7	5,062
1974	2,384	94.3	4,659
1975	3,607	106.0	3,920
1976	4,749	122.3	2,302
1977	5,255	120.8	1,566
1978	6.340	130.1	1,186
1979	7,936	130.3	796
1980	10.290	121.7	960
1981	14,485	120.7	947
1982	16,284	107.5	954
1983	18,881	100.5	762
1984	18,381	90.8	598
1985	18,604	96.6	685
1986	17,218	100.0	929
1987	16,331	95.9	1,661

Source : International Financial Statistics; Yearbook, 1990, pp. 704-705.

The second period (1973 to 1977) saw the largest number of Haitians leaving for Canada as was the case with the Jamaicans. A total of 16,342 immigrants left, which accounted for 39 per cent of the total Haitian migration from 1968 to 1987. It cannot be denied that the poor economic, social and political situation could possibly have played a large role in the Haitians' decision to leave during this period. However, there is no solid evidence to support the idea that these factors were the primary cause for Haitians leaving for Canada at such a high percentage. A large percentage of the workers leaving during this period had intended to be employed in fabrication and assembly repair jobs. Those seeking clerical and service jobs also accounted for a substantial percentage of the workers leaving for Canada.

The last two periods (1978 to 1982 and 1983 to 1987) witnessed a decline in the number of Haitians leaving Haiti. The decline was much less than what Jamaica, Trinidad or Barbados experienced. The last two periods accounted for 42 per cent of the total migration in the twenty-year period. It is arguable that economic, social, demographic and political push factors increased during these two periods, especially the political push factor. It was during this time

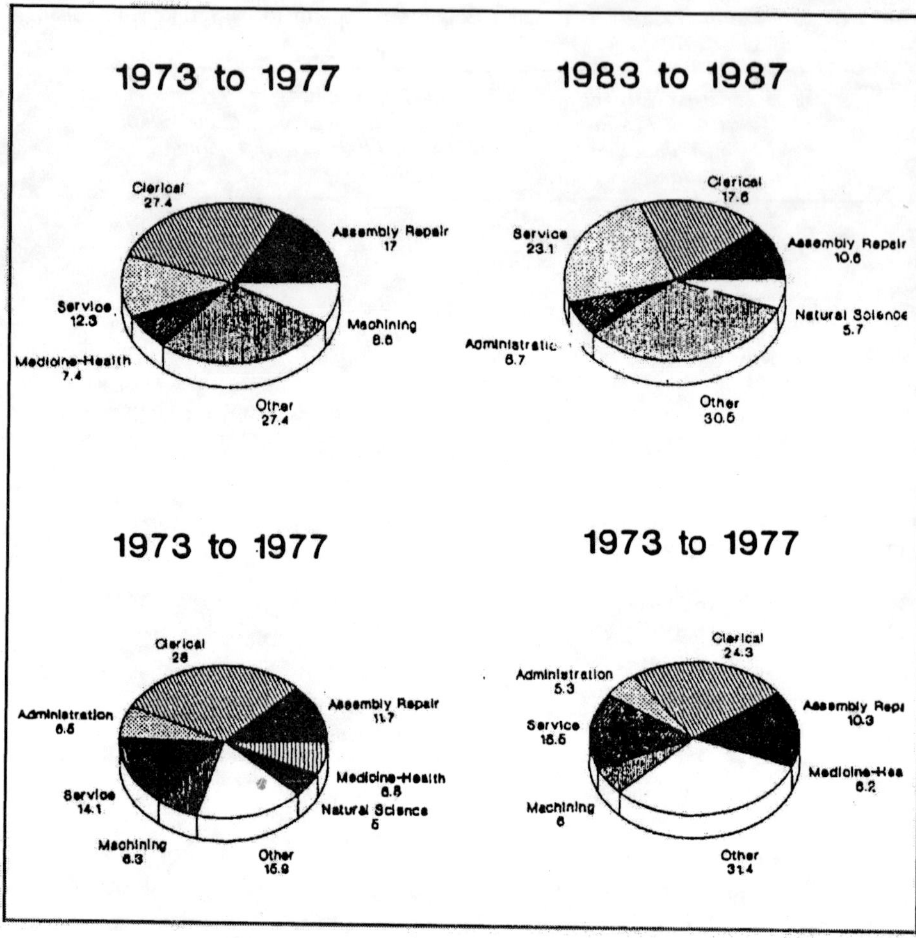

Source : Tabulated from Employment and Immigration Canada, 1990.

Fig. 19.3 : Intended Occupations of Trinidadian Immigrants to Canada : Percentage to Total for Selected Periods

that the Duvalier regime became more corrupt and repressive forcing Haitians to risk life and limb to escape the clutches of the brutal dictator. In 1981 numerous boat voyages to the continental U.S.A. attested to the political strife that existed in Haiti.

During the third period, 1978-1982, the number of Haitians emigrating to Canada declined. However, near the end of the period, from 1980 to 1982 the number of Haitians leaving the country rose markedly. This increase in Haitian migrants was largely due to the deepening repressive nature of the Duvalier regime. Stepick states that the Duvalier regime exacorbated the poor economic and social conditions and the dangerous political atmosphere that had established the mechanisms for migration.[12] The Haitian immigrants, who were leaving during this period, intended to work mainly in fabrication and assembly repair, service and clerical jobs. The intended employment of Haitians in this period was much the same as the previous period with a decline in the percentage choosing fabrication and assembly repair jobs.

The fourth period (1983 to 1987) saw a decline in the number of Haitian immigrants entering Canada. The numbers were similar to those of the third period when 9,586 Haitians migrated to Canada. Haitians were still eager to leave the country because the political climate was still unstable and dangerous. Employment was difficult to find and social conditions were unbearable, therefore, those who had the opportunity to leave did so. The Haitians who left during this period intended to be employed in the clerical, service and fabrication and assembly repair jobs. From 1973 to 1987 the Haitians appeared to have lost many of their skilled workers. The data does not indicate any substantial loss of professional workers. The average percentage of Haitians intending to be employed in fabrication and assembly repair jobs during this period was lower than the previous two periods. However, the average percentage was much higher than any other Caribbean nation. Unlike the three other major source countries, from 1973 to 1987 most Haitians preferred to be employed in fabrication and assembly repair jobs (Fig. 19.4). The large percentage of Haitians intending to be employed in fabrication and assembly repair jobs may have come from the dynamic Haitian assembly plant industries sector. Although a large number of skilled workers left Haiti in this period the 'brain drain' did not seem to have occurred in Haiti; rather, it appeared whoever could leave, did so.

The push factors that had an impact on the Haitians leaving were over population, unemployment, poverty and especially an unsafe political climate. The political terror that influenced emigration was not a new phenomenon in Haiti. When Francois 'Papa Doc' Duvalier's repressive regime took power in 1958, emigration escalated.[13] When Francois Duvalier's son, Jean Claude 'Baby Doc' Duvalier assumed power in 1971, he perpetuated the repressive regime that his father had established.

Emigration also increased under 'Baby Doc's' oppressive government. More than any other nation in the Caribbean, the political push factor influenced the migration of Haitians.

Barbadian Emigration

Barbados supplied the fourth largest number of West Indian immigrants to Canada. During the twenty-year period from 1968 to 1987, 11,631 Barbadians departed for Canada. The Barbadians comprise 5 per cent of the West Indians

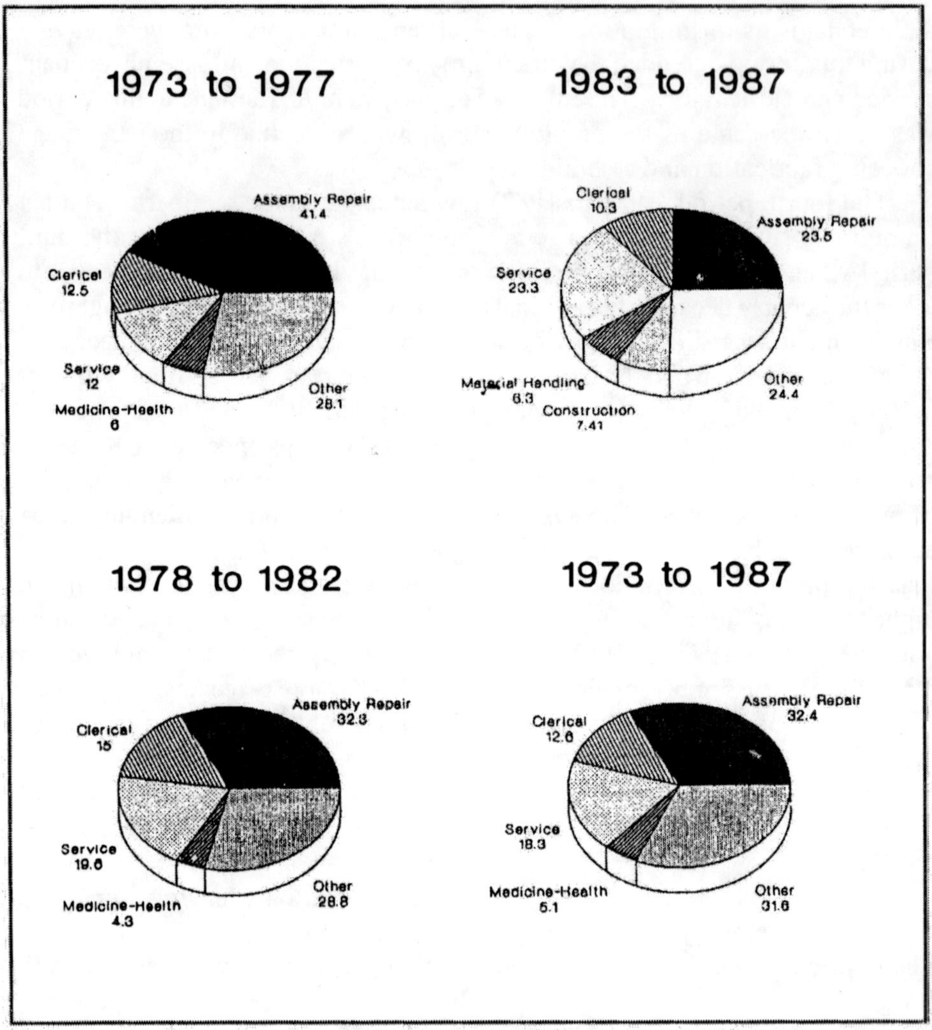

Source : Tabulated from Employment and Immigration Canada, 1990.

Fig. 19.4 : Intended Occupations of Haitian Immigrants to Canada : Percentage to Total for Selected Periods

who migrated to Canada from 1968 to 1987 but they only make up 0.8 per cent of the total Caribbean population. As was the case in Trinidad and Tobago, most (74 per cent) of the Barbadians emigrated during the first half of the twenty year period. The first period (1968 to 1972) accounted for 40 per cent of the total of Barbadians. The push factors in Barbados had a greater impact on the number of people leaving the island than in Jamaica and Trinidad. During the 1960s Barbados had a chronic problem of overpopulation which induced the government to actively participate in the process of sending immigrants abroad to live and work.

Thirty-four per cent of the Barbadians who emigrated to Canada from 1968 to 1987 did so during the second period (1973 to 1977). Apparently population increase and high unemployment contributed to the Barbadians' departure. Most of the Barbadians who left for Canada had intended to be employed in clerical, service and fabrication and assembly repair jobs. A substantial percentage of Barbadians also intended to be employed in medicine and health. The Barbadians exhibited similar intended occupations to the other major source countries in that they intended to be employed primarily in service, clerical and fabrication and assembly repair jobs. It was during this period that most of the Barbadians chose the skilled and professional fields as jobs of potential employment.

The third period (1978 to 1982), did not witness much migration from Barbados to Canada. The improved performance of the Barbadian economy may have been a factor in keeping Barbadian emigration at a low level. The workers who left during this period did not appear to be skilled, due probably to the Barbadian government's promotion of migration of its unskilled workers during the late 1970s and early 1980s. The Barbadian government did not mind its population leaving as long as the people leaving were unskilled. Clerical and service jobs were still well represented as fields of intended employment, but medicine and health along with fabrication and assembly repair, dropped markedly. A disproportionate percentage of the Barbadians intended to be employed in other jobs. Such a high percentage of workers in the 'other' category indicates a lack of workers in the more skilled and professional areas.

The low emigration rate that had begun in the third period continued in the final period (1983 - 1987). The Barbadian economy was still healthy with respect to earlier years. Again, this may have acted as an incentive for Barbadians to stay home. The workers who were leaving had intended to be employed in service and clerical jobs. As was the case in the previous period, a large percentage of workers was in the 'other' category. No evidence of 'brain drain' was noticed during this period.

From 1973 to 1987 the Barbadians intended to be employed mainly in clerical and service jobs. Of the four major source countries, Barbados has the weakest argument for the 'brain drain'. A large percentage of Barbadians did not expect to be employed in skilled and professional jobs such as machining, construction, management and administration, to name a few (Fig. 19.5).

As stated earlier, the Barbadian government has actively encouraged, and even promoted, emigration of its citizens. Over the years the Barbadian government has provided loans for non-sponsored immigrants and has instituted programmes to train its citizens in skills that are needed abroad.[14] This however, practice, did not seem to have been emphasised during the late seventies to the

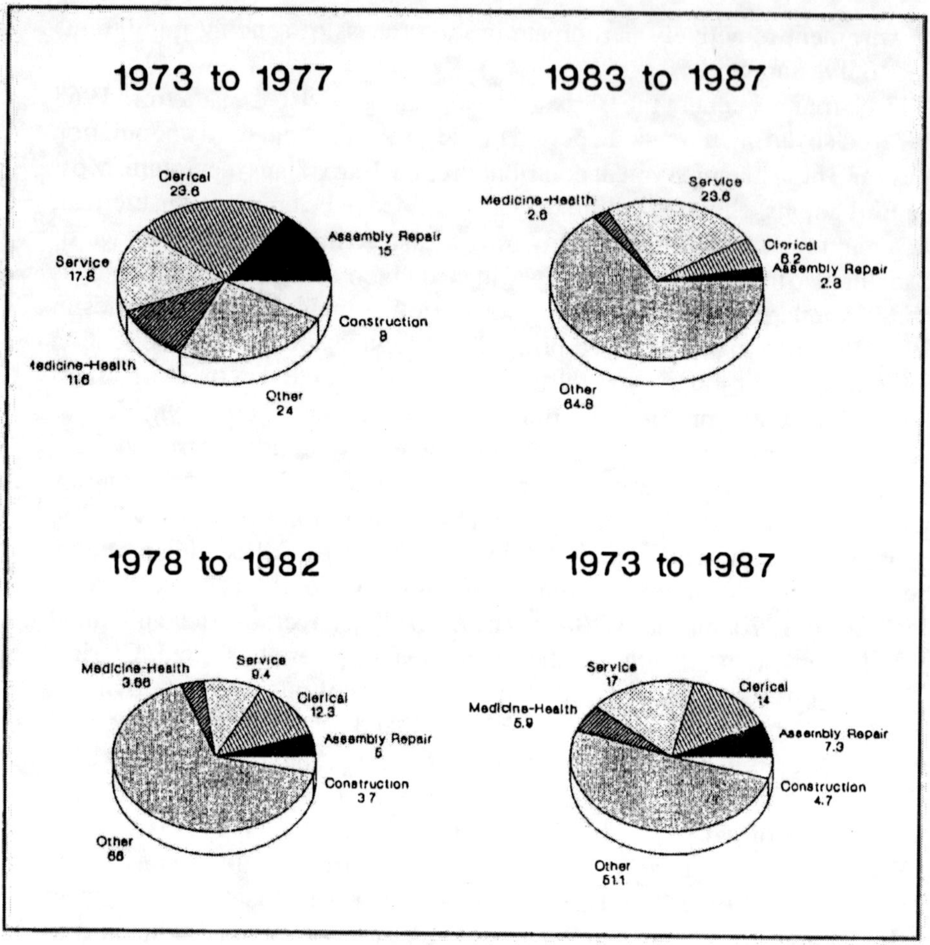

Source : Tabulated from Employment and Immigration Canada, 1990.

Fig. 19.5 : Intended Occupations of Barbadian Immigrants to Canada : Percentage to Total for Selected Periods

mid-eighties. This may have been the result of an improved Barbadian economy. The Barbadian government was more intent on keeping its skilled and professional workers at home to help develop its growing economy. Diaz-Briquettes states that the Barbadian government encouraged emigration of unskilled workers, but was concerned with the loss of skilled workers.[15]

Other Caribbean Nations

The other twenty Caribbean nations contributed only 10 per cent of the total migration to Canada from 1968 to 1987. Most of these immigrants came from the island of Grenada and St. Vincent. No strong evidence of a 'brain drain' among the other Caribbean nations exists. The push factors that affected emigration were similar to those found in the four major contributing countries. The workers from the other Caribbean nations intended to be employed mainly in clerical and service jobs.

Although the occupations discussed in this chapter are intended occupations, they are quite similar to the actual occupations of West Indians presently living in Canada (Table 19.2).

It can be argued from the above table that for the most part the West Indians were employed in their intended occupations, *i.e.*, clerical, service, and fabrication and assembly repair jobs.

Table 19.2 : Occupational Structure of the West Indians in Canada, 1985

Occupations	*Percent*
Clerical and Related Fields	22
Professionals	18
Services	14
Assembly and Repair	14
Processing	8
Other	26

Source : Statistics Canada, 1986.

Conclusion

Migration has always been a part of history. Most migrants have economic motives though political conflicts, religious persecutions, and natural disasters have also caused people to migrate. The recent migration of Caribbean people to Canada is all economically motivated. It accentuated when the receiving country, Canada, needed a workforce at different levels to boost its economy. The migrants, almost all, were destined to urban areas and not all could be absorbed in the occupation for which they had a technical background. After coming to Canada, the Caribbean migrants, who themselves were of mixed ethnic, racial and religious origins, created their own cultural nitches and added further diversity in the pre-existing, European-based, diverse Canadian culture.

References and Notes

1. Dutt, Ashok K. and Kalbach, Warren (1970). *The Impact of Immigration on Canada's Population*, Ottawa : The Queen's Printer for Canada, ref. p. 11; Noble, Allen G. (1982). 'The Cultural Geography of India in Perspective', in Allen G. Noble and Ashok K. Dutt (eds.), *India: Cultural Patterns and Processes*, Boulder: Westview Press, pp. 367-373.
2. Ramcharan, Subhas (1982). *Racism : Non-whites in Canada*, Toronto : Butterworth & Co., ref. p.12.

3. *See* ref. 2, p.19.
4. *See* ref. 2, p. 19.
5. Pastor, Robert (1985). 'Introduction: The Policy Challenge', in Robert A. Pastor (ed.), *Migration and Development in the Caribbean : The Unexplored Connection*, London Westview Press, ref. p. 118.
6. Cooper, Michael (1985). 'Migration from Jamaica in the 1970s : Political Protest or Economic Pull', *International Migration Review*, 19, 4, pp. 728-745.
7. *See* ref. 5, p. 120.
8. Diaz-Briquets, Sergio (1985). 'Impact of Alternative Development Strategies on Migration : A Comparative Analysis', in Robert A. Pastor (ed.) *Migration and Development in the Carribean : The Unexplored Connection*, London : Westview Press. p. 57.
9. *See* ref. 8, p. 57.
10. Henry, R. & Johnson, K. (1985). 'Migration Manpower and Underdevelopment of the Commonwealth Caribbean', in Robert A. Pastor (ed). *Migration and Development in the Caribbean : The Unexplored Connection*, London : Westview Press, pp. 273-288.
11. *See* ref. 8, p. 58.
12. Stepick, Alex (1987), 'The Haitian Exodus : Flight from Terror and Poverty', in Barry B. Levine (ed.), *The Caribbean Exodus*, New York : Praeger, pp. 131- 151.
13. *See* ref. 12, p. 136.
14. *See* ref. 8, p. 57.
15. *See* ref. 8, p. 57.

20

AIDS DIFFUSION PATTERNS AND CONTAINMENT POLICY

U.S.A. and India

ANUPA MUKHOPADHYAY AND ASHOK K. DUTT

Introduction

Pioneering works on medical geography have been done by May, Howe, Pyle, and Hunter.[1] Learmonth another pioneer, studied Indian medical geography.[2] Others, who have worked on disease patterns of India, are Akhtar, Chakravarti, Dutt, Dutta and Dutt, Hazra, Hyma and Ramesh, Mishra, and Singh and Dutta.[3] Mukerji has also made a significant contribution in this field by analysing the disease ecology of a micro-region, the Chandigarh Dun, an area that has attracted his attention for indepth research for a long time.[4] This study on global AIDS pattern, examines the policies to limit the spread of the disease; using India and USA as case studies and focussing on socio-cultural constraints that affect AIDS containment policies.

AIDS

Acquired Immunodeficiency Syndrome (AIDS) has emerged as a leading cause of morbidity and premature mortality in cities around the world in a short period of over a decade. In North America AIDS had become the principal killer of young adult males and it has outstripped overdose, suicide, homicide, and hepatitis B.[5]

Despite the still relatively small reported totals, international concern about AIDS, a disease which is primarily transmitted sexually, with no known cure and an ever increasing mortality rate, is immense. The mean incubation period following infection with the causative agent of AIDS, the Human Immunodeficiency Virus (HIV), is estimated at 8-12 years. Consequently, current global AIDS statistics are effects of earlier HIV spread processes.[6]

Despite efforts to curb the dissemination, HIV has spread rapidly around the world. By 1994, WHO estimated the number of HIV infections worldwide at 17 million. Such figures suggest that, within the next few decades, AIDS will indeed come to rank with the world's other great pandemic. In fact, WHO estimates that by 2000 A.D. about 40 million men, women and children are likely to be infected by HIV; cumulative number of AIDS cases will be around 10 million. Heterosexual and perinatal transmission will be the most common modes of transmission of HIV infection. Of these infected persons, 75 per cent will be living in the developing countries. As information and estimates become available from countries which previously have withheld or purposely under-reported HIV infection rates and AIDS cases, it is very likely this estimate will be revised upwards significantly. In economic terms, the costs of the first decade of the disease have already been staggering, with global cumulative direct and indirect costs estimated at US $240 billion.[7]

Because of its diverse social, ethical, and legal implications and its variable effects on different segments of society, AIDS has been characterised as a disease of the social system.[8] Not only does AIDS reflect social characteristics, as in promiscuous homosexual and heterosexual behaviour, drug use, and prostitution, but its social impact has been felt differentially by segments of society. 'AIDS is clearly affecting mortality, though differentially, by age, gender, racial, and geographic groups'. Further, the effect of epidemics such as AIDS 'extends far beyond their medical and economic costs, shaping the nature of cultural and social life'.[9]

Discovery of AIDS

Two major opportunistic diseases associated with AIDS : *Kaposi's Sarcoma* (KS), equally rare form of pneumonia, and *Pneumocystis Carinii Pneumonia* (PCP) were found among small groups of homosexual males in California and New York.[10] The common origin of these two rare and distinct clinical manifestations were determined and the acronym, AIDS, was applied to the syndrome.[11]

In 1983, researchers at the Pasteur Institute in Paris isolated a new human retrovirus from a lymph node biopsy of a New York homosexual and labelled the agent 'Lymphadenopathy Associated Virus' (LAV), and subsequently the National Institute of Health in Bethesda, Maryland also isolated the same class of virus and further determined that it was 'attracted to' or had a selective tropism for a cell critical in the human cell-mediated immune response system, the T4 or T-helper/inducer cell.[12] Therefore, they labelled the virus the Human T-Cell-Lymphotropic-Virus III (HTLV-III).[13] Another variant of HIV virus has been isolated in West African AIDS patients, this is now known as HIV 2, and the 'original' virus is now labelled HIV-1.[14]

Though first identified in the United States, evidence in the form of blood samples, first patients point to a large region of Central Africa (including Zaire, Zambia, Uganda, Rwanda, and the Central African Republic) as the possible index location of the origin of the AIDS pandemic.[15] For a while, the occurrence of KS and PCP in Haitian immigrants to the United States postulated the belief that Haiti was the origin of this pandemic.[16] The substantial biological, spatial and temporal evidence now points to Central Africa as the 'index' location for AIDS. Further, as research suggests HIV-1 may have evolved along with HIV-2 from a common ancestor as recently as forty years ago.[17]

Evidence suggests that HIV-1 infection has been occurring in an epidemic form in Central Africa since the late 1970s.[18] Blood samples suggest the probable existence and stability of the HIV-1 and AIDS related disease for a long period in the more remote areas of Central Africa.[19] Further support for probable African origin of HIV-1 derives from the presence of close relatives of HTLV-1 and HIV-1 in the non-human primates found there such as Simian Immunodeficiency Virus (SIV) in wild African green monkeys.[20] The HIV-2 was first found in West Africans including residents from the Ivory Coast.[21]

There are unsubstantiated statements about non-African origin of the AIDS. Hunt[22] argued 'that AIDS did not originate in Central Africa because the known AIDS virus in the U.S.A. is at least two decades older than that of Africa. Hunt also argues that if AIDS originated from green monkeys in Africa it must have been a disease in rural areas. Today it is the urban areas of Central Africa that have much higher incidence of AIDS compared to the rural areas. In actuality, there are two AIDS viruses— HIV-I and HIV-II. HIV-I was discovered earlier and HIV-II at a later time. HIV-II, which Hunt does not discuss, originated in West Africa and then came to Europe. In North America this virus is still not widespread. The U.S. origin of the disease from preserved blood samples of 1950s, cannot in itself be justified. The fact is that preserved blood samples from Africa are not available prior to the 1970s. Therefore, it is impossible to determine whether the HIV-I virus was present among Africans in the 1950s or earlier. Shannon and Pyle confirm that African origin of AIDS, Smallman-Raynor and Cliff also confirm that HIV-II has its origin in West Africa. Hunt's argument of non-African AIDS origin does not follow the mainstream thinking held by the world scientific community and its validity is questionable.'[23]

AIDS was discovered in the United States in the early 1980s, and by the middle of 1993, about 315,390 cases of AIDS had been reported in this country alone. The number of reported AIDS cases represents only the 'tip of the iceberg' with regard to the prevalence and incidence of AIDS, as well as the potential future numbers of persons infected with the purported etiological agent of AIDS, the HIV.

The pandemic is coursing through the world's population, and there is little the medical and public-health communities can do to directly prevent the spread of the infection or cure those infected. Recent evidence indicates that life expectancy is decreasing for HIV-infected persons who develop an aggressive form of *Kaposi's Sarcoma* (KS), a soft-tissue cancer. Until such time as effective vaccines or cures are developed and made available on a global scale, the only viable source to slow the pandemic is preventive educational programmes directed toward changing human behaviour patterns and cultural traditions known to facilitate infection with the HIV.

Educational programmes directed toward safer sexual practices among male homosexuals in the United States have contributed to at least a reduction in the incidence of HIV infection among this group. In the long run, however, if not already, this group may ultimately represent only a small fraction of the number of people infected with the HIV. Globally, the majority of persons infected with the HIV and succumbing to AIDS presently and in the foreseeable future will be heterosexuals.

At present, Africa is the geographic area with highest rate of transmission and infection with the HIV. Male to female ratio of infected persons in African nations is approximately 1:1. Clusters of heterosexual females infected by one promiscuous HIV-infected male have already been identified in Europe and the United States. For many world regions and the millions of people living in these areas, there prevails scant knowledge concerning the status of the HIV epidemic and the number of AIDS cases. To date, the people and countries of sub-Saharan Africa appear to be the greatest risk for the potential destructive force of the pandemic.

The primary mode of HIV transmission is through intimate contact with infected blood and blood products, semen and cervical/vaginal fluids, or perinatal events. In developed countries those people particularly at risk to infection have been homosexual males, hemophiliacs, parenteral drug users, and children of infected mothers. In underdeveloped countries especially in Africa and the Caribbean, the infection appears to spread predominantly by heterosexual intercourse.

Global Diffusion of AIDS

It has been hypothesized that quite possibly Central Africa experienced the HIV infection first in the early 1970s. Today Africa has the highest rate of transmission of HIV. This is primarily due to lack of screening of blood used in transfusion and usage of unsterilized needles for injections for medical, scarification and tattooing. Also, greater frequency of anal intercourse among heterosexuals has contributed to increase of HIV infection among women. It has been further postulated that infected immigrants from Central Africa as well as visitors to Central Africa carried the HIV infection to other parts in Europe such as Belgium, and America during the later 1970s. By 1981, AIDS was found in Mexico and Canada as well. The disease further diffused to Brazil, Australia, and New Zealand. During the mid-1980s, HIV infection was identified in China, Japan, and Philippines. During the same period the west European countries such as, Denmark, France and Netherlands had emerged as AIDS clusters (Fig.20.1).

Four broad distinct patterns of infection have been distinguished.[24] These patterns are as follows

Pattern I

This pattern is found in North America, Western Europe, Australia, New Zealand, and urban areas of Latin America. HIV infections first started spreading in the late 1970s in these areas. Exposure to HIV was due to homosexual, bisexual, and intravenous drug use activity.[25] Heterosexual transmission occurs from infected males, predominantly intravenous drug users or bisexuals. Increasingly, women are being infected by heterosexual and bisexual males and this has resulted in an increase in perinatal transmission.

Pattern II

The pattern is found in the sub-Saharan Africa and in Latin America, especially the Caribbean. HIV infection in these areas presumably began in the mid-1970s. The predominant transmission of HIV is due to heterosexual activity. The male to female ratio of cases is 1:1. Exposure to HIV via intravenous drug use exposure is limited, however, other skin-piercing instruments for medical and ritual purposes are a source of transmission. Infection through tainted blood products also plays an important role in HIV transmission.

Pattern III

This pattern is seen in Eastern Europe, the Middle East, North Africa, and most countries of Asia and Oceania. HIV infection was probably introduced in the early to middle 1980s. AIDS cases are mostly due to transfusions of blood products and sexual contact with both heterosexual and homosexual populations from Pattern I and Pattern II countries.

Pattern IV

This pattern is seen in Brazil and Honduras which show predominant mode of

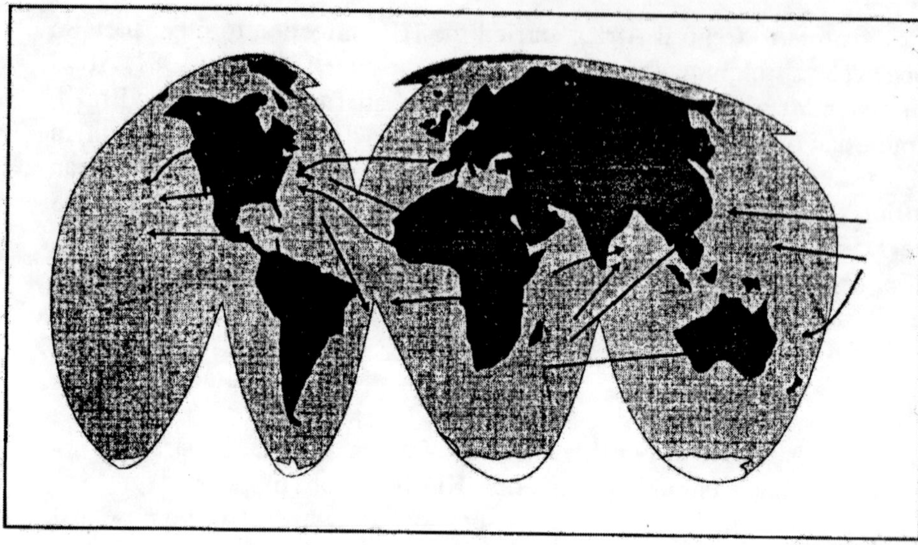

Fig. 20.1 : AIDS : Global Diffusion Pathways

HIV transmission due to bisexual activity. Bisexuals have infected large segments of the heterosexual population, especially the women.

AIDS Diffusion Patterns in the United States

During the 1980s, AIDS was a disease of major urban areas which were first to receive the infection. Concentration of AIDS victims were within urban places with the largest number of male homosexuals and intravenous drug users, such as in New York City, Los Angeles, San Francisco and southern Florida.

With more contact of the infected individuals with the susceptible, AIDS spread like any sexually transmitted disease, spreading outward from larger to smaller metropolitan areas and from large cities to adjacent sub-urban areas following a hierarchical diffusion pattern. Radial diffusion pathways of the disease were also established whereby, concentrations of populations at risk were selectively located in increasing distances from points of infection origin. Mid-west-Chicago shows a radial diffusion of AIDS, whereas, New York to Chicago follows a hierarchical pattern along routes of transportation.[26]

More than half of all the persons diagnosed with AIDS in the United States since 1981 have died. AIDS continues to increase annually in importance in terms of potential years of life lost. The AIDS epidemic in the United States is rapidly increasing within the lower socio-economic status and minority population groups. Increasingly HIV infected individuals and persons with AIDS are the poor members of ethnic minorities, often concentrated in large cities.[27]

These population groups are the most difficult to reach with educational and prevention programmes.

The majority of the patients infected with the HIV and dying of AIDS belong to realtively young ages. Female prostitutes in the United Stated are increasingly at risk to infection with the HIV.[28]Others such as, health care workers, physicians and dentists are also at increased risk to HIV infection. Furthermore, heterosexual exposure route of transmission is accounting for increasing number of AIDS cases each year especially among the women.

Geographically, wherever high concentrations of population are found, those places become the source nodes of HIV infection. From these nodes, mostly urban areas, the disease then spreads outward to the sub-urban areas and non-metropolitan areas. The spread of AIDS is totally based on sociological interactions between humans and thus education to change social behaviour is the only remedy. Educational programmes promoted to practice safer sexual practices has brought down the rate of HIV infection among homosexuals in the United States. However, heterosexuals and drug users are still at very high risk in this nation and everywhere else. An emerging group of victims are young infants and mothers, especially in the non-metropolitan areas.[29]

AIDS in the Metropolitan Statistical Areas (MSAs) of the United States

By the early 1980s three centres of diffusion were noticed for the MSAs (Fig.20.2), New York in the northeast, Miami in the southeast and Los Angeles and San Francisco in the west. The bulk of the early cases were first identified in Los Angeles, San- Francisco, and New York. From these large metropolitan centres the disease was discovered to affect population groups in other large metropolises across the United States. A pattern of diffusion of AIDS cases emerged by the mid-eighties, it quickly became evident that places such as Dallas, Denver, and Chicago which were highly connected, accessible, points of transfer for air plane passengers and points of transhipment of goods and people-were emerging as major centres of AIDS concentration (Fig. 20.2).

In the United States there were less than 7 cases per 100,000 population in 1982 and less than 20 cases per 100,000 in 1993 for any MSA respectively. During the mid-1980s the northeastern MSAs experienced an increase in rates of AIDS cases. Similarly, the southeastern and the western metropolitan areas also experienced an increase in rates especially San Francisco which had 38.6 cases per 100,000 population.[30]

In 1985 AIDS had spread in a contiguous pattern from New York to Boston in the north and Washington in the south, in Miami it spread north to Jacksonville, and in the west AIDS had spread from San Francisco to Los Angeles following the linear routes of transportation lines. Also, centres of

AIDS concentration were found in Houston, Dallas, Austin, Denver, Kansas City and Atlanta, which clearly indicated that an expansion-hierarchic diffusion pattern of AIDS was in progress. The emergence of the two diffusion patterns of linear contiguousness and expansion-hierarchic became evident in the diffusion pattern of new AIDS cases in the subsequent years.[31]

As the hierarchical and linear diffusion routes of AIDS spread strengthened during 1985-1987 so did the contiguous spread into the heart of the country (Fig. 20.2). By 1988 linear spread of AIDS in the northeastern coastal MSAs, southeastern coastal MSAs and western MSAs were experiencing an increase in AIDS case rates as well as influencing the interior MSAs in the hinterland (Fig. 20.2).

The late 1980s and early 1990s clearly demonstrate the pattern of AIDS spreading to the interior MSAs.[32]

Difficulty in the AIDS Vaccine Discovery

Efforts to discover AIDS vaccine have been made since the end of 1980s in the US and France in particular. Institutions supported by the World Health Organization are also quite active in this regard. Difficulties in the development of the vaccine surfaced as the multifaceted characteristics of the AIDS virus were unearthed in the process. Koff[33] summarized the difficulties as, 'The first generation AIDS vaccines were developed mainly based on the envelop glycoprotein of HIV.[34] Such vaccines were aimed at preventing AIDS in North America and Europe by targeting a sub-type of HIV-I that prevailed in those regions of the world.[35] Potential success of the candidates quickly waned as the following information on the immunobiology and pathogenesis of HIV became available: (a) the capacity for multiple sub-types of HIV to circulate concomitantly in different parts of the world,[36] (b) the capacity of the virus to infect by means of cell-free and cell-associated forms,[37] (c) the potential for selected regions of the envelope glycoprotein to induce immunosuppressive, immunopathologic, or infection-enhancing responses,[38] and (d) the inability of the first generation vaccines to stimulate or maintain the levels of immune responses likely to be effective against HIV.'[39]

In the US, the Federal Co-ordinating Council on Life Sciences and Health has affirmed that one should have a set of criteria for an ideal AIDS vaccine. They are the following :

1. The vaccine is safe.
2. The vaccine elicits a protective immune response in a high proportion of vaccinated individuals.

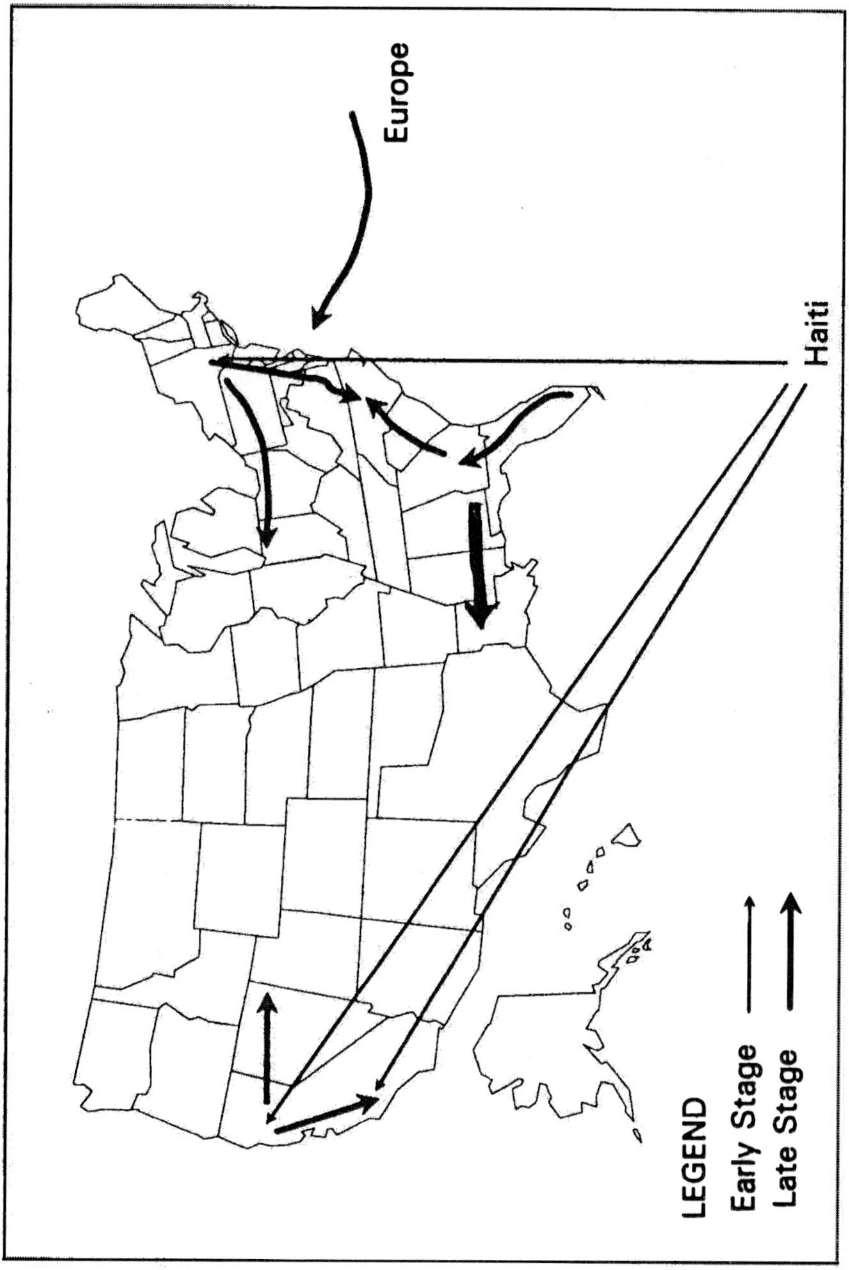

Fig. 20.2 : AIDS : Diffusion Pathways in the Unnited States

3. The vaccine stimulates both the cellular and antibody components of the immune system because HIV is transmitted by different routes (sexual, intravenous, and perinatal) and by different modes (as cell-free and cell- associated viruses).
4. The vaccine protects against different subtypes of variants of HIV because several variants of HIV are now circulating simultaneously in different geographic regions of the world.
5. The vaccine induces long-lasting protection because exposure to HIV may occur at long intervals after immunization, throughout an individual's sexually active lifetime.
6. The vaccine induces local immunity in the mucosa of the genital tract or rectum, which may be important to impede infection at the site or sites of sexually transmitted HIV infection.
7. The vaccine is practical (for example, with regard to the number of immunizations and cost) for world wide delivery and administration.[40]

It may not be possible to realise all the aspects of the criteria when the AIDS vaccine is finally discovered but there ought to be a continued incentive for AIDS vaccine development. In the US, the private sector is shy in investing for an AIDS vaccine development because of certain disincentives, such as, legal liability, low profit margins for the manufacturer and potential price controls. Any large-scale field trials of AIDS vaccine needs to be a specially considered strategy generated by the US Government and there needs to be an international regulatory consensus as well, in order that, the poor people of Africa and Asia, where the HIV infection is most rampant, may not be debarred from using the vaccine because of the costs. There is also a necessity to improve information transfer to vaccine manufacturers. This would not only expedite the discovery but minimise the duplication of the different aspects of research.[41]

Koff rightly asserts that there is an urgency to develop a safe and effective vaccine and estimates that for the world as a whole 100 million people will need to be vaccinated and the cost is expected to exceed 30 billion US dollars. Inspite of the enormity of the costs and difficulties the scientists and the various governments are bent upon discovering a vaccine. The fact remains that hardly 10 per cent of the current money spent on AIDS research goes to the vaccine development. This has to change.

AIDS Diffusion Patterns in India

India first confirmed evidence of AIDS infection in April 1986 when six prostitutes from Tamil Nadu were found to be HIV positive.[42] HIV cases testing positive for intravenous drug users are increasing at an alarming rate especially for the northeastern states of Manipur, Assam and Nagaland. Available statistics on AIDS and HIV infection in India do not convey the seriousness of the epidemic. By December 1990, 57 cases of AIDS and 4,696 cases of HIV positive status were reported to the World Health Organization (WHO).[43] But the WHO already estimates that 500,000 people in India may already be infected with the virus.

The major mode of transmission of HIV infection in India is through heterosexual promiscuity. A progressive rise in number of cases among promiscuous heterosexuals and blood donors has been observed since mid-1980s. Though the number of AIDS cases were not very high, but cases were showing up among the different population sub-groups. The very first detections were made in April 1986, when seropositivity was detected in an Indian prostitute; in May 1986, an AIDS patient was identified; in August 1986, seropositive Indian male in a STD (Sexually Transmitted Disease) clinic was identified; September 1986, a seropositive pregnant woman was identified; December 1986, a patient of AIDS in the homosexual exposure category was identified; July 1987, a seropositive blood recipient (from Indian blood donor) was identified, spouse to spouse transmission identified; October 1987, seropositive infant identified. By the end of 1987 screening activities had been intensified and it soon became evident that though promiscuous heterosexuals and blood donors were experiencing higher rates of HIV infection it was not limited to them alone.[44]

Sources and Diffusion of AIDS in India

Clinical surveillance of seropositive status was initiated in October 1985. The first patient with AIDS was reported in 1986 in Pune and the source of infection was traced to blood transfusion given in the United States. Similarly, the second case was of a blood recipient who received infusion of blood products in the United States. The rest of the cases which followed were primarily due to heterosexual promiscuity. First seropositive blood donors were identified in Vellore, Tamil Nadu in 1987.

During early 1988, several cities in Maharashtra and Tamil Nadu had identified seropositive blood donors (Fig.20.3). Subsequent seropositive monitoring shows that seropositivity rates among drug users and prostitutes are much higher than those among other risk groups. Different seropositive groups are predominant in different geographical areas of the country. Prostitutes are the predominant seropositive group in Chennai (Madras), prostitutes and blood donors in Mumbai (Bombay), and intravenous drug users in the northeastern region of the country. Studies carried out by the National Institute of Virology (NIV), Pune, Christian Medical College (CMC), Vellore and centres in Mumbai have shown that HIV-2 infection also exists in India.[45]

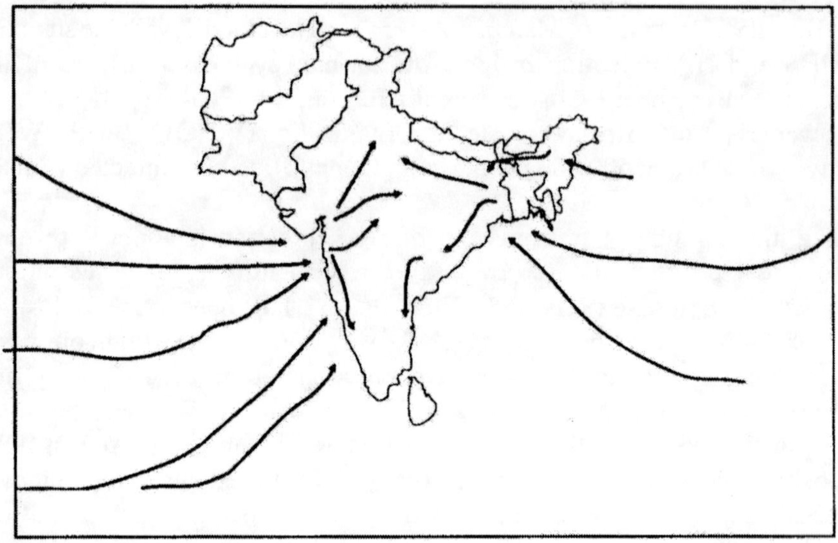

Fig. 20.3 : AIDS : Diffusion Pathways in India

Sero-surveillance has shown that HIV infection is present in all states both urban and rural areas. Predominant modes of transmission are heterosexual activity and perinatal. However, it is difficult to estimate the infection load as : (i) most of the data collection has been done in 4-5 states and there are large areas in the country from where adequate data and sero-surveillance are not available; (ii) almost all the data has been collected in urban areas; (iii) there are substantial differences in the reported seropositivity rates in similar groups from different regions; (iv) precise estimates of the size of the high, intermediate and low risk groups in the country are not available. Estimated figures of HIV infected persons range from 0.2 to 1 million.[46]

It has been estimated that there might be 400,000 to 1 million HIV infected persons in India and of these between 50 to 75 thousand will develop AIDS by 1995.[47] The Union Ministry of Health and Family Welfare, Directorate-General of Health Services (DGHS), and State Health Authorities have already initiated steps to strengthen the existing health care facilities, especially the Departments of Internal Medicine and STD Clinics so that adequate health care for seropositive individuals and AIDS patients can be provided in major cities of India.

AIDS Containment Policies in India

Since the late 1980s National AIDS Control Programme has made efforts to strengthen health education for HIV prevention in India. Poverty, illiteracy and traditional reticence in discussing STD, pose major obstacles for mounting a mass media based health education campaign.[48] It is envisaged that health education and HIV prevention programmes will become a part of the educational system in the nation to make youths aware.

Since late 1980s the Indian Council of Medical Research along with DGHS has proposed and vigorously pursued several containment policies to combat the transmission of HIV infection. The following policies are in effect

I. Strengthening of screening of high risk groups especially promiscuous individuals and providing counselling and care of seropositive individuals are major thrust areas in AIDS control plan. Screening programmes were initiated and launched in Mumbai (January 1989), Chennai (February 1989), Delhi (March, 1989) and Kolkotta (April 1989).
II. Health education not only for high-risk groups, but also the general public to prevent and control HIV infection and AIDS.
III. Sentinel surveillance for HIV infection; monitoring of different population groups such as pregnant women, blood donors, and high-risk groups. A network of five reference centres and 43 surveillance centres all around the nation were established in mid-1980s.
IV. Clinical facilities for counselling and care of seropositive individuals are to be established.
V. Appropriate hospital policies for care of HIV infected persons including AIDS patients are being instituted.
VI. Diagnostic facilities for HIV antibody detection are being made available in more centres.

The National AIDS Control Programme responsible for containment of HIV/AIDS is being carried out within the existing health infrastructure. Organization of such programmes have posed many challenges to the health care delivery system even in developed countries. However, DGHS sero-surveillance programme to track HIV infection early on, even before AIDS cases registeries were developed, has been notable. Today it is a known fact that seropositivity data are a better tool for estimating future AIDS cases among the different population groups. In the light of such facts India has shown the capability to seriously develop a programme to contain and control HIV infection.

National AIDS Control Programme—Broad Prevention Approaches and Education

Until a vaccine or cure for AIDS is found, the only means at present available is health education to enable people to make life-saving choices (*e.g.*, avoiding indiscriminate sex, using condoms). There is, however, no guarantee that the use of condoms will give full protection. Intravenous drug users should be informed that the sharing of needles and syringes involves a high risk. Women suffering from AIDS or who are at high risk for infection should avoid becoming pregnant since infection can be transmitted to the unborn or newborn. Educational material and guidelines for prevention should be made widely available. All mass media channels should be involved in educating the people on AIDS, its nature, transmission, and prevention.

Prevention of Blood-Borne HIV Transmission

People in high-risk groups should be urged to refrain from donating blood, body organs, sperm or other tissues. All blood should be screened for HIV-1 and HIV-2 before transfusion. Transmission of infection to hemophiliacs can be reduced by introducing heat treatment of factors VIII and IX. Strict sterilization practices should be ensured in hospitals and clinics. Pre-sterilized disposable syringes and needles should be used as far as possible. One should avoid injections unless they are absolutely necessary.

Specific Prophylaxis

At present there is no vaccine or cure for treatment of HIV infection and AIDS. However, several researchers are working on drugs to interfere with HIV's production cycle at one stage or the other. Anti-viral chemotherapy with the chemical compound zidovudine (AZT) while not a cure, has proved to be useful in prolonging the life of severely ill patients. The AZT, however, neither restores the immune system, nor does it destroy the HIV virus already installed in cells.

Primary Health Care

Because of its wide-ranging health implications, AIDS touches all aspects of primary health care, including mother and child health, family planning and education. It is important, therefore, that AIDS control programmes are not developed in isolation. Integration into countries' primary health care system is essential.

Lastly, but not the last, is the policy of AIDS surveillance—reporting and record keeping of cases for scientific purposes. A network of 150 surveillance centres were to be established by the end of 1990 which are intended to monitor the high-risk groups.

Universal Social and Economic Implications of AIDS

AIDS has and will result in the death of men and women in the reproductive age groups and in the working age groups. Population balance will be seriously jeopardized. The elderly and the young will be left without support. Health services and resources will be consumed by AIDS at the cost of other diseases. Indirectly, AIDS is going to lower productivity of the workforce.The sickness seen in urban areas will filter down to the rural agricultural areas and eventually cripple the economy completely. The negative costs will finally erode the socio-economic system.

Behaviour Change or Risk Everything

The World Bank (1993) asserts that the AIDS epidemic was pretty bad in the early 1990s and is getting worse. Nine million people in the world carried the AIDS virus in 1990, by the year 2000 it will rise to 40 million. AIDS, then will contribute to about 3.5 per cent of global burden of disease and 2 million people would die of this disease each year. As it takes a short time for the infection rates to double in many developing countries, particularly in Asia and Africa, the projected figures will be surpassed. It was estimated that in 1990, 80 per cent of the AIDS infected people lived in developing countries. In the year 2000 it will rise to 95 per cent. 'In Thailand one in fifty adults is infected. In sub-Saharan Africa one in forty adults is already infected and in certain cities of Africa the prevalence of infection is as one in three. In some of these high-prevalence communities AIDS is already starting to reverse long-term declines in child mortality'.[49]

Two areas, sub-Saharan Africa and Asia, where the infection rates are the highest, there are two alternative assumptions regarding HIV infection. The worst scenario is that in Asia about 4 million and in sub-Saharan Africa about 2 million people will have new infections per year in the year 2000 (Fig. 20.4). Such a possibility is devastating, and will eventually bring about mortality and disabilities in the same way bubonic plague did in Europe during the Middle Ages. Until the vaccine is discovered, AIDS can be contained to certain level only by behaviour change, which involves individual, social, and community factors. Models have been constructed depicting individual factors affecting behavioural change. These models assume that individuals make decisions about, and potentially have control over, their behaviour.[50] The central understanding of the social community factors revolves around social pressures, cultural expectations, the conventional sexual negotiation, drug and alcohol use culture, costs involved in safer sex behaviour, laws of certain social groups, political and religious ideology and 'limitation of public and private resources that impairs the quality and adequacy of AIDS intervention programmes'.[51]

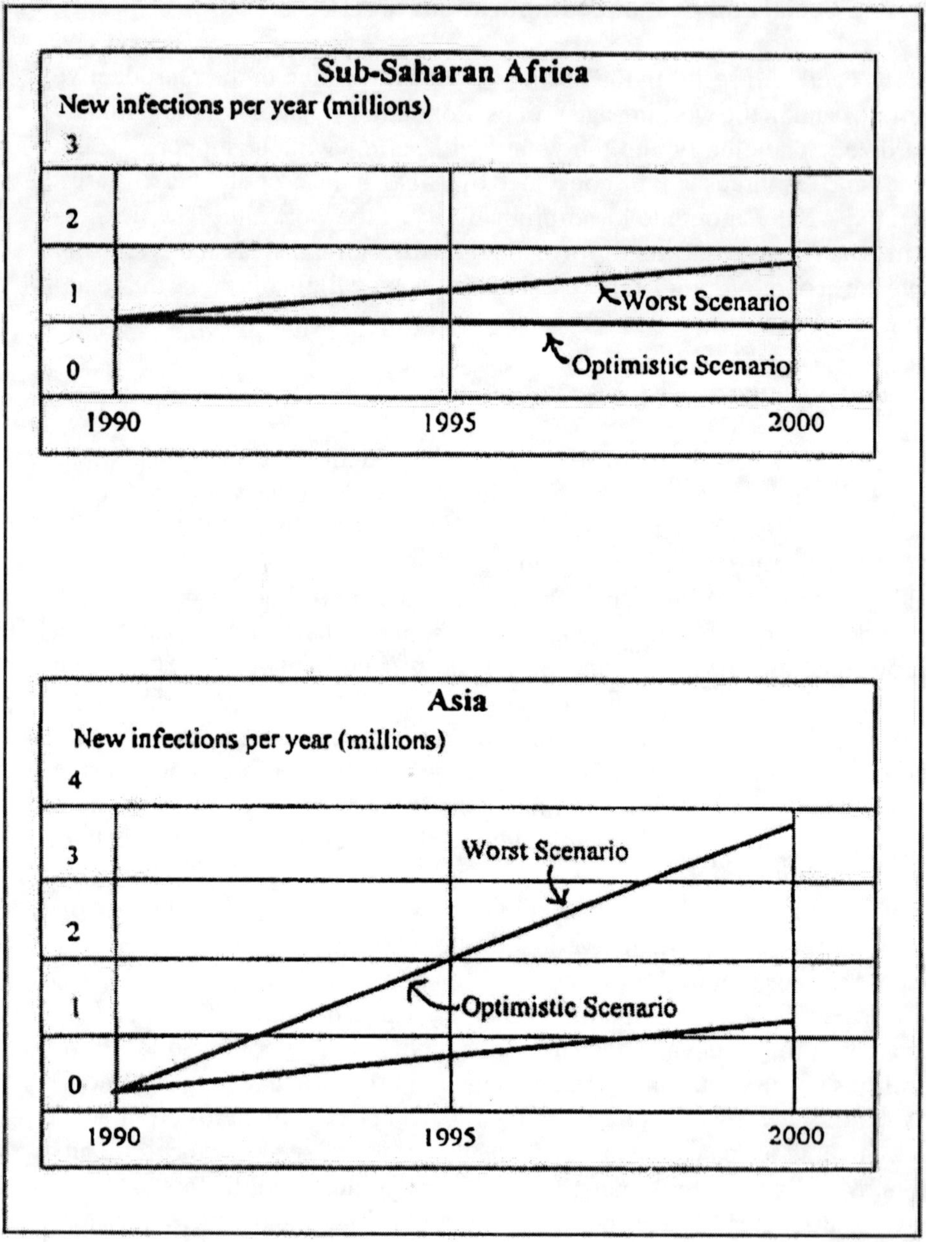

Source : World Health Organisation and World Bank Data (World Health Report), World Bank, p. 101.

Fig. 20.4 : Trends in New HIV Infections for Sub-Saharan Africa and Asia : Worst and Optimistic Scenarios, 1991-2000

An intervention in the form of modifying individual behaviour is essential, because in the absence of a vaccine, this is the only way to limit the fast development of AIDS (Fig.20.5). The WHO, which has had a conservative AIDS infection projection of 19.5 million in the year 2000, calculates that such intervention can half the number of infections. The idea is that in a decade or so, with appropriate intervention the impact of the disease can be significantly minimised. For that purpose the most economic and rightful strategy is to find out the main target groups of the countries from where AIDS diffuses. In the case of USA the target groups in the 1990s were the homosexuals and intravenous drug users. In the case of India, prostitutes, truck drivers, and single males who visit large cities like Mumbai, often are the most likely to get infected and spread the disease to their spouses and children. The young adults in both countries, ought to be targeted as well.Thus, a behaviour change strategy for controlling AIDS before the discovery of a vaccine, in particular, is most rational for the world as a whole. Target-group oriented education campaign is a necessary strategy for containing a fast rate of increase in AIDS incidence.

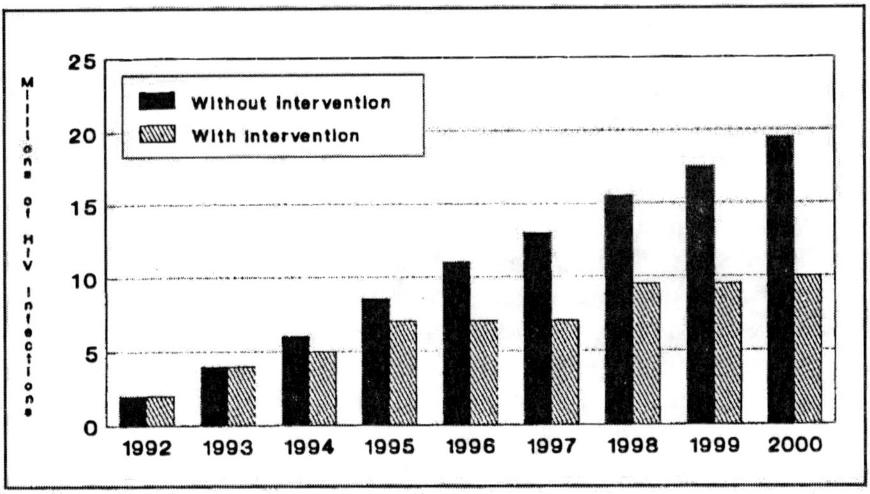

Fig. 20.5 : Project Resulting from with or without Intervention : Cumulative Adult HIV Incidence from 1992 to 2000

Source : World Health Organization (as quoted in *science*. V265, Jul., 94)

Comparative Analysis of AIDS in USA and India

The United States and India have radically different medical conditions and social environments that provide very different epidemiological climates in each nation. In the United States, AIDS has radicalized a public debate on homosexuality, extramarital sex, and contraception that had already been opened up by the great social movements of the sixties and seventies. In India poor medical facilities, infected blood products, and inadequate AIDS preventive information pose a more immediate threat to all Indians. Thus, the key question in the Indian view of AIDS policy has to consider basic right of access to humane medical treatment, respect for privacy, and the promotion of preventive education. AIDS policy should include safe sex education that targets both men and women, and that allows women in particular more control over their lives.

On the other hand in the United States, the general population has information about AIDS and is fully aware of all the precautionary measures to avoid the disease. In the case of India a social change is very necessary to implement educational and medical policies to disburse information about AIDS. Containment of activities of high-risk, irresponsible individuals in the United States is top priority whereas, fighting ignorance of the people and providing medical care is the top priority in India.

Conclusion

As noted by Dutt, the AIDS virus has spread from Central Africa primarily through heterosexual contacts and has diffused throughout the United States.[52] This pattern of diffusion is true for the European nations who were infected by migrating infected individuals, through whom the virus was transmitted via sexual practices, drug abuse, or perinatal transmission sources. Dutt points out that for United States, the AIDS virus has been limited to homosexual/bisexual males and intravenous drug users appearing originally within metropolitan areas of New York, San Francisco, and Los Angeles. From the urban centres the disease has now diffused to other sub-urban areas and is becoming increasingly a heterosexual disease, while the incidence of AIDS is declining among the homosexual community due to their safe sexual practices. Alarmingly the disease has entered the infant groups through infected mothers and increasingly females are the victims of AIDS.[53] Dutt, suggests that the rising number of victims will be from the heterosexual groups in the future, as the disease remains uncontrolled.

The only way to prevent this epidemic is for the society to change their social and cultural behaviour. AIDS has diffused throughout the world and nearly 17 million (WHO estimates for 1994) people are infected. Presumably half of them will develop the disease within the next decade. Due to lack of curative medicine the final level of diffusion is difficult to predict except that if present trends continue the whole world, developing and developed nations will face enormous destruction of life with developing nations being the hardest hit, as they lack the resources to screen blood, institute social and cultural changes, and encourage the usage of sterilized needles. AIDS in the diffusion process has gone through the infusion stage, it is now in the inflection stage and will be saturated when everyone is infected, whether it will wane after that is doubtful keeping in mind the medical progress the scientists have had in checking the virus.

Due to the nature of the HIV and its infectious process, should a vaccine be developed, it will not be a panacea.[54] The number of cases will continue to rise as medicine battles it. The only effective avenue for dealing with the problem, for the foreseeable future, is prevention at all levels: primary (avoidance of transmission), secondary (minimize complications in infected persons through aggressive therapy), and tertiary (minimise pain and suffering in persons with AIDS).

Hopefully, the ultimate answer will come from the medical-biological research in the form of vaccines to prevent infection and drugs to rid the body of the virus. Meanwhile, it is imperative for social scientists to collaborate their efforts on understanding of the virus and its mode of transmission. Based on such understanding they should examine the socio-behavioural context of HIV infection and AIDS and should provide ideas for development of programmes appropriate to controlling the infection process and caring for those infected.

References and Notes

1. May, Jacques (1950). 'Medical Geography, Its Method and Objectives', *Geographical Review*, Vol. 40, pp. 9-41; Howe, G. Melvyn (1936). *Man, Environment and Disease in Britain*, New York : Barnes and Noble Books; Pyle, Gerald F. (1969). 'The Diffusion of Cholera in the United States in the Nineteenth Century', *Geographical Analysis*, Vol. 1, pp. 59-75; Hunter, John M. (1974). 'The Challenge of Medical Geography', in John Hunter, ed., *The Geography of Health and Disease*, University of North Carolina at Chapel Hill, Department of Geography, No. 6, pp. 3-20.
2. Learmonth, A.T.A. (1958). 'Medical Geography in India and Pakistan : A Twenty Year Data for the Former British India', *The Indian Geographical Journal*, Vol. 33, pp. 1-59.
3. Akhtar, Rais (1978). 'Goitre in the Kumaon Region of India', *Social Science and Medicine*, Vol. 12D, pp. 157-163; Chakravarti, Aninda K. (1982). 'Diet and Disease: Some Cultural Aspects and Health and Disease in India', in Allen G. Noble and Ashok K. Dutt (Eds.) *India: Cultural Patterns and Processes*, Colorado: Westview Press; Dutt, Ashok K. Jyotshna Jaiswal (1985). 'Two Approaches to Medical Geography', in S.C. Mukhopadhyay (ed.) *Geographical Mosaic*, Calcutta: Modern Book Agency Private Ltd., pp. 498-511; Dutta, Hiran, M., Dutta, Ashok K., Vishnukumari, G. (1979). 'The Resurgence of Malaria in Tamil Nadu', *Social Science and Medicine*, Vol. 13D, pp. 193-196; Hazra, Jayati (1994) ed. *Health Care Planning in The Developing World*, Calcutta : Jayati Hazra ; Hyma, B. and Ramesh, A. (1997). *Cholera and Malaria Incidences in Tamil Nadu, India*, Department of Geography Publication Series No. 9, University of Waterloo, Waterloo, Ontario ; Mishra, R.P. (1970). *Medical Geography of India*, Delhi: National Book Trust ; Singh, Sitasaran, Dutta, Hiran. M. (1981). 'Smallpox Pattern and its Correlates', *Geographical Journal*, Vol. 5.1, pp. 77-82.

4. Mukerji, A.B. (1980). 'The Disease Ecology of a Small *cul de sac* : Chandigarh Dun', *Social Science and Medicine*, Vol. 14D, No.3, pp. 331-36.
5. Shannon Gary,W., Gerald F. Pyle and Rashid L. Bashshur, (1991). *Geography of AIDS : Origins and Course of an Epidemic*, New York : The Guilford Press.
6. Smallman - Raynor, M.R., and A.D. Cliff ,(1990). 'Acquired Immunodeficiency Syndrome (AIDS) : The Global Spread of Human Immunodeficiency Virus Type 2 (HIV-2),' in R.W. Thomas, ed., *Spatial Epidemiology*, London : Pion Limited.
7. See ref. 5.
8. Willis, D.P. (1989). 'The Social Consequences of AIDS', in *New Perspective on HIV - Related Illnesses : Progress in Health Services Research*, pp. 79-83, National Centre for Health Services Research and Health Care Technology Assessment, Public Health Service, U.S. Department of Health and Human Services.
9. See ref. 8.
10. Centres for Disease Control (1982). 'Opportunistic Infections and Kaposi's Sarcoma Among Haitians, in the United States', *Morbidity and Mortality Weekly Review*, Vol. 31, pp. 465-467.
11. Volberding, P. (1987). 'AIDS—Variations on a Theme of Cellular Immune Deficiency, in J.C. Gluckman and E. Vilmer, ed. *Acquired Immunodeficiency Syndrome*, pp. 191-98, Amsterdam : Elsevier.
12. Fauci, A. (1988). 'The Human Immunodeficiency Virus : Infectivity and Mechanisms and Pathogensis', *Science*, Vol. 239, pp. 617-622.
13. Shannon Gary, and Gerald F. Pyle, (1989). 'Origin and Diffusion of AIDS', *Annals of the Association of American Geographers*, Vol. 79, March, No.1, pp.1-24.
14. Clavel, F. K. Mansinho and Chamanet, S. (1987). 'Human Immunodeficiency Virus Type 2 Infection Associated with AIDS in West Africa', *New England Journal of Medicine*, Vol. 316, pp. 1180-85; Denis, F., Gershy-Damet G. *et al.* (1987). 'Pervalence of Human T-Lymophotrophic Retroviruses Type III (HIV) and Type IV in Ivory Coast', *The Lancet*, Vol.I, pp. 408-411; Mabey, C.C, Tedder, R.S. *et al.* (1988). 'Human Retroviral Infections in the Gambia', *British Medical Journal*, Vol. 296, pp. 83-86.
15. Clumeck, N., Dewit, S. (1988). 'AIDS in Africa', in M.A. Sande and P. Volberding (eds.), *The Medical Management of AIDS*, Philadelphia : Saunders
16. Pape, J.W., Liautaud, B. *et al.* (1983). 'Characteristics of the Acquired Immuno deficiency Syndrome (AIDS) in Haiti', *New England Journal of Medicine*, Vol. 309, pp. 945-950; Pitchenik, A.E., Fischi, M.A. *et. al.* (1983). 'Opportunistic Infections and Kaposi's Sarcoma among Haitians : Evidence of a New Acquired Immunodeficiency State', *Annals of Internal Medicine*, Vol. 108, pp. 277-284; Brunet J.B. and Ancell R.A. (1985). 'The International Occurrence of the Acquired Immunodeficiency Syndrome' *Annals of Internal Medicine*, Vol. 103, pp. 670- 674.
17. Smith, T. F., Srinivasan, A. *et. al.* (1988). 'The Phylogenetic History of Immunodeficiency Viruses', *Nature*, Vol. 333, pp. 573-575.
18. Kreiss, L.I., Lee, S. *et al.* (1986). 'AIDS Virus Infection in Nairobi Prostitutes', *New England Journal of Medicine*, Vol. 312, pp. 414-418; Mann, J.M. (1987). 'The Epidemiology of LAV/HTLV-III in Africa', in J.D. Gluckman and E. Vilmer, (Eds.), *Acquired Immunodeficiency Syndrome*, pp. 131-136, Amsterdam : Elsevier.
19. Nahmias, A.J., C. Reynolds - Kohler, *et al.* (1988). 'Human Immunodeficiency Syndrome Virus Detected in Bowel Epithelium from Patients with Gastrointestinal Symptoms', *The Lancet*, Vol. 1, pp. 259-262; Nzilambi, N., DeCock, K. *et al.* (1988). 'The Prevalence of Infection with Human Immuno deficiency Virus Over a 10 Year Period in Rural Zaire', *New England Journal of Medicine*, Vol. 318, pp. 276-279.
20. Zuckerman, A.J. (1986a). 'AIDS and Insects', *British Medical Journal*, Vol. 292, pp. 1094-94; Zuckerman, A.J. (1986b). 'AIDS in Primates', *British Medical Journal*, Vol. 292, pp. 158; Fukusawa, M., Maria, T. *et. al* (1988). 'Sequence of Simian Immunodeficiency Virus from Africa Green Monkeys, a New Member of the HIV/SIV Group', *Nature*, Vol. 333, pp. 457-461; Kanki, P., Alroy, J.J., Essex, M. (1985). 'Isolation of T-Lymphotrophic

Retrovirus Related to HTLV-III/LAV for, Wild Caught African Green Monkeys', *Science*, Vol. 230, pp. 951-954; Montagnier, L. and Alizon, M. (1987). 'The Human Immune Deficiency Virus (HIV) : An Update', in J.C. Gluckman and E. Vilmer, (eds.), *Acquired Immunodeficiency Syndrome*, pp. 13-21, Amsterdam: Elsevier.

21. Montagnier, L. and Alizon, M. (1987). 'The Human Immune Deficiency Virus (HIV) : An Update', in J.C. Gluckman and E. Vilmer, (Eds.), *Acquired Immuno-deficiency Syndrome*, pp. 13-21, Amsterdam: Elsevier.
22. Hunt, Charles W. (1993). 'The Social Epidemiology of AIDS in Africa : Migrant Labour and Sexually Transmitted Disease', in Peter Conard and Eugene B. Gallacher (Eds.), *Health and Health Care in Developing Countries : Sociological Perspectives*, Philadelphia : Temple University Press, pp. 1-37.
23. Dutt, Ashok (1995). 'A Book Review', Peter Conrad and Eugene B. Gallagher (Eds.) 'Health and Health Care in Developing Countries : Sociological Perspectives', *Social Science Medicine* , Vol. 40, No.1, pp. 137-140
24. Mann, J.M. (1987). 'The Epidemiology of LAV/HTLV-III in Africa', in J.D. Gluckman and E.Vilmer, (Eds.), *Acquired Immunodeficiency Syndrome*, Amsterdam: Elsevier, pp. 131-136; Torrey, B.B., Way, P.O. and Rowe, P.M. (1988). 'Epidemiology of HIV and AIDS in Africa : Emerging Issues and Social Implications', in N. Miller and R. Rockwell (Eds.), AIDS in Africa: The Social and Policy Impact, Lewiston, NY : E. Mellon Press, pp. 31-54 ; Shannon Gary and Gerald F. Pyle, (1989). 'Origin and Diffusion of AIDS', *Annals of the Association of American Geographers*, Vol. 79, March, No.1, pp. 1-24; Shannon Gary, W., Gerald F. Pyle and Rashid L. Bashshur, (1991). *Geography of AIDS : Origins and Course of an Epidemic*, New York: The Guilford Press.
25. Torrey, B.B., Way, P.O. and Rowe, P.M. (1988). 'Epidemiology of HIV and AIDS in Africa : Emerging Issues and Social Implications', in N. Miller and R. Rockwell (Eds.), *AIDS in Africa: The Social and Policy Impact*, Lewiston, NY : E. Mellon Press, pp. 31-54.
26. *See* ref. 13.
27. *See* ref. 5.
28. Centers for Disease Control (1986). 'Acquired Immunodeficiency Syndrome (AIDS) among Blacks and Hispanics—United States', *Morbidity and Mortality Weekly Report*, Vol. 35, pp. 655-666; Centers for Disease Control (1987). *Morbidity and Mortality Weekly Report*, Vol 36, p. 157; Centers for Disease Control (1987). 'Human Immunodeficiency Virus (HIV) Infection Codes : Official authorized addendum ICD-9-CM effective January 1, 1988, *Morbidity and Mortality Weekly Report*, Vol. 36, pp. 1-24.
29. Dutt, Ashok K., Miller, D., and Dutta, H.M. (1990). ' Reflections on the AIDS Distribution Pattern in the United States of America', in R.W. Thomas (ed.), *Spatial Epidemiology*, London: Pion Limited, pp. 183-96.
30. Mukhopadhyay, Anupa (1994). 'Spatio-Temporal Diffusion of AIDS in the United States', Dissertation Submitted to the Graduate Faculty at the University of Akron.
31. *See* ref. 30.
32. *See* ref. 30.
33. Koff, Wayne C. (1994). 'The Next Step Toward a Global AIDS Vaccine', *Science*, Vol. 266, pp. 1335-1337.
34. McAleer, W. J., *et al.* (1984). *Nature*, Vol. 307, p. 178.
35. Myers, G., *et al.* (1993). 'Human Retroviruses and AIDS: A Compilation and Analysis of Nuclei and Amino Acid Sequences', *Theoretical and Biology and Biophysics*. Los Alamos National Laboratory, Los Alamos, NM.
36. *See* ref. 35; Mascola, J.M. *et al.* (1994). *Journal of Infectious Disease*, Vol. 169, p. 48.
37. Phillips, D.M. (1994). *AIDS*, Vol. 8, p.719.
38. Golding, H.R., *et. al.* (1989). *Clin. Invest.*, Vol. 83, p. 1430
39. Fast, P.E. and Walker, M.C. (1993). *AIDS*, Vol.7, (suppl.1), S147.
40. *See* ref. 33, p. 1335.
41. *See* ref. 33.

42. Park and Park, (1992). *Preventive and Social Medicine*, India.
43. *See* ref. 42.
44. Indian Council of Medical Research, (1989). *ICMR Bulletin*, Vol. 19, No.11, November.
45. Indian Council of Medical Research, (1991). *ICMR Bulletin*, Vol. 21, No. 12, December.
46. *See* ref. 45.
47. *See* ref. 45.
48. *See* ref. 45.
49. World Development Report (1993). *Investing in Health*, New York: Oxford University Press.
50. Aggleton, Peter, *et al.* (1994). 'Risking Everything? Risk Behavior, Behavior Change, and AIDS', *Science*, Vol. 265, pp. 341-345.
51. *See* ref. 50, p. 343.
52. Dutt, A.K., Monroe, C. and Prince, B. (1987). Geographical Patterns of AIDS in the United States', *The Geographical Review*, Vol. 77, pp. 456-71.
53. *See* ref. 29.
54. Osborne, J. E., Jenson, A. R., Cooke, M. *et al.* (1986). 'AIDS : Science, Ethics and Policy', *Issues in Science and Technology*, Winter, pp. 40-54.

21

COMMERCIAL ACTIVITIES IN HILLY REGIONS

Tradition *vis-a-vis* Options— A Case of a Tea Garden

NASREEN AHMAD AND RAZIA SULTANA

Introduction

The hilly region of Bangladesh comprising the Greater Districts of Sylhet, Chittagong and the Chittagong Hill tracts, covers only 10-11 percent of the total area in the north-eastern and eastern part of the country bordering both India and Myanmar. This region is distinctive both in terms of its economy which is based on minerals, stones, timber and tea, and its population composition which primarily include the cultural or ethnic minority populations the Khasias, Manipuris, Santhals, Oraons, Garos, Hajong, Tripra, Chakmas, Mogh, Murang, Kukis and others.[1] These people attach strong values of tradition and customs to the activities pursued by them. In the districts of Greater Sylhet most ethnic groups/minorities work on their own *jhum* of betel leaf/nut or in the tea gardens, which constitute the principal industry in Bangladesh and are largerly concentrated here. Of a total of 156 tea gardens, 134 are located in Greater Sylhet and of the total land devoted to tea in Bangladesh (47, 239 ha.), 93 per cent of the land (43, 902 ha) is in Sylhet. Tea in Sylhet is grown on the *tilas* and hillocks (Fig.21.1). Although Bangladesh contributes only 2 per cent to the world production of tea, after meeting the domestic demand, tea comprises 6 per cent of all exports of the country. The industry further absorbs about 106,000 workers[2] *i.e.* 0.67 per cent of the total labour force employed in the industrial sector,[3] majority of the them belonging to ethnic/cultural groups.

The tea industry has a history of its own . With the first tea garden established in 1854, the pioneer planters, the British, realised that tea plants/gardens needed constant care and attention. Therefore, to build up a permanent work force which would reside and work within the estate, tea labourers were drawn from far off places in India — Madras, Madhya Pradesh, Orissa, Bihar and others.[4] Lured by *sardars*/contractors (also known as *arkathis, maghi*) and sold to the British planters, the first batch of labourers (who were gradually joined in by others) were thus trapped in an alien and remote environment. This work force no matter how dissatisfied, had no means of going back to their ancestral homes and were thus, compelled to go on living and working in these isolated regions.

Over the years this isolation, presumably also bred a sense of insecurity that barred them from venturing outside their new, but by now, familiar domain. The isolation of the work force also led to the preservation of their language (which is today a mixture of Hindi, Uriah, Bangla), religion, rituals and attitude towards life and living. Here the tea-garden workforce is referred to as 'people' who indulge in an activity respecting the customs and practices of the past *i.e.* people who have 'tradition' that guides them in their choice of activity.

Objectives

This study through a case in point of an ethnic minority dominant labour force in a Tea Garden argues, that it is not 'tradition' but 'lack of options and opportunities' that bind people to an activity *i.e.* activities that their forefathers had indulged in. This study also tries to look into the effects of the present day change in activities, if any, of the labour population of the tea industry in general and the consequent change in attitude regarding commercial activitiy of the garden population in particular.

This study is based on data collected through a questionnaire survey of the household heads in all the labour households of the study garden. Informal discussion and long personal experience with the people and the region also forms a basis of relevant information regarding present activity status of the garden population and their future aspirations.

Study Area and Population

Nestled amidst the foothills of the Khasia Jaintia hills in the north eastern corner of Bangladesh (Fig. 21.1), Sreepore Tea Estate is a small proprietory garden with a total land grant of 415 ha. of which only about 80 ha. is under tea. The estate supports 53 labour households while about 40-45 households of non-labourers reside illegally within the Estate property. The study garden, in existence for about 88 years, has changed ownership thrice but the labour population has remained mostly unchanged, with a majority being descendants of the original batch of labourers brought for this particular plantation from India.

To the north of the study garden, there is a small Khasia settlement or *pungee* comprising 22 households who own and work in their *jhums* of betel leaf and nut. A stone quarry in operation since the early sixties, but which has grown to its present peak status in the last seven years, is also located within the garden area.[5]

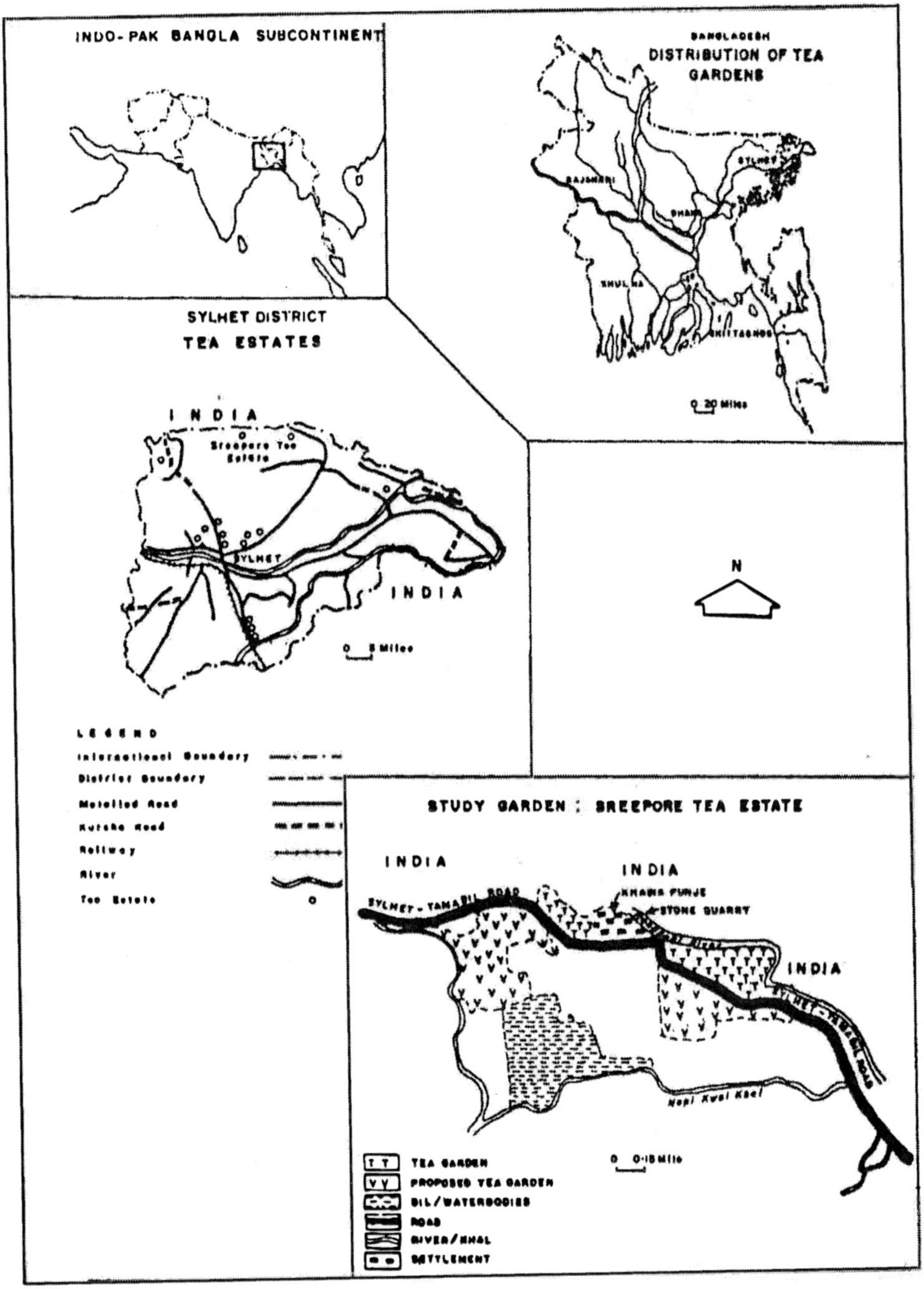

Fig. 21.1 : Study Area

Garden Work Force : Their Activities and Aspirations

Of the 53 labour households surveyed, all, except 7 households (13.2 per cent) are ethnic minorities/caste groups including the Raothias, Uriahs, Santhals, Goalas, Karmakar, Ghatwals and Ghonrs (Table 21.1). Practicing Hinduism and adhering to the caste system these groups are further divided into sub-groups based on social status and ceremonial observances. Thirteen per cent of households belong to Muslims who came in from adjoining Kachar district (province of Assam) and Agartala (Tripura) following communal disturbances during the sixties and after.

Table 21.1 : Origin of Different Groups of Labour Population

Origin Place/DISTRICT	*Caste Groups*
Ranchi, BIHAR	Ghatwal, Mura, Munda
Dumka, Santal, Pargana	Santal
Deogmar, ORISSA	Uria, Rawthia
Bilashpur, W. BENGAL	Ghonr
Bikrampur, DHAKA	Dey Karmoker
Manikgonj, DHAKA	Ghoala

Registration of at least one member of a household as a worker in the garden is mandatory to ensure that households get all facilities provided by the management, *viz* housing, sanitation, free medical aid and free primary education for all children, subsidised ration etc., (originally a convention but now mandatory over the years). Thus, of a total household population of 269, only 76 persons (which is 51 per cent of the total employed people) are registered garden labourers, while the rest (193) are employed elsewhere. Hence securing the extra 80-100 labourers needed during the peak plucking season is difficult and the plantation suffers.

Of the total 269 household members, 147 (54.6 per cent) are gainfully employed, with only the very young (0-10 years) and the very old not working (Table 21.2).

Besides working in the plantation (51.7 per cent) the primary occupation for about 46 per cent of the currently employed labour population is to work in the stone quarry, while about 3 per cent work in the Khasia *jhum* (Table 21.3).

Work at Stone Quarry : A New Option

The stone quarry within the garden area which engages 46 per cent of the employed came into full operation in the late 1980s providing a viable and

Table 21.2 : Distribution of All Members by Age and Sex

Age	*Male*	*Female*	*Total*
0-4	15	17	32
5-9	20	17	37
10-14	24	19	43
15-19	9	12	21
20-24	10	3	13
25-29	6	5	11
30-34	10	11	21
35-39	17	12	29
40-44	2	4	6
45-49	5	7	12
50-54	9	8	17
55-59	9	3	12
60-70	7	8	15
Total	143	126	269

Table 21.3 : Occupation of All Employed Persons

Occupation	*Primary*	*%*	*Secondary*	*%*
Registered with Garden	76	51.7	—	
Stone quarry	67	45.58	19	
Jhum	4	2.72	1	
Business/shops	—	—	3	
Transport	—	—	1	
Construction	—	—	1	
Poultry/Live stock	—	—	1	
Total	147		27	

competitive area of employment *vis a vis* the plantation. Attracted by the cash earned, people from the tea estate, nearby areas and also from all over Bangladesh especially from the districts of Comilla, Chittagong, Mymensingh and Tangail, prefer to work in the quarries. The excess garden population, who previously could not be persuaded to work in other tea estates, are now actively engaged in the stone quarry.

For the registered households, with basic amenities and security of a home ensured, the purpose and prospect of earning extra cash, encourages them to send their offsprings and/or dependents to this new activity abandoning their 'generation old' work (Table 21.4). This new option of quarrying near their homes amidst familiar surroundings has also evidently lured 50 per cent male and 7 per cent female registered garden workers to be involved with quarrying (as a secondary occupation) neglecting their garden work (Fig. 21.2). This has given rise to absentism and low labour productivity in the garden under study.

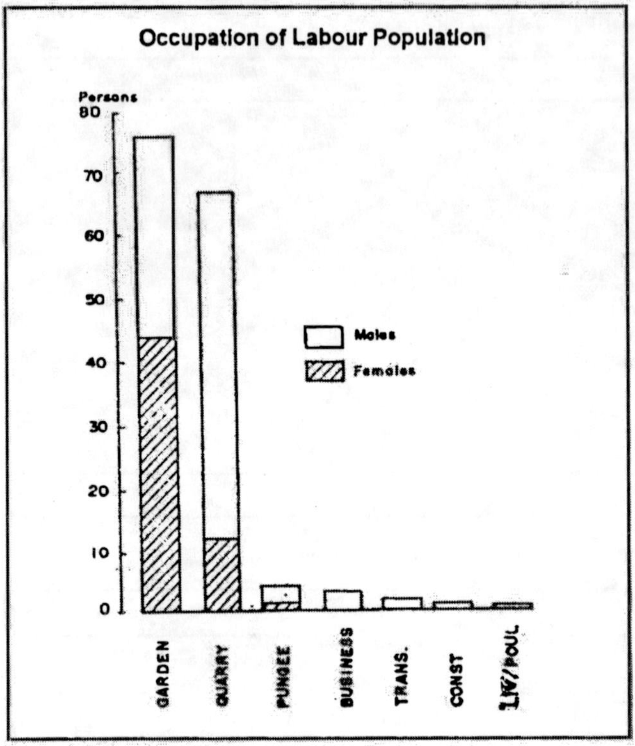

Fig. 21.2 : Occupation of Labour Population

Table 21.4 : Relationship of Workers at the Quarry with Household Heads

Relationship	*Number*	*% age*
Son	31	46.3
Husband	10	14.9
Daughter	5	7.4
Wife	4	6.0
Himself	8	11.9
Herself	2	3.0
Nephew	1	1.5
Brother	2	3.0
Sister	2	3.0
Brother-in-law	1	1.5
Sister-in-law	1	1.5
Total	67	100.0

Preference for working at the quarry is mainly due to the fact that the various types of jobs in the quarry (*viz*. extraction of stone and sand, sorting shingles and gravels, breaking of boulders and loading the stones/sand onto trucks to be transported to different parts of Bangladesh), can make use of as many men and women (however young or old they may be) who are capable of and willing to do the job. As a result, children little over 10 years of age can be seen working besides a 60 year old and it is estimated that about 1500-2000 persons are employed daily in the peak production season between mid October-March at the quarry.

Further, the quarry offers these jobs at various rates on a daily payment basis. Sorting at Taka 30 for females, Taka 40 for males; breaking stones at the rate of Taka 4.00 per cft. ; loading at the rate of Taka 1200-1500 per truck. Depending on the types of jobs done, a person can earn from a minimum of Taka 30 to a maximum of Taka 200 per day. Compared to the garden that pays cash, on a weekly basis, at a fixed rate of a minimum of Taka 20 per day per head (excluding bonuses/provident fund, other facilities/benefits), the cash received at the quarry evidently is more attractive.

The New Option and Consequent Change in Commercial Activity

Working in the stone quarry amidst people coming from all over the country has brought over a change among the garden labourers regarding commercial activities pursued. Moreover to cater to the needs of the people working at the quarry, different commercial activities have been introduced. The quarry has helped to install the local people in commercial ventures that were hitherto unknown to them. Small shops selling tea, betel leaf, areca nut and other items have come up. While two labour households have invested in boats for the sole purpose of transporting stones down the Rangpani river, another has taken on the vocation of construction and maintenance of the road within the garden used by the stone laden trucks. Further, the illegal stills/ distilleries have also increased from one to five in the last few years.

The New Option and Consequent Change in Attitude

The subsequent contact with outside people has helped to widen the vision of the garden population regarding work. This is reflected in their aspirations for their children and dependents. For people who have been occupied all their lives in one activity, to seek a different activity for their offsprings certainly indicates attitudinal change and a broadening of vision. But based on the respondent's personal experience and perceptions, the aspirations for jobs for the future generations are restricted to immediate vicinity, thereby somewhat curtailing the 'broadening of vision'. Evidently social status, power and security that the activity generates is an important factor, but linked with it all is the cash money to be ultimately derived.

Preferred Future Activity/Aspirations

An overwhelming majority of the garden labourers feel that education is the key to all success, but in reality however, only 15 per cent of households send their children to school. Among the total number of school going children (5-14 years), only 30 per cent males and 3 per cent females are presently school going. Labour households do not seem to take advantage of the free education programme for children provided by the management, as contributing to household income by working at the quarry takes precedence.

The location of the garden on the India-Bangladesh border has, since partition in 1947, exposed the population to the presence of the paramilitary forces [the Bangladesh Rifles (BDR) and the Border Security Force (BSF)] of both Bangladesh and India. As they wield considerable power here the respondents have thus shown (Table 21.5) a keen preference for their male child to get educated and join the BDR and police force or else take up office jobs as custom officials. As regards choice of activity, the next for a job as a driver of a motor vehicle, preferably a truck, followed by work in the garden. Work at the quarry comes next. Only 5 respondents showed a preference for independent commercial activity such as owning shops/tailoring etc.

Although sex roles are more fluid in garden culture, but for the girl child, the security of a marriage (like in rest of the country) is the first choice of most parents (35 per cent). As far as occupation is concerned, a slight shift can be noticed. As evident from Fig. 21.3, although more women (57.9 per cent) than

Table 21.5 : Future Occupation Preferred for Children

Job Preferences	*Females*		*Males*	
	No.	*%*	*No.*	*%*
Education & joining BDR/Custom	4	11.9	30	33.7
Driver	X	X	18	20.2
Work in Garden	3	8.8	14	15.7
Work in Quarry	5	14.7	11	12.3
Own Business	X	X	4	4.5
Mechanic	X	X	4	4.5
Household	3	8.8	4	4.5
Teach	1	2.9	1	1.2
Tailor	5	14.7	1	1.2
Midwife	1	2.9	X	X
Farm	X	X	2	2.2
Marriage	12	35.3	X	X
Total	34	100.0	89	100.0

(Multiple preferences were voiced for the son. Response was often nil for the girl child).

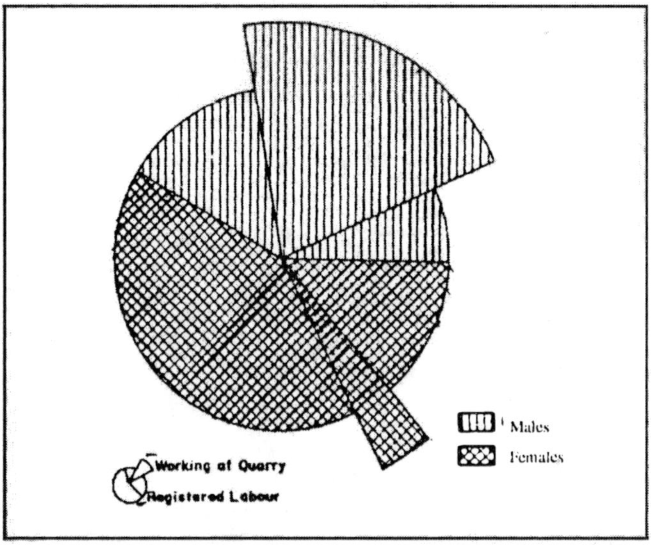

Fig. 21.3 : Registered Workers Involved in the Stone Quarry

men are registered workers in the study garden, households now would like to see their daughters work in the quarry or master a skill rather than work in the tea gardens.

By and large while the older population (55 and above) still prefer to carry on with their traditional activity, it is the younger group of respondents who express their preference for activities other than working in the tea gardens. This trend is apparently natural, response to a tendency to reject rules and a preference for a consumer life style, today's people want independence to control ones work.[6] This tendency also indicates that with time and growing options the garden population, which had so far confined itself to a single activity within a limited space is apt to spread out in various fields encompassing a larger area.

Conclusion

A deep sense of insecurity perpetuated by the knowledge of having been uprooted from their ancestral homes once and kept in isolation as also an awareness of their cultural/minority status kept the work force in the tea gardens doing the same work for generations. However, with growing options and opportunities and consequent exposure, the bindings of respect for past practices have slackened to enable the people to accept activities that can be termed as non-traditional to the tea garden community. In other words, options and benefits have helped the garden population to overcome the controls created by an environment of physical barriers and behavioural inhibitions.

Tea is a labour intensive industry, but this industry which could so long be complacent about its source of labour force, often faces a problem of labour shortage. In fact, many tea gardens in Bangladesh, where employment opportunities exist nearby, are presently experiencing such a problem. It is here that Management Development Programmes in the context of human resource management have come into focus on the motivational aspect of how to retain the labour force and increase labour productivity. As traditions can no longer ensure holding back the labour population and as individuals have a right to pursue an activity of their choice, the Management of tea plantations have to think of ways and means to accomodate and accept the changed circumstances.

References and Notes

1. Rashid, Haroun (1965). *East Pakistan—A Systematic Regional Geography and its Development Planning Aspects*, Lahore.
2. Bangladeshiyo Cha Sangsad (BCS). (1992). *Annual Report, 1991*, Chittagong: Bangladesh.
3. Bangladesh Bureau of Statistics (BBS), (1992). *Statistical Pocket Book of Bangladesh 92*, Statistics Division, Ministry of Planning, Government of Bangladesh.
4. Choudury, Md Safwan (1990). *Working Conditions in Bangladesh Tea Plantations, Problems and Possible Solutions*, University of Oxford, Department for External Studies (unpublished project paper).
5. It may be mentioned that Bangladesh has so long been self-sufficient in its requirement of stones/boulders for construction work. It is only from the year (1993-94) that boulders are being imported from India. Most stone quarrying is done till now in Greater Sylhet in the *thanas* of Jaintiapur (Sreepore/Jafflong), Goinghat (Bisnakandi), Kanaighat (Lubhacherra) and Companygonj (Bhalaganj).
6. Chruden, H.J., Sherman, A.W. (1984). *Managing Human Resources*; Ohio : South-Western Publishing Co.
7. Armstrong, M. (1987). *A Handbook of Personal Management*, London : Kogan Page Ltd.
8. Sargent, A., (1989). *The Missing Work Force*, London: Institute of Personal Management.

22

TIBETAN LIFE IN EXILE-MOMENTUM OF CHANGE AND SOCIO-CULTURAL ADAPTATION

CHAMPA MITRA AND MAMTA DESAI

Introduction

Refugees are a significant social category of the twentieth century, the first half of which has been characterised as 'the age of the homeless'. As such, they have always evoked humanitarian concern dealing with human sufferings, deprivation, their rehabilitation, as also theoretical interest among the social scientists. At a point in transition, their adaptation to ecological, socio-cultural, and economic changes have been of special interest.

Tibet, in the northern slopes of the Himalaya, is the original homeland of the Tibetans, whose religion is Lamaistic Buddhism, in which His Holiness, the Dalai Lama is believed to be a reincarnation of Buddha. No one, with the exception of the Chinese was permitted to enter their country. Being a landlocked territory Tibet was dependent on China to a certain extent. In 1951, the Chinese and the Tibetan Government signed an agreement wherein the existing political system was not to be altered and Tibet was granted regional autonomy under the central Peoples Government of China. However, the agreement was repudiated when China became agressor, and the Tibetan head of state, His Holiness, the *Dalai Lama* and a large number of Tibetans were forced to flee from Tibet and seek refuge in India. Ever since 1959, the *Dalai Lama*, as also a large number of Tibetans settled in India and other parts of the world. As a result of this influx, several resettlement camps were opened by the Government of India to rehabilitate the Tibetans. The *Dalai Lama* himself moved to and continues to stay at Dharamsala along with a number of refugees.

This abrupt change from one kind of social milieu and ecological set-up to a strikingly different one, was a very difficult experience for the Tibetans. Not only were they faced with drastic climatic change, but their occupations, food habits, everything changed. Many lost families and relatives and reached India completely destitute, having no other support, except the hope of security of life and freedom of religion. India endeavoured to do its best in alleviating the sufferings and hardships of the Tibetan refugees. By now perhaps more than 80,000 Tibetan refugees have managed to follow His Holiness into exile. Seeking to save his people and culture, His Holiness began a peaceful struggle to preserve Tibets' unique identity and regain the country's independence.

Consequently, soon after His Holiness, the *Dalai Lama's* escape to India in 1959, the Central Tibetan Administration (CTA) was established in 1960. It is presently located at '*Gangchen Kyishong*' a Tibetan phrase, meaning 'happy valley of snow' midway between Dharamsala and McLeodganj. The different departments constituting the 'CTA' function according to the 'Charter of Tibetans in Exile', which is, in effect, a constitution, based on modern democratic principles and works towards the Tibetan people's struggle for independence and survival. For the Tibetan refugees who migrated into Indian territory following the Chinese takeover of Tibet, India pledged to provide all facilities. As such resettlement camps were established in different parts of India and refugees were provided accommodation and land along with bullocks, agricultural equipment, seeds and fertilizers. These camps, located in different parts of India, differed in their ecological settings. While the headquarters at Dharamsala, Rabongla in Sikkim, and Kalimpong in Darjeeling district were similar in their physical environment to Tibet, the transit camp in New Delhi or agricultural camps in Ganjam district of Orissa or Bylakuppa in Karnataka were entirely different in terms of relief and climate.

In the context of their relocation, and focusing on the original Tibetan society, the social and cultural changes that were subsequently affected have been the focus of this study. Data pertaining to their class, occupation, economic status, ethnic background in original Tibetan society and their occupation, economic and social status in exile was collected and changes in their religio-cultural setting, and their socio-economic status and education, were studied and analysed. After a preliminary enquiry about all Tibetan settlements in India two settlements *i.e.* New Delhi and Sikkim were selected for the purpose of the present study for which purposive sampling was used. The samples were selcted on the basis of the following criteria

1. The pattern of acculturation and adjustments of a settlement composed of more than one ethnic group;
2. Locational factor *i.e.*—rural, urban and rurban;
3. Occupation types, whether influenced by the existing ecological set-up or whether imposed by the concerned authority.

A hundred respondents, including the agricultural labourers/peasants, religious personnel, personnel from CTA and some traders were studied

(Table 22.1). Of these while 40 per cent were associated with agricultural activities, 9 per cent belonged to the elite society (holding dignified posts), 25 per cent were religious personnel, 22 per cent were actively engaged with the CTA and the remaining 4 per cent belonged to the traders' community. In view of the traditional occupational pattern of the Tibetans in their homeland, the occupational status of the respondents in Tibet was found to include 45 per cent of them having been associated with agriculture, 8 per cent constituting the elite society, 25 per cent were religious personnel, 10 per cent herders, and the remaining 12 per cent were traders, thereby covering well the livelihood pattern of Tibetan society in their homeland, Tibet.

Table 22.1 : Percentage Distribution of Tibetan Refugees (interviewed) in Accordance with Their Livelihood Pattern.

S. No.	*Occupational status*	*No. of respondents*	*Percentage*
1.	Attached to agriculture	40	40
2.	Elite	9	9
3.	Religious personal (monks)	25	25
4.	CTA	22	22
5.	Traders	4	4
		100	100

Table 22.2 indicates the occupational status of the respondents, in Tibet. From the view point of the Tibetan traditional occupational pattern, the Table is well represented, as it covers the livelihood pattern of Tibetan society in their homeland *i.e.* Tibet.

Table 22.2 : Percentage Distribution of Tibetan Refugees (interviewed) in Accordance with Their Livelihood in Tibet.

S. No.	*Occupational status*	*No. of respondents*	*Percentage*
1.	Attached to agriculture	45	45
2.	Elite	8	8
3.	Religious personal (monks)	25	25
4.	CTA	10	10
5.	Traders	12	12
		100	100

Social Structure of Tibetan Refugees—Past and Present

It is an established fact, that environment plays an important role in determining the style of man's socio-cultural life. In fact, man and his environment are interrelated. The dynamism of this interrelationship is one of the chief characteristics of culture.

In the past, Tibetan society was a rank startified society, which was divided into two main groups *i.e.*—the lay society and the monastic society. The lay society consisted of elitic society, followed rank-wise by merchants and farmers, small traders, carpenters, tailors, masons and lastly, the smiths, tinkers, butchers. Apparently there was no middle class in the past.[1]

The noble men, belonging to the elitic class, attributed their descent to three main sources. First, the family of a commoner in which a *Dalai Lama* was reincarnated, and which therefore, received large estate from the Government, by virtue of which it came to be among the nobles in society. Thus, families of the *Dalai Lamas* were treated as nobles of lay society. Secondly, families who were recognised and awarded by the Government for their services to the state, and were given the status of nobles. Thirdly, families who traced their ancestry to the early monarchs, who reigned Tibet, prior to the rule of the *Dalai Lama*, were also treated as nobles. As per existing rules, nobles could marry only within the noble society. Both polygamy and polyandry were practised by them. The main objective was to save the property from being divided. The nobles earned their livelihood from three different sources firstly, from their family estate. Though one or two members of a family served as Government officials, the salaries they earned were negligible. The main sources of income were the family estates, and the services rendered by the officials were granted as '*Kral*' or tax. Secondly, every estate had some pastoral subjects. According to Bell, one of the leading families, the *Pa - Lhas* paid 2,200 pounds to the Government as tax.[2] Thirdly, many of the nobles had rug or blanket factories or some other trade, although the main source of income remained landed property.

Next was the peasantry, which constituted five - sixths of the total population. The long winters, harsh climate, limited cropping season, simple and primitive methods and implements for farming were responsible for very low crop yields per acre. Moreover, due to the poor quality of soil, the land holdings were large, which required more hands to look after.[3] As such, among the peasantry too, both polygamy and polyandry were practised to keep the family land intact. The most able son inherited the land, while other brothers shared his wife and land. In the absence of a male heir, daughters were able to inherit the property and a young man was adopted as her husband. The head of the family was the tax payer. The system was a unique example of social adaptation to the existing ecological set-up.[4]

The herdsmen too played a vital role in the lay society and the mixed economy practised. Yak and sheep were the most important domestic animals kept by the Tibetans. The herdsmen with their herds had to live within well defined territory and routes fixed by the Government. The butter prepared from sheep or yak milk, was sent to the Government treasury as part of the revenue, and the district administration in turn sent a portion of it to the monasteries for the burning of lamps in the temples. The social life of the herdsmen was controlled by religion and they too followed the different sects of Buddhism. They also sent their sons to the monasteries to make them monks.

Similarly, trade and traders played an important role in Tibetan economy and society. In fact, every Tibetan household, whether rich or poor, produced its own food and clothing. However, items such as salt, utensils, cotton goods, tea, silk etc. had to be purchased from India and China. In the absence of any manufacturing industry, the main export items were wool, yak tail, firs, borax and herbs. The barter system was the only existing system of trade. Mules, yaks and ponies were the main means of transportation. All sections of people from nobles to peasants were somehow engaged in trade.

Craftsmen of different categories did not hold any high position in society. With the exception of Nepali metal workers, other craftsmen belonged to low class Tibetans. However, only the butchers, regarded as sinners and outcasts, were not allowed to enter priest-hood for performing religious rites. The lowest strata of society comprised the beggar community responsible for carrying away corpses to particular places where they were given over to vultures. It is evident therefore, that Tibetan society was based on a well defined system of stratification although any religion based class system was absent. The officials had to wear clothes and jewellery in accordance with their ranks; even their hair styles were strictly framed by the authority.

Tibetan society, on the whole, was controlled by different religious segments, as a result of which priests played a very important role in the socio-cultural life of the Tibetans. They were engaged to perform religious rites for the families, to cure illness, to perform rites after birth and death. The lay society therefore, was to a great extent dependent on the monastic society and an understanding of the structure and organisational influences of the monasteries thus, becomes imperative to the understanding of the socio-cultural context.

There were 1,800 monasteries of different categories, belonging to the ruling Geluk-Pa and other sects. The *Dalai Lama* held the highest position in the monastic society and had a well organised administrative body under him. In each monastery, there was a three tier organisational set-up, known as the 'university board' (*La-chi*), and 'the college' (*Khamtsten*). The chief administrator of the college was known as '*Khen-Po*' under whom there were '*Kham-tehens*'. The '*La-chi*' or the university board was composed of '*Khen Pos*'. All members of the administrative boards had to render some services allocated by the higher administration, or else pay exemption fees.

The persons who wanted to be an incarnate *lama*, had to apply to the *Dalai Lama*, pay a handsome donation and organize a feast for the whole congregation. ·Often novices from ordinary families could not afford to arrange all these.

Once admitted to a monastery, a person became entitled to an allowance in cash and kind. Accommodation however, was alloted, according to the wealth and rank of the monks. Only very few monks who succeeded in the learning of Buddhism, became high officials of the monastic administration.

The *Panchen Lamas,* the royal incarnations and abbots of the three great monasteries also held a very high position similar to the *Dalai Lama*. Equal in status to the *Dalai Lama* in the temporal sphere, the *Panchen Lama* held a subordinate position as he controlled only three districts.

In the absence of wheeled vehicles in Tibet, the cost of transportation was rather high. With the exception of people attached to the monasteries all had to pay '*ula*' or transport charges. Many *lamas* and monasteries were engaged in trade and money-lending business. According to Bell, the total income of the monastic estates was higher than the Lhasa Government.[5] Individual *lamas* could also earn through handicraft business, other trades and by charging fees as priests. Though the structure and administration of the monastic society was entirely different, the lay society was controlled by monastic influence.[6]

In fact, in Tibet day-to-day activities, thought, and culture were all controlled by and centred around religion. The monk and monastery were held in great respect by the Tibetans. The political structure too was controlled by the monasteries. His Holiness, the *Dalai Lama* was the head of the administrative set-up, both sacred and secular. As such the head of the Government was always an incarnated monk, and the concept of '*Cho - Si - hui - dan*' is (where '*Cho*' is means religion, '*Si*' politics, '*Nui*' means number '2' and '*dan*' is the possessive suffix'). Two separate civil services existed in tandem, means lay and the ecclesiastical. While the financial office comprised people from lay society, the monks constituted the ecclesiastical court.

Model of Stratified Structure of Tibetan Society in the Past

The Dalai Lama / Panchen Lama

The Lay Society	The Monastic Society
The Nobles	La - chi or University Board
The Big Merchants and Landlords	Dra Tshong (the college)
Rich Traders Peasantry Herdsman	Kham tsten
Craftsman, Ordinary Traders	Ordinary Monks
Beggers, Butchers	

Religion and Education

The influence of religion in moulding human attitudes and behaviour, in the past as also its influence on contemporary society is well recognised. In Tibet too since centuries its influence was all pervading. The Tibetan social structure, socio-cultural activities, even the economy were controlled by religion which played a vital role in Tibetan life till 1950, when China strengthened its aggressive actions over Tibet. Religion was the centre of Tibetan culture and civilization. This typical Buddhist religion, a peculiar blend of Lamaistic Buddhism and animism also isolated Tibet from the rest of the neighbouring states.

The social structure was completely dominated by religion. The powerful religious institutions or monasteries were the centre of bureaucratic activities. As such the political history of Tibet is actually the history of its religion, and one can hardly be discussed without the other.

Tibetans, before the inception of Buddhism, practised a kind of nature worship which after the arrival of Buddhism, did not disappear, but was integrated with Buddhism. The prime reason for the success of Buddhism in Tibet was its capacity of absorption and assimilation.Thus, Tibetan Buddhism, the product of long continued historical processes and political development, differs markedly from Buddhism in the rest of the world. The Tibetan political, economic, and social institutions enriched by religious tradition led to a very different type of socio-cultural organisation. Thus, Buddhism in Tibet underwent a varied degree of adaptation, adjustment, compromise, in accordance with Tibetan wisdom and rationality.

In traditional Tibetan society the monasteries were responsible for education and few general schools existed. The *lamas* were the teachers and the monasteries were designated universities, colleges or schools, according to their status and prosperity. Alongwith religious education, general and vocational training were imparted by the monasteries. As a result some of the Tibetan monasteries developed into world famous Buddhist Universities. Traditional Tibetan sciences such as medicines, astronomy and astrology, were also taught in the monasteries. A large number of *lamas* also specialised in performing rituals and controlling supernatural powers. Even special training was given for police or security staff, who were trained not only to perform sentinal duties at the monasteries, but took part in the national militia.

Social life in Tibet was directly and completely controlled by the monasteries. It was the duty of the monasteries to make the common people conscious about the religious code, and to regulate their lives according to the religious beliefs. All natural phenomena, rain, snow, wind, were indications of the manifestations of God and divine anger was expressed through hail storm, crop damage, illness or any type of disaster. Potala was the principal centre of religious-cum-political activities. In traditional Tibetan society, almost everyone had to take refuge in *Buddha, Dharma, Sangha* and the *Lamas*, who symbolized the first three. Supporting a monk or donating a monastery was a holy and meritorious task.

As mentioned earlier polygamy and polyandry both were practised in Tibet, particularly a fraternal polyandry. The sharing of the same wife by all brothers, led to an almost stable population in Tibet and there was never any population explosion. Thus marriage was a significant social system. Any form of marriage was good to the Tibetan provided it prevented fragmentation of landed property. Marriages were arranged by parents strictly according to the rules of endogamy and exogamy, the main objective being to keep family property intact, to maintain purity of the family stock, and to retain class distinction. Hence, the Tibetan marriage system was a complex one. In order to avoid inbreeding, relatives were not allowed to marry except after seven generations. However, there was no bar to marriage along the kinship relationship through women, except marriage between maternal uncle and niece. The maternal uncle played a very important role at a girl's wedding, as he was the supreme authority regarding blessings.

In a society where polyandry was practised, disparity in the age of marriage is expected. However, the youngest brother became entitled to have any sexual relationship with the woman only at the age of eighteen. Thus the marriage age for males was eighteen, while it was fifteen for females.

Divorce was not a common feature. But in the event of a divorce, the woman's rights were well protected. A divorced wife with her children was entitled to get one-fourth of the family property. If she had a daughter, then she was entitled to one-sixth of the property. In case of one of the children becoming a monk or nun, he/she remained entitled to a share of the family property while the parents had to provide him/her with clothing and provisions. Widow re-marriage was common as to a Tibetan a fatherless family was like a cart without a horse. At the same time, however, freedom in man-woman relationship was not permissible.

Present Day Socio-Cultural Life in the Camps in India

As mentioned earlier, the Chinese aggression and atrocities on the Tibetans, in their motherland forced them to flee to India where they reached completely destitute, and having lost their relatives, with no support. Only few Tibetans were able to come along with their families. As a result, their entire socio-cultural life underwent great transformation *e.g.* the polyandry system gradually disappeared, and came to be replaced by the nuclear family system.

Similarly, the traditional class system gradually waned as separated from their families, individuals had to select partners, ignoring the traditional class system. Thus Tibetans in India came to adopt the monogamous family form. Although trying to maintain traditional marital alliances, marriages of refugees have taken place outside the traditionally accepted groups. Infact, Tibetans marrying Indian, American, Canadian or Swiss is not rare, especially the aristocrat who prefer marital relations with rich foreigners rather than poor Tibetans and in the process have become aloof from the masses. Previously marriages were mainly arranged by the parents through a mediator, but in the new set-up, the younger generation prefers to select partners of their choice. Rituals and ceremonies, deeply connected with the social structure are gradually loosing their value due to the changing socio-economic structure of society. In the traditional class stratified society, the marriage ceremony was an occassion for display of family fortune and presitge, but now there is very little scope to maintain old traditions.

The old orthodox hierarchical system no longer exists among the Tibetan refugees in India due to the changing mode of occupation. Services and business have emerged as the two new and principal occupational divisions. Even in the administrative establishments of the Dalai Lama, various service institutions have emerged to cater to the newly educated generation. At the same time the proportion of peasants has not declined. It is observed that the number of farmers is greater than Tibetans in service because of the greater number of agricultural camps for refugees than the administrative establishments.

Following migration it were the Tibetan *lamas* and other religious personnel who experienced the most drastic change, as it was not possible for them to continue their traditional monasterial way of life in the new ecological set-up.[7] Thus, rehabilitation for the large number of refugee *lamas* emerged as one of the biggest problems for the refugee community and the host administration. Realising fully the value and need of preserving the Tibetan religio- intellectual tradition, some religious institutions were established by the Central Tibetan Administration (CTA), where highly intellectual *lamas* were rehabilitated. Thus, the CTA was able to create the traditional religious environment in the different refugee camps. In India wherever Tibetan settlements occur the refugees have built *gomphas* or monasteries, wherein the traditional rituals are performed by the *lamas*, who continuously recite prayers, roll prayer wheels, and recite *mantras*.

The rituals related with birth, death, and marriage which were traditionally practised in Tibet, are generally followed with some modifications by the refugees in exile. For example, butter oil lamps have been replaced by electric bulbs. Everywhere in India, wherever a Tibetan settlement is located, one can recognise it by the prayer flags flying atop trees, electric poles or hill tops.

Religious paintings, astronomy and magic are still practised by them. Thus, the Tibetan refugees in India have been able to maintain the religious essence in their social life despite the drastic environmental changes. Even in an urban metropolitan city like New Delhi, it is observed that they have been able to recreate a nucleus of the old order of monastic society (near Ladakh road). A number of monks and *lamas*, belonging to different sects have been resettled by their followers in different parts of India. One of the most important and oldest sects, the *Sakya*, has built a monastery in Dehradun. Similarity, the Reumtek monastery in Sikkim has been built by the *Kagya* sect just as the monastery in Kalimpong has been built by the chief of the *Nyingmapa* sect. The followers of the *Panchen Lama*, belonging to the *Tashilhunpo* sect have established a monastery in Bylakuppa in Karnataka. Each of these monasteries belonging to different sects have their own abbots, college council etc., in accordance with the traditional Tibetan system. As the host governments or the foreign donors had very little concern about religion, the refugees themselves co-operated fully with the CTA to complete the task of religio-cultural rehabilitation despite tremendous odds.

However, it is observed that Tibetans still fear the erosion of religious values. Their fear is not totally baseless. The tradition of sending one family member to the monasteries has been weakened, which is affecting the recruitment of the monastic order.

In Tibet *lamas* or monks were sponsored by their respective families, which is not possible in India due to economic reasons. Moreover, the younger generation members do not wish to study in the religious institutions, and prefer public schools. In traditional Tibetan society, education was in the hands of the monasteries. Thus, one of the major problems faced by the refugees in India, was related to education of the younger generation. Realising the seriousness of the problem, with the help of the Ministry of Education, Government of India, His Holiness the *Dalai Lama* established the Tibetan School Society. In all seven residential schools and one technical school were established in Kalimpong, Dalhousie, Dharamsala, and Panchmari. Apart from these residential schools, there are day schools, located at different resettlement camps. All these schools, emphasised teaching Tibetan religion, language, and culture. For the purpose of teaching, initially the planners had to depend entirely on Indian teachers. Now after thirty years in exile, one can expect trained Tibetan teachers from the refugee community, but educated Tibetans do not want to be engaged in low salaried jobs in the educational institution and are temped to get into petty business. As a result there is general tendency of decline in educational standards.

Although India has become a second home to Tibetans, they continue to maintain their distinct identity and culture, and fervently hope to return some day to an independent Tibet. However, being refugees, they are treated as second class citizens, and most of the service opportunities available to the general Indian public, are not available to the refugees. Thus, they face lack of equal opportunities for higher education. Getting admission into an Indian college or training centre is extremely difficult, in some cases impossible for a refugee.[8] To be eligible to get admission, a Tibetan refugee needs a domicile certificate, for which he has to be an Indian Citizen. However, getting Indian citizenship is a very sensitive issue.[9] The main reasons for seeking refuge in India, was not because Tibetans were scared for their lives under Chinese aggression, but they fled their country, only to save their religion, unique culture and identity. Had the *Dalai Lama* and 1,00,000 refugees not fled their country, perhaps today Tibet would have been an old name for a Chinese province. But if the Tibetan Government in exile cannot create more openings for the young generation, the younger generation Tibetans would have to take Indian citizenship to get admission into higher education centres or to get a chance in Government offices. In fact many Tibetans in India and abroad have taken citizenships, but they have not given up their religion or culture. Most of them are living with their families within the Tibetan colonies. Most of them have proved that they are still Tibetans by maintaining their unique culture, religion and identity. They have proved that the question of identity remains with traditional culture or religion and not with citizenship.

References and Notes

1. Chand Attar (1982). *Tibet Past and Present : 1660-1981,* New Delhi.
2. Bell Charles A. (1968). *Tibet, Past and Present* , Oxford.
3. See ref. 2.
4. Carraseo, Pedro (1959). *Land and Policy in Tibet*, Seattle.
5. See ref. 2
6. Ray Burman Bina (1979). *Religion and Politics in Tibet*, New Delhi.
7. Sithar Kunchock (1989). 'Tibetan Monastic Life in the Past and Present',*Tibetan Review* , Vol. XXIV, No. 3, March 1989.
8. Tenzin Kesang (1986). 'Education of Tibetans in India : Problems of Tibetan Education', *Tibetan Review* ,Vol. XXI , No. 8 .
9. Tsering Thondup (1990). 'Should Tibetan Students in India Accept Indian Citizenship', *Tibetan Review*, Vol. XXV, No. 2, Feb. 1990.

SECTION-IV

CULTURE TRAITS AND DISTRIBUTION

23

REGIONAL PREFERENCES FOR FOOD

Some Aspects of Food Habit Patterns in India

A.K. Chakravarti

Introduction

India is rich in a variety of languages, religions, castes, cultures, and ethnic groups. There is also a wide range of food habits which vary from place to place within the country. Two common problems related to Indian food habits, however, are the regional preferences for certain cereals which constitute the bulk of the diet, and the serious protein deficiency (nutrition) and the quantity of food supply in India,[1] but, as far as the author knows, an analysis of the spatial variation of food preferences and the pattern of food habits in India has not been conducted. Such an analysis is deemed significant for at least two principal reasons. First, regional food shortages in India are related to a large extent to the rigid food habits of the people.[2] Secondly, there are certain malnutritional problems prevalent in parts of India which are not due to scarcity of food but to the unique food habits of the people of the region. Similarly, some diseases are endemic in certain regions because of ignorance and the consumption of certain preferred foods by the people.[3] A study of the spatial variation of food habits in India will be helpful in gaining a better understanding of the regional problems of food shortages, in planning nutritional education, and in making efforts to change the food habits for maximum utilization and equitable distribution of food resources within the country. The objectives of this study, therefore, are to investigate the nature and complex relationships of Indian food habits, to analyse the areal pattern of food habits in relation to the varieties of cereals which constitute the bulk of the diet, and to examine the types of vegetarianism / non-vegetarianism practised in different parts of the country with special reference to the influence of their religion on the vegetarianism practised by Hindus in India.

* Reprinted from *The Canadian Geographer*, Vol. 18, 4, 1974.

Data and Methodology

Extensive diet surveys covering different parts of India were conducted from 1935 to 1948 by the Nutrition Research Laboratories and various reports have been published since. Further, during the recent period diet surveys were conducted in collaboration with the Nutrition Sections of the Public Health Directories in some of the states and the results have been published.[4] These publications contain detailed and significant information on the composition of diets for different age, income, occupation, and rural-urban groups, but they do not include any information on the food habits by religion, caste, and cultural group in different regions of India. Therefore, the necessary information for this study has been gathered through field studies and indirect sources. During 1970 and 1972 field work was conducted by the author in the major linguistic (cultural) regions of India.[5] The field work involved the collection of relevant published materials from regional sources, and the interviewing of representatives of various cultural and government organizations regarding the food habits with particular reference to the vegetarianism practised by different Hindu castes of the region. It is admitted, however, that some conclusions drawn in this study are no more than generalizations of dominant trends of complex food habits.

General Nature of Food Habit in India

With its year-round growing season and wide range of climates, a large variety of food is produced and consumed in India. The Nutrition Research Laboratories list 856 food items, with their scientific as well as regional names, which are consumed in different parts of India.[6] Though different varieties of food are consumed in different regions, there are some common features of the Indian food habit. An important characteristic of an Indian meal is the preponderance of foodgrains, which supply as much as 70 to 90 per cent of the total caloric requirement.[7] Cereals, such as rice, wheat, millets, barely, ragi, and maize, comprise the bulk of the diet, but a variety of pulses and such foods as cooked vegetables, meat, fish and some dairy products are taken in very small quantities, primarily as an aid in consuming the bulk of staple cereals.

In a densely populated agricultural country such as India, it is natural for foodgrains to constitute the bulk of a diet since they provide a cheaper source of calories than foods of animal origin. When crops are fed to animals instead of being eaten directly by human beings, they lose a large proportion of their caloric value before they re-emerge in the form of meat or milk, about seven calories being required to derive one calorie from animal products.[8] As a consequence, the per capita consumption of food of animal origin, such as meat, fish, eggs, and dairy products, in India is extremely low and far below nutritional requirements (Table 23.1). Although there may be some bias against consuming flesh foods or eggs, most people of all parts of India like to consume milk and other dairy products if they can get them. However, since very few homes have refrigerators, usually the only dairy products prepared and consumed are those which can be kept for a long time without spoiling. Those who can afford it consume boiled butter (called *ghee*), curd, or yoghurt, and a large variety of sweets and candies made out of condensed milk and fresh cheese. Further, fresh milk is consumed warm after boiling, the same milk being heated each time before consumption in order to prevent coagulation.

Table 23.1 : Composition of An Indian Diet (for an adult male in ounces per day)

Class of food	*Average quantity consumed*	*Adequate quantity recommended for a balanced diet*	*Minimum quantity recommended for an improved diet without heavy expenditure*
Cereals	16.6	14	14 (includes 7 oz of millets)
Pulses	2.4	3 (includes nuts and oil seeds)	3
Leafy vegetables	0.7	4	4
Other vegetables	3.2	6 (includes 3 oz of root vegetables)	3
Fruits	0.2	3	2
Milk	2.8	10 (includes milk products)	6
Sugar and jaggery	0.7	2	2
Oils and fats	0.5	2 (includes vegetable oils and ghee)	1
Meat, fish, and eggs	0.5	4.5 (includes one egg)	1

Source : Compiled from Aykroyd *et al.* (see Ref. 6), 28-35. For diets recommended for different age, sex, occupation, and vegetarian/non-vegetarian groups, see "Recommended Daily Allowances of Nutrients and Balanced Diets" (Nutrition Research Laboratories, Indian Council of Medical Research, Hyderabad, India, 1968).

Indian food is prepared principally to satisfy the taste, with little care or concern for nutritional aspects. Even people from well-to-do classes are indifferent and do not have a properly balanced diet.[9] Most housewives spend several hours each day in the kitchen to prepare a variety of fried, roasted, and curried dishes of vegetables, meat, or fish, richly seasoned with spices. Heavily spiced food with hot chillies is favoured in most parts of the country, though the mixture of spices used varies from region to region (there is nothing like a common curry powder). Various pickles and condiments are also used widely, the most common types being those of mangoes, lemons, and hot peppers. As far as is known, these have no nutritive value but they do enhance the palatability of foods. During hot and humid seasons, which last for several months each year, many people suffer from lack of appetite and chronic digestive disorders. The spices containing essential oils and carminative properties help digestion by stimulating the appetite and the secretion of digestive juices, and certain spices such as garlic and asafoetida have some active properties which correct digestive disorders by inhibiting the growth of putrefactive bacteria in the intestinal tract.[10] Further, for millions of people who cannot afford vegetables, meat, or fish, spices and pickles are inexpensive and effective aids in consuming the bulk of plain bread or boiled rice.

Seasonal Nature of Food Habit

An interesting feature of the Indian food habit is that it changes significantly with the seasons. There is little refrigeration for perishable commodities either for storing the surplus produce of the season from a region or for transporting it to distant markets. Canning facilities are also insignificant and, with the limited market, canned foods are too expensive for most consumers. Therefore, most seasonal and perishable commodities such as varieties of vegetable and fruits have to be consumed soon after harvesting. For example, vegetables such as tomatoes, cauliflowers, and cabbages are in good supply and available at very low prices only through the winter, whereas *parval* (*Trichosanthes dioica*), *tinda* (*Citrullus vulgaris*), and baby okra (*Abelmoschus exculentus*) are available only during the summer and have to be consumed in that season. Similarly, some very popular fruits such as mangoes and *lichi* (*Nephelium*) are in abundant supply and inexpensive for a few weeks in summer only, whereas oranges and guavas are plentiful in the winter and can be afforded by people with limited means. Some off-season vegetables and fruits can be bought, however, in exclusive markets in big cities at exorbitant prices.

Another reason for seasonal consumption or avoidance of many foods in India is traditional belief or superstition.[11] Generally, there is a common belief that certain foods are heat-producing or 'hot' and should not be consumed in summer, and other foods are 'cold' and should not be consumed in winter. According to one case study of a village near Delhi, the villagers listed some of the 'hot' foods as egg, peanut, almond, and varieties of millets, and some 'cold' foods as orange, sweet lime, carrot, radish, buttermilk, and barely.[12] Thus many villagers do not eat oranges or other citrus fruits in winter because they believe that they will cause cough, cold, pneumonia, and pain in body. As a result, they deprive themselves of a rich source of vitamin C in winter, and in summer egg-eaters do not eat eggs, believing that they have a 'heat'-producing effect which is harmful for the body. Nutritionists calim that there is no scientific basis for such seasonal discrimination with regard to human health, and find this dietary belief to be one of the major hurdles against the success of their applied nutrition programme.[13]

Relationships of Food Habit to Other Factors

Most food items, including varieties of foodgrains, vegetables, meat, fish, and dairy products (Table 23.1), have been in use in India for centuries, dating back to 2000 B.C.[14] With the passage of time and subsequent influences from the many cultures, races, and religions which came to India, food habits became more refined, and certain regions developed their own food habits. By the early period between 75 A.D. and 300 A.D., inhabitants of northwest India were known for their special liking for butter, milk, curd, wheat, groats, and meat, and the people of east India were recognised for their deep preference for fish, rice, and pungent spices.[15] These food habits are still prevalent in these areas.

An examination of the general food habits prevalent in the country reveals that complex inter-relationships exist between the food habit pattern and various cultural, religious, and physical environmental factors. In order to understand this complexity, a basic model has been prepared to show the types and main relationships of the food habit in India (Fig. 23.1). Culture, religion (including Hindu castes), economic status, sex, and physical environment, all play a role in determining the food habit of a region. Food habits in India can, however, be placed in a two-fold classification, first on the basis of the type of cereals consumed, and secondly, on the basis of flesh foods consumed. In the first category there are rice, bread, and rice-and-bread consuming groups, and in the second category there are vegetarians and non-vegetarians, and there are also various types of vegetarians and non-vegetarians (Fig. 23.1).

Food Habit Pattern for Cereals

The relationships between the food habit and the influencing factors noted above vary greatly from region to region. This has led to characteristic spatial patterns of food habits in India. Further, the regional preference and firm habits of consuming particular types of cereal (bread cereals or rice) on the one hand, or following a vegetarian or non-vegetarian diet on the other (Fig. 23.1), constitute significant barriers to the free flow and optimum utilisation of the total food resources available in the country.

A large variety of cereals are consumed individually or in various combinations. Most cereal consumption is in the form of boiled rice or bread made from other grains, so cereal consumers in India can, effectively, be grouped into rice eaters, bread eaters, and rice-and-bread eaters (Fig. 23.1). The people of a region generally tend to consume habitually those cereals which are favoured

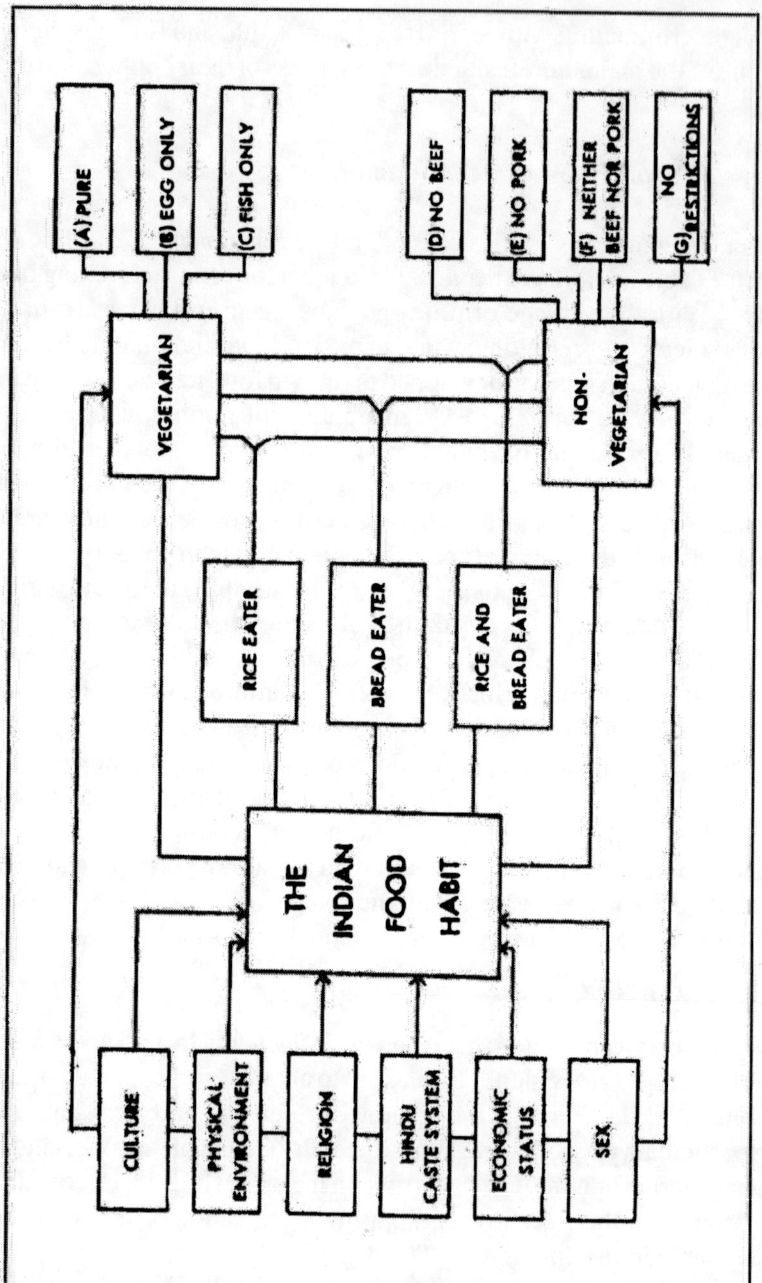

Fig. 23.1 : A Basic Model of the Types and Relationships of Food Habits iin India

by the physical environment and which are the principal foodgrains produced in the area.

Furthermore, a cultural region in India is recognised on the basis of its language; indeed, the whole country is divided into states on the basis of language. The people in a linguistic state not only speak the same language but may have similar ethnic composition, wear similar dress, and eat similar kinds of food. Thus, for example, Bengalis in Bengal are rice eaters and Punjabis in Punjab are predominantly bread (wheat) eaters. On the basis of the predominant cereal consumption and culture of an area, therefore, India has been divided into rice, bread, and rice-and-bread consuming regions—this is the major food habit pattern[16] (Fig. 23.2). There may be islands with different habits in such regions, however, which are generally large urban-industrial complexes such as Bombay and Calcutta. For example, in Calcutta there are millions of labourers, white collar workers, and business men from other states who consume bread, and who maintain a continuous demand for wheat in the predominantly rice-consuming state of West Bengal. Nevertheless, for areal generalization, the dominant pattern of the larger rural areas, which include 80 per cent of the total population of the country, has been taken as the norm (Fig. 23.2).

The main reason for choosing the particular division into bread and rice areas is that, whereas in the former areas a variety of cereals can be prepared into breads, either alone or mixed, in the latter dependence is on rice only. People in the rice areas are reluctant to consume any other cereal brought in from other areas as their staple food. This dependence on one cereal is a critical food problem not only because India is deficient in rice production but also because crop failure may lead to starvation and famine in rice-consuming areas.[17] In the bread-consuming areas, on the other hand, about half a dozen major cereals, namely wheat, *jowar, bajra*, maize, barley, and *ragi*, are consumed individually or in mixes. These cereals are ground into flour, and may be mixed for economy of more expensive cereals while maintaining the flavour and the baking quality of the dough. The dough, rolled into round, flat, and unleavened cakes, is baked over an open fire and called *chappati* or *roti*. Only a few, rich people, mostly in the cities, consume western style breads at breakfast or as snacks.

Generally, wheat is preferred over other cereals for making breads. Wheat does not contain more food value than many millets,[18] and the choice would seem to be made chiefly because it is more glutinous, makes better dough, and bakes into lighter and finer *chappatis*. Other bread cereals, also known as coarse cereals, such as varieties of millets, barley, and maize, are less in demand, get lower prices, and constitute the bulk of the poor man's food. Data on the consumption pattern of these cereals are not available, though people tend to use for breads those cereals which are the predominant foodgrains produced in

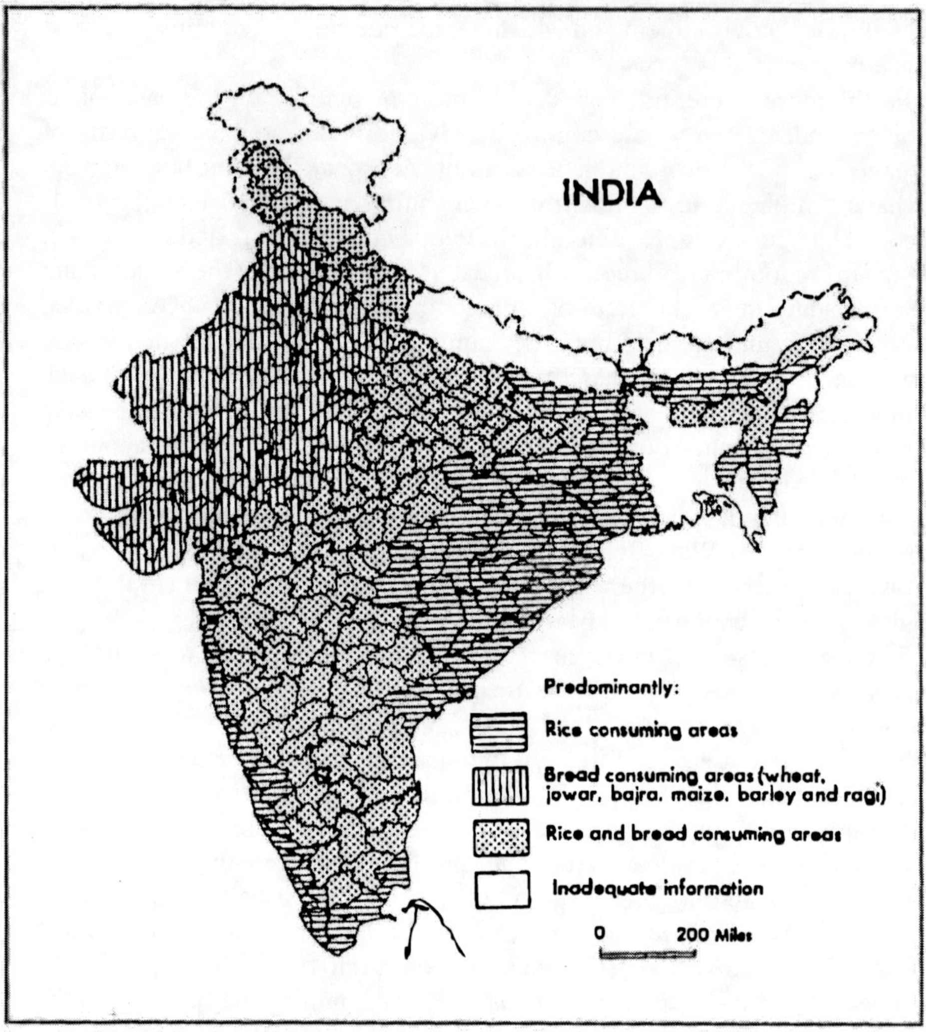

Fig. 23.2 : Food Habit Pattern for Cereals

their area. In recent field investigations, however, it was noted that with the introduction of high-yielding varieties of wheat and better farming conditions, many farmers who used to sell wheat and consume coarse cereals are consuming more wheat than they did previously.

The rice-eating population is mostly confined to the warm and moist areas, where more than four-fifths of the land under foodgrains is given over to rice alone (Fig. 23.2). These are densely populated areas and rice is the most appropriate staple food crop. Rice gives comparatively heavier yeilds than other grains, and as many as two or three crops may be raised in a year. For transplanting the seedlings, from the nurseries to the fields, the large requirement for cheap manual labour is met by the dense population. In the hot, moist climate, rice can be stored for a longer time than other cereals without being spoiled. In fact, older rice is preferred over newly harvested grain, since in cooking it expands more and becomes fluffier. Also rice needs little preparation, fewer utensils, and less fuel to cook than bread. The rice is simply boiled and is eaten for all the meals of the day. Some of these qualities have made it difficult for rice eaters to substitute other cereals as a staple food. In the Bengal famine of 1943, for example, when one and a half million people perished, substitution of other cereals for rice was reported to be the most difficult task by the Famine Inquiry Commission: 'During famine, large quantities of wheat and millets are sent to Bengal ... but efforts to persuade people to eat them in their homes in place of rice met with little success ... In numerous grain stores the wheat flour was deterioriating for lack of demand ... The poor rice eater does not possess the necessary iron utensils and if he did, would not know how to use it. Further, *chappatis* as a food differ in bulk and consistency from a bowl of rice. The rice eater is accustomed to bulky meals of soft consistency which give him a feeling of repletion, and does not relish more concentrated food which needs chewing'.[19]

The Commission also observed a similar condition in Travancore (now Kerala State) and concluded that the problem of how to wean rice eaters from their determined preference for a food in short supply, and from their reluctance to turn to alternative grains, is of importance to the whole of India. Among the predominantly rice eating areas, however, Kerala is the only one where a large amount of manioc is consumed as a supplementary food. This root crop is cultivated in the poor laterized soils and is consumed chiefly by the poor who cannot afford a full rice meal.

In between the predominantly bread and rice areas, there is an extensive zone where both bread and rice are consumed (Fig. 23.2). A variety of cereals are here used for bread, depending on the crops produced in each area. For example, in eastern Uttar Pradesh, wheat, maize, barley, and *bajra* are common, and in large parts of Maharashtra, Andhra Pradesh, and Madhya Pradesh, *jowar* consumption is significant together with wheat, *bajra* and, of course, rice. In Mysore and adjoining districts of Tamil Nadu and Andhra Pradesh, however, *ragi* becomes a significant cereal for breads. In general, the bread-and-rice consuming areas, because of their intermediate location and mixed food habit, have seldom experienced any difficulty in substituting for a particular cereal during an emergency period.

A group of leguminous crops supply a variety of pulses (seeds) which are cooked into a thick soup called *dal*. Among the common varieties are Bengal gram (*Cicer arietinum*), black gram (*Phaseolus mungo*), cow gram (*Vigna catiang*), green gram (*Phaseolus aureus*), *khesari* (*Lathyrus sativus*), lentil (*Lens esculanta*), moth beans (*Phaseolus acontifolius*), peas (*Pirum sativum*), and red gram (*Cajanus cajan*). These are cooked individually or in combinations and their preparation varies from family to family and from one region to the other. There is a tendency, however, to prefer and consume those pulses which are the major produce of an area. Since the pulses are used in small quantities as a supplementary food, and since their method of preparation is similar in most parts of the country, regional preferences and the inter-regional transfer of surplus varieties do not pose any serious problem.

The pulses are of great significance in an Indian diet since they are the principal source of protein for the majority of the people who are either vegetarians or too poor to buy enough meat, fish, or eggs. Besides, they are a good source of B vitamins and, when sprouted, a rich source of vitamin C. Owing to lack of knowledge, taste, or both, however, people of only a few areas consume sprouted pulses, and few take pulses in an adequate amount (Table 23.1).

The above discussion refers only to the most popular methods of consuming cereals and pulses.There are also other methods which are used to add variety. One familiar method is parching or popping them like popcorn. In most villages an ingenious method is used for roasting the grains with heated sand, and popcorn or parched grams are common snacks.

There are three major problems in those areas which are accustomed to using mostly one kind of cereal. First, during crop failures it is not only difficult to arrange for supplies of the foodgrain that has failed, but it is also difficult to induce people to eat other cereals. Secondly, it is the opinion of the nutritionists that a mixture of cereals and pulses provides a more balanced combination of amino-acids than the consumption of only one type of cereal and pulse.[20] Thirdly, the continuous consumption of one type of cereal and pulse in an area has given rise to some serious endemic diseases. For example, beri-beri is common in rice-eating areas, though it is unknown in wheat areas.[21] A major occurrence of beri-beri has been recorded in Bengal and some districts of Andhra Pradesh as a result of the continuous use of highly polished milled rice by people of low income. Pellagra occurs in parts of the central Indian plateau owing to consumption of *jowar* (*Sorghum vulgare*) as the staple food; and lathyrism, causing paralysis of the lower limbs, is common in parts of Madhya Pradesh, Uttar Pradesh, and Bihar states as a result of excessive consumption of the pulse *khesari* (*Lathyrus sativus*).[22]

The Vegetarian and Non-Vegetarian Food Habit Pattern

Vegetariansim and non-vegetarianism constitute a very complex issue in the food habit of the people in India. There is no unanimity about the definition of a vegetarian, and for the majority of the population, *viz.*, Hindus, the practising of vegetarianism or non-vegetarianism is purely a personal choice affected more by family tradition and the cultural environment of the region than by religion.[23] According to existing food habits in India, however, the vegetarian habit has been divided into three groups and the non-vegetarian into four groups (Fig. 23.1).

So many Hindu leaders and writers have eulogized the philosophy of vegetarianism that most Hindus are erroneously considered to be vegetarians. According to the National Sample Survey, approximately 70 per cent of the population in India is non-vegetarian.[24] Among the religions in India, Hindus comprise the major religious group and Christians, Sikhs, Buddhists, and Jains together constitute a small segment of the total population.[25] Buddhists, and Jains together constitute a small segment of the total population.[25] Buddhists and Jains are vegetarians (Type A, Fig. 23.1). If all the Moslems, Christians, Sikhs, and even the non-classified others (totalling 16 per cent) are taken to be non-vegetarians, then at least the remaining 54 per cent of non-vegetarians are Hindus. Since Hindus constitute 82.7 per cent of the total population, approximately 65 per cent of the Hindus in India may be assumed to be non-vegetarians. Hindu non-vegetarianism is a complex concept, However. Hindus do not take beef, of course, but the consumption of other meat and of fish or eggs is discriminative and varies with the culture and region. For example, fish may be considered part of a vegetarian diet in West Bengal but not in Gujarat, and eggs may be consumed by many professed vegetarians though it is a non-vegetarian item for others (Fig. 23.1).

It has been documented that beef was consumed in ancient India by the Hindus, including Brahmins, and by some Jains, who are now stirct vegetarians.[26] It was argued that those who took life were at fault, but not the persons who ate flesh. It appears, however, that vegetarianism became popular and the symbol of a high culture with constant preaching and adoption of the practise by many illustrious Buddhists, Jains, and Vaishnav and Saiva saints.[27] The mission behind the strong fervour for vegetarianism has been one of respect for life, kindness to animals, a sense of responsibility, and an endeavour after higher life, and not any diehard religious belief or taboo. There has been, of course, a great feeling for the cattle in India, for in an agrarian society cattle still represent the wealth and major investment of a farmer, and help him at each stage of his farming activity, by providing transportation and even fuel (dried dung) for cooking food. It is interesting to observe that the Hindus of Bali island and the Buddhists of Sri Lanka (Ceylon) eat beef.

Non-vegetarianism in Hindu society is related to the caste system, but with a great degree of flexibility. Among the principal castes,[28] the Brahmins and Vaisyas, originally the priest and business communities respectively, are generally pure vegetarians, (Type A) and most of them may not consume even onion and garlic; the Kshatriyas and Kayasthas, originally the warrior and service communities, are usually non-vegetarians (Types D and F, Fig. 23.1). In the latter two castes, the women may be mostly vegetarians but the male members of the family relish non-vegetarian dishes These rigid habits are rapidly breaking down, however, with more education, travel, and rapid industrialisation and urbanisation. In many urban areas it is not uncommon to find male Brahmins and Vaisyas eating non-vegetarian dishes. Some vegetarians realise the nutritional value of non-vegetarian food and admit that they avoid it because of old habits and not because of any belief. Further, in many such families, the children and particularly the boys are allowed to consume eggs with the permission of the strict vegetarian parents.

A unique aspect of vegetarianism in India is that there are some cultural regions where all Hindus, including even devout Brahmins, are non-vegetarians. For example, Hindu Bengalis, including Brahmins and Vaisyas, comprising over 78 per cent of the total population in West Bengal, are extremely fond of eating fish and also take goat meat. Similarly, most Hindus of Assam, Kashmir, and the hilly districts of northern Uttar Pradesh are non-vegetarians, (Fig. 23.3) and in the Maithili-speaking area of Bihar adjoining West Bengal, Hindus (including Brahmins) will consume fish and also goat meat if the animal is sacrificed before the appropriate goddess. Most fish-consuming people of these areas, however, prefer freshwater fish, although marine fish can be purchased at lower prices. They consider marine fish less tasty and of inferior quality.

It is difficult to provide any specific explanation for the non-vegetarian food habits among the majority of Hindus in the areas shown in Fig. 23.3. West Bengal has a long sea coast together with a large number of rivers and ponds which abound in fish; but unlike the Bengali Brahmins, the Brahmins and other orthodox Hindus of Andhra Pradesh, Tamil Nadu, and Kerala, with similar environmental facilities, do not consume fish. Judging by the typical location of these Hindu non-vegetarian areas, however, it is likely that their food habit has been influenced by their predominantly non-vegetarian neighbours of Mongoloid origin and Islamic faith along the northeastern and northwestern frontiers of India (Fig. 23.3). Another explanation regarding Bengali non-vegetarianism is based paradoxically on the Hindu religion itself. From very early times, Hindus in Bengal had a special liking for sensual modes of *tantric* worship, and were worshippers of a particular bacchanalian type of *tantric* deity, the goddess *Varahi*. The majority of the people in Brahmanical Hindu society in Bengal have a special relish for partaking fish because the image of the goddess *Varahi* is depicted holding a fish in one of her hands.[29] Recently a stone image of the goddess *Varahi* holding a fish, dating back to the ninth or

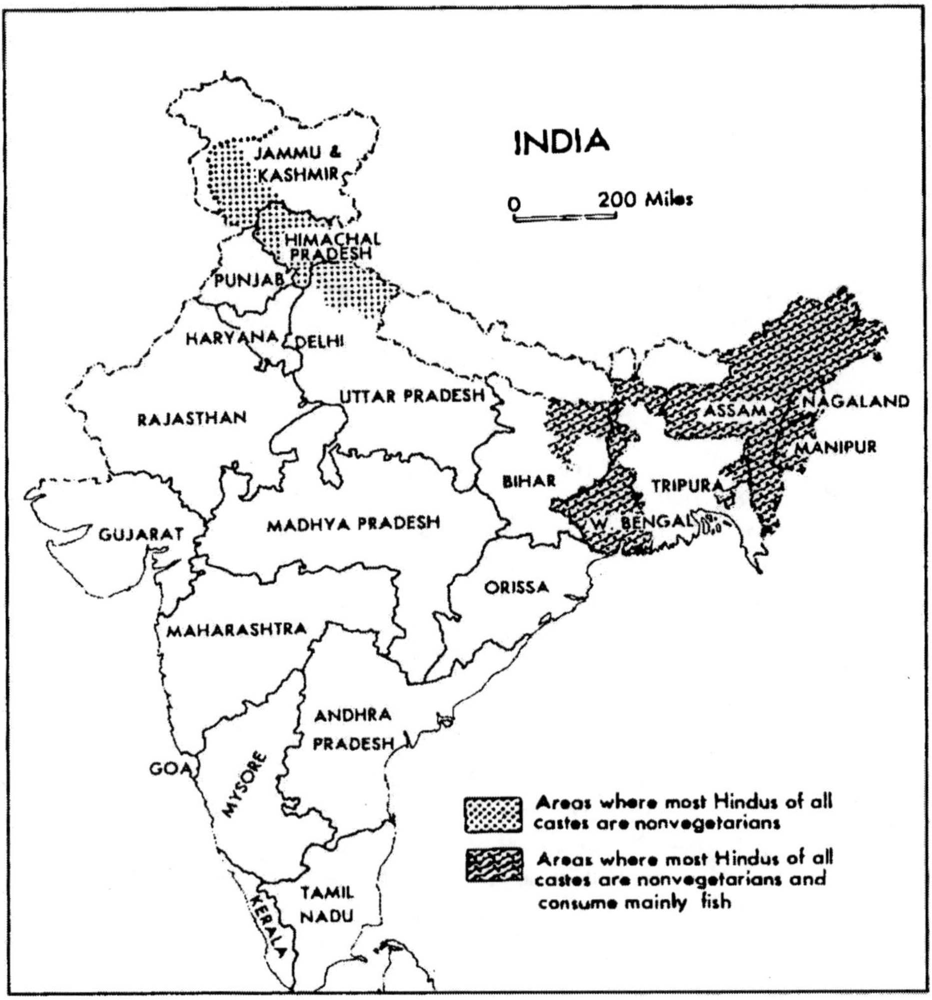

Fig. 23.3 : Non-Vegetarian Hindus

tenth century A.D., was discovered in a village in West Bengal. There are, however, Hindus who are worshippers of Lord *Vishnu* or *Krishna* and who style themselves as *Vaishanavas* (similar to the followers of the *Hare Krishna* movement in North America and Europe). These *Vaishanavas*, comprising only a negligible minority in West Bengal, are pure vegetarians (Type A, Fig. 23.1), as is also true in other parts of India. Another important exception to fish, meat, and egg consumption by Bengalis is the Bengali widow. After the death of her husband, a Bengali woman observes strict vegetarianism and does not consume fish, meat, and egg, or even onion and garlic.

Most non-vegetarian Hindu and particularly the Kayasthas will consume poultry meat and eggs, but in West Bengal many Hindus will not eat chicken meat or chicken eggs although they may take duck eggs. In many villages in all parts of the country, poultry-keeping is considered a side occupation of low-caste Hindu or Moslem farmers, although this prejudice is disappearing gradually. There is also a taboo against consuming pork by Hindus. Excepting some modern farms which raise pigs for the large urban market, only scheduled or low-caste Hindus raise pigs and consume pork. These pigs act as scavengers for the villages, and most Hindus consider pork impure. Among the scheduled castes, however, certain groups within a single caste are vegetarians, some are non-vegetarians, and some may consume any kind of meat including beef and pork. Members of one caste called *Chamar* skin the dead animals for hides, and some of them are so poor that they will consume the meat of a dead cow. The situation has improved since the independence of India (1947), and many such scheduled caste people refuse to work as scavengers and do not eat beef in order to improve their position in Hindu society.

Moslems, who are non-vegetarians (Type E, Fig. 23.1), comprise one-fifth to one-half of the total population in the northern, northeastern, and southwestern parts of India, and provide a large potential market for beef consumption in those areas.

Beef is the cheapest meat and there is a vast potential for beef supply, yet Moslems and other smaller groups such as Christians and Jews in India may not consume the available amounts. Putting the entire responsibility on the Hindus for making the cow inviolable is an over simplification of the issue. There is no homogeneous set of values but a gradation from a completely sacred to a completely secular attitude towards cattle. Although the entire country is committed to the 'sacred cow' policy, a large segment of the population is indifferent and some groups are even dissenting. The Moslems have most violently opposed the 'sacred cow' policy, an issue around which Hindu sentiment has been aroused from time to time by Hindu leaders for political advantage.[31] In large parts of India, nevertheless, when the oxen become too old they are sold to the Moslem butchers. No sentiments are involved in the slaughter of buffaloes, goats, and sheep, however.

Utilisation of cattle for meat is also affected adversely by food preferences and the social status of the beef-consuming population. Many Moslems and Christians prefer goat meat and poultry to beef since the latter is coarse and fibrous and does not make a good curry—the commonest method of meat consumption in India. There are two other reasons for the smaller consumption of beef. The beef is usually of very poor quality since only old and unwanted cattle are slaughtered; most well-to-do families who could consume beef therefore avoid buying this poor quality meat because it is consumed mostly by the poor and shows poor taste. The culture and sentiments of the communities who are in the majority in a village may also influence the food habits of the minorities. For example, in a predominantly Hindu village, Moslems, may develop the same attitude towards beef eating as their Hindu neighbours.[32]

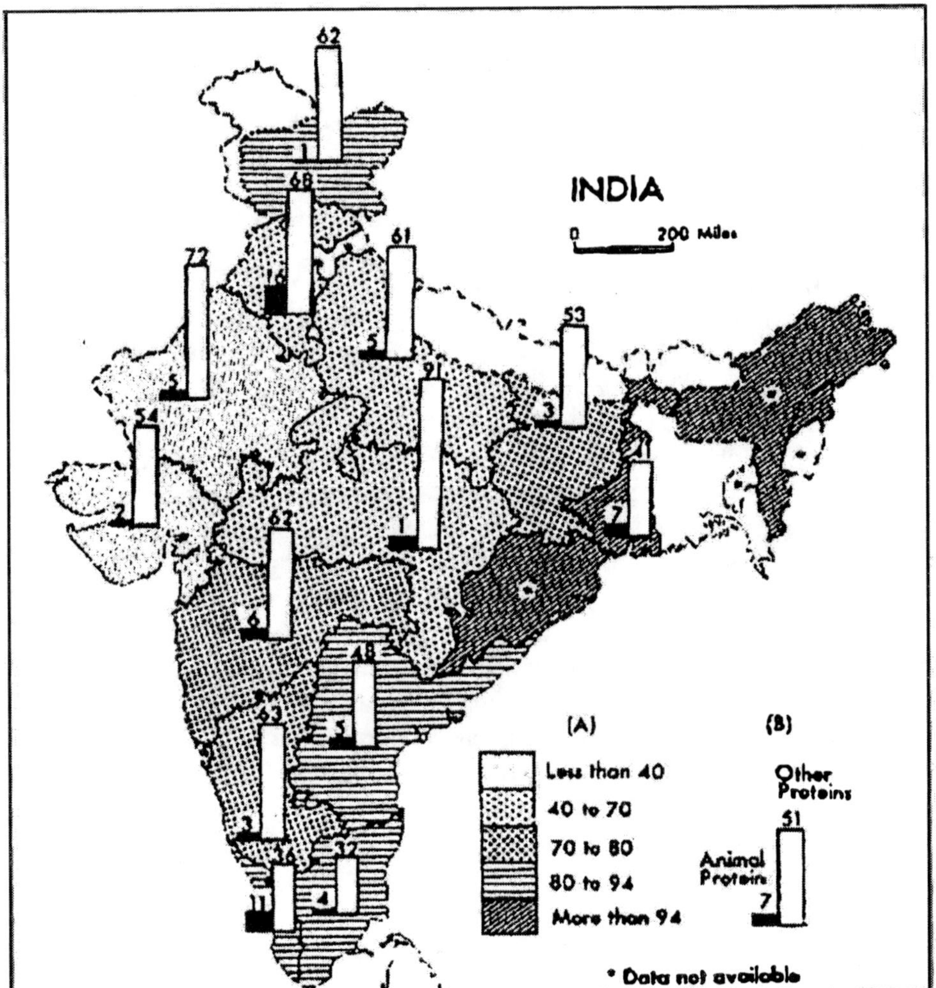

Source : The National Sample Survey Number 99 and 107 (Ref. 24); and Gopalan *et.al.* (Ref. 4). 119 Separate Estimates for the Urban Population are not shown Since the Urban Population Occupies very small Areas, and is Less than 20 per cent of the Total Population of India.

Fig. 23.4 (a) : Non-Vegetarian Population as a Percentage of the Total Population
(b) : Average Intake of Proteins per Capita per Day, in Grams

As is evident from the above discussion, except for Jainism and Buddhism, with very small numbers of followers, religions in India only guide the selection and do not prohibit the consumption of flesh foods. Thus the areal distribution of the non- vegetarian population has little relationship with the distribution of major religions in India. For example, in West Bengal and Orissa, 78 and 96 per cent of the total population respectively are Hindus,[33] yet the proportion of non-vegetarians is as high as 94 per cent (Fig. 23.4). On the other hand, in some states such as Kerala, Punjab, and Jammu and Kashmir, where the proportion of Hindus is as low as 59, 38, and 30 per cent of the total population respectively, the proportion of non-vegetarians is comparatively low (Fig. 23.4) Paradoxically, these states have some of the highest concentrations of Sikhs (60 per cent in Punjab), Moslems (66 per cent in Jammu and Kashmir), and Christians (21 per cent in Kerala), who are mostly non-vegetarians.

The practice of vegetarianism or non-vegetarianism bears no relationship with the amount of intake of animal protein in any state. For example, in Jammu and Kashmir state, where over 80 per cent of the total population is non-vegetarian, there is only one gram of animal protein consumption per capita per day; whereas Punjab, with a much lower proportion of non-vegetarians, has the country's highest per capita consumption of animal protein (Fig. 23.4). It appears that the economic condition is a very significant factor in determining not only the amounts of non-vegetarian foods consumed but also the practice of non-vegetarianism, since most people of low-income groups are perhaps vegetarians because of economic necessity.[34] In one investigation in South India, it was observed that 'only 15 per cent of the people object to flesh foods on religious grounds; most people will eat them it they can get them'.[35] Furthermore, many people do not eat an adequate amount of animal protein because of ignorance. As has been noted, many people in North India do not eat eggs in summer because they think they are harmful and have a 'heat-producing effect on the human body'. Thus during summer, when the egg supply goes down, the prices decline too since most people stop consuming eggs.[36]

Conclusion

The food habit pattern in India is determined by a combination of regional or local factors including religion, physical environment, economic conditions, and even ignorance. Since the influence of these factors individually or in combination varies from place to place, no single factor can be held responsible for the food habits of the people. Regarding the major source of calories, people consume or do not consume certain foodgrains because these may or may not be the major produce of the area. Thus people have formed rigid food habits, erecting regional barriers against the consumption and free flow of certain surplus foodgrains, and hampering the optimum utilisation of the total resources within the country. Also preference for and the continuous consumption of certain grains have resulted in malnutrition and endemic diseases in some parts of India.

A major problem with the Indian food habit is that taste is given more importance than the nutritional aspects of a diet. As a result, a large number of people who can afford to consume a balanced diet do not do so, and cases of serious malnutrition and deficiency diseases are common even among the children of richer people. People with limited means can eat a more balanced diet if they have the knowledge and the interest. There are seasonal surpluses of some protective foods, and there are plenty of edible green leaves rich in vitamins growing the year round, yet few care to consume them. It is possible, therefore, for many families with low incomes to eat a nutritionally more balanced diet if they have the knowledge and have developed the taste. The consumption of mixed cereals, germinated pulses, more leafy vegetables, and some seasonal fruits may not involve extra expense for improved diets.

The role of the Hindu religion and the caste system has previously been over-emphasised. People consume less animal protein primarily because of (a) economic necessity, (b) regional or seasonal prejudice, and (c) social pressure of the majority community over the minority, rather than because of their religious beliefs.

It is evident that no single effort can solve the nutrition problem in India. It is certain, however, that much of the malnutrition is related to deep-rooted food habits. These habits have a spatial pattern with distinctive and characteristic featuers. It will need detailed and large-scale studies to understand the food habit pattern of each region fully, and to suggest methods for eradicating the regional habit-barriers to the optimum utilisation of the total food resources. This study is a preliminary attempt in that direction.

References and Notes

1. Patwardhan, V.N. (1952). *Nutrition in India.* Bombay: Hind Kitab Ltd.; Sinha, R.P. (1961). *Food in India—An Analysis of the Prospects for Self-Sufficiency by 1975-76* . Oxford University Press.
2. Woodhead, Sir John (1945). *The Famine Enquiry Commission: Final Report 1945* , Delhi, India : The Manager of Publications, p. 199.
3. See, for example, the geographic distribution of protein caloric malnutrition in children and some endemic diseases in India, Gopalan, C. and Vijaya Raghavan, K. (1969). *Nutrition Atlas of India* , Hyderabad: National Institute of Nutrition, Indian Council of Medical Research, pp. 56-7 and 84-5.
4. Indian Council of Medical Research Special Report Sl. No. 20 (1951). '*Results of Diet Surveys in India, 1935-48,*' ; Gopalan, C., Balasubramanian, S.C., Rama Sastri, B.V. and Visweswara Rao, K. (1971). *Diet Atlas of India,* Hyderabad: National Institute of Nutrition, Indian Council of Medical Research.

5. A cultural region in India is recognised on the basis of its language, and the country is divided into states primarily on the basis of 18 major languages; see Spate, O.H.K. and Learmonth, A.T.A. (1967). *India and Pakistan: A General and Regional Geography*, London: Methuen & Co. Ltd., pp. 153-7.
6. Aykroyd, W.R., Gopalan, C., and Balasubramanian,S.C. (1966). *The Nutritive Value of Indian Food and the Planning of Satisfactory Diets*, New Delhi : Nutrition Research Laboratories, Indian Council of Medical Research, Appendix IV, pp. 154-257.
7. Indian Council of Medical Research Special Report Sl. No. 20 (1951). '*Results of Diet Surveys in India, 1935-48*,' p.13.
8. *World Food Surveys* , Washington, D.C. : Food and Agricultural Organization of U.N.O., July 1946, p.19.
9. Roy S.M., (1952). 'Food Consumption in India,' *Agricultural Situation in India,* New Delhi: Ministry of Food and Agriculture, Vol.7, p. 89.
10. Swaminathan, M. and Bhagwan, R.K. (1969). *Our Food* , Madras: Ganesh & Co. Private Ltd., 6th ed., p. 48.
11. An analysis of food concepts and seasonal diets in North India has been made by Planalp, J.M. (1971). *Heat Stress and Culture in North India*, Natick, Mass.: U.S. Army Research Institute , pp. 139-203.
12. Sound, K.K. (1967), ' Changing Food Habits in a Rural Community—A Case Study', *Agricultural Situation in India* , New Delhi : Ministry of Food and Agriculture, Vol. 22, pp. 379-81.
13. See ref.12, p. 381.
14. Prakash Om, (1961). *Food and Drinks in Ancient India from Earliest Times to 1200 A.D.*, Delhi: Munshiram Manoharlal, pp. 1-33.
15. See ref. 14, pp. 7-33 and 165-6.
16. The term 'predominant' refers to a major trend and not to any precise quantity, and is based on the report on types of cereal consumption in different parts of India by Aykroyd, W.R. (1948). *Diet Survey in India* ,Coonoor, South India: Indian Research Fund Association, Special Report No.16, pp. 6-35; the district level production data and cereals in *Season and Crop Report* issued annually by each State Agricultural Department of India; Bhatia, S.S. (1965). 'Patterns of Crop Concentration and Diversification in India', *Economic Geography*, 40, pp. 39-56. The above published materials were used together with the information collected from the major linguistic-cultural regions in delineating the food habit patterns for cereals in India. The recent diet surveys have included all the varieties of cereals consumed in an area irrespective of the amount and the frequency of consumption of a cereal. Therefore, the map in the Diet Atlas showing the areal pattern of consumption of varieties of cereals in selected parts of India is unbalanced. See ref. 4, pp. 19 and 118.
17. Chakaravarti, A.K. (1970). 'Foodgrain Sufficiency Patterns in India', *Geographical Review*, 60, pp. 210-211 and 221.
18. *See* ref. 6, pp. 51-52.
19. *See* ref. 2, p. 177.
20. Sound, K.K. (1968). 'Food Culture in a Rural Community: A Case Study', *Punjab Medical Journal* , Medical College, Amritsar, India, 17, p. 507.
21. Gopalan, C. (1966). 'Major Nutritional Problems of India and Southeast Asia', *Proceedings, Seventh International Congress of Nutrition, Hamburg*, ed. H.D. Cremer,(1966). *Nutrition Under Various Geographic and Climatic Conditions*, New York:Pergamon Press,Vol.3, p.325.
22. Swaminathan, M. (1968). 'Diet and Nutrition in India', *Journal of Nutrition and Dietics*, Coimbatore, India: Home Science College, 5, p. 234; See ref. 3, p. 85.
23. Gandhi, M.K.(1960). *The Moral Basis of Vegetarianism* vs *Non-vegetarianism*, Bombay: Popular Book Depot.
24. *The National Sample Survey Number 99*, (1962). Tables with Notes on Consumer Expenditure in Urban Areas, Draft, Fourteenth Round, July 1958-June 1959, Calcutta: India Statistical

Institute, July 1962, Table 1.13.0; *The National Sample Survey Number 107*, Tables with Notes and Consumer Expenditure in Rural Areas, Draft, Fourteenth Round, July 1958-June 1959, Calcutta: Indian Statistical Institute, October 1962, Table 13.0. According to these surveys, the proportion of non-vegetarians is higher in the urban areas and lower in rural areas in some states, but the reverse is true in others. Since rural areas are extensive and contain more than 80 per cent of the total population of India, the analysis and the following Fig. 23.4 are based on the conditions in the rural areas. The data on non-vegetarians by religion are not available, however.

25. Hindus form 82.7 per cent, Moslems 11.2, Christians 2.6, Sikhs 1.9, Buddhists and Jains 1.2, and others 0.4 per cent of the total population. *Census of India 1971*, Series 1-India: Paper 2 of 1972: *Religion*, Delhi, 1972, p. 6.
26. Prakash Om (1961). Food and Drinks in India from Earliest Times to 1200 A.D., Delhi: Munshiram Manoharlal, pp. 64-65; Simoons, F.J. (1961). *Eat Not This Flesh: Food Avoidances in the Old World* , Madison, Wisconsin: Univ. Wisconsin Press, pp. 48-49. An opposite but not convincing argument has been given by Gita Press, *A Review of Beef in Ancient India*, Delhi, 1971.
27. Bapat, P.V. (ed.) (1964). *2500 Years of Buddhism*, New Delhi: Ministry of Information and Broadcasting, pp. 311-312.
28. For a description of the caste system in India, see Ghurye, G.S.(1957). *Caste and Class in India* ,Bombay: Popular Book Depot, pp. 41-111.
29. Chakravarty, T. (1959). *Food and Drink in Ancient Bengal* , Calcutta: Gossain & Co., (Printers) Private, Ltd., pp. 52-53.
30. See ref.25, Map 2, p. viii.
31. Roy, P. (1955). 'The Sacred Cow in India', *Rural Sociology*, 20, pp. 8-15.
32. Wiser, C.V. (1955). 'The Foods of an Indian Village of North India'. *Annals Missouri Botanical Garden*, 42, p. 346.
33. See ref.25, pp. 2-5.
34. See ref. 4, p. 48.
35. Burgess, A. and Dean, R. F. A. (eds.) (1962). *Malnutrition and Food Habits, Report of an International and Interprofessional Conference* , London, p. 35.
36. Gupta, S. B. L. and Tewari N.P. (1969). 'Marketing of Eggs in Varanasi City', *Agricultural Situation in India*, New Delhi: Ministry of Food and Agriculture, 24, pp. 621-624.

24

GEOGRAPHIC PERSPECTIVES ON LITERACY AND EDUCATIONAL LEVELS IN NORTH-EAST INDIA

H.N. Sharma

Introduction

Literacy and education are essential to the development of modern society. Their relevance to India as an impetus for development and social change is imperative. This is particularly true for the North-Eastern region comprising the states of Arunachal Pradesh, Assam, Manipur, Meghalaya, Mizoram, Nagaland and Tripura, (Fig. 24.1) which is relatively backward and characterized by very low levels of literacy and education. A distinct entity derived from historical and politico-administrative factors, the region has immense physical, racial and socio-cultural diversities and a literacy and education pattern which is highly varied both in spatial and social terms. Thus, this study is focused on a geographic study of the problem in order to reveal the various patterns of literacy and education in the region and the processes associated with them.

The objectives of the study are thus, to outline the pattern of literacy in various parts of the region including the rural-urban differential; to identify the sex disparity in literacy at the rural-urban level; to examine the pattern of educational levels and the degree of its correspondence with that of the literacy levels; and finally to derive a composite social profile for each state of the region towards assessing their relative position in totality. An exposition of the processes responsible for the present situation of literacy and educational levels in the region and its varied and complex character is another objective of the study.

'That literacy and education are major *dessiderata* to significant economic development and social change',[1] underlines the significance of the problem. 'This is the premise for which marked disparities in economic development across regions in India are in consonance with the inequity in educational development and the levels of responses to educational opportunities in them'.[2] This is particularly true of the North-East, where in addition to the general factors, its peculiar geographical setting and cultural attributes also contribute

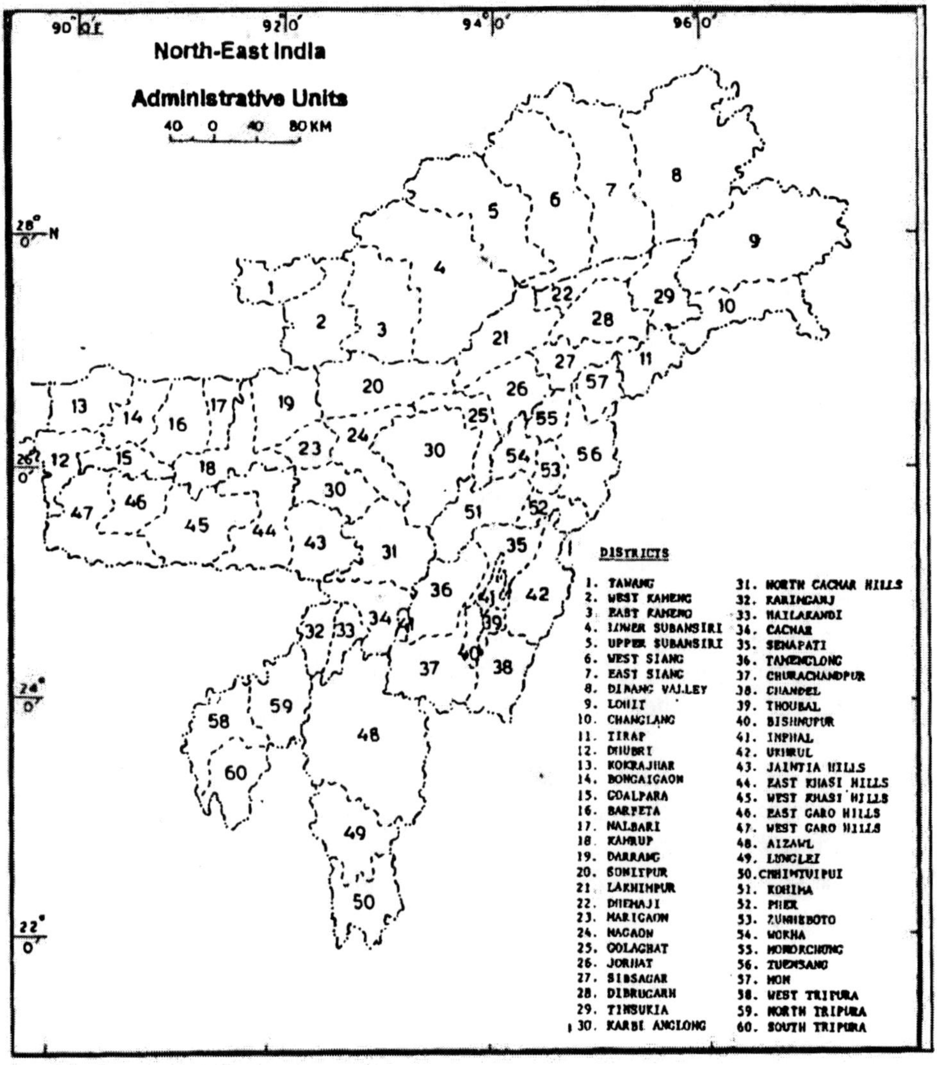

Fig. 24.1 : North-East India Administrative Units

to this situation. Therefore, a proper understanding of the patterns and processes of literacy and educational levels, disparities and distortions in this region are crucial, not only to lend impetus to the pace of socio-economic development in the region but also to make an integrated and less heterogeneous India, an urgent need for national unity at present. In the absence of any pragmatic geographic study for this region, this study becomes significant.

The main sources of data include different census publications and Government Reports,[3] although other relevant literature and various reports have also been used. Conclusions of the study are based both on the results derived from data and personal experience gathered through field observations. For the analysis of literacy patterns simple percentage or ratios, as also Sopher's Disparity Index,[4] have been used and the results represented through a number of maps and graphs.

Regional Background

The geographic background of the North-East has a great bearing on the spatial pattern of its literacy and educational levels, and the socio-economic development of the region. An understanding of the region's distinctive geographic character from the core areas of the country is significant and helps to place the problem in proper perspective.

Located between $21^0 59' 40''$ N and $29^0 40'$ N latitudes and $89^0 51'$ E and $97^0 25'$ E longitudes, North-East India with an area of 2,55,036 square kilometres and a population of 31,386,911 (1991 census) represents 7.75 per cent of the land area and 3.72 per cent of the population of the country. Bearing strategic importance it is truly a frontier tract being hemmed in by foreign countries—Bhutan, China, Myanmar and Bangladesh on all sides except for a narrow corridor, only 60 km wide, called the Bengal Duars, which connects the region with Bengal and hence the rest of the country.

Its locational peripherality and highly inaccessible character, on the one hand, and the distinctive character, of its population composition resulting from different streams of peopling at various stages of its history have an important bearing on the problem under study. Added to these, are its bewildering physical, racial and socio-cultural diversities, as already mentioned, a combined result of many geographic factors including historical and politico-administrative ones.

The region's peripheral location in the extreme North-East virtually separated from the rest of the country, isolates the region from the mainstream. This has a significant bearing on the difference in the character of literacy, education, and development in the region from other parts of the country. Secondly, the primarily rugged terrain comprising the Arunachal Himalayas to the north and their eastern appendages of the Patkai Barail, Naga-Manipur - Mizo ranges makes the region inaccessible, leaving only the intermontane valleys as in-roads for development. The Meghalaya-Karbi plateaus covering almost 12 per cent of the region's total area add to the problem of inaccessibility in the region, although to a lesser degree and extent. The Brahmaputra Valley, and the relatively small Barak Valley, confined only to the state of Assam, are the most expansive of the plain areas. The intermontane Manipur valley and the piedmont plain of Tripura in the hills are the other significant plain areas. Adding to the physical adversity of the area is the tropical monsoon climate with exceedingly heavy rainfall during the monsoon months, and recurring devastating floods. The region thereby emerges as a *cul-de-sac* where social transformation can have, but only, a slow pace.

Culturally the region is distinctive being a veritable cauldron of diverse racial, linguistic, and socio-cultural groups since pre-historic times.[5] Such an inflow of people brought into the region diverse genetic strains, languages, cultural traditions and societal patterns as also varied types of economic organisation. This led to the emergence of different types of politics and political organisations in different parts at various historical periods on the basis of factors of tribal origin and identity, spread of Hindu ideology and caste, warfare, conquest and amalgamation of territories into kingdoms under powerful rulers.[6] As for instance the matriarchal form of society of the Khasis and Jaintias who are linguistically Austro-Asiatic speaking people and the earliest settlers of the region, has in essence profound influence on the present character of male-female literacy and educational levels in Meghalaya. The mongoloids (belonging to the Tibeto-Burman group) and the Caucasoids (the Indo-Aryan language speaking people) settled mostly in the plain areas of the Brahmaputra valley and a small portion in the Barak Valley, being culturally advanced inherited a better socio-economic position, as is presently apparent. On the other hand, the tribal people who generally occupied the inaccessible terrain with a tribal economy and life had lower socio-economic position, compared to their counterparts in the plains.

However, the socio-economic structure of the region underwent a metamorphosis following its annexation with British India in 1826. As different peoples from different parts of the country, Bangladesh, and Nepal, moved into the region, its original ethnic structure underwent considerable transformation including marked changes in the literacy pattern of the plain and hill areas.

Isolation, inaccessibility and non-contact of both the tribals and non-tribals with the modern world have been responsible for the low social profile of the region which had low levels of literacy and education till the beginning of the last century. Extension of western influence, especially with the mass conversion of tribals to Christianity, as also the spread of education initiated by Christian missionaries, has led to the tranformation of the region's traditional social surface towards a new order. This process of westernisation, modernisation, and urbanisation received further impetus after independence resulting consequently in what is seen today. Thus, the present pattern of literacy and educational levels in the North- East has to be viewed and understood in the context of such a regional backdrop.

Spatial Pattern of Literacy

North-East India has at present 44.44 per cent literates as against an average of

43.29 per cent for the country (Table 24.1). As in most parts of the country the progress of literacy in the region was very slow prior to Independence and it was only after Independence that increase in literacy gained momentum. However, the level of literacy and education in the region even now is lower than many parts of the country. More particularly, its position in the field of higher education is discouraging. The literacy growth rate is also highly varied within the region [Table 24.2] (Fig. 24.2) attributable primarily first, to the initiative of Christian Missionaries notably in the Christian majority hilly areas especially prior to Independence and second, the overhelming influence of modern forces of westernisation and urbanisation. Further physical and social constraints, including past legacy, retard the literacy growth in most parts of the region especially in rural areas, and also the common factors which are operating against literacy development in the country as a whole.[7]

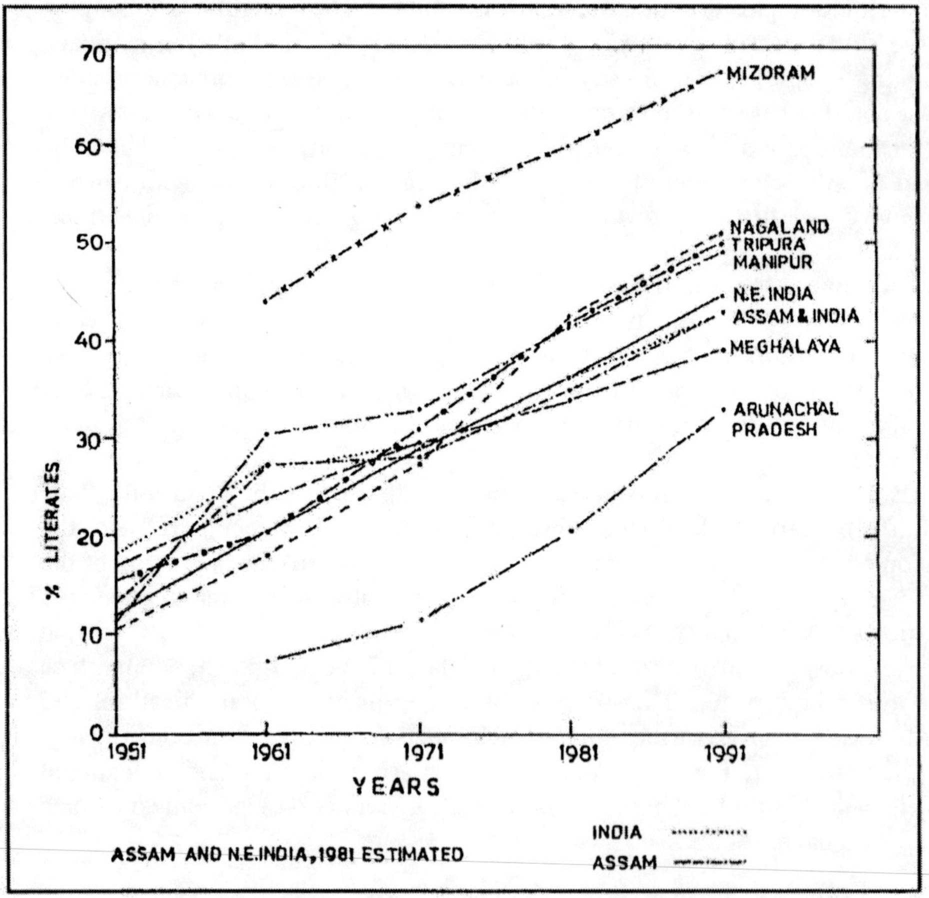

Fig. 24.2 : Pattern of Growth of Literacy in North-East India, 1951-1991

Table 24.1 : Literacy Pattern in North East India, 1991

States		*Literacy Rate (in percentage)* Persons	Males	Females
1. Arunachal Pradesh	T	32.87	41.34	23.03
	R	29.41	37.95	19.68
	U	57.73	64.16	49.20
2. Assam	T	43.20	50.62	35.18
	R	39.96	47.53	31.88
	U	69.19	74.17	63.26
3. Manipur	T	49.01	58.24	39.40
	R	44.76	53.73	35.39
	U	60.11	70.10	49.81
4. Meghalaya	T	39.16	41.72	36.45
	R	32.60	34.96	30.14
	U	67.68	70.34	64.74
5. Mizoram	T	67.36	70.36	64.11
	R	58.39	62.40	54.01
	U	77.81	79.42	75.76
6. Nagaland	T	51.09	56.05	45.52
	R	47.60	52.82	41.87
	U	67.82	70.49	64.42
7. Tripura	T	49.86	58.23	41.01
	R	45.83	54.93	36.18
	U	72.24	76.67	67.61
N.E. India	T	44.44	51.61	36.72
	R	40.55	47.94	32.67
	U	68.60	73.61	62.84
India	T	43.29	53.11	30.57
	R	36.70	47.36	25.37
	U	62.30	69.30	54.48

Source : Census of India, 1991, Series I, Paper-2, Provisional Population Totals.
T—Total, R—Rural, U—Urban

Table 24.2 : Growth Rate in Literacy in North East India, 1951-1991.

States	*Percentage of Literates to Total Population*				
	1951	*1961*	*1971*	*1981*	*1991*
1. Arunachal Pradesh	—	7.2	11.3	20.8	32.9
2. Assam	18.3	27.4	28.2	35.0*	43.2
3. Manipur	11.4	30.4	32.9	41.4	49.0
4. Meghalaya	13.2	26.9	29.4	34.1	39.2
5. Mizoram	—	44.0	53.8	59.9	67.4
6. Nagaland	10.4	17.9	27.4	42.6	51.1
7. Tripura	15.5	20.2	31.0	42.1	49.9
N.E. India	11.7	20.6	28.9	36.2*	44.4
India	16.7	24.0	29.5	36.2	43.3

Source : Census of India, 1951-1991. * Estimated.

Marked variation exists in the literacy rates among the states of the region (Fig. 24.3). Mizoram records the highest literacy rate of 67.36 per cent second in the country after Kerala with 78.09 per cent. It is distantly followed by Nagaland with 51.09 per cent, Tripura with 49.86 per cent, Manipur with 49.01 per cent, Assam with 43.20 per cent and Meghalaya with 39.16 per cent. Arunachal Pradesh has still the lowest literacy rate with 32.87 per cent among all the states of the country except for Rajasthan (31.03 per cent) and Bihar (21.10 per cent). It is thus evident that the states of Assam, Meghalaya and Arunachal Pradesh which comprise 72.38 per cent of the total area and 79.37 per cent of the total population of the region have literacy rates below the national average.

Further, the rural-urban differential in literacy is generally high in the region.[8] Thus the states of Nagaland, Manipur and Tripura which show higher literacy rates have, in fact, significantly lower levels of rural literacy (below 50 per cent) compared to their urban counterparts. In Assam and Meghalaya the rural literacy rate falls below 40 per cent and below 30 per cent in Arunachal Pradesh. Mizoram is perhaps the only exception with high rural (58.39 per cent) and urban (77.81 per cent) literacy rates (Fig. 24.4).

With the exception of Meghalaya and Mizoram, male-female disparity in literacy is well marked, especially in Arunachal Pradesh. This disparity, however, is gradually decreasing indicating thereby the spread of female education in the region (Table 24.3). Disparity in literacy rates on the basis of gender are more pronounced in the rural areas (Fig.24.5). This is particularly so in Arunachal Pradesh (0.3969), Tripura (0.3223), Manipur (0.3261) and Assam (0.2867) where the influence of Christianity is much less compared to the other states of the region.

Table 24.3 : Sex Disparity in Literacy in North-East India (1971-1991)

States	*Male-Female Disparity in North-East India (1971-1991)*		
	1951	*1981*	*1991*
1. Arunachal Pradesh	0.6683	0.5038	0.3719
2. Assam	0.4044	0.3150*	0.2761
3. Manipur	0.5567	0.4442	0.3312
4. Meghalaya	0.1989	0.1515	0.0962
5. Mizoram	0.2978	0.1740	0.1234
6. Nagaland	0.3439	0.2900	0.1834
7. Tripura	0.4302	0.3568	0.3031
N.E. India	0.4097	0.3200*	0.2673
India	0.4530	0.4220	0.4103

Source : Calculated by the author on the basis of Census of India, 1971-1991.
* On the basis of estimated literacy rates.

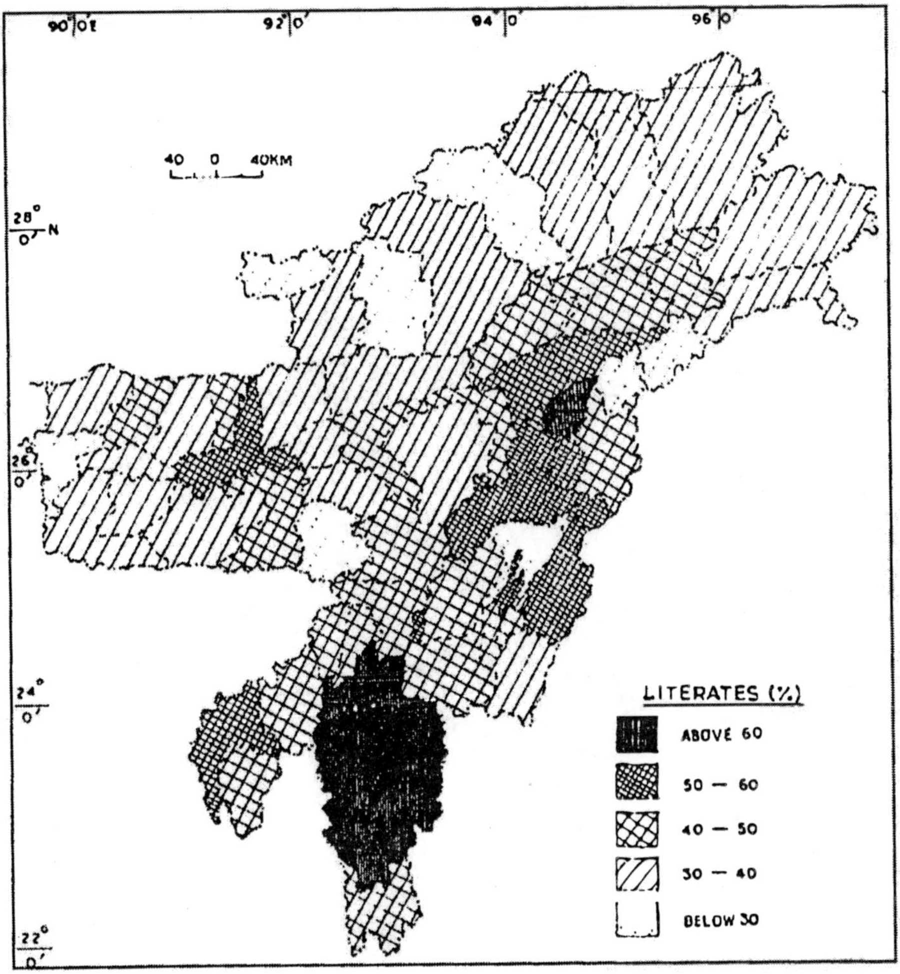

Fig. 24.3 : North-East India : Literacy, 1991

A variety of physical factors such as the isolated location and inaccessible terrain, especially in Arunachal Pradesh, the legacy of a non-conducive educational atmosphere in the region, only recent contact with the modern world, a varied population composition such as the Christian majority areas in the hilly states, Hindu dominated plain areas, tribal religion practising areas of Arunachal Pradesh, tribal areas of Assam and Tripura, are factors responsible for causing variation in the topography of literacy of the region. Within the states, marked variations exist between areas occupied by relatively advanced social groups and those by backward ones. Thus, the areas occupied by the *Adis* of Arunachal Pradesh, *Aos*, *Angamis* of Nagaland, *Hmars* in Manipur highlands, *Mizos* in Mizoram, *Khasis* and *Jayantias* of Meghalaya, caste Hindus

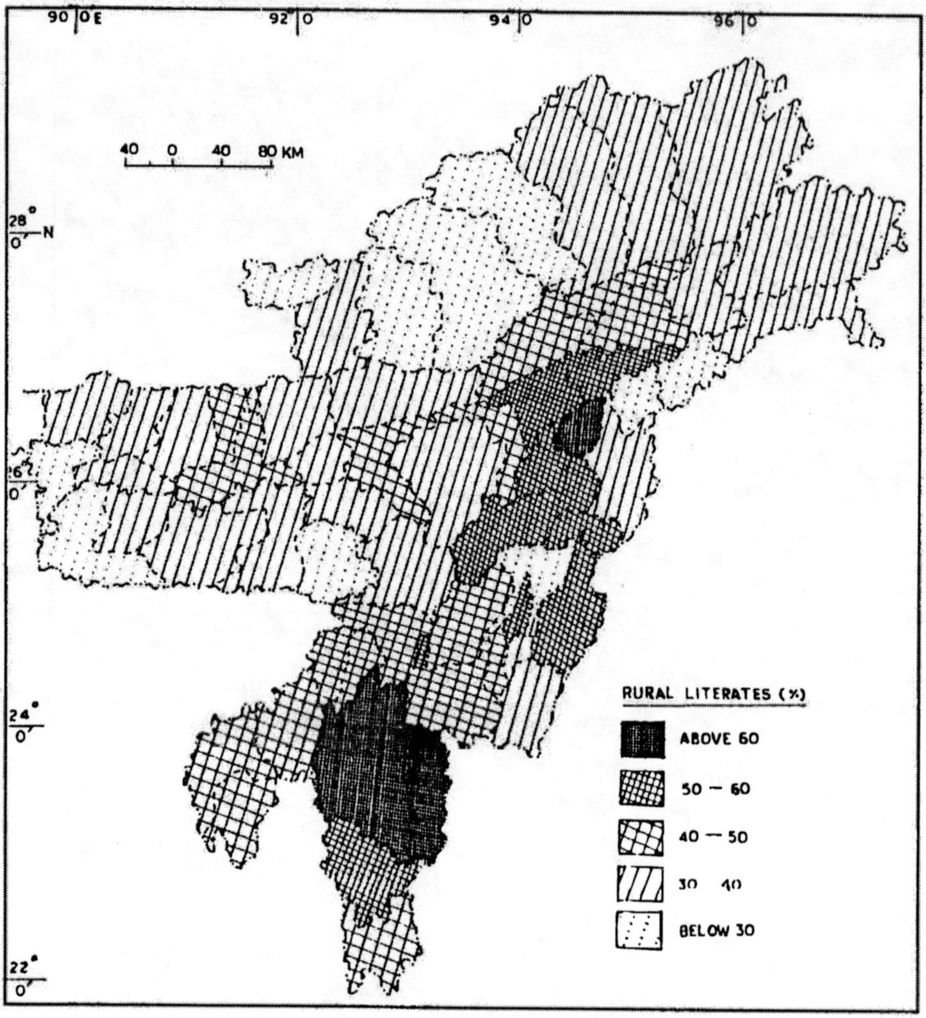

Fig. 24.4 : North-East India : Rural Literacy, 1991

of Assam and Tripura, *Meitis* of Manipur plain are in a better position in this respect.[9]

Literacy by age-group which has great significance for assessing the potential for literacy development has not been covered here because of non-availability of relevant data. On the basis of earlier census data it is seen that the literacy in any particular age group except the age group of 5-9, is higher than the percentage of literacy in any preceeding age group. The lower literacy in the age group 5-9 is attributable to the fact that children enter school after the age of 5, especially in the rural areas and because children do not immediately become literate upon entering school.[10] That the literacy rate peaks at the age-

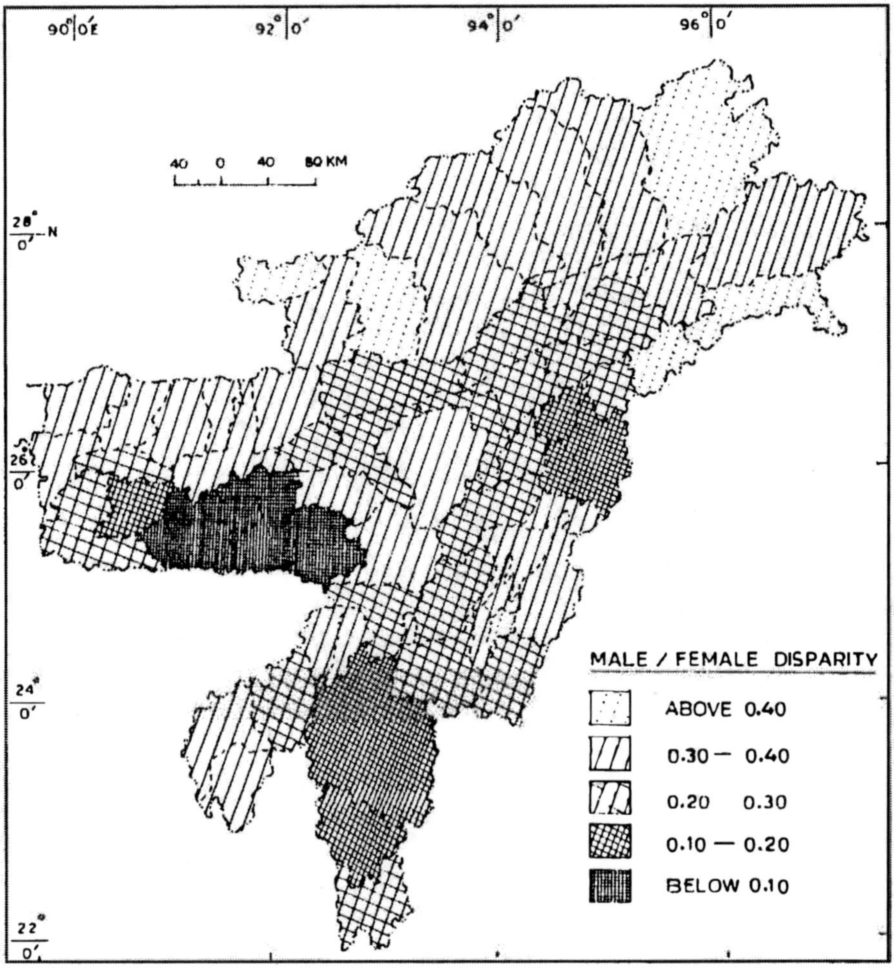

Fig. 24.5 : North-East India : Sex Disparity in Literacy Rural Population, 1991

group 10-14 is encouraging for it would mean a persistent trend of increase in the level of literacy over time. This is mainly because of literacy elevation programmes implemented during the different plan periods after independence. Further as clear from the data literacy is spreading fast and literacy rates improving among the new generation, especially in Arunachal Pradesh and Nagaland (Table 24.2).

Spatial Pattern of Educational Levels

Although the level of literacy in the region is not too low, the per- centage of persons educated above secondary level is very low (Table 24.4). Interestingly enough, Mizoram which has the highest literacy level in the region and the second highest in the country, has only 1.39 per cent of its population educated to the secondary and higher secondary levels and 0.33 per cent to the graduate and above level. The corresponding figures for Assam were 1.82 per cent and 0.41 per cent and for Manipur 3.01 per cent and 0.67 per cent respectively.[11] While the pattern in Nagaland and Mizoram was similar, Meghalaya's position was slightly better but Arunachal Pradesh was the worst. Tripura was close to Assam. During the last two decades the number of educated persons above the primary level has increased significantly due to the establishment of educational institutions and various development measures taken by the Government (Table 24.4). But even in 1981 although Mizoram still ranked second in terms of literacy in India, it had the lowest percentage of persons at the graduate and above level and second lowest at the secondary level. Besides rural urban differential and sex disparity in the higher educational levels are even more glaring in the region.

Educational development at the district level is better as in the districts of Aizwal and Lunglei in Mizoram, Imphal in Manipur, Kohima and Mokokchung in Nagaland, Jorhat, Sibsagar, and Kamrup in Assam, West Tripura in Tripura, and East Khasi Hills in Meghalaya. In fact, in these areas not only is the literacy rate higher, but male-female disparity in literacy is considerably low. This is primarily attributed to the legacy of generally higher literacy as also the location of district headquarters here.

During the last two decades the number of educated persons above the primary level has increased significantly in the region. Following the establishment of educational institutions in various areas and the literacy elevation drives taken by the Government, the number of persons educated above the primary level has increased significantly in the region although the higher educational level scenario, especially in the remote areas, is far from satisfactory. The dearth of persons with higher qualifications is pervasive in most parts of the region.

Mere literacy without any educational level could be considered a liability to society as such literates are not well-equipped to meet future challenges and are more likely to be susceptible to frustration and discontent. Contrastingly, persons who have education at least above the primary level are likely to be better equipped for meeting the various needs of an emerging society.[12] Therefore, rapid increase in literacy rates without simultaneous increase in educational levels could be misleading. In reality, despite the higher literacy levels, in Mizoram and Nagaland, stability and progress seem to have a better foundation, in Manipur, Assam and Tripura.

Table 24.4 : Pattern of Educational Level in North East India, 1981

States	Literates			Primary			Middle			Secondary			Higher Secondary			Technical Diploma or Certificate above			Graduate and above		
	T	*M*	*F*	*T*	*M*	*F*	*T*	*M*	*F*	*T*	*M*	*F*	*T*	*M*	*F*	*T*	*M*	*F*	*T*	*M*	*F*
1. Arunacha Pradesh	20.79	28.94	11.32	4.80	8.88	2.39	2.33	3.39	1.11	2.15	3.18	0.94	0.69	1.01	0.31	0.14	0.25	—	0.89	1.34	0.37
2. Assam*	35.00	44.73	30.00	14.75	17.65	11.54	6.37	8.70	3.80	4.98	6.93	2.83	2.05	2.80	1.23	1.07	1.97	0.08	1.52	2.12	0.89
3. Manipur	41.35	53.29	29.06	9.71	12.30	7.05	8.48	11.65	5.21	4.90	8.93	2.82	2.08	2.87	1.26	0.05	0.09	—	1.98	2.93	1.00
4. Meghalaya	34.08	37.89	30.08	9.73	10.51	8.93	8.00	9.34	8.59	2.87	3.56	2.14	0.90	1.28	0.81	0.04	0.08	—	1.22	1.57	0.85
5. Mizoram	59.88	64.45	54.91	13.16	15.36	10.77	5.65	6.72	4.49	2.78	3.65	1.84	1.63	2.31	0.90	0.08	0.13	—	0.73	1.16	0.25
6. Nagaland	42.57	50.06	33.89	11.88	14.12	9.30	5.73	7.32	3.88	3.92	5.39	2.22	0.93	1.45	0.33	0.08	0.14	—	0.88	1.34	0.34
7. Tripura	42.12	51.70	32.00	10.70	12.87	8.40	5.55	7.24	3.75	3.11	4.34	1.80	1.78	2.34	1.13	0.09	0.16	—	1.04	1.50	0.55
N.E. India	39.70	47.80	31.00	13.57	16.25	10.62	6.33	8.57	3.97	4.59	6.37	2.63	1.90	2.59	1.15	0.82	1.51	0.06	1.46	2.02	0.83
India	36.23	46.89	24.82	10.95	13.60	8.11	6.34	8.89	4.15	4.19	5.93	2.34	1.47	2.16	0.75	0.20	0.31	0.09	1.41	2.05	0.72

Source : Census of India,1981, Series 1, India Part IVA, Social and Cultural Tables
+ Estimated Figures

Social Profile

The literacy pattern, rural-urban literacy differential, male-female disparity in literacy and the educational level lend a complexity to the social profile of the region, which varies both structurally and spatially within the region (Table 24.5 and Fig. 24.6).

Table 24.5 : Ranks of Social Variables in Order of Importance for the States of North East India, 1991

State	*Liteeracy Percentage*	*Rural-Urban Differential in Literacy Stage*	*Male-Female Disparity in Literacy*	*Educational Level above Secondary*
Arunachal Pradesh	7	5	7	7
Assam	5	6	4	2
Manipur	4	1	6	1
Meghalaya	6	7	1	6
Mizoram	1	2	2	5
Nagaland	2	3	3	4
Tripura	3	4	5	3

Source : Census of India, 1981 and 1991.
Note : Ranks on the basis of calculated data by the author.

At the sub-regional level, Mizoram, characterised by higher literacy, lower sex disparity in literacy, and rural-urban differential in literacy, in terms of higher educational level, is found to be in the worst position indicative of a social situation not desirable for its social development. The impetus brought forth by Christian Missionaries for the spread of literacy has not been reinforced, especially in the field of higher education even after Independence mainly pernaps due to the persistence of political instability. Also, inadequate infrastructural facilities for higher education, particularly in the rural areas, and the prevalence of primitive economy, have been other negative forces hindering promotion of higher education. Urbanisation, instrumental in alleviating higher education is confined to few pockets only. As such, the social profile, though not very well defined as of now, *vis-a-vis* education, holds promise.

Nagaland, with a relatively high literacy level in the region and a rapid literacy growth rate during the last few decades, has high gender disparity in literacy as well as rural-urban differential, while its position regarding higher education is virtually the worst. In fact, its literacy position today is what was Mizoram's in the 1960s. Here, the literacy drive made by the Missionaries gained momentum only after the declaration of Nagaland as a state in 1982. Continued political instability further retarded the spread of literacy and social development process. The impulse of literacy spread is yet to reach the interior rural areas, especially the womenfolk, and hence, the prevalence of high gender and rural-urban disparity in literacy rates. In respect of higher education it is at par with Mizoram. Therefore, although stunted the social profile of the state, with rapid social transformation and extension of infrastructural facilities in the recent years, is expected to witness all round social development.

Meghalaya, another hilly and Christian majority state on the other hand, marked by lower literacy rate and very high rural-urban differential in literacy, has the least gender-disparity in literacy and also a better position in respect of higher educational level than Nagaland and Mizoram. Development here is primarily confined to Shillong and its surroundings. Better educational levels, the result of Missionary influence and subsequently the establishment of the capital of Assam in Shillong in 1874 by the Britishers, helped transform the social landscape of the area, although the impact of this development was largely restricted to Shillong. This continued even after Independence, although now change in the areas around a number of other urban centres is evident. The problem of low rural literacy level has been further accentuated by Nepali immigrants among whom the literacy rate is very low. Hence, the highest rural-urban literacy differential is found here. However, disparity in literacy on the basis of gender is mainly attributable to the existence of the matriarchal system, wherein females play a dominant social role. Thus, with the exception of the area around Shillong, the social profile of Meghalaya is low and it will have to strive hard even to come to the level of Mizoram and Nagaland.

Arunachal Pradesh, known for its extremely hilly terrain and socio-economic backwardness bears a social profile with the lowest scores in almost all the social variables considered in the study. Inaccessibility and prevalence of tribal religion have been the major impediments in its progress of literacy. In fact, till recently Arunachal Pradesh, bordered primarily by China, a country of very different political ideology, was almost closed to any outside influence. It was after the Chinese aggression in the early 1960s, that it came in to the limelight and in-roads of development started taking place. Therefore, literacy, educational level, and total social development are at the lowest. However, the last few decades have witnessed an increase in literacy, a trend indicative of radical change in the near future.

The social profile of Manipur is characterised by relatively high literacy, low rural-urban differential in literacy and relatively high proportion of people in different educational levels. The gender disparity in literacy here is, however, quite high, but overall it ranks high at the social level. This is primarily the result of two different cultural forces acting parallel to each other in the Imphal valley, on the one hand, and the hilly areas around it, on the other. In the Imphal valley, apart from the effects of modernisation and urbanisation which have

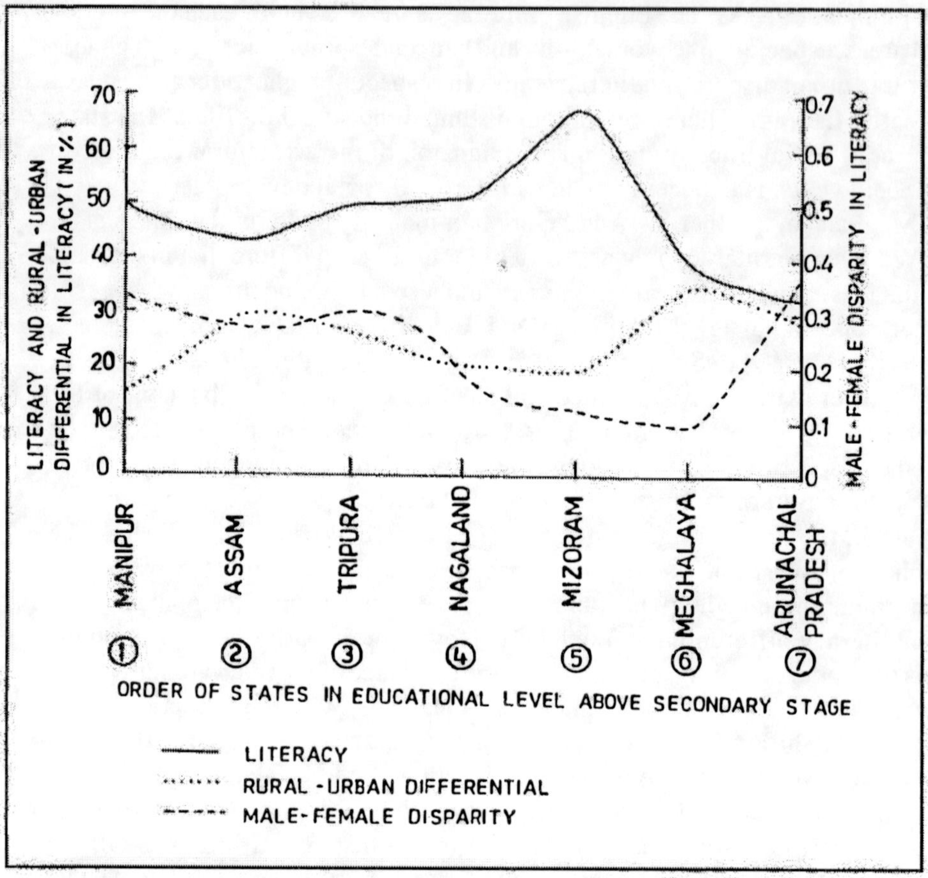

Fig. 24.6 : Proflies of Literacy, Sex Disparity in Literacy and Rural-Urban Differential in Literacy in North-East India, 1991

been quite vigorous in the last few decades, the rich Vaishnavite cultural heritage of the valley people, who are mostly Hindus, has contributed significantly to the higher literacy and educational levels. At the same time the hilly areas are also progressing conspicuously in this respect mainly due to the influence of Christian Missionaries and the growing impact of urban development. The Churachandpur area provides a case in point which is undergoing notable social changes at present because of the interplay of these forces. Disparity in male-female literacy is higher primarily because of the prevalence of illiteracy among the Hindu females in the Manipur Valley as well as the Nepali females of the hilly areas. On the whole, the state is well placed as an emerging prospective social landscape.

Tripura stands out differently altogether. Relatively high literacy, but high gender disparity, and high rural urban differential in literacy point to a pattern of literacy development having prominent areal and social group-wise differences. The majority of the people here being Bengali Hindus, who are mainly concentrated in the plain areas in and around Agartala and other urban centres, have greatly contributed to the fairly high literacy level in the state. Contrary to this, the Tripuries and other tribal groups including the tea garden labourers of highlands and the hilly areas with a low literacy level, especially among females, has widened the gender and also the rural-urban differential in literacy. The absence of Missionary influence, except in case of a few small tribal groups, has a bearing on the low literacy level in the hilly areas. Nevertheless, the state has a relatively high proportion of persons with different educational levels mainly due to the presence of a culture in the plain area congenial to the growth of higher education. The social profile of the state, therefore, is characterised by sharp differentiation between the plains and the hill areas.

With an average literacy level and the second highest proportion of persons of different educational levels, Assam stands out prominently in the level of social development among the states of the region. As already mentioned, the advantageous physical conditions, the past legacy of a rich socio-cultural background, the presence of culturally advanced groups of people and a large number of educational institutions and a wider contact with the rest of the country, among other factors, have contributed largely towards achieving this level. Notwithstanding this, relatively high gender disparity in literacy and rural-urban differential in literacy prevail. The population composition of the state, as referred to earlier, is such that a large section of people, such as the tribals, plains tribals, scheduled castes, immigrant Muslims, Nepalis and tea garden labourers inhabiting mostly the rural areas are generally illiterates especially the females, a factor largely responsible for low literacy and a rural-urban differential in literacy. Further, the social profile of the state is highly uneven because of sharp differentiation in the levels of literacy and education among the various social groups over its different parts. Thus, are seen the socially advanced areas such as the Upper Assam Valley, certain enclaves of old Kamrup district in Lower Assam, and Silchar area of the Barak Valley as against the social deeps present in the tribal and immigrant dominated areas. Irrespective of these and other conditions of instability prevailing at present, with the increasing degree of urbanisation and modernisation, the sub-region is in a state of flux heading towards a prospective level of social development.

Conclusion

What emerges from the study is that not only literacy and educational levels are low in the region, but great areal variations exist in them in different forms.

This is clearly borne out by the social characteristics—literacy levels, rural-urban differential in literacy, gender disparity in literacy and the educational level dealt with in the study. Further, the social profile of the region, apart from being uneven, is complex in character, a result of the interplay of a host of physical and cultural attributes, as discussed above. The complexity has got accentuated because of the diversities prevalent among the various ethnic and religious groups at different stages of socio-cultural transformation. Viewed in totality, the region emerges as a social deep, although not without the scattered presence of certain social ridges. Again the social peaks apparent in the hilly areas seem to be hardly generative in character and, therefore, have contributed only marginally towards strengthening the educational base in such areas. But despite all constraints, both physical and cultural, significant changes in the educational development have taken place in the region since Independence and are becoming more pronounced during the last few decades. Various schemes, including universalisation of educational programmes, are doing much in enhancing the level of literacy and education in the region although, literacy and educational development of the region is as yet far from satisfactory. Therefore, efforts are needed to speed up the process of spread and improvement of literacy and educational levels in the region to achieve desirable development. Existing disparities between relatively advanced and backward areas including the urban and rural sectors, among the various social groups, and between males and females of all the states need to be addressed and removed.

References and Notes

1. Schuth, Sister Katarina (1980). 'Village Literacy and Its Correlates : A Mysore Case Study', in David E. Sopher (ed.), *An Exploration of India: Geographical Perspectives on Society and Culture*, Ithaca: Cornell University Press, p. 191.
2. Ahmad, A. (1985-86). 'Literacy in India in the Regional Context', *Abstract of UGC National Lecture* at the Department of Geography, Gauhati University.
3. *Census of India* (1991). Series 1, India : Provisional Population Totals, Paper 1; Provisional Population Totals : Rural-urban Distribution, Paper 2; The definition of literacy used in the present study is the one taken by the Census of India 1981-91 which accounts as literate a person who has the ability to read and write with understanding in any language; *Census of India* (1981). Series-1, India, Part IV A, Social and Cultural Tables; *Census of India* (1971). Series 3, Assam, Part II-C (ii), Social and Cultural Tables; North Eastern Council (1990), *Basic Statistics of North Eastern Region*, Shillong
4. Sopher, David E. (1974). 'A Measure of Disparity', *Professional Geographer*, Vol.26, pp. 389-392. The disparity index DIS, has the formula

$$\log_{10} \frac{p(1-q)}{q(1-p)}$$

where p and q are the probabilities that members of the two groups, respectively, have the property involved. The probability is the number of persons in the group who have the property divided by the population of the group.

5. Taher, M. (1977). 'Tribes of North-East India: A Diagnostic Survey in Spatial Pattern', *The North Eastern Geographer*, Guwahati, Vol. IX, No.1 and 2.
6. Bhagabati A.C. (1992). 'Social Formations in North-East India', *Bulletin of the Department of Anthropology*, Gauhati University, Vol. VI, pp. 9-27.
7. Gosal, G.S. (1979). 'Spatial Perspective on Literacy in India', *Population Geography*, Vol.1 and 2, June-December, pp. 41-67.
8. *Census of India* (1991). Series I, India: Provisional Population Totals : Rural Urban Distribution, Paper 2.
9. Sharma, H.N. (1985). 'Sex Disparity in Literacy and Social Topography of Assam', in A.B. Mukerji and Aijazuddin Ahmad (eds.), *India : Culture, Society and Economy—Geographical Essays in Honour of Prof. Asok Mitra*, New Delhi: Inter-India Publications, pp. 379-401.
10. Sharma, O.P. and Rutherford, R.D. (1987). *Recent Literacy Trends in India*, Occasional Paper No.1, Census of India, New Delhi, pp. 9-11.
11. Census of India (1971). Series 3, Assam, Part II-C (ii), *Social and Cultural Tables of Assam.*
12. Roy Burman, B.K. (1970). *Demographic and Socio-Economic Profiles of the Hills Areas of North-East India*, New Delhi: Census of India, 1981, p. 22.

25

CASTE AND TERRITORY

Boundaries of Socio-Cultural Regions

IRINA EFREMOVA

Introduction

In recent years many aspects of social and cultural geography of India have attracted the interest of Indian and foreign scholars, although the social and cultural fabric of India and its constituent regions have been themes of specialised study of several generations of European, American and Japanese scholars. It cannot be denied that an understanding of the social and cultural imperatives of a people within a territory/region has a significant impact on the formulation and effective implementation of economic and political programmes.

In India, wherein cultural imperatives are so deep rooted whether in matters of social, religious, occupational or even territorial stratification, identification of socio-cultural space niches becomes important as these factors exercise considerable control on the stereotypes of economic and political behaviour of the people who belong to a particular region. Caste differences are not new and go back centuries to the division of society on the basis of *varnas* and later castes. Some castes then emerged as the dominant groups in specific territories which came to be identified with them. Successive waves of settlers into the Indian sub-continent took this process forward.

These processes operating in India throughout its history have been very strong and were determined very largely by the natural environment. The settlers encountered a variety of relief divisions and generally the strongest, most powerful and dominant groups settled areas which were akin to their own habitat and which allowed them to practice the mode of living with which they were most familiar, making their ventures economically viable. Thus, different groups associated with different and distinctive modes of living settled different ecological niches leading to the evolution of characteristic cultural groups, each identified with a specific territory and occupation. Thus, cultural regions of sorts came to be identifiable on the landscape. Geographical barriers, thus, played a significant role in the emergence, evolution, and development of socio-cultural regions.

With time, a convergence seemed to emerge between the specific occupational groups and associated religiosity culminating in the development of the caste system and a definitive caste structure recognised in the villages. This was then perpetuated by the symbiotic relationship between the service classes (*varnas*) which simultaneously came to be bestowed with ritualistic meaning as well. Thus, 'caste structure' in a village can be recognised both horizontally and vertically, bound together by a system of inter-caste serving called the *jajmani system*. While a certain trade specialisation exists of villages, it is not consistently reflected in all villages, although within small territorial clusters a combination of villages having diverse functions is relatively constant.

In this study an attempt has been made to identify and demarcate social and cultural regions, which coincide to a great extent to different social and cultural complexes. Since data regarding caste composition of population is available only at the district level, the latter has been treated as the basic unit and Census of India data, 1931 has been used for this study. For the delineation of the socio-cultural regions, the Index of Dissimilarity of caste structures, first used by David Sopher in Sociology (to reveal variations in caste composition of the districts of United Provinces of Agra and Oudh (UP)) has been used.[1] In this the Index of Dissimilarity (G) was calculated by applying the given formula *i.e.*

$$G = \frac{\sum_{j=1}^{m} (xij - xkj)}{2}$$

i = 1,2,3............., n

neighbour districts that have mutual borders,

k = 1,2,3...............,n

xij — quota of a particular caste in the population of district i

xkj — quota of the same caste in population of district k,

m — number of castes in the country

Those castes, irrespective of their differences of religion, occupation status or language, were considered for the calculation of the Index of Dissimilarity, G, which comprised atleast 1 per cent of the total population of a district. Only neighbouring districts were taken into consideration. The value of G ranged from '0', indicating a complete identity of caste composition, to '1' which meant absolute dissimilarity. Within this spectrum of the value of G from 0 to 1, the equivalent of $G = 0.224$ and $G = 0.280$ was considered significant, for while the former value led to an identification of socio-cultural units within the distirct, the latter value was indicative of the existence of essential differences in the socio-cultural complexes within a district.

In order to make the results more reliable, differences in the organisation of socio-cultural space in the North and in the South have been taken into account. The caste composition in the North is relatively complex in its expression on the landscape through the organisation of socio-cultural space. Territorial communities in the North comprise a single caste as also its sub-groups or clans, lending an inbuilt heterogeniety. On the other hand, in the South, the territorial communities comprise groups having common ethnic identity, onto which caste structures (as perhaps a trait borrowed from the North) have been superimposed as a secondary and subsequent development.

Socio-cultural boundaries often coincide with natural frontiers especially between varying relief zones[2] (Fig.25.1). The edge of the Western Ghats between the humid and wet Malabar coast and the Deccan tableland represent one such border between different socio-cultural communities. Similarily, the edge of the Eastern Ghats, when moving from the plain districts of coastal Andhra to the adjacent hilly plateau districts, also comprise a boundary coinciding with a socio-cultural border. In fact, an overlapping of relief and socio-cultural borders are also evident between the plains and the Himalayan foothill zone in Darjeeling, Dehradun, and Nainital districts of Uttarakhand and Uttar Pradesh. Marked variation in social and cultural complexes also exist between plain districts of coastal Andhra and the Telangana plateau, plain districts of Saurashtra and hilly districts of former Rajputana Agencies, the North-Indian Plains and the Gondvana plateau.

Highlands or mountain areas reveal a greater degree of diversity and the existence of smaller, closed territorial communities when compared to the plain regions. This is attributable to inaccesibility and limited cultural contacts. Moreoever, the population here is essentially tribal comprising more than 50 per cent of the total population of the district. As a result the mountains are characterised by extremely high figures of the index of dissimilarity of caste structures.[3]

Rivers too have significantly influenced the spread of population across the sub-continent. Most rivers in India flow from west to east. The thrust of the waves of mirgants, especially the Indo-Aryans was eastwards along the Ganges to Bengal. The Ganges acted as a unifying force binding together districts belonging to culturally different regions, with socio-cultural barriers between culturally different regions *e.g.* the Ballia district in Uttar Pradesh and the Saran district in Bihar, fading away, leading to an integration of caste profiles. However, in their ingress southwards to Maharashtra, through the gates of Gwalior and to Telengana through Bundelkhand and Baghelkhand, the rivers such as Krishna, Tungabhadra and Godavari (also flowing from west to east) served as natural barriers and as in the case of the Godavari river, in its middle part, the Balaghat mountains along the right steep bank became, a cultural barrier for the districts Parbhani and Bidar as well.

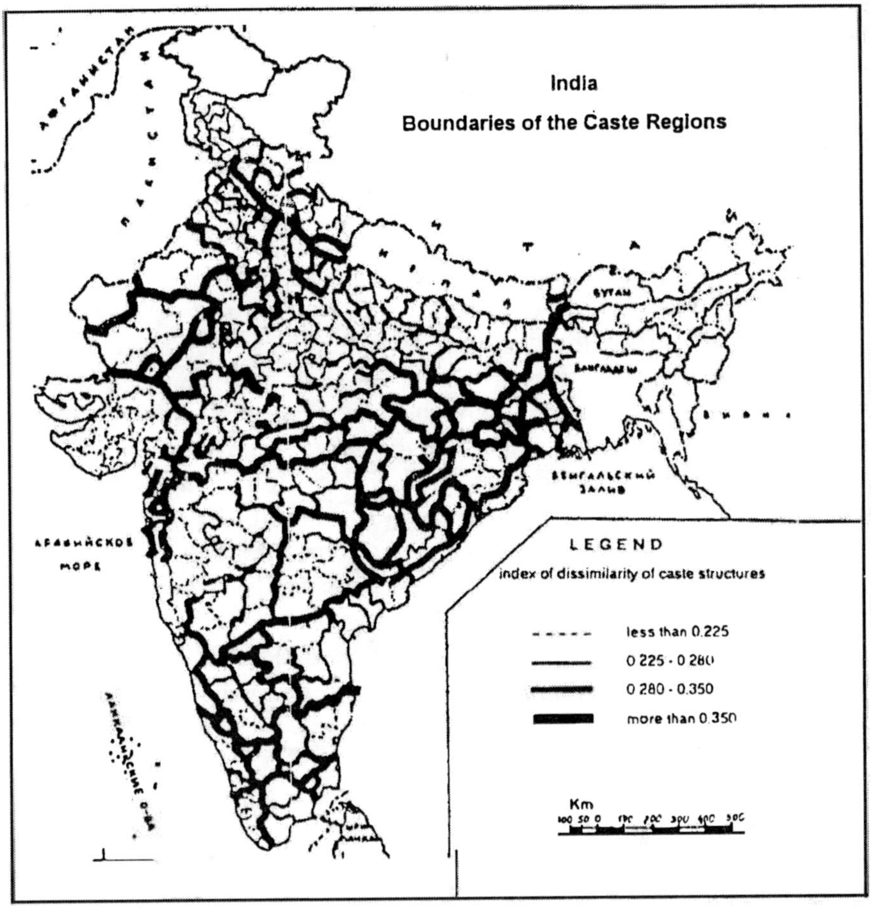

Note : For this map Data of Census of India 1931 were used. For Madras Presidency Data of Census of India 1911 were used.

Fig. 25.1 : India : Boundaries of the Caste Regions

In some cases however, the socio-cultural borders and natural borders do not coincide. As in the case of the fertile alluvial Ganga Plain. Primarily agrarian in context, the agro-climatic conditons change steadily from the south-west to the north-east following the gradual increase in precipitation. But here the socio-cultural borders instead of following the same arrangement, intersect the fertile plain from north to south perpendicular to the main stream of the Ganga.

These socio-cultural borders were perhaps synonymous to the political frontiers between ancient states, frontiers which continued to influence the network of subsequent administrative territorial divisions, such as the borders between Rohilkhand and Oudh and between the eastern part of Uttar Pradesh and Bihar. Thus, as suggested by Sopher[4] even though both Rohilkhand and Oudh comprised areas of strong Muslim influence, it was the political frontier between them that led to the marked difference in the caste composition of their population. Thus, while Rohilkhand was dominated by Pathans, Sheikhs etc., a high proportion of Oudh's population was that of the Pasi, a low caste, primarily involved in breeding pigs, thereby indicating a non-Islamic context.

The core areas of the socio-cultural regions comprise districts with little variation in their caste composition. In fact, such areas have, through the centuries, evolved and emerged as centres of regional political power on the North Indian Plain. Despite variations through time both in the area under their influence and in their configuration, they have remained consistent as cores of political formations. This is essentially related to their geographical position, physical aspects and historical associations. Magadha, for example, from ancient times and through the early Middle Ages, was one such area. Being the eastern, most part of the area under the influence of the Aryans, it was on this fertile plain that their political and administrative acumen was sharpened and concept of statehood born. Magadha's political centre corresponds to the core of the socio-cultural region.

Historical processes and the evolution of statehood in India gradually led to the shifting of the centre of political power from the east to the west *i.e.* from the area of Magadha through Kashi and Kaniya-Kubdja to Agra and finally to Delhi.

Under the Mughals, however, the capital was not at the centre of any political area, rather it was established at the juncture of cultural and political borders, a principal that enabled the Mughals to exercise control over vast and socio-culturally diverse territories.

The socio-cultural areas of eastern Uttar Pradesh and Oudh with Kashi and Kaniya-Kubdja as their centres and cores, in the analysis of caste composition, are not well defined by rigid borders. The flexibility of these borders is perhaps the result of the large scale cultural interminglings that took place on the North Indian Plain and reflect consequent smooth and gradual changes of caste profiles. This was the area which through history came under the influence of the waves of Indo-Aryans and later Muslim invaders with each wave making an impact on the political formations. In fact, the northern cultural area proved to be a testing ground for social innovations with every wave of migrants. A synthesis of cultures was taking place consistently between the groups moving in and the indigenous inhabitants leading to a greater degree of socio-cultural integration in the North as compared to the South. Further, in the absence of any physical barriers along the Northern Indian Plain, and with its continuity being enhanced by the west to east flowing Ganges, providing an inherent unifying force, the territories different in their caste compositions came to be less defined in the

North than in the South. Almost at the same time the caste system was taking root with the caste regions later evolving as the bases of the socio-cultural regions in the North.

Almost similar processes were afoot in the south corridor of Aryanization. For instance, in Maharashtra (as in Bihar) the border between the core region, Desh and Marathwada was quite distinct. However, unlike the north which was constantly exposed to varied influences, the south remained as *cul-de-sac*. Cultures, already transformed in the north and spilling over to the south met with little resistance and were incorporated into the already developed ethnic structures *per se*. Thus, the caste system which in the north had evolved alongside the formation of the socio-cultural communities came to be accepted in the south within secluded cultural systems as a borrowing of a ready structure already formed in the north. In fact, the borders of socio-cultural regions in the south were distinctive and marked by sharp differences in their caste structures. They often coincide either partially, (as in the case of Rayalsyma) or completely, (as in the case of Tamilnadu) with contemporary linguistic borders.

Further, the Berar and Telengana social and cultural communities are associated with the second trunk of Aryanization. Distinctive and developed caste regions can be identified within areas forming relatively small states in west and south India, the result of the disintegration of the erstwhile large kingdoms, which took place about the Middle Ages. For example, by the end of the fifteenth century, the states of Bijapur, Golkonda, Ahmadnagar, Bidar and Berar had emerged on the ruins of the former Bahmani kingdom. Irrespective of the territorial size of these new states, the local caste profiles revealed correspond to the original, small geopolitical areas which existed earlier and which were annexed and incorporated into the big kingdoms. Here too, territorial analysis of caste structures reveals that the distinctive, socio-cultural borders coincide with linguistic borders. As within the erstwhile state of Hyderabad, the boundary between Golconda, including the Telugu speaking areas of Telangana and Bidar, and the Kannada speaking area, is clearly evident.

On the basis of local caste structures, Saurashtra forms a separate region. However, in the case of the erstwhile Bombay Presidency, an analysis of the caste profiles is extremely difficult, for one, data on caste composition is available only for the agencies of British administration and two, the wide variation in the territorial size of neighbouring agencies makes a comparative study not only impossible but also meaningless. Further, the existence of a large number of petty states, limited in territorial size, further complicated the matter. Any attempt at analysis could well lead to a false picture of caste gradients and misleading socio-cultural micro regions.

In the case of the large, but politically scattered and culturally parcelled Rajputana Agency, all political and administrative borders are evidently synonymous to the socio-cultural borders. Establishing the authencity of the borders in these large territorial units is problematic. Existence of a large number of small territorial communities conforming to their own evolved caste systems and the consequent fragmentation of the socio-cultural space, reveals a hidden territorial and political isolation. A marked contrast to the surrounding regions, the former could be considered as a single whole in the hierarchical spectrum.

To conclude, it can be stated that the degree of disintegration of the Indian socio-cultural space is high revealed by the high number of Indices of Dissimilarity of the caste structures. On the other hand, the borders of territorial units having similar caste structures are not only rare and in isolation, but also realtively ill-defined. This testifies that the method of analysis used is effective in the conditions of Indian society and is able to reveal the socio-cultural territorial communities as they exist in reality. The caste regions thus revealed enable us to evaluate their extent and also verify the disputable position of some socio-cultural boundaries.

Small distinct areas (as those inhabited by tribes) ensconed within a larger area reveals not only the high degree of disintegration of the socio-cultural space but also the socio-cultural characteristic of the territory. Also revealed are the zones of transition between the large territories, each with its distinctive hierarchical structure.Thus, the Bundelkhand - Baghelkhand region appears as a buffer between the North and the South where the gradual intermingling of the socio-cultural complexes of the North and South is taking place.

The map prepared (Fig.25.1), while being informative on the aspect of boundaries of caste regions, demands further detailed analysis. An understanding of the hierarchy of the revealed boundaries is necessary. The boundaries, of socio-cultural regions which are not well defined, as in the North, seem to indicate a hidden and complex hierarchical organisation of the socio-cultural space, which needs further investigation. The map may be treated as a blue print to be used further as a base for the preparation for the map of socio-cultural regions of India.

References and Notes

1. Sopher, David (1977). 'Rohilkhand and Oudh: An Exploration of Social Gradients Across a Political Frontier', in R.G. Fox (Ed.) *Realm and Region in Traditional India*, New Delhi: Vikas Publishing House Private Ltd., p.286.
2. The investigation did not touch the North-East and Kashmir because these areas are different in socio-cultural terms from the rest of the sub-continent. A desire to present a comparison of similar things and a genuine interest to reveal the boundaries of socio-cultural regions not apparent *a priori* prompted the decision to preclude these areas.
3. Following conversion to Hinduism, tribes are being incorporated into the caste system. With only few exceptions, as a rule they are treated as lower castes, usually untouchables.
4. *See* ref. 1 p. 290.

SECTION-V

CULTURAL ECOLOGY

26

THE NILGIRIS OF TAMIL NADU, INDIA, AS A DISTINCTIVE UPLAND ISLAND

William A. Noble

Introduction

As in every region of the world, it is the magnificent and interconnected web of life forms and attendant environment features which cause each region to have its share of uniqueness. The Nilgiri Upland Island, that part of the Nilgiri massif rising above 1,200 metres, or 4,000 feet, and culminating in the 2,637 m peak of Doda Betta, lies within the Nilgiri District of extreme northwestern Tamil Nadu (Fig. 26.1). The Island is extraordinarily rich in life forms. Unique links in the context of lands surrounding the Nilgiri massif also contribute to the distinctiveness of the Nilgiris. In and close to the Nilgiris there are endemics and species whose relatives live far away. Some life forms are thus, disjunctly separated by as much as 1,500 km of space from the same species, or sub-species, or related species elsewhere. We now know, too, that tropical rather than temperate vegetation dominates in *sholas*, the upland forests, mostly lying above 1,800 metres. They and grasslands, certainly spread in areal extent by humans in more recent times, once contributed greatly to the distinctiveness of the Island. Thanks to the detailed work by Blasco and his colleagues,[1] the uniqueness of the high rainfall *sholas* and grasslands of the far western Nilgiris are being revealed for the first time. We now have some inkling of what happened in the Nilgiri Upland Island during the Pleistocene—the Ice Age—and after, and this region was made even more distinctive by the changes which occurred. Upper level soil cores reveal changing conditions of the past, and intervening environments away from the Nilgiris were sometimes suited to the spread of species to the Nilgiris. Changes resulting from a human presence in the Nilgiris seem not to have occurred until long after the Pleistocene. After humans came, their impact accelerated through time. While all the factors contributing to the distinctiveness of the Nilgiri Upland Island are in reality closely interreleated, for the sake of simplicity, there is in this contribution a focusing upon (1) Life Forms, (2) *Sholas* and Grasslands, and Their High Rainfall Varients, and (3) The Pleistocene and Its Aftermath.

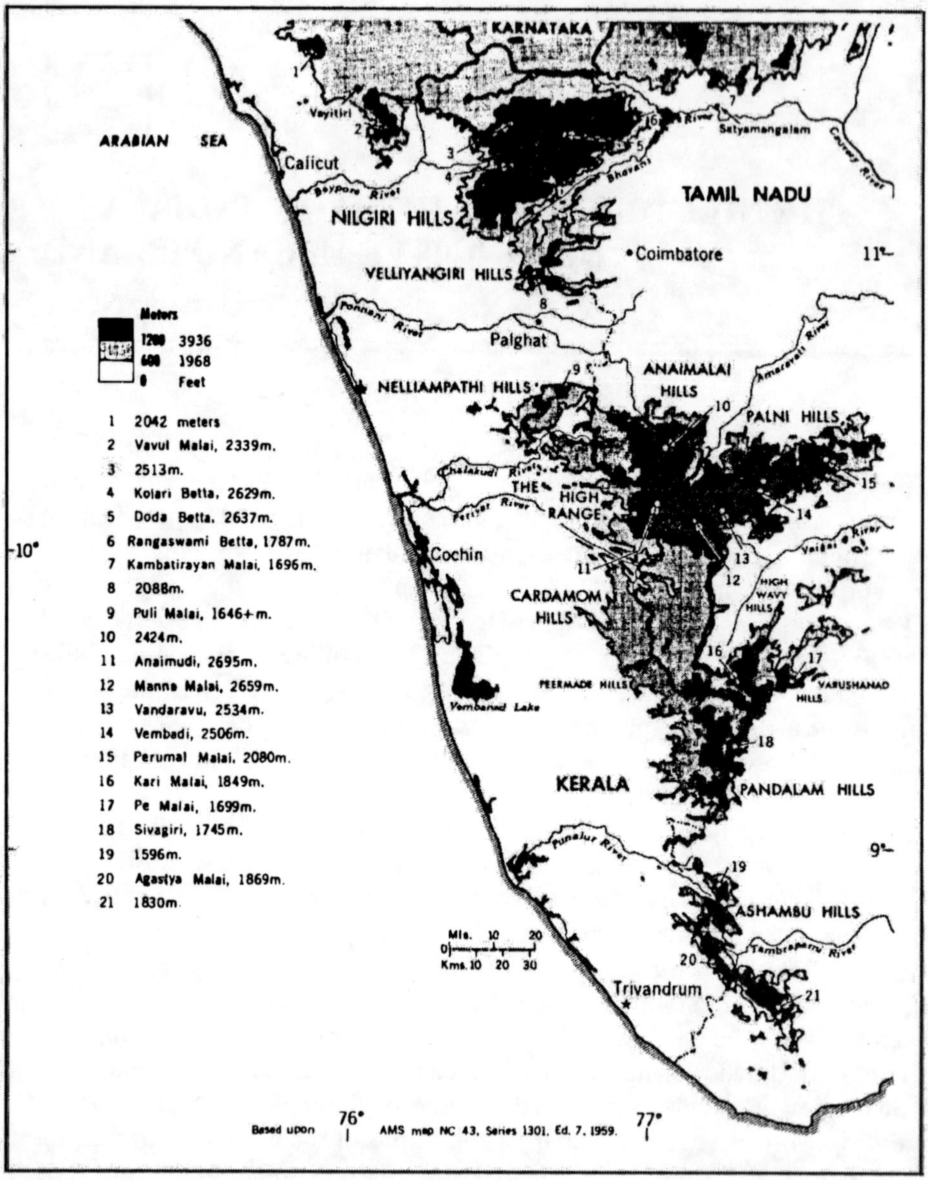

Fig. 26.1 : Indian Peninsula Southernmost Ranges and Peaks

Life Forms

The ideas of Sunder Lal Hora[2] *versus* those of M.S. Mani[3] assist us to focus upon different examples of life forms. Hora intuitively postulated the Satpura Hypothesis.[4] He was especially struck by the presence of related fishes in extreme southwestern India and Southeast Asia. He offered all manner of proof for the spread of creatures over a continuous upland which followed the Satpura trend from the Khasi and Garo Hills of Assam, to the Rajmahal Hills on the other side of the Ganges and Brahmaputra distributaries, to the end of the Satpura Range near the mouths of the Tapti and Narmada rivers in Gujarat. From there species could make their way down the Western Ghats. Other scholars, including the great Salim Ali,[5] followed Hora's lead, and there eventually were an impressive number of supporting publications. Alas, the pronounced break in terrain between the Rajmahal and Garo Hills, which might date back to Archaean times, has continued to be an insurmountable bit of evidence against Hora's hypothesis. Thus it has gone out of style.

Mani aruges for general distributions in earlier times with, then, a host of factors leading to the disappearance of species over vast tracts and, ultimately, their survival in relict refugiums.[6] Environmental changes through time and many different evolving life forms in competition are among the factors involved. The drift of the Peninsula Shield, the formation of the mighty Himalayas, climatic changes during the Pleistocene, and increased post-Pleistocene desiccating trends between refugiums were involved. So, too, was the extermination of species and the destruction of habitat by humans. Some progenitors became extinct, while their descendants survived in refugiums far from the habitat of their ancestors. We are all faced with the overwhelming realization that astounding changes have taken place in geologic time, that evolution has routinely been associated with the extinction of species, and that fossilized remains often offer very limited proof of species living in the past.

Example Fauna Relating to the Nilgiri Upland Island

Caecilians : The distribution of caecilians in the most primitive genus *Ichthyophis* defies explanation (Fig. 26.2). With the exception of *Icthyophis sikkimensis*, endemic to the eastern Himalayas, other Indian species survive in two disjunct portions of the Western Ghats and nearby lowlands.[7] Two of the species also live in Borneo and Java, but one has thus far been found in only the Nilgiris (*I. beddomei*).[8] The caecilians are seldom seen, because they live underground in moist soil. A past link with Gondwanaland seems to have ideal potential in explaining their distribution, but they do not live in Madagascar. Hora's focus upon uplands of spreading, along with different climatic conditions in the past, comes out ahead in the quest for a logical explanation of spread. Mani admits to some possibility of a past land bridge by north to south trending Dharwar rocks of Monghyr, between the Darjeeling region and the Rajmahal Hills. We now also know that so-called Malayan fishes migrated westward along the rising Himalayan arc and that some species appear to have moved

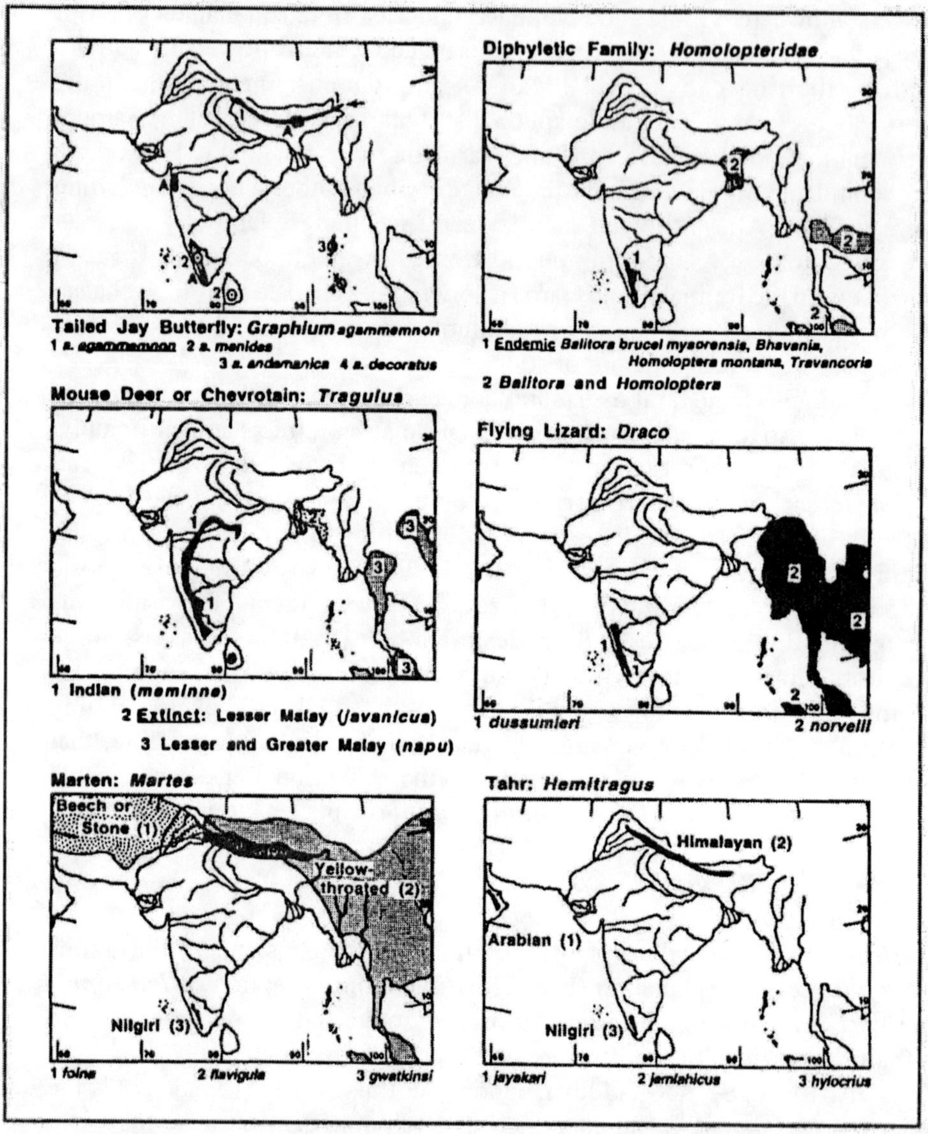

Fig. 26.2 : Example Fauna Relating to the Nilgiri Upland Island

southward to the Indian Peninsula from the more westerly Himalayas in an early period.[9] The once higher northerly to southerly trending Aravallis, formed in an earlier Gondwanaland, later smashed into the Himalayas.[10] Thus they formed a land bridge, but at a time too early to explain the spread of fish and other species existent today in the Western Ghats. Even though it is the remaining stub of a far higher Aravalli Range, the Delhi Ridge, just to the west of New Delhi reminds us of past possibilities.

Butterflies : How is the distribution of the sub-species of the tailed jay butterfly (*Graphium agammemnon*) to be explained? (Fig. 26.2) This butterfly is clearly a wandering and hardy survivor, both overland and in long flights over water. It has already spread to southern China, the Solomon Islands, and Australia, and after moving through Myanmar (Burma), one sub-species of the butterfly (1) is still spreading westward in the Himalayas.[11] Other sub-species live in (2) southern India, including the Nilgiris, and Sri Lanka, (3) the Andaman Islands, and (4) the Nicobar Islands[12] (Fig.26.3). The sub-species of southern India can be seen in Parson's Valley and between Coonoor and Kallar, the Nilgiris, in September and October. While it apparently had a far more limited range in the 1920s (including the Nilgiris), it has spread widely and is now quite commonly seen in the cities of Bangalore, Chennai (Madras) and Mysore. Could its ancestral stock have wandered or been blown into southern India and Sri Lanka from the northeast, perhaps after being carried afar in a storm? More likely, still, seems the possibility of both males and females being blown westward in a storm system affecting the Andaman and/or Nicobar islands.

Fishes : Is it likely that the ancestors of *Homolopteridae* fish in extreme southwestern India spread overland, perhaps in areas of higher terrain existing now and/or existent in the past, in the intervening regions (Fig.26.2)? If this was the case, the late dying out of species occurred on a large scale. Two endemic sub-species, and especially two endemic genera, indicate that local evolution in southwestern India probably went on for a long time. *Homoloptera montana* is an endemic of the Anamalai Mountains, just to the south of Nilgiris. *Balitoria brucerei mysorensis*, named in part after the former state of Mysore, is endemic to the region just to the north of the Nilgiris. It is fascinating to discover that three other sub-species live in (1) the eastern Himalayas and Assam, (2) Myanmar (Burma), and (3) an area encompassing the Myanmar (Burma) Thailand border.[15] Another species, *B. maculata*, is restricted in its distribution to the Darjeeling area of the Himalayas. Because fish of the genus *Bhavania* were first collected from a tributary of the Bhavani River in the Nilgiris, the endemic genus is named after the river. The endemic genus *Travancoria* is named after the former state of Travancore that covered southern Kerala, but the species also lives farther to the north.

Mouse Deers : The mouse deer or chevrotain of southern India prefers thickly forested areas, including tropical rain forests, on or near mountain slopes. Its survival may thus, in part, be related to its shyness and withdrawal from areas colonized by humans. Because *Moschiola* is a generic name also used for this animal, there is some indication that it has for a long time evolved separately from its Southeast Asian relatives(Fig. 26.2). However, because humans may have brought about the extinction of lesser Malay mouse deers in overcrowded Bangla Desh by eating them, there is a hint that mouse deers once lived over

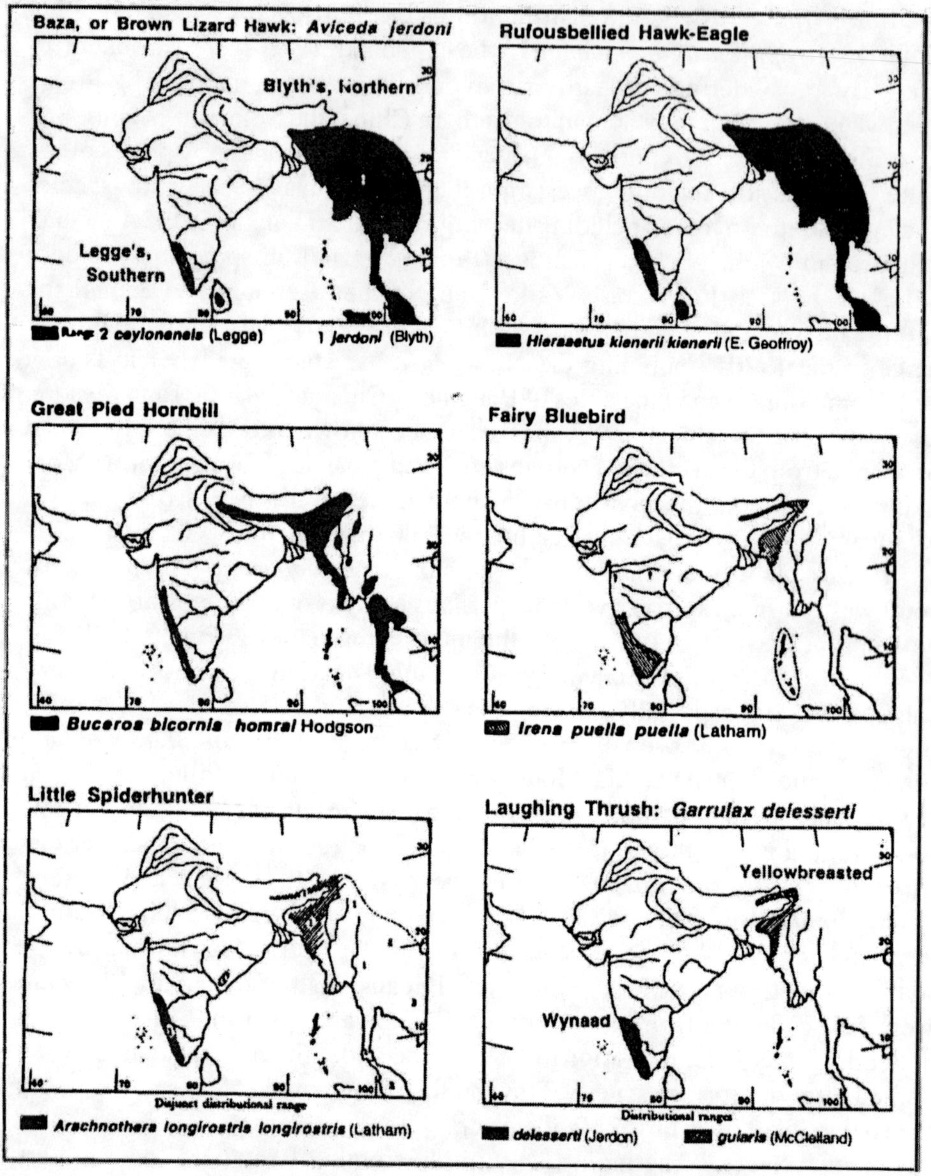

Fig. 26.3 : Example Fauna Relating to the Nilgiri Upland Island

contiguous land areas (and all the way across Myanmar ?) from Southeast Asia into southern India. Reza Khan interviewed over 100 people who remembered eating the meat of an animal fitting the description of the mouse deer, but he did not manange to collect a single remnant of it.[14]

Flying Lizards, Chameleons : Flying lizards of genus *Draco* spend most of their lives in trees. Six extended ribs to each side of the body, enfolded by skin, form rudimentary planes which enable the lizard to glide from one tree to the next.[15] The single species living mainly on mountain slopes in tropical and sub-tropical moist evergreen forest in southwestern India, offers a fine example of disjunct distribution (Fig. 26.2). The flying lizard in the Nilgiris lives as well in the Silent Valley National Park, located in Kerala and in the extreme southwestern part of the Nilgiri massif. From northeastern India, through Southeast Asia, and as far as the Philippines there are about 40 species of *Draco.*[16] We again face the most intriguing question of how so many life forms came from the east and survived in the extreme southwestern refugium of India (and especially in the mountains there). Hora's focus upon mountains and similar climatic conditions in the past again becomes an inescapable guide in any quest for a logical explanation. On the other side of the Nilgiris, in savanna woodland and in close to broken thorny thickets of the Mysore Ditch, the single species of chameleon (*Chameleon zeylanicus*) living in India is common. By contrast with the migration of species affiliated with Southeast Asia, the chameleon is affiliated with the over 80 species of chameleons in Africa (the Ethiopian Realm) and Madagascar. The Indian species might be descended from ancestors moving through similar climate (tropical steppe) and vegetation all the way from Africa and across the southern portion of the Arabian Peninsula, but it is tempting to suggest that ancestors of Nilgiri chameleons travelled northward on the Peninsula Shield after it broke away from Madagascar.

Nilgiri Martens, Nilgiri Tahrs : Nilgiri martens and Nilgiri tahrs inhabit the upper Nilgiri (Fig.26.2). They once were inhabitants of Coorg to the north, but are now probably extinct there. The status of the marten is in doubt in the Nilgiris and in other mountainous regions to the south. While the tahr is endangered, it survives in mountains from the Nilgiris south to the Ashambu Hills.On the Nilgiri Upland Island, most tahrs live within Mukerti National Park in the extreme western Nilgiris, and this is the region where martens are most likely to be seen. Martens probably evolved in the Palearctic Realm[17] and they flourish in the boreal forest of Eurasia and North America. The Nilgiri marten is most closely related to the yellow-throated marten of the Far East. It is an outstanding example of an animal whose ancestors migrated far to the south and into the Tropics. In that martens prefer to live above 2,000 m in the Nilgiris, their ancestors probably spread slowly south over mountainous areas during the Pleistocene. Arabian, Himalayan, and Nilgiri tahrs, forming three distinct species, are widely separated survivors (Fig. 26.2). Because the Nilgiri tahr has 58 chromosomes and the Himalayan tahr has only 48, they clearly had different evolutionary paths.[18] The Himalayan tahr has a ruff of long hair on the neck and shoulders, and the Nilgiri tahr has shorter hair all over. The female

Himalayan tahr has four mammae, whereas the Nilgiri tahr has two. The Himalayan tahr lives in terrain with precipices up to the timberline, which may be as high as 3,000 to 3,600 m, whereas the Nilgiri tahr prefers cliffs next to grasslands that are often 1,800 to 2,400 m in elevation. There is evidence for a tahr species (*H. stehlini*) coming into Europe in the early Pleistocene and another tahr species (*H. bonali*) moving in during the late Pleistocene, in periods when the climate was colder and drier.[19] European tahrs became extinct between 17000 and 10000 BP. Considering the past and widespread distribution of tahrs, it is suggested that unknown ancestor of Nilgiri tahrs evolved far away, and that they migrated to the southern part of the Indian Peninsula during the Pleistocene.

***Birds* :** Links between extreme southwestern India and Southeast Asia again become evident in the distribution of bird species. Using the disjunct distribution of four very different kinds of birds, the rufousbellied hawk-eagle, the great pied hornbill, the fairy bluebird, and the little spiderhunter (Fig.26.3), we find that breeding populations of the same species are separated by over 1,500 km of intervening land. Two separate sub-species of the baza and a laughing thrush are separated by equally long distances, and we thus have proof of the passage of sufficient time for single species to evolve into sub-species. Legge's baza prefers rain forest on lower mountain sides (150 to 900 m), but the rufousbellied hawk-eagle prefers higher elevations (to 1,200m) in rain forest and moist deciduous forest (jungle). While the great pied hornbill (to 1,500 m), the fairy bluebird (to 1,800 m), and little spiderhunter (to 2,100 m) inhabit the Nilgiri upland island as well, the first two species are far more likely to be seen in rain forest at lower elevations (in the Silent Valley, for example). Turning again to the why of distribution, both Hora and Mani are apparently right in their own way.[20] It seems inconceivable that most species just covered could have slowly spread so far south without the existence of varied mountain niches along most of the migration routes. If continuous mountain arcs never existed, there is some likelihood that gaps with vegetative associations (including tropical evergreen forest) different from today might have enabled migrations to take place in any case. Ice masses moving far down Himalayan valleys during the Pleistocene, and then retreating, must have caused climatic changes to the south. There is, for example, some evidence indicating the climate and vegetation (including evergreen forest) were very different 2,000 years ago from the tropical steppe and related xerophytic plants surviving presently in the *Brij* region, around Mathura, in northern India.[21] Since that time, too, we must remember that humans scouring landscapes in their perpetual quest for fuel have destroyed most, if not all, natural forests in the *Brij* region, and thus have contributed to the desiccation of the region.

Laughing Thrushes : Laughing thrushes (genus *Garrulax*) offer an outstanding example of evolution in disjunct regions.[22] There are 55 species of laughing thrushes in northern South Asia, and they are spread outside all the way from Tadzhikistan and Afghanistan to China and North Vietnam. The giant laughing thrush also lives in the Tsangpo Valley of Tibet, and as high as 4,900 m in the Szechwan Basin of China. Apart from being spread from Hunza and Gilgit in Pakistan and over the Himalayan Range, there are laughing thrushes in the Garo and Khasi Hills, in Nagaland and Manipur, India, and in the Chittagong Hill Tracts of Bangla Desh. The southern laughing thrushes provide us with some insight into evolution. Apart from the Wynaad laughing thrush, so closely related to the yellowbreasted laughing thrush of northeastern India (Fig. 26.3) there are five other laughing thrushes. One (*G. cinereifrons*) lives in Sri Lanka, and three closely related sub-species live in Coorg (*G. jerdoni jerdoni*), Kerala (*G. j. fairbanki*), and southern Kerala (*G.j. meridionale*). The fifth, the Nilgiri laughing thrush (*G. cachinnans*), is closely related to the three. It lives almost exclusively on the Upland Island, up to the tops of the highest peaks, and is also the only Nilgiri endemic bird.

More Birds, Animals : Another noteworthy bird is the woodcock (*Scolopax rusticola L.*), wintering on the Nilgiri Upland Island. Before and after seasonally staying in the south, this bird flies from the Himalayas to and from the Nilgiris, Palnis, and other mountainous areas of the far south in non-stop flights. While they live elsewhere in southern India as well, these birds named after the Nilgiris also contribute to their distinctiveness: Nilgiri blackbird, Nilgiri flowerpecker, Nilgiri house swallow, Nilgiri pied bush chat, Nilgiri pipit, Nilgiri plain wren-warbler, Nilgiri quaker babbler, Nilgiri small skylark, Nilgiri thrush, Nilgiri verditer flycatcher, Nilgiri white-eye, and the Nilgiri wood pigeon. The Nilgiri langur (*Presbytis johni*), a black monkey occassionally seen on grasslands, is mostly confined to upland island *sholas* of the Nilgiris. It more often lives at lower elevation in other mountainous regions of southern India.[23] As one goes up the ghat road to Coonoor, the bonnet macaque (*Macaca radiata*) is often sighted. There are urban troops of this monkey in Coonoor and Kotagiri. The famous wildlife photographer Krishnan recorded a miniature bonnet macaque ('hardly half the size of the commoner species') in the Mudumalai Wildlife Sanctuary at the northwestern base of the Nilgiri massif.[24] One of the main surviving habitats for the endangered liontailed macaque (*Macaca silenus*) is the rain forest on western Nilgiri slopes, in the southwesterly Silent Valley, and on slopes down to the Upper Bhavani River.[25] The Hanuman or common gray langur (*Presbytis entellus*, sub-species *priam*) is a main attraction among the tourists visiting the Mudumalai Wildlife Sanctuary. Elephants (*Elephas maximus*), gaurs (*Bibos gaurus*), and spotted deer (*Axis axis*) are frequently seen in the Sanctuary, and wander widely in the Mysore Ditch to the east. The elephants form one of their main surviving populations in India. In the savanna woodland and in close to broken thorny thickets spreading from the northeastern and eastern base of the Nilgiri massif, the black buck (*Antilope cervicarpa*) is making a comeback after having greatly declined in numbers. The four-horned antelope (*Tetraceros quadricornis*) is also present in small numbers within the Mysore Ditch.

Sholas and Grasslands, and Their Varients in High Rainfall Areas

Sholas and Grasslands **:** A 'wilting effect due to plants exposed to the morning sun being unable to draw from the frozen soil' was used by Ranganathan as an explanation for why grasses grew together in such stark contrast from the nearby *sholas*.[26] There can be no doubt that fires followed by grazing have had a profound impact, as Bor concluded, 'I believe that the *shola* forest is the relic of an evergreen forest climax which has been pushed back to its last stronghold by fire and grazing. The grassland I consider to be a biotic climax rendered stable by fire and grazing and only one more proof of the stability of grassland under such condition'.[27] Among the complex influences upon vegetation which exist in the upper Nilgiris, it seems reasonable to draw a parallel with the treeless balds of the Appalachians in the United States, as conceptualized by Billings and Mark , 'The balds are then essentially forest dominants'.[28] Quite apart from the shock of widespread morning frost, the tolerance limits of upper Nilgiri plant species are severely tried by fast-moving winds sometimes gusting to over 110 km per hour, heavy rains, drought, sometimes prolonged for weeks, exposure to a bright tropical sun that can soon burn the human skin, and cooling shade from enfolding, clinging damp mist (the last two sometimes experienced by humans within 30 minutes). Along with the spread of glaciers to lower elevations in the Himalayas during the later Pleistocene, we now know that there was sufficient climatic change farther to the south to cause grasslands to dominate in the Nilgiri Upland Island.

Before briefly covering the nature of *sholas* and grasslands, a comment regarding the writer's use of the term grassland is in order. Blasco uses 'common savanna' and 'high rainfall savanna' instead of grassland, and quite correctly so in terms of his view that grasses on the Nilgiri Upland Island are typically accompanied by a host of herbs and shrubs.[29] While recognising this fact, the writer prefers to use grassland by using the term downs, the Nilgiri downs reminding them of open grassy tracts in their homeland. Secondly, remembering the original tropical *sabana* of Cuba, from which the word savanna is derived, the more specific use of savanna for a landscape with grasses and scattered trees seems more appropriate. While rhododendron and other tree species scattered on grassland form limited savanna areas, it is the overwhelming presence of low-lying grasses over larger tracts (even over 90 per cent of the terrain) that visually gains the upper hand. Grasses have for long formed a significant part of the vegetation in the Nilgiri upland island, and the degree of endemism among them may reach over 10 per cent. While the writers' reference to grassland equates with Blasco's common savanna, his use of the term high rainfall grassland follows Blasco's lead in regard to areas with a unique ecology next to far western Nilgiri escarpments.

Due to the nature of plant associations, and because so many revisions were made while writing about them, documentation became impossible. The writer would feel remiss, however, if these basic works of Champion, Fyson, Gamble, Meher-Homji and Shankarnarayan were not mentioned.[30]

Sholas : Above 1,800 meters there are scattered low canopy evergreen forests called *sholas*, with most trees less than 20 metres in height. The *sholas* are separated by grasslands and agricultural clearings, and typically lie within more sheltered valleys, depressions, and on the leeward sides of ridges. Most tree species in *sholas* are distinctly tropical in type. The continuous canopy forests are cool, dark, and sometimes open near the surface. Linked to the great frequencey of clouds during the year and due to the resultant drenching of trees, many *sholas* fulfill the common criteria of tropical cloud forests by having a host of epiphytes growing on tree trunks and branches. These are common *shola* trees, all of which are tropical: big cinnamon (*Cinnamomum wightii Meissn.*), Nilgiri mock-olive (*Elaeocarpus oblongus* Gaer.), (*Gordonia obtusa* Wall.), spiraea, sacred Toda *tir* (*Meliosma wightii* Planch.), Nilgiri magnolia (*Michelia nilagirica* Zenck.), Schefflera (*Schefflera racemosa* Harms.), sweet-leaf (*Symplocos foliosa* Wt.), black plum, jambu (*Syzygium spp.*), and Turpinia (*Turpinia nepalensis* Wall.).

***Grasslands* :** Grasslands above 1,800 metres were once generally associated with grasses kept short through annual fires and frequent grazing by livestock, but there are now vast tracts where firing is prohibited and where the grasses as a result have grown into large clumps. Along with introduced species, there are over 120 species of grasses growing on the upper Nilgiris. One species of grass (*Chrysopogon zeylanicus*) is so common over large tracts that a grassland type can be named after it. Another grass (*Andropogon polypticus*) dominates in some boggy depressions and frost pockets. Introduced Kikuyu grass (*Pennisetum clandestinum*) from Africa has also become widespread. In areas the least disturbed biotically, some grass species (*Cymbopogon polyneuros, Heteropogon contortus, Ischaemum indicum,* and *Themeda* spp.) will flourish. Other species (*Andropogon* spp., *Bothriochloa* spp., and *Eragrostis* spp.) indicate increased biotic disturbance, while others (*Cynodon dactylon, Digitaria* spp., and *Eragrostis poaeides*) indicate intense biotic disturbance and soil erosion. Besides grasses, a host of other plants, ranging in size from minute herbs to short trees, grow on the upland grasslands. Blooming *kurunji* (*Strobilanthes* spp.) periodically colours entire slopes with flowers. While there are balsams growing within and in the fringes of *sholas*, many form herbaceous clumps on open grasslands. Bracken, probably introduced, sometimes grows in colonies over burned areas on lower slopes. Imported white and yellow broom, common gorse, and goat weed have become weeds, not infrequently forming large colonies. Because goat weed thrives where structures built by humans have disintegrated, it can be of great assistance to archaeologists. Some shrubs grow singly on the grasslands, or form colonies (*Berberis, Dodonaea, Gaultheria, Hypericum, Mahonia, Oldenlandia, Rhodomyrtus* [also a tree], and *Sophora*). Rhododendron (*Rhododendron nilagiricum* Zenk.) trees—highly resistant to fire—are capable of growing singly and far out on the grasslands, but also form loosely knit colonies (and, thus, savannas). There are also heaths composed of rhododendron mixed with shrubs, and these may form a transitional ecotone between grassland and *shola*.

High Rainfall *Sholas* and Grasslands

The high rainfall *sholas* and grasslands of the far western Nilgiris are little known scientifically, but are ecologically distinctive. They for the most part are located in a band running from a northerly to southerly direction, following the western escarpment, and then, for a shorter distance, the southerly escarpment. Included are the Pandiar cliffs, all the area surrounding Nilgiri Peak, the Mukerti Peak area, the three Western Catchments, the Bangitapal Rest House and adjacent area, the zone close to the cliffs in the Nadgani and Sispara areas, and then the *sholas* and grasslands alongside steep southerly slopes to and somewhat eastward from the Bison Swamp area. In 1961, an exceptional year, high rainfall grasslands at Mukerti Ridge Top and in Western Catchment Three had the distinction of having the two highest annual rainfall figures ever recorded in southern India: 10,608 mm (414", or 34.5') of rain fell at Mukerti Ridge Top and 10,867 mm (424", or 35.3') of rain fell in Western Catchment Three.[31] Mukerti Ridge Top receives an average of over 5,000 mm (200") of rain during the westerly monsoon. This station edges the western rim, and a part of the ridge is actually an overhang. The June, July, and August median rainfall values for Mukerti Ridge Top are 1297, 2159, and 1186 mm (51, 84, and 46 inches) respectively. The heaviest rainfall of 713 mm (28") ever recorded in a 24-hour period at Western Catchment Three, on July 7, 1958, and the heaviest rainfall of 1283 mm (50") ever recorded in a 48-hour period at Mukerti Ridge Top, on July 12 and 13, 1943, further demonstrate how high the rainfall is at recording stations thus far set up. The harsh environmental conditions of the high rainfall grasslands during westerly monsoon rains, with their numbing wet cold, regularly prompted past Toda and Badaga pastoralists to move back eastward from them and adjacent grasslands before each westerly monsoon set in . In contrast to the rainfall and great frequency of clouds, mainly in the period of the westerly monsoon, but during the easterly monsoon as well, there are also periods of drought and varying lengths of intense exposure to the Sun's rays in the high rainfall grasslands.

While there are *sholas* in protected valleys and depressions, there are areas where high rainfall grasslands dominate up to both banks of streams and over low-lying terrain. Associated soils are often leached by the high rainfall. A description by Rao helps us envision how Nilgiri wet *shola* and grassland landscapes compare with the very high rainfall Cherrapunji area of northeastern India... 'the situation at Cherrapunji, until recently famed as the wettest place on earth, needs special notice. The area looks disappointingly bleak and bare of wooded vegetation, due to the poor soil cover; all the soil being leached out by the heavy rains, leaving behind smooth, bare rocks. For vast distances all around, only dwarf grass-growth is visible. It is only in the comparatively sheltered depressions, as at Mamloo and Moswmai, where there is a deposit of soil and humus, that there are small islands of wooded vegetation, in an otherwise vast sea of grassland.'[32]

High Rainfall Sholas : According to Blasco and Lengerke, high rainfall *sholas* are composed of a dense growth of trees, often dwarfed, forming a shorter canopy to about 10 m from the ground.[33] The dominant tree species is often a black plum (*Syziygium calophyllifolium* Walp.[trop.]). Other trees, all tropical species, are cinnamon (*Cinnamomum macrocarpum* Hook. and *perottetii* Meissn.), another black plum (*S. arnottianum* Walp.), apple bladder-nut (*Turpinia cochinchinensis* Merr.), white milk tree (*Isonandra condolleana* Wt. and *perottetiana* Wt.), and mock-olive (*Elaeocarpus ferrugineus* Wt.). The less frequent Nilgiri magnolia [trop.], spindle (*Euonymus crenulatus* Wall.) [temp.], and cryptocarya (*Cryptocarya lawsoni* Gamb.)[trop.] trees also grow in these *sholas*. Epiphytes are more common in the high rainfall *sholas*. In addition to a variety of orchids, pepper-elder (*Peperomia reflexa* A. Dietr.) is a very common epiphyte. As an evolutionary adaptation to the rigorous climatic conditions, two balsams are often epiphytic (*Impatiens orchioides* Bedd. and *laticornis* C.E.C. Fischer) on boulders and trees, whereas one (*neo-barnesii* C.E.C. Fischer) has become a full-fledged epiphyte on trees. The last typically grows on branches with moss and pepper-elder, on trees often located close to the western escarpment and above 2,400 m.

High Rainfall Grasslands : The high rainfall grasslands have *Isachne kunthiana* and *Themeda trianda* grasses, and *Arundinella fuscata* Nees is the commonest grass in some areas. Among the plants most frequently growing in these grasslands are common anemone (*Anemone rivularis* Ham.), balsams, creat (*Andrographis lawsonii* Gamb.), everlasting (*Anaphalis* spp.), *kurunji* (strobilanths), teasel (*Dipsacus leschenaultii* Coult.), wood germander (*Teucrium*), black plum (*Syzygium*), Nilgiri privet (*Ligustrum*), and some scattered rhododendrons. The high rainfall grasslands offer the ideal habitat for Nilgiri tahrs, animals which prefer living on and close to cliffs. The grasslands are sometimes still burned over in the dry season, despite regulations to the contrary, and it is noteworthy that tahrs relish young grasses sprouting after firings. Thus, they often tend to congregate in such areas. Attempts by the Forest Department to plant trees in high rainfall grasslands have typically ended in failure. Row upon row of dead young plants now stand out on them.

The Pleistocene and Its Aftermath

Possible Changes Over 40,000 Years

The palynological evidence provided by Vishnu-Mittre and Gupta[34] and Gupta[35] with a prime focus upon pollen grains from soil profiles at Kakathope, near Ootacamund in the Nilgiris, tentatively suggest past vegetational changes over a long period. A cooler and drier climatic regime probably existed from about 40 to 35,000 years BP, and the evidence indicates that there was widespread grassland on the upper Nilgiris. Along with grasses, there were sedges and a variety of other herbs and shrubs. While it is usually impossible to identify actual species from the pollen grains gathered from different depths in soil profiles, it is possible to identify plant genera which were present. Herbs of the genera *Campanula, Chenopodium, Geranium, Justicia,* and *Lilium,* shrubs of the genus *Berberis*, and a few sporadic trees of the genera *Ilex* and *Rhododendron* appear to have spread in from the north. What is even more significant are species that did not move down into the Peninsula from the north. Thus, the oaks (*Quercus*) and pines (*Pinus*) of Meghalaya never managed to spread southward, whereas *Ilex* and *Rhododendron* did. Latitudinal variation and insufficient change in climate as a result of glaciation appear to have jointly operated to prevent the spread of most temperate species. The *Berberis, Ilex* and *Rhodedendron* genera that came are currently hardy species, quite capable of surviving as lone sentinels on grassland. *Impatiens* herbs and *Strobilanthes* shrubs (more recently, however, divided into several genera) appear to have always been present on the Nilgiris in small numbers. This may be the result of their ancestors coming northward on the Peninsula Shield from Madagascar. Along with the high degree of endemism among Nilgiri grasses, the highest degree of plant endemism has occurred among the *Impatiens* and *Strobilanths*.

Between 35,000 and 15,000 BP there was an increase of shrubs and the establishment of thickets over grassland. Cool and dry climatic conditions probably persisted, with a shift toward greater aridity. The summer westerly monsoon may have greatly weakened (or not have existed?), and did the fall easterly monsoon become stronger?[36] There was a decline in the pollen grains of grasses and sedges, while those of the *Compositae*, *Dipsacaceae*, *Leguminosae*, *Malvaceae*, and *Rosaceae* families, and the genus *Jasminum* (or *Ligustrum*, instead?) notably increased. Pollen grains identified with the genus *Portulaca* were present, but no related species survives today. While pollen grains of the genera *Campanula, Geranium,* and *Justicia* declined, those of *Chenopodium* and *Lilium* continued to be present in about the same quantity. The pollen grains of *Berberis, Ilex*, and *Rhododendron* are also present, with the last declining as time went by. As shrubs, there is also some showing of the genus *Sarcococca*, and a better representation of the genus *Lonicera*. Sporadic pollen grains of the genus *Artemisia*, a herb or shrub, also occur. Ultimately, there was thus an indication of widespread heaths with a few scattered trees.

The climate probably became warmer and more humid between 15,000 and 7,000 BP. There is some evidence for a strengthening of summer westerly monsoons, with them reaching a peak at about 11,000 BP, and then for a progressively drier period to 6,000 BP.[37] The greater presence of tree pollen grains in related soil horizons indicates the spread of *shola* with evergreen tree genera *Elaeocarpus, Euonymus,* and *Gordonia*. The spread of *sholas* tends also to be supported by the marked increase of pollen grains of the genus *Peperonia*, now a common epiphyte in *sholas*. It seems unlikely, however, because of a strong continued presence of grass and sedge pollen grains, that the Nilgiri Upland Island was ever covered entirely by *sholas*.

From about 7,000 BP and continuing into the Post-Contact Era, starting in the early 1820s when the British came, a decline in the pollen grains of *shola* species indicates a reduction in the areal extent of *sholas* and an increase in grasslands. There are intervals with a marked decline of *Peperomia* pollen grains, which contrast with a marked increase of *Impatiens* pollen grains. While pollen grains of the genus *Jasminum* (or *Ligustrum*?)—so abundant in the earlier heaths—are conspicuously absent, those of the families *Compositae, Dipsocaceae, Malvaceae*, and the genus *Geranium* are present. In contrast to the earlier heaths, too, there is a lack of *Berberis, Ilex*, and *Rhododendron* pollen grains. Might this be due to fires on open grasslands increasingly destroying the related tree species? Climatic change may, in part at least, be responsible for the increasing shift from *sholas* to grasslands. A stage to at least 6,000 BP, and probably beyond, may be due to increased drying. A weakened westerly monsoon may have again been established by 6,000 BP. By 600 BP, and from about 1200 to 1400 AD, there appears to have been a warmer and moister period than what exists at present.[38] It is thus, perhaps, no coincidence that there is an abundance of archaeological evidence (partially seen by the author during 1994 field work in the area) for the presence of more people living in the Mysore Ditch, just to the north of the Nilgiri, in that period. The possible roles played by lightening strikes or spontaneous combustion during dry seasons in starting fires and thereby increasing the areal extent of grasslands, leading to an increase of grass pollen grains in Nilgiri waterlogged soils, have not been investigated. Human hunters-gatherers setting fires for millennia might have contributed to *shola* destruction (but there is no archaeological proof of this).

The Past 3,000 Years

A more conservative time span of 3,000 years provides us with a still more revealing and possibly more accurate glimpse of the past. The pioneering effort of Vimala Menon is most useful because of its zone by zone analysis of a soil profile in a bog at Pykara.[39] The profile extends to a depth of 1.7m and is divided into 15 zones. From zone 1 at the bottom to zone 15 at the top, these are the percentages of grass pollen grains: 83, 61, 54, 53, 60, 49 (the lowest), 55, 53, 59, 78, 83, 79, 85 (the highest), 80, and 71. The evidence (count of 2368 pollen grains) thus indicates the long continued dominance by grasses in the area. The pollen grains of ferns are the next largest group (spore count of 340). Although many ferns prefer shadowed, cool, and moist environments, suggesting *shola*, some thrive in shade provided by the rocks and plants of open grassland. We must remember, too, that ferns give off many spores, which can profoundly effect the rate of their accumulation in a soil profile. Sedges (*Cyperaceae*, pollen count of 159) come next. Represented families which follow, with pollen counts, are the *Liliacae* (123), *Compositae* (100), *Balsamina-ceae* (96), *Scrophula-riaceae* (63), *Xyridaceae* (62), *Chenopodia-ceae* (21), *Gentianaceae* (15), *Umbelliferae* (12), *Cruciferae* (12) and *Caprifoliaceae* (11). These generally indicate the presence of grassland with herbs and shrubs, while pollen grains from trees are conspicuously lacking. The occasional pollen grains of these families probably represent species living in *sholas*: *Araceae, Ericaceae, Euphorbiaceae, Meliaceae, Orchidaceae, Thymelaea-ceae, and Urticac*eae.

Blasco and Thanikaimoni studied pollen grains from profiles in sediments down to 3.8 m at Pykara and down to 2.1 m at Parson's Valley.[40] They divide the pollen grains examined into four separate ecological groups: (1) herbaceous (grass mainly, but other herbs as well), (2) forest border (forest—grassland ecotone), (3) tree/shrub savanna (trees/shrubs growing on the grassland), and (4) forest (*shola*). With one exception, a down trend of herbaceous pollen grains and the rise of forest border pollen grains from the 1.3 to 1.9 m depth at Pykara (during the wetter period and tree spreading, c. 600 BP?), the graph profiles of herbaceous pollen grains consistently ranged far higher than the graph profiles for the three other groups. Next came the forest border group, then the tree/

shrub group, and, lastly, the forest group. The pollen grains of *shola* species in the families *Celastraceae, Lauraceae*, and *Myrtaceae* were relatively rare, and those of *Araliaceae, Elaeocarpaceae, Sapotaceae, Symplocacaeae*, and the genus *Ilex* were also recorded. While the pollen grains from both places in general ranged from 65 to 80 per cent for plants of the open grassland, the pollen grains of forest species were at a low of about 5 per cent. Thus the overall findings parallel those of Menon.[41]

Unfortunately, Menon, Blasco, and Thanikaimoni have no radiocarbon dates for any parts of their profiles. However, Blasco and Thanikaimoni used comparable evidence from the Palnis to devise a dating system. Interpolating with a 1770 AD date for a sample at about 30 cm below the surface and a 1050 BP date for a sample at about 130 cm below the surface in the same locality, they postulate a going back to some 3,000 to 4,000 years BP for layers lying at 3 to 4 m below the surface in the Nilgiris and Palnis. Considering the evidence covered, we are confronted with the possibility of there being extensive natural grasslands on the Nilgiri Upland Island for thousands of years, and more certainly for at least the past 3,000 years. So strong is the evidence that we may not need to use the annual burning of grasslands any longer to explain their presence. Even if this should be the case, however, there seems no denying that the annual burning of grasslands must have contributed to their spread. While we have no archaeological proof for the burning of grasslands over thousands of years, their seeming presence for so long now raises the possibility of there being humans on the upper Nilgiris for far longer than hitherto thought.

While a characteristic polished Neolithic pointed butt stone implement found in the Mysore Ditch (and photographed by the author in 1994) indicates the presence of humans there by 1000 BC, there is no such indication for the Nilgiri Upland Island. From the evidence yielded thus far from the well known pre-historic sites, among which are the walled stone circles on the tops of peaks, we have no proof for the presence of humans prior to 1 AD.[42] The most logical choice of a people who could have been around for long are the Todas, who are a main element making the Nilgiri Upland Island distinctive. After watching a Toda gather the roots of so-called Toda potatoes (*Ceropegia pusilla* Wt.), eating some of them myself in 1994, and realising the variety of other food stuffs which can be gathered by the Todas, one concludes that their ancestors were probably gatherers on the Nilgiri Upland Island and outer slopes. Toda males give ceremonial, non-functional bows and arrows to determine parenthood. There are also references in Toda folk tales to some deities using bows, arrows, and spears in the 'Dream Time', when deities dwelt among humans, so Toda ancestors might also have been hunters. However, from 1603, when the priest Yacome Finicio wrote his account of a visit to the Nilgiris, we know that the Toda were buffalo herders. They remained so into post-contact times after the coming of the British in the early 1820s.[43] The historic record shows that they burned the grasslands each year. Fires set for even 300 years must have contributed to the spread of grasslands, to the detriment of *sholas*. We have no ideas as to when the distinctive Kota farmers settled on the upper Nilgiris. As artisans who can still make knives and pots, there is some potential for their ancestors being the makers of the effigy pots and knives buried in the centres of archaeological features. Most ancestors of Badaga farmers, the third distinctive group on the upper Nilgiris and now far outnumbering the other two, may have mainly migrated to the Nilgiris after the fall of the Vijayanagar Empire in 1565 AD. The Kotas to a lesser degree and the Badagas also once set fires and lived seasonally with their buffaloes and cattle out on the grassland.[44] The Forest Department eventually prohibited the firing of grasslands and most migrations into the grasslands, and these activities for the most part came to an end in the 1960s. It also engaged in the planting of trees, and by now black wattle (*Acacia mearnsii* Willd.), blue gum (*Eucalyptus globulus* Labill.) trees, indigenous to Australia, and Mexican pines (*Pinus patula* L) grow over thousands of hectares once dominated by grasses. By 1987, eucalyptus alone covered about 12,000 ha.[45] Mexican pines have typically been planted in frost pockets where the other two species failed.

References and Notes

1. Blasco, F. and Hans J. von Lengerke (1989). 'The Nilgiri Environment', in Hockings, Paul (ed.) *Blue Mountains : The Ethnography and Biogeography of a South Indian Region*, Delhi : Oxford University Press, pp. 54-61.
2. Hora, Sunder Lal (1949). 'Satpura Hypothesis and the Distribution of the Malayan Fauna and Flora to Peninsular India', *India, National Institute of Science, Proceedings 15B* : pp. 309-313; Hora, Sunder Lal (1950). 'Hora's Satpura Hypothesis,—An Aspect of Indian Biogeography', *Current Science 19* : pp. 363-370.
3. Mani, N.S. (1974) 'Biogeographical Evolution in India' in N.S. Mani, *Ecology and Biogeography in India*, pp. 698-724; Mani, N.S. (1974) 'Biogeography of the Peninsula, in N.S. Mani, *Ecology and Biogeography in India*, pp. 614-663.
4. *See* ref. 2
5. Ali, Salim (1948). 'The Gujarat Satpuras in Indian Ornithology', *Gujarat Research Society Journal*, 10 : pp. 35-45.
6. Mani, N.S. (1974). 'Biogeographical Evolution in India', in N.S. Mani, (ed.) *Ecology and Biogeography in India*, pp. 698-724.
7. Gorham, Stanley W. (1962). 'Gymnophiona, *Das Tierreich*, 78, Berlin : Walter de Gruyter, p. 25 ; Jayaram, K.C. (1974). 'Ecology and Distribution of Fresh-water Fishes, Amphibia and Reptiles,' in N.S. Mani (ed.), *Ecology and Biogeography in India*, pp. 517-584.
8. Gorham, Stanley W. (1962). 'Gymnophiona', *Das Tierreich*, 78, Berlin : Walter de Gruyter, p. 10.
9. Sahni, Ashok (1982). 'The Current Status of Hora's Satpura Hypothesis', in K.S. Valdiya, V.K. Gaur and S.B. Bhatia (eds.), *Geology of Vindhyanchal*, Delhi : Hindustan Publishing Corporation (India), pp. 143-147.
10. Pascoe, E.H. (1965). *A Manual of the Geology of India and Burma, Vol.1*, Calcutta :

Government Press, pp. 246, 344-352; Pascoe, E.H. (1959). *A Manual of the Geology of India and Burma, Vol.2*, Calcutta : Government Press, pp. 533, 549-561.

11. See ref. 6, p. 632.
12. Evans, N.H. (1927). *The Identification of Indian Butterflies*, Madras : Diocesan Press, for the Bombay Natural History Society, pp. xi, 302, xxxii end plates.
13. Jayaram, K.C. (1974). 'Ecology and Distribution of Fresh-water Fishes, Amphibia and Reptiles in N.S. Mani, *Ecology and Biogeography in India*, pp. 528-529.
14. Khan, Mohammad Ali Reza (1984). *Endangered Mammals of Bangladesh*, Oryx 18, 3 : pp. 152-156.
15. Hawkins, R.E. (ed.) (1986). *Encyclopedia of Indian Natural History*, Bombay : Bombay Natural History Society, p. 620.
16. See ref. 13, p.558, Fig. 122.
17. Anderson, Elaine (1970). 'Quaternary Evolution of the Genus Martes,' *Acta Zoologica Fennica*, 130, Helsinki : Helsinglfors, p. 132.
18. Benirschke, Kurt and Arlene T. Kumamoto (1980). 'The Chromosomes of the Nilgiri Tahr', in P.J. Olney (ed). *International Zoo Yearbook, Vol. 20*, London : Zoological Society of London, pp. ix, 520.
19. Schaller, George B. (1977). *Mountain Monarchs : Wild Sheep and Wild Goats in the Himalayas*, Chicago : University of Chicago Press, p. 48.
20. See ref. 2 and 3.
21. Rhandawa, M.S. (1944-1945). 'Progressive Desiccation of Northern India in Historical Times', *Bombay Natural History Society Journal*, pp. 558-565.
22. Ali, Salim and S. Dillon Ripley (1987). *Compact Handbook of the Birds of India and Pakistan, Together with Those of Bangladesh, Nepal, Bhutan and Sri Lanka*, Delhi : Oxford University Press, 10 Vols. xlii in p. 737 and 104 end plate.
23. Kurup, G.U. (1975). 'Status of the Nilgiri Langur, *Presbytis johni*, in the Anamalai, Cardamom, and Nilgiri Hills of the Western Ghats, India', *Bombay Natural History Society Journal* 72, pp. 21-29.
24. Krishnan Madhaviah. (1971). 'An Ecological Survey of the Larger Mammals of Peninsula India', *Bombay Natural History Society Journal*, 68, pp. 503-555, ref. p. 537 and photograph 47a, opposite p. 539).
25. Kurup, G. U. (1978). 'Distribution, Habitat and Status Survey of the Liontailed Macaque, *Macaca silenus* (L.)', *Bombay Natural History Society Journal*, 75, pp. 321-340.
26. Ranganathan, C.R. (1939). 'Studies in the Ecology of the *Shola* Grassland Vegetation of the Nilgiri Plateau', *Indian Forester*, 65, pp. 523-541.
27. Bor, Noel L. (1938). 'The Vegetation of the Nilgiris', *Indian Forester*, 64, pp. 600-609.
28. Billings W.D. and A.F. Mark (1957). 'Factors Involved in the Persistence of Montane Treeless Balds'. *Ecology*, 38, pp. 140-142.
29. See ref. 1, pp. 59, 61.
30. Champion, Harry G. (1935). 'A Preliminary Survey of Forest Types of India and Burma', *Indian Forest Records, New Series, Silviculture, Vol.1*, No.1, Delhi : Government of India, 1961 reprint; Fyson, P.F. (1932). *The Flora of the South Indian Hill Stations*, Madras: Superintendent, Government Press 2 Vols., V.1 description with p. 697, V.2., p. 611 illustration; Gamble, J.S. (1967) *Flora of Madras*, Calcutta : Botanical Survey of India, second reprinted edition in 3 Vols; Meher-Homji, V.M. (1965). 'Ecological Status of the Montane Grasslands of the South Indian Hills : A Phytogeographical Reassessment', *Indian Forester*, 91, pp. 210-215 ; Meher Homji, V.M. (1967). 'Phytogeography of the South Indian Hill Stations', *Torrey Botanical Club, Bulletin*, 94, 4, pp. 230-242; Meher Homji, V.M. (1972). 'Himalayan Plants on South Indian Hills : Role of Pliestocene Glaciation *versus* Long Distance Dispersal', *Science and Culture*, 38, 1, pp. 8- 12 ; Meher Homji, V.M. (1975). 'On the Montane, Species of Kodaikanal, South India', *Phytocoenologia* 2, 1-2, pp. 28-39; Meher Homji, V.M. (1987-88). 'Ecological Diversity in the Nilgiri District', *Journal of Palynology*, 23-24, pp. 159-166 ; Shankarnarayan, K.A. (1958). 'The Vegetation

of the Nilgiri', *Journal of Biological Sciences*, (Bombay), 1, pp. 90-98; Shankarnarayan, K.A. (1959). 'The Vegetation of the Nilgiris', *Journal of Biological Science*, (Bombay), 2, pp. 14-18.

31. Lengerke, Hans J. Von (1976). *Climatological Survey of the Nilgiris Area (South India)*, Heidelberg : University of Heildelberg, A Dissertation, pp. 174, 179 and 188.
32. Rao, A.S. (1974). 'The Vegetation and Phytogeography of Assam-Burma', in N.S. Mani (ed). *Ecology and Biogeography in India, Monographic Biologicae*, Vol. 23, The Hague : Dr. W. Junk b.v. Publishers.
33. See ref. 1, pp. 54-61.
34. Vishnu-Mettre and Gupta H.P. (1970). 'The Origin of *Shola* Forest in the Nilgiris, South India', *The Palaeobotanist*, 19, pp. 110-114.
35. Gupta, H.P. (1971), 'Quaternary Vegetational History of Ootacamund, Nilgiris, South India', *The Palaeobotanist*, 20, pp. 74-90; Gupta, H.P. (1989). '*Sholas* in South Indian, Montane : Past, Present and Future', *Palaeobotanist* 38, pp. 394- 408.
36. Sukumar, R. *et al.* (1993). 'A C13 Record of Late Quaternary Climate Change from Tropical Peats in Southern India', *Nature*, 364, pp. 703-706.
37. See ref. 36, pp. 705.
38. See ref. 35, pp.705.
39. Menon, Vimala K. (1966-67). 'Pollen Analysis of the Post-Glacial Deposits from Pykara, Ootacamund, India', *Journal of Polynology*, 1 and 2 , pp. 1-10.
40. Blasco, F. and Thanikaimoni G. (1974). 'Late Quaternary Vegetational History of Southern Region [India]', in K.R. Surange, R.N. Lakhanpal, and D.C. Bhardwaj, *Aspects and Appraisal of Indian Palaeobotany*, 632-643, Lucknow : Birbal Sahni Institute of Palaeobotany, pp. 674.
41. See ref. 39.
42. Noble, William A. (1989). 'Nilgiri Prehistoric Remains', in Paul Hockings (ed.) *Blue Mountains : The Ethnology and Biogeography of a South Indian Region*, Delhi : Oxford University Press, pp. 102-132 and end plates 13,14 and 15.
43. Rivers, William H.R. (1906). *The Todas*, London : Macmillan & Co. Ltd., pp. xviii, 755, Geneology lists.
44. Noble, William A. (1977). 'Settlement Patterns and Migrations Among Nilgiri Herders, South India', *Journal of Tropical Geography*, 44, pp. 57-70 and plate1.
45. Meher-Homji, V.M. (1987-88). 'Ecological Diversity in the Nilgiri District', *Journal of Polynology*, 23-24, p. 159.

27

CONSERVATION PRACTICES AND RELIGIOUS IDIOM

A Case Study of Bishnois of India

M.H QURESHI AND SURESH KUMAR

Introduction

Human well-being significantly depends upon the availability of natural resources. But, the natural resource-base in many countries of the world is rapidly depleting and creating very serious problems for the sustenance of every life-form. Mismanagement of natural resources is the main reason for this state of affairs. This mismanagement has occurred due to a variety of reasons, some of which are, unfortunately, justified by present day technique of neo-classical economic analysis.[1] Thus, neo-classical economics, which has a profound and pervasive effect on development policies the world over, miserably failed in sustainable use of natural resources. Indian policies after independence were also moulded by Nehru's vision of market economy, while Gandhi's ideas of India as a country of self-reliant villages was not given a trial. However, the failure of most of the development projects, search for viable and sustainable alternatives to current models of resource use, and new found pride in traditional values and institutions have led to renewed interests in the local level community based resource use and management systems.[2] Many leading international, governmental, and non-governmental organisations have shown tremendous interest in it.

Many community based resource conservation/rational utilisation systems have collapsed mainly due to increased participation in market economies, breakdown of traditional value systems, population growth, increasing centralisation of power and application of inappropriate pricing, subsidies, legislation or other governmental incentives.[3] However, there are still many communities the world over which are protecting their natural resources through their own indigenous methods. The well known examples are the 'Tuarege' of Niger, the 'Kuna' of Panama, the 'Amish' in the Appalachians, the 'Bishnoi' of India and the 'Kayapo' of Brazil. Unfortunately, perhaps due to the vested interests of people in the 'iron triangle', such conservation efforts by local communities remained isolated and were not taken note of. But, now it is widely recognised that for the very survival of humanity, it is essential to understand the conservation efforts of local communities in order to resurrect their traditional resource use and management systems.

The present study is an attempt in this direction, as it examines the evolution of environmental conservation concerns in the 'Bishnoi' community of India. It also empirically evaluates the present perception about, and actual position of age-old conservation ethos among the Bishnois located in the arid and semi-arid parts of Rajasthan and Haryana states.

Evolution of Conservation Ethos

In most parts of the *Thar* desert of India, herding of cattle, camel and small ruminants is the basic reliable source of sustenance. But, the excessive browsing and grazing, cutting of trees and hunting put a question mark not merely to desert ecology but also to the survival of this source of their livelihood especially during periods of drought. Long back, during the late fifteenth century, this fact was realised by a young man named Jambheshwar, the son of the headman of Pipasar village in the Nagaur district of Rajasthan. He was a cattle herder and lived in the lap of nature. It is believed that during one prolonged drought when people and animals starved to death, he realised that people themselves created their plight by cutting and over-browsing trees. This led him to give up his inheritance and undertake a mission to teach the people to care for their health and environment.[4] In 1485 A.D. he prescribed twenty-nine tenets which his followers had to strictly follow.[5] The followers by virtue of their accepting the twenty-nine (*Bees+nau*) tenets came to be known as Bishnois. Among these tenets two are specifically oriented towards the conservation of environment. These include one which forbids Bishnois from cutting trees and another, which forbids them from killing the animals.[6] The rest of the tenets pertain to morally accepted conduct which is more universal in the Indian context.

The preachings of this young man had quick and profound effect on the people and he enjoyed considerable following during his life time. Today, this religious sect comprises of about tens of thousands practising Bishnois[7] and is mainly confined to the ecologically fragile region of north-western Rajasthan and adjoining parts of Haryana and Punjab. In modern times, when forest and wildlife have become truism, this community, under the influence of its religious tenets, has been protecting the trees and wildlife with the same enthusiasm and commitment.[8] As a consequence, in the midst of the vast *Thar* desert, the Bishnoi villages are refreshingly greener and appear to be sanctuaries of wildlife.[9] Both the cultivated and uncultivated lands in Bishnoi villages abound in *Khejari*

(Prosopis cineraria) trees. The flocks of deer and bluebulls can be seen freely roaming in the fields of the Bishnois.

Bishnois are so committed to protect the green trees, especially the *Khejaria*, and wild animals, that they not only themselves avoid their felling and killing, but if someone tries to cut the tree and hunt the animals, they always protest and sometimes there are violent protests as well. There have been many incidents in which they individually or jointly have protested against the killing of wild animals and the felling of trees.[10] In these confrontations, they have assaulted and killed many poachers and sometimes have even sacrificed their own lives to save the environment. The highly tragic and heroic incident, in which the Bishnois demonstrated their convictions to the protection of environment in a truly remarkable way was that of Khejadli village in the Jodhpur district of Rajasthan in 1730 A.D.[11] It is recorded that during 1730 A.D. Maharaja Abhay Singh of Jodhpur decided to build a fort and was in need of firewood to get the lime-kilns burning for making mortar. There were very few trees around Jodhpur except in the Bishnoi village of Khejadli. Hence, the Maharaja's Minister, Girdhar Das, sent his axemen to get the Khejari trees of this village felled. But, the Bishnois under the leadership of Bishnoi Ardoji forbade them to cut the trees. In those days of monarchy, disobeying of king's order meant very harsh punishment. The villagers, being aware of the fact that Girdhar Das would send troops along with axeman, immediately convened a meeting of the *Sarpanchas* (village headmen) of the surrounding 84 villages to seek their help to save the Khejari trees. Back in Jodhpur, when Girdhar Das was told that the villagers did not allow the trees to be cut, he was so infuriated that he personally proceeded to Khejadli village along with troops and axemen to teach the villagers a lesson. However, as the axemen moved to cut the tree, a Bishnoi woman named Amrita Devi with her three daughters hugged the trees to save them. The Minister was so angry that he ordered his men to cut the women along with the trees. The troops obeyed and did so, but it motivated others to rush forward to hug trees and get killed. It is said that 363 women, men and children were killed in this incident. One couple who got married on the same day also got killed while hugging a tree. The people were inspired by the slogan propagated by Guru Jambheshwar '*sir shante aur rukh rahen tohi sasto jan* (if the trees can be saved by getting oneself beheaded, even then the bargain is cheaper). Stones mark some of the sites in the village where trees were felled and people killed. In the ecological history of India and in the evolution of the concept of *chipko* (saving trees from being felled by embracing them) this has played a leading role.[12]

Hearing of this incident the king was shocked and rode to Khejadli village. He apologised to the Bishnois and proclaimed that no trees would be felled and no wild animal would be killed in Bishnoi villages. His proclamation was inscribed on a copper plate (*Tamra Patra*) which the villagers still hold. It authorised the Bishnois to challenge the hunters, including members of the Royal family.[13] Today, the site of the mass graves of the Bishnois killed in 1730 in Khejadli is known as '*Jal Nadia*'. A small shrine of Guru Jambheshwar has been built and 363 trees have been planted there in the memory of the deceased. Every year on the full moon (*Purnima*) of the month of *Bhadon*, a five day fair in the village is held to commemorate the historic sacrifice by the community.

Present Scenario of Conservation Ethos

In order to capture the present scenario of conservation practices and behaviour among Bishnois, an empirical exercise was conducted to collect data from the historical village of Khejadli in Jodhpur district of Rajasthan during July 1995. For a detailed household survey, a sample of 49 households (about 11 per cent of the total) were selected using stratified random sample design. The village community was divided into the Bishnoi and non-Bishnoi strata. However, keeping in view the purpose of the study, a higher representation was given to the Bishnoi than the non-Bishnoi households in the sample. Thus, a sample of 20 per cent was taken from Bishnoi households, while only about 6 per cent from the non-Bishnoi households. The household level information, including sources of motivation for conservation and its intensity, level of awareness, extent of commitment, changes therein etc. were collected from the sample households by canvassing a semi-structured questionnaire. This was then taken as the basis for analysis.

Table 27.1 : Level of the Awareness of the Historical Incident in which the Bishnois Sacrificed Their Lives to Save Trees

Response	*Number of households and their percentage to the total households*					
	Bishnois		*Non-Bishnois*		*Total*	
	No.	*Per cent*	*No.*	*Per cent*	*No.*	*Per cent*
Fully aware	21	70.00	8	42.10	29	59.18
Partially aware	2	6.67	2	10.53	4	8.16
Not aware	7	23.33	9	47.37	16	32.66

To begin with, the empirical data gathered reveals the level of awareness of this incident in Khejadli village during 1995 as indicated in Table 27.1, which shows that about 77 per cent (70 per cent fully and 7 per cent partially) of Bishnoi households were aware of the episode as their historical heritage. The level of awareness in non-Bishnoi respondents was, as obvious, low. But as seen from the table a considerable proportion of Bishnoi households (23 per cent) were not aware about this incident.

Table 27.2 : Preferences of Reverence to Wildlife

Wildlife	*Number of Households and their percentage to the total households*											
	First preference						*Second preference*					
	Bishnois		*Non—Bishnois*		*Total*		*Bishnois*		*Non—Bishnois*		*Total*	
	No.	*Per cent*	*No.*	*Per cent*	*No.*	*Per cent*	*No.*	*Per cent*	*No.*	*Per cent*	*No.*	*Per cent*
Deer	29	96.67	19	100.00	48	97.96	1	3.33	—	—	1	2.04
Peacock	—	—	—	—	—	—	7	23.34	8	42.11	15	30.16
Bluebull	—	—	—	—	—	—	11	36.67	2	10.52	13	26.53
Rabbit	—	—	—	—	—	—	10	33.33	9	47.37	19	38.78
Others	1	3.33	—	—	1	2.04	1	3.33	—	—	1	2.04

As has been emphasised earlier, Khejari is the most revered plant for Bishnois. The deer and the bluebull are their sacred animals and get their maximum protection.[14] Table 27.2 presents preferences of sample households for reverence to wild animals in Khejadli village. Deer is a highly revered animal in the village as all households except one assigned first preference in showing reverence to it. Moreover, that one household also assigned second preference to it. The peacock, bluebull and rabbit are other wildlife which were assigned second place in terms of reverence by the people.

The concern towards their religious belief of not cutting trees and killing wildlife is very deep among Bishnois. About 97 per cent of Bishnoi households believed that cutting a tree was a sin (Fig. 27.1). Moreover, all sample households of Bishnois confirmed their reverence towards wildlife (Fig. 27.2). Interestingly, non-Bishnoi households of Khejadli village, with marginal variation in degree, were following similar norms towards trees and wildlife. It is clearly the policy of accepting the norms followed by the majority community of the village.

The extent of commitment to preserve vegetation cover is very high among Bishnois of Khejadli village. A large majority of Bishnoi households (80 per cent) refrained from cutting even the branches of trees (Table 27.3). However, 10 per cent of sample households of Bishnois refrained from cutting trees although they normally chop-off the branches. In a clear departure from the community's accepted norms, about 10 per cent of the households stated that they could cut trees if the need arose. The proportion of households who refrained from cutting both trees as well as branches is lower among the non-Bishnoi stratum of the village.

Table 27.3 : Refrain from Cutting Trees and Their Branches

Refrain from Cutting	*Number of households and their percentage to the total households*					
	Bishnois		*Non-Bishnois*		*Total*	
	No.	*Per cent*	*No.*	*Per cent*	*No.*	*Per cent*
Trees but not branches	3	10.00	8	42.11	11	22.45
Both trees and branches	24	80.00	9	47.37	33	67.35
None of the above	3	10.00	2	10.52	5	10.20

Most of the modern ecology movements in India have emerged as the people's response to the threat to their survival and as a demand to save their vital life support system.[15] But the source of motivation for conservation is spiritual in nature in the case of Bishnois. About 82 per cent of the total

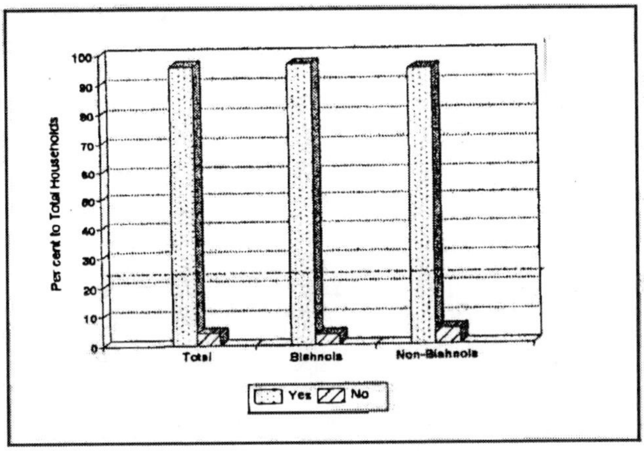

Fig. 27.1 : Pattern of Belief that Cutting Trees is a Sin

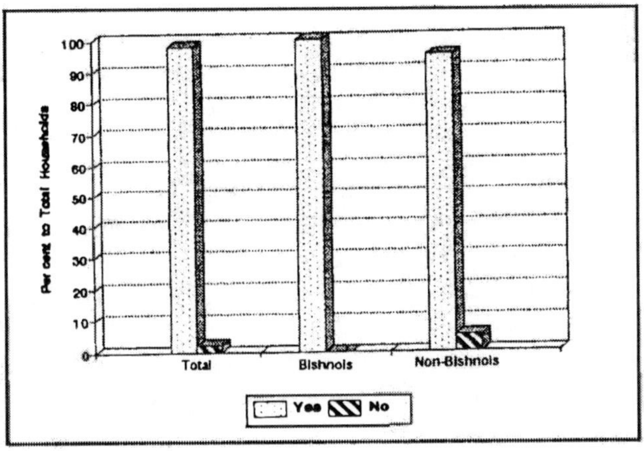

Fig. 27.2 : Pattern of Reverence to Wild Life

households admitted religious idiom as the source of motivation for tree conservation (Fig. 27.3). The proportion of such households was higher (97 per cent) among the Bishnois than non-Bishnois (58 per cent). In other words, only 3 per cent of Bishnoi households accepted tradition as the source of motivation for tree protection, while corresponding proportion is higher (42 per cent) in case of non-Bishnoi households.

Similarly, for the protection of wildlife, the community is motivated neither by any on-going national of international conservation movements nor by any governmental policy. In fact, it is their religious belief which has motivated them for it. All the sample Bishnoi houses of Khejadli village averred religion as a source of motivation (Fig. 27.4). However, the opinions of non-Bishnoi households about the source of motivation for wildlife conservation come from both the religious beliefs as well as traditional norms. This perhaps emanates from the fact that they are following the beliefs, norms and values observed by the majority community without going into details of their sources of evolution. When confronted with further supplementary question as to whether or not they would have observed similar norms in case they had been located in a non-Bishnoi village, they expressed the view that, in such a situation, they would not have hesitated in removing trees and chasing wild animals. Thus, simply following the majority community is an obvious conclusion.

The responses about benefits provided by trees and wildlife (Tables 27.4 and 27.5) indirectly suggests that it was not their demand for food, fodder, fuelwood, timber etc. which motivated them for tree and wildlife conservation but their religious beliefs. Khejari, the most revered plant of the Bishnois, is an evergreen and long lasting friend of the locals. It is said to be the *Kalpa variksha* (the mythological tree of munificence) of the Indian desert, as almost every part of the tree is used in one form or the other. It is an important source of very nutritious fodder known as '*loong*' (dry leaves of the tree). Its fruit is also used as vegetable (*sangri*). Its long roots, by drawing moisture from deeper layers of the soil, improve soil moisture and do not adversely affect the crop under its shade. It is observed that the density of crop plants under a Khejari tree is more than in other parts of the field. Also falling leaves seem to provide a very effective mulch effect and protect the crop from direct heat and the desiccating effect of wind. Its fruit attracts a large number of birds whose droppings fertilise the soils. However, the knowledge of the villagers about the indirect benefits which they receive from the tree is limited. The main benefits perceived by both the Bishnoi and non-Bishnoi households from the trees is their fruit and shade as indicated by a large proportion of total benefit frequencies assigned to them in the study village. The second level benefits perceived by them include clean air and fuelwood which recorded about 13 and 8 percent of total benefit frequencies (Table 27.4).

In case of wildlife, more than 70 per cent households across both social categories were of the view that they obtained no benefit from wild animals (Table 27.5). They considered wildlife as a burden in economic terms and added that it was only for the sake of their religion that they were protecting them. About 27 per cent of the Bishnoi and 16 per cent of non-Bishnoi households perceived some minor benefits provided by wildlife. Beauty provided by the peacocks and rabbits, the manure from wildlife droppings and a comfortable place in heaven, were the perceived benefits mentioned by the respondents. These benefits are more psychological than economic in nature.

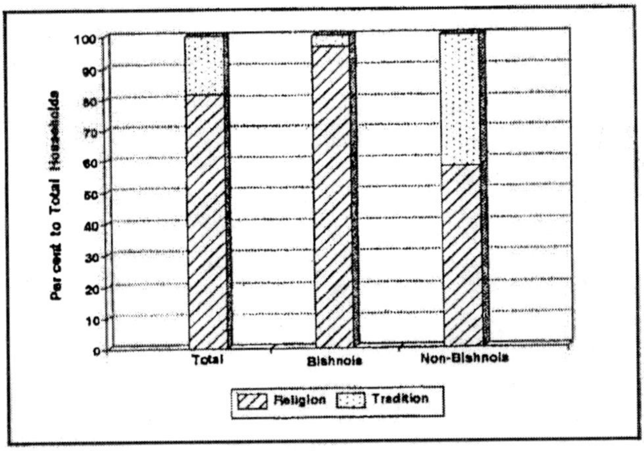

Fig. 27.3 : Sources of Motivation for Tree Protection

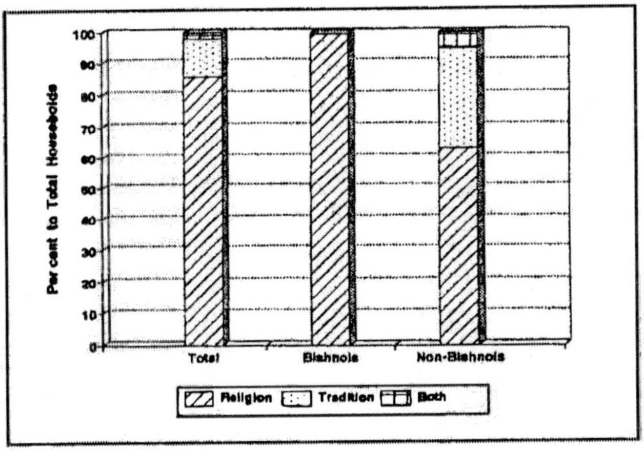

Fig. 27.4 : Sources of Motivation for Wildlife Conservation

Table 27.4 : Perception of Benefits from Trees

Benefits	*Number of frequencies and their percentage to the total frequencies*					
	Bishnois		*Non-Bishnois*		*Total*	
	No.	*Per cent*	*No.*	*Per cent*	*No.*	*Per cent*
Shade	29	22.83	16	26.23	45	23.94
Fuelwood	6	4.72	9	14.75	15	7.98
Clean air	16	12.60	8	13.12	24	12.76
Fruits	41	32.28	14	22.95	55	29.25
Fodder	4	3.15	5	8.20	9	4.79
Others	31	24.42	9	14.75	40	21.28

Table 27.5 : Perception of Benefits from Wildlife

Response	*Number of households and their percentage to the total households*					
	Bishnois		*Non-Bishnois*		*Total*	
	No.	*Per cent*	*No.*	*Per cent*	*No.*	*Per cent*
Some benefits	8	26.67	3	15.79	11	22.45
No benefits	21	70.00	15	78.95	36	73.47
Not response	1	3.33	1	5.26	2	4.08

Changes in the Conservation Ethos

Many ecologically and socially sound traditional norms or religious beliefs of local communities have been withering away mainly because of either the greed and compulsions of the community as a whole or of its certain sections, or due to outside intervention. The growth of population and integration of the local community with the outside world, *i.e.*, mainly market economy, during the last few decades are some of the formidable reasons for it.[16] The views of the sample households regarding the effects of increasing population pressure, and interaction with the outside world, on their conservation beliefs during the last one decade, were collected.

It is known that population pressure compels people to expand the net sown area as well as the area under settlements. This leads to the shrinkage in the land under forests, pastures and fallows which are the available habitat for wildlife. Population pressure also forces the people to overuse or misuse the natural resource base. Perhaps it was only due to their strong convictions that above 70 per cent of the respondents in Khejadli village perceived that mounting population pressure had no impact on the conservationist attitude towards trees and wildlife (Table 27.6). Interestingly, some households believed that increasing population pressure had a positive effect on their conservation ethos. In their opinion with large number of people they could now easily save trees and wildlife from outside attacks. Thus, very few households, felt that population pressure had negatively affected their conservation policy during the last decade.

The modern high-tech transportation and communication facilities have significantly enhanced the interaction among different communities. With this increased interaction, many age-old perceptions of the people about their beliefs, norms or values have undergone a remarkable change. Surprisingly, more than 75 per cent of the Bishnoi households in the Khejadli village did not feel that their interaction with the outside world had adversely affected their attitude towards tree and wildlife conservation (Table 27.7). The proportion of such households is slightly low amongst non-Bishnois. Moreover, some households averred that their interaction with other communities had in fact, reinforced

Table 27.6 : Effects of the Population Pressure on the Strategies of Tree and Wildlife Conservation

Response	*Number of Households and their percentage to the total households*											
	First preference						*Second preference*					
	Bishnois		*Non-Bishnois*		*Total*		*Bishnois*		*Non-Bishnois*		*Total*	
	No.	*Per cent*	*No.*	*Per cent*	*No.*	*Per cent*	*No.*	*Per cent*	*No.*	*Per cent*	*No.*	*Per cent*
Positive change	1	3.33	1	5.26	2	4.08	4	13.33	1	5.26	5	10.20
Negative change	2	6.67	—	—	2	4.08	4	13.33	1	5.26	5	10.20
No change	21	70.00	16	84.2	37	75.51	19	63.34	16	84.21	35	71.43
No response	6	20.00	2	10.53	8	16.33	6	20.00	2	10.53	8	16.33

Table 27.7 : Influence of Interaction of Community with outside World on the Strategies of Tree and Wildlife Conservation

Response	*Number of Households and their percentage to the total households*											
	Tree Protection						*Wildlife conservation*					
	Bishnois		*Non-Bishnois*		*Total*		*Bishnois*		*Non-Bishnois*		*Total*	
	No.	*Per cent*	*No.*	*Per cent*	*No.*	*Per cent*	*No.*	*Per cent*	*No.*	*Per cent*	*No.*	*Per cent*
Positive change	1	3.33	1	5.26	2	4.08	1	3.33	—	—	1	2.04
Negative change	1	3.33	—	—	1	2.04	1	3.33	—	—	1	2.04
No change	25	83.33	13	68.42	38	77.55	23	76.67	12	63.21	35	71.43
No response	3	10.00	5	26.32	8	16.33	5	16.67	7	36.84	12	24.49

their dedication towards trees and wildlife conservation. However, one respondent was of the opinion that their interaction with communities, which were economically sound but were not concerned about ecology, had a negative effect on their attitude to environment and wildlife conservation. He regretted that the educated youth, owing to their greater interaction with the outside world, were not so committed to their religious tenets.

Conclusion

The Bishnois represent one of those communities the world over who are motivated by their religious idiom to conserve environment. The level of awareness leading to high degree of commitment to the cause of environmental conservation is observed even amongst the illiterate villagers. The household level analysis in Khejadli village reveals that more than three-fourth of the sample Bishnoi households were aware of the historical incident of mass sacrifice by the community. A substantial majority of Bishnoi households in this village not only believed that cutting a tree was a sin but actually refrained from cutting even their branches. Moreover, all the Bishnoi households expressed their very strong reverence towards wildlife. The deer is a highly revered animal among the Bishnoi as well as non-Bishnoi households of the village. A large proportion of Bishnoi households believed that the religious tenets propounded by Guru Jambheshwar in the 15th Century had been the main source of motivation for tree and wildlife protection. Their limited knowledge about the benefits of tree wealth further confirmed that it was not the economic consideration but the conservation ethos which inspired them to preserve the trees and wildlife. The conviction is so strong that above 70 per cent of the respondents believed that increasing population pressure and interaction of the community with the outside world during the last decade have had no impact on their commitment to the conservation ethos.

Religious beliefs and traditional norms, normally, wither with progress and development. However, the Bishnois are by no means a primitive community, yet they continue to follow their religious tenets of environmental protection with a lot of enthusiasm. Some deviation is observed but we can hope for the continuity of their conservation idiom in the future too with perhaps some minor changes in emphasis.

References and Notes

1. Goodland, R. *et al* (1989). 'Meeting Environmental Concerns Caused by Common-Property Mismanagement in Economic Development Projects', in F. Berkes (ed.), *Common Property Resource : Ecology and Community Based Sustainable Development*, London: Belhaven Press, pp. 148-163.

2. Berkes F. and Farvar, M.T. (1989). 'Introduction and Overview', in F. Berkes (ed.) *Common Property Resources : Ecology and Community Based Sustainable Developments*, London Belhaven Press, pp. 1-17.
3. Bromley, D.W. and Chapagain, D.P. (1984). 'The Village Against the Center : Resource Depletion in South Asia', *American Journal of Agricultural Economics*, Vol. 66, No. 5, pp. 868-873 ; Berkes, F. (1985). 'Fisherman and the Tragedy of the Commons', *Environmental Conservation*, Vol.12, No. 3, pp. 199-206 ; Gadgil, M. (1989). *Deforestation : Problem and Prospects*, The Foundation Day lecture delivered on May 12, 1989 at the Society for Promotion of Wasteland Development, New Delhi.
4. Sankhala K.S. and Jackson, P. (1985). 'People, Trees and Antelopes in Indian Deserts', in J.A McNeely and D. Pitt (eds.) *Culture and Conservation : The Human Dimension in Environmental Planning*, London : Croom Helm, pp. 205-210.
5. The 29 Doctrines of Bishnoi Religious Sect which had to be strictly observed by the followers are:
   (i) The lady should take rest for thirty days after giving birth to a child (*sutak*).
   (ii) The lady who is in period should take rest and should not perform household duties for five days.
   (iii) All people should take bath in the morning.
   (iv) All men should always observe the best principles of character, reflection and satisfaction.
   (v) People should offer prayer (*sandhya*) in the morning as well as in the evening.
   (vi) In the evening, people should meditate over the virtues of God.
   (vii) To realise the Purusharth *i.e. Dharma*, wealth (*Artha*), sex (*Kam*) and salvation (*Moksha*). *Havana* should be performed.
   (viii) One should not steal.
   (ix) One should not do back biting.
   (x) One should not indulge in baseless talks and unnecessary debate.
   (xi) One should not lie.
   (xii) For the purpose of internal purification, one must observe fast on the day of *Amavashya* (moonless night of the month).
   (xiii) People should have compassion towards living beings.
   (xiv) People should not cut the green trees.
   (xv) One should drink water and milk only after filtering it with cloth.
   (xvi) The firewood should be purified before using it for *Yajna* and cooking.
   (xvii) All should always speak politely.
   (xviii) Desire, anger, greed and delusion (*Moh*) are formidable enemies and these must be kept under control.
   (xix) Maintaining tolerant attitude, one should practice forgiveness.
   (xx) *Pahal* ceremony is a must for all the people of the sect. *Pahal* is water purified by recitation of some *mantras*. It is administered to those who want to join the sect of Bishnois, a sort of baptism.
   (xxi) All people should worship and praise the omnipresent *Vishnu*.
   (xxii) Animals should not be sold to butchers.
   (xxiii) Bullocks should not be castrated.
   (xxiv) Opium should not be taken.
   (xxv) Tobacco should not be used in any form.
   (xxvi) Cannabis (*ganja and bhang*) should not be used.
   (xxvii) One should never drink wine.
   (xxviii) One should not eat meat.
   (xxix) One should never wear blue clothes.
6. *See* ref. 5, Nos. 14 and 22.
7. *See* ref. 4.
8. Qureshi, M.H. *et al.* (1995). *Bishnoi Community and Environmental Conservation: A Study of Their Perception, DSA Working Paper No.3*, New Delhi: Centre for the Study of Regional Development, Jawaharlal Nehru University.

9. Gadgil M. (1985). 'Social Restrains on Resource Utilization : The Indian Experience', in J.A. McNeely and D. Pitt. (eds.) *Culture and Conservation : The Human Dimensions in Environmental Planning*, London : Croom Helm, pp. 135-154,. ref. p. 144.
10. Dharniya, S.R. (ed). (1985). *Panchshati Smarika*, Mukam (Bikaner) : Akhil Bhartiya Bishnoi Mahasabha, (in Hindi).
11. Dharniya, S.R. (ed.) (1985). *Panchshati Smarika*, Mukam (Bikaner) : Akhil Bhartiya Bishnoi Mahasabha (in Hindi) ; Sankhala, K.S. and Jackson, P. (1985). 'People, Trees and Antelopes in Indian Deserts', in J.A. McNeely and D.Pitt. (eds), *Culture and Conservation : The Human Dimension in Environmental Planning*, London : Croom Helm, pp. 205-210 ; Gadgil, M. (1985). 'Social Restrains on Resource Utilization : The Indian Experience', in J.A. McNeely and D. Pitt. (eds.) *Culture and Conservation : The Human Dimensions in Environmental Planning*, London : Croom Helm ; Bahuguna, Sunderlal, (1989). 'Environmental Conservation for Survival', *Indian Journal of Public Administration*, Vol. 35, No. 3, pp. 374-379 ; Gadgil, M. and Guha, R. (1992). *This Fissured Land : An Ecological History of India*, New Delhi, Oxford University Press ; Qureshi, M.H. *et al* (1995). *Bishnoi Community and Environmental Conservation : A Study of Their Perception, DSA Working Paper No.3*, New Delhi: Centre for the Study of Regional Development, Jawaharlal Nehru University.
12. Bahuguna, Sunderlal (1989). 'Environmental Conservation for Survival', *Indian Journal of Public Administration*, Vol. 35, No. 3, pp. 374-379; Shiva V. *et al* (1991b). *Ecology and the Politics of Survival : Conflicts Over Natural Resources in India*, Tokyo : United Nations University Press and New Delhi : Sage Publications India Pvt. Ltd. pp. 19-22.
13. Sankhala, K.S. and Jackson, P. (1985). 'People, Trees and Antelopes in Indian Deserts', in J.A. McNeely and D.Pitt (eds.) *Culture and Conservation : The Human Dimension in Environmental Planning*, London : Croom Helm, pp. 205-210; Dharniya, S.R. (ed). (1985). *Panchshati Smarika*, Mukam (Bikaner): Akhil Bhartiya Bishnoi Mahasabha.
14. *See* ref. 8.
15. *See* ref. 12.
16. Berkes, F. and Farvar, M.T. (1989). 'Introduction and Overview,' F. Berkes (ed), *Common Property Resources : Ecology and Community Based Sustainable Development*, London Belhaven Press, pp. 1-17 ; Gadgil, M. (1989). 'Deforestation Problem and Prospects,' The Foundation Day lecture delivered on May 12, 1989 at the Society for Promotion of Wasteland Development, New Delhi ; Gagdil, M. and Guha, R. (1992). *This Fissured Land : An Ecological History of India*, New Delhi : Oxford University Press.

28

CULTURAL ADAPTATION IN THE RIVER ISLANDS (*CHAR* AREAS) OF THE BRAHMAPUTRA (ASSAM)

Abani K. Bhagbati

Introduction

The bed of the great river Brahmaputra, for a length of 720 km is dotted with innumerable river islands, locally known as *chars* or *chaparies*. These *chars* vary greatly in size, shape and physical vulnerability. However the Assam State *Char* Area Development Authority has identified some 1,200 *chars* having some degree of physical and cultural significance. Characterised by frequent physical changes resulting from seasonal fluctuation of the fluvial process, these *chars* are capable of supporting only some varieties of riverine grasses, bushes and certain aquatic plants and thus, constitute a unique, but highly sensitive natural habitat for birds and wild animals. These natural habitats, till late, were not under direct human occupance and in fact, considerably away even from any remarkable human influence. It was only at the beginning of the 1900s that people, especially those who migrated from erstwhile East Bengal, started transgressing the hitherto uninhabited *chars* of the Brahmaputra. In fact, the similarity of the Brahmaputra riverine ecology with that of East Bengal easily attracted the land-hungry peasants from the latter. Concomitantly, some Nepali graziers, after being informed primarily by the Nepali soldiers recruited by the British Government, about the presence of luxuriant grasslands along the Brahmaputra river, also slowly and silently started stepping into the *char* areas. Thus, the process of humanisation in the Brahmaputra *chars* began first at the western end of the valley and gradually spread upwards to cover its central and, more recently, the upper part as well. Interestingly most of the inhabited *chars*, have been named by immigrants after a feature or a species or an event of local ecological significance.

At present, almost all the permanent *chars* of the Brahmaputra are under direct human activities. Sedentary as well as temporary, seasonal dwellers of these chars have already adopted economic pursuits, especially agriculture, suited to the peculiar physical make-up of their habitats. Simultaneously roads, periodic markets (*hats*), schools and other public institutions have also come up in some of the larger *chars*. The settlements, agricultural patches and other secular institutions have resulted in a new cultural landscape quite distinct in identity and attributes from the other areas of Assam.

Although recent in evolution, the cultural landscape of the *chars* is being rapidly transformed. The transformation has been basically directed by the peculiar physical make-up of the *chars* and articulated eventually by the ability of the dwellers to adapt to the situation. More recently, the growing interaction between the *char*-dwellers and the neighbouring indigenous people and increasing relation with urban culture has perceptibly accelerated the process of ecologico-cultural transformation in the *char* areas. Consequently, the *char* areas have at present assumed special importance as a promising field of study as to how people have been exploring the possibilities of living provided by nature and the manner of their adaptation to the physical set-up of these areas. It is towards this end that this study is oriented.

Field investigation and interviews carried out in some selected *chars*, such as Baladmari *Char* in Goalpara District, Kalar *Char* and Bhanganmari *Char* in Nalbari District (Fig. 28.1) and Burhachapari in Sonitpur and Nagaon districts form the basis of this study. Relevant information has also been collected from written records at the Sub-Deputy Collectors' office, Block Development offices and the Assam State *Char* Areas Development Authority

Physical Context

The *chars* are an integral part of the very complex fluvial regime of the Brahmaputra river. The Brahmaputra, originating from the Tamchuk Khambab glacier in the Kailas Range of the Himalayas at an elevation of 5,300 metres flows through China, India (Arunachal Pradesh, Assam and West Bengal) and Bangladesh for a length of 2,880 km. In Assam, the river flows through a highly braided channel with numerous lateral and mid-channel bars (*chars*). These bars are composed of both bed load and suspended load and are generally triangular in shape. Their height, naturally, can never be greater than the height of the highest flood since they are built under flood environment.[1]

During floods the *chars* get submerged and some may even get completely wiped away by erosion. Generally, erosion is the dominant process in the upstream parts while deposition is characteristic at the downstream end. As a result, the *chars* migrate downstream. It is pertinent to note that sediment transport in the Brahmaputra increases with the progress of the flood cycle during the rainy season. In this season, the thalweg starts to change position resulting in a change in the geometry and location of the *chars*.[2] Thus, many of the *chars* are quite temporary or semi-permanent and only a few are permanent.

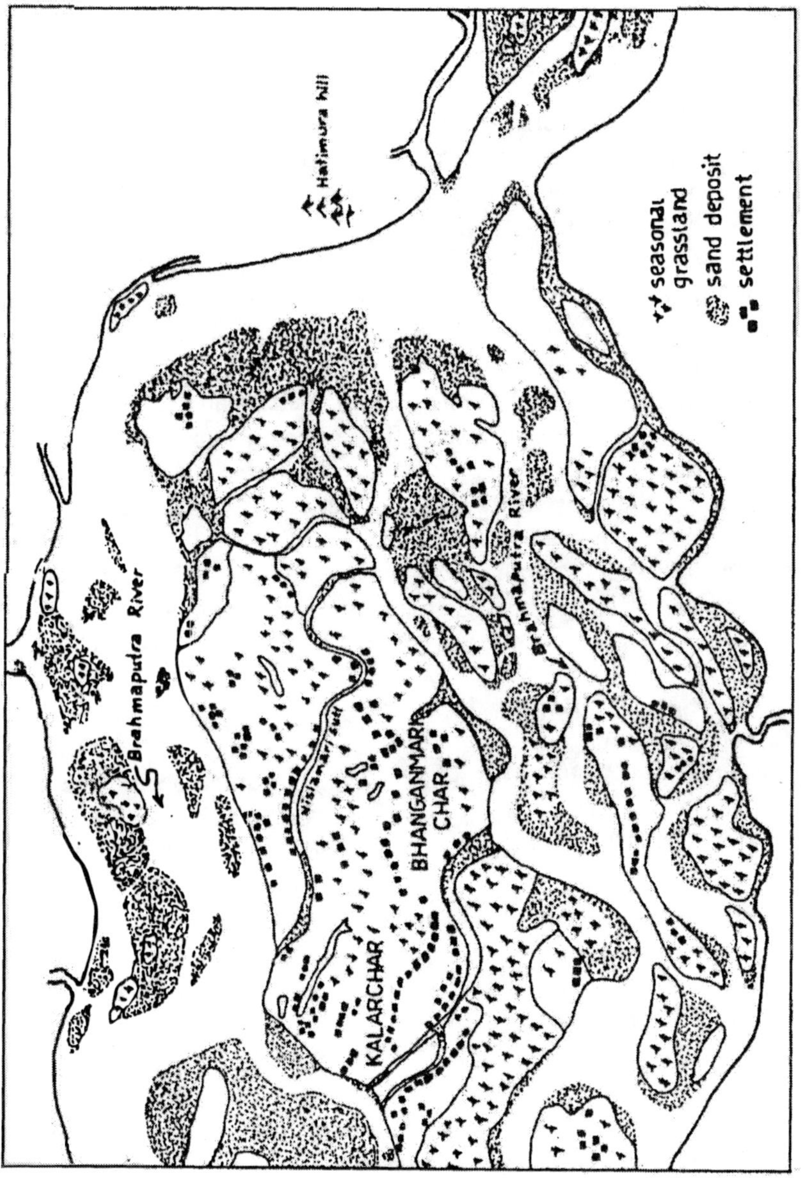

Fig. 28.1 : A Sketch of Kalarchar and Bhanganmari Char in the Brahmaputra River

Undergoing a distinctive process of soil formation, the soil type of the *chars* is different from that of other areas. The soil here is a mixture of fine sands, silt and clay freshly laid by the river. The uppermost layer of the *chars*, if regularly innundated, remains sandy with little humus content. But the sub-surface layer in which vegetal remains are abundant, is rich in humus and favours luxuriant growth of those plants which can extend their roots down to this layer. In most of the semi-permanent *chars*, the soil is normally azonal in character as it does not get sufficient time to develop a well-structured vertical profile. But the silt continuously deposited by the river make the soil very suitable for some fast growing seasonal plants.

A variety of shruby bushes and grasses characterise the natural vegetation of the Brahmaputra *chars*. As most of the *chars* are semi-permanent in nature, matured tall trees can rarely be found. The sandy soil, enriched with regular silt supply, favours growth of tall and sharp grasses often up to a height of 3-4 metres. Grasses such as *nal, Khagari, birina, kahua* and *kher* not only add to the scenic beauty of the *chars*, but also fulfil the demands for building material and fodder for domesticated animals. However, tall trees such as the *simul* and bamboo are found in the relatively bigger and permanent *chars*. Another type of hard-stemmed bushy vegetation called *jaoban* grows abundantly in the *char* areas. The waterlogged and marshy areas on the other hand, are the breeding grounds of water hycinth and some other aquatic flora, a number of which are, edible.

Background of Humanisation

Compared to other parts of the Brahmaputra Valley, human occupance in the riverine tract, especially the *char* areas, is of recent origin. Chronically affected by floods and poorly endowed by nature, the *chars* are, in fact, the last of the ecological niches to attract human habitation. Despite this, due to certain external as well as internal factors people have gradually moved into and settled down in these areas during the recent past. However, the settlers of the *chars*, with the exception of very few, are not of indigenous origin. They are mostly immigrants from erstwhile East Bengal (present Bangladesh) and Nepal. Thus, the present population inhabiting the *chars* of the Brahmaputra comprises immigrant Muslims, Bengali refugees, Nepali immigrants and a very small number of indigenous people, each with their distinctive cultural background creating their own characteristic cultural landscapes.

Immigrant Muslims : During the 15th Century, Assam, including the Brahmaputra valley, was regarded as a region having vast virgin tracts of forests and rich cultivable lands. The state, in fact, remained an area of low population density compared to the nearby province of Bengal (present Bangladesh).

In 1911, for instance, while the Brahmaputra Valley had a population density of 63 persons per sq.km, East Bengal recorded a density as high as 243 persons. The relatively low density areas of the valley naturally attracted a large number of landless peasants from the neighbouring areas, especially East Bengal, and as mentioned in the Line System Committee report, 'The influx into Assam of thousands of immigrants from the districts of Eastern Bengal particularly Mymensingh, in search of land on which to settle and make their homes, is a problem of no recent origin and one which has come under the consideration of Government on many occasions during the last twenty years'.[3]

The East Bengal Muslim immigrants mostly from the districts of Mymensingh, Rongpur, Pabna and Bagra entered Assam in thousands and started settling since about 1905 on the uninhabited riverine tracts.[4] The tempo of this movement of population gradually increased to reach culmination during and after the liberation war of Bangladesh in the early seventies. The immigrants thus moved further east along the Brahmaputra in order to reduce the unbearable pressure in the lower part of the valley. Thus, the Muslim immigrants have been able to reclaim large areas of cultivable waste and fallow lands including the reverine grazing lands and have now consolidated their position economically and culturally in the riverine tracts, especially of the lower and central Brahmaputra valley districts of the state.

Bengali Refugees : The migration of Bengali Hindu refugees from erstwhile East Bengal to Assam in huge numbers was more a consequence of political events than economic ones. The refugee migration was mainly triggered by communal conflicts that rocked East Bengal just after the partition of the country, and, on humanitarian grounds, many refugee colonies had to be opened. For instance, the Baghbor colony covering about 2,800 hectares in the former Barpeta sub-division, was established in 1950 while Gobindapur and Theka colonies in the same sub-division were established in 1965.[5] Besides the specified colonies, the Hindu refugees started occupying the uninhabited areas of the lower Brahmaputra valley including the *chars* close to the indigenous Hindus.

Another major influx of refugees took place during and after 1971 when East Pakistan became the independent State of Bangladesh. The policy of Islamisation of the then Bangladesh Government helped generate tense situations and some forms of communal conflicts and consequently many of the Hindus had to migrate mainly to Assam. Many of the refugees being poor and helpless, however, had no other way but to occupy the virgin *char* areas for their settlement and sustenance. They retained their cultural heritage even in the new land and started taking up both farming and fishing as their prime occupations.

Nepali Immigrants : The process of Nepali immigration to Assam has a long and interesting historical background. There is no denying that in the remote past India and Nepal have shared closed socio-cultural links. The spread of the

Sanskrit language and culture in Nepal and the movement of Nepali scholars to higher learning centres such as Nalanda in India are good examples of Indo-Nepalese cultural relations in the past. Such type of exchange formed the basis of movement of Nepali people to India including to its remote north-eastern part.

Recruitment of Nepali soliders and workers to maintain external defence and internal law and order, and to guard offices and administrative buildings during the British period, gave further impetus to the process of Nepali immigration. These pioneer immigrants passed on information regarding the presence of extensive grassland and culturable wasteland in Assam to the poverty-stricken people of Nepal and subsequently streams of Nepali graziers and cultivators, along with their herds of buffalos and cattle, began to arrive in the state. It is interesting to note that the buffalo *khuti* (buffalo keeping yard) existed in Nongthymmal near Shillong even as far back as 1874.[6] The availability of grasses in the riverine tracts and the *char* areas of the Brahmaputra and the gently sloping marginal foothills facilitated the Nepali graziers to combine dairy farming and crop farming which automatically encouraged them to settle in these selected belts of the State.

Settlement and House Type

Settlements and house types are significant indicators of man's adaptation to ecological conditions. Important as expressions of culture, their study is of particular relevance in the context of the *char* areas as they represent a harmonious living with nature under a tradition-guided, but newly arrived cultural environment.

In the *chars* there is considerable scarcity of habitable land. The slightly convex central part alone provides a certain degree of habitability as generally the *chars* are not above the highest flood level. In other words, with increasing distance from the core of the *chars*, there is a general increase in water depth during floods, making the process of settlement increasingly difficult away from the core. Thus, the initial selection of site for settlement is greatly controlled by the depth of flood water during summer. As the *chars* are susceptible to being entirely submerged during high floods, no alternative is left for the inhabitants than to raise earthen heaps for building their dwellings. Consequently, in the *char* areas, one frequently encounters scattered heaps raised particularly by the Muslim immigrants and the Hindu refugees for erecting houses and homesteads (*bastis*).

As has been observed in Burhachapari, the Nepali immigrants, prefer *Chang ghar* (tall house over wooden posts) instead of raising earthen heaps. Thus, unlike the fertile flood-plains occupied by the indigenous people, the settlement sites, including the homesteads in the *char* areas, are extremely narrow and discontinuous. Relatively dense settlements are seen only in the central parts of the *chars* along natural levees of small channels which criss-cross them. Perpendicular in alignment to the channels, the plots of land occupied by the *bastis* enable the households to enjoy the facilities for bathing, washing, boating and fishing provided by the river.[7] Thus, the settlement pattern generally appears to be linear, an arrangement that does not obstruct the flood water current. Once sites for settlement along the channels are no longer available, settlements are sited randomly in the midst of agricultural fields forming a chaotic settlement pattern.[8] Aligned often to the serpentine foot paths and the narrow zig-zag *kaccha* roads, houses comprising a settlement form a highly meandered linear settlement.

The *char*-dwellers are economically poor, culturally backward, and lead a somewhat nomadic life.[9] Heavy and unpredictable erosion of the Brahmaputra force frequent shifts of settlement sites, with people moving by small country boats alongwith their belongings in search of habitable and cultivable land from one area to the other, reflecting thereby, a semi-nomadic character on account of which they are called 'river nomads'. Both environmental constraints and poverty negate any construction of modern, permanent dwellings. Thus small, temporary huts and sheds made of locally available building materials constitute the dwelling units for the *char*-dwellers.

The Muslim immigrants normally construct their houses similar to the type in their ancestral land, *i.e.* Bangladesh. The roofs of the houses are made of locally available thatch and grasses with a steep slope so as to enable the heavy rainwater to rush down quickly. The frame of the walls is made either of bamboo or dry jute stems over which a thin layer of mud is plastered. With the exception of the *matabbors* (group leaders) who may have two to three relatively good houses sometimes with tin roofs, all other families reside in single thatched houses along with a small open shed for housing the domesticated animals. Houses located in the peripheries of the *chars* are, however, damaged either completely or partially by floods every year. Therefore, most of the houses in this zone are dismantled before the arrival of high floods and are erected temporarily over the embankments or the raised platforms built under the initiative of the State Government. At times the people prefer to stay over the boats or banana rafts placed either inside or close to the house for days together along with the animals during days of floods.

The houses of the Nepali immigrants on the other hand, are relatively long and spacious with provision of separate rooms for youngsters and elders. They often build houses over a wooden or bamboo-made platform about two metres above the ground in order to get rid of flood water. The open space below the platform is generally used for the domesticated animals during the dry season.

Agricultural Practices

As mentioned earlier, the *chars* of the Brahmaputra with their physical vulnerability and cultural nascency present a geo-environmental situation quite distinct from the rest of the Brahmaputra Valley. The *char*-dwellers, therefore, have to adopt cropping practices suitable to the peculiar environment of their habitats.

The farming system in the *chars* is basically characterised by multiple cropping and interculture. These methods have been developed traditionally to suit the local ecological conditions similar to what is elsewhere called 'ecofarming'.[10] The plant biomass, weeds, animal dung and rain water are constantly integrated into the farming process following a unique crop-calendar developed by the farmers themselves in consonance with the seasonal rhythms of the Brahmaputra (Fig. 28.2). Floods, which are the main natural constraint, are perceived by the *char*-dwellers as part of life while deciding the cropping pattern. For them, a timely flood of low magnitude is not a hazard, but a useful phenomenon which makes soil more fertile and fish abundant. As most of the *chars* get inundated during the summer, the periods of sowing and harvesting are selected in such a manner that damage by floods is minimised. In the *char* areas, the peak flood period characterized by several waves (3-5 major waves) from July to September is an agriculturally lean period. However, for most parts of the Brahmaputra plain away from the river, agricultural activity continues with land being covered with *sali* rice, the dominant *kharif* crop of Assam.

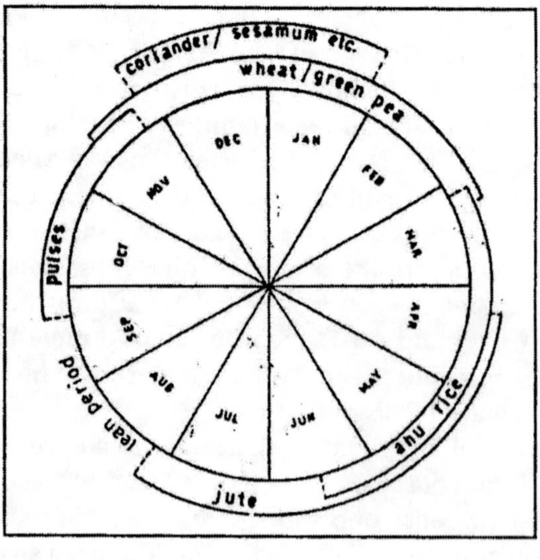

Fig. 28.2 : Crop Calendar of the *Char* Areas

Just after the recession of flood waters the *char*-dwellers start preparing land for cultivation of *rabi* crops such as wheat, lentil, blackgram, mustard, rapeseed, seasamum and a variety of vegetables. From late October, to late November and early December, the cultivation of these crops continues. The main *rabi* season, however, spans from November to March, wherein wheat is sown in November and harvested in March. When wheat attains sufficient height, green pea is sown as an intercrop. Other *rabi* crops, especially a variety of vegetables, are cultivated during the period from November to early February. The cultivation of *ahu* rice, the major grain crop of the *chars*, starts in March and is harvested in the month of July just before the occurrence of the first wave of floods. Jute takes the period from April to early August during which the washing and drying of fibres is also completed.

It is interesting to note that the *char*-dwellers do not generally grow the same crop repeatedly in a plot of land. They follow a suitable system of crop rotation. For example, a plot given to *ahu* rice cultivation during the current year is given over to cultivating *rabi* crops for more than a year before rice is again cultivated in that plot of land. *Sone*, *khesari* and some variety of pulses which have special ability to fix nitrogen in the soil, are also grown as early *rabi* crops as they considerably help enhance the yield of the subsequent crops. During an agricultural year. however, raising *rabi* crops on a plot of land is the focus of the peasants as most of their lands remain idle during the *kharif* season due to the rise of the Brahmaputra waters in summer.

The peasants of the *char* areas mostly use crude farming technology and traditional variety of seeds that are cheap and easily available. The traditional practice has relevance to the skill of the peasants, their life, environment and mode of production and is horizontally integrated with the total way of life.[11] They are not generally interested in cultivating the high yielding variety crops which are expensive and highly vulnerable to insects and diseases. In fact, the traditional varieties of rice such as the *hasan gaonburha*, *kalamanik*, *ghuni* and *saithadhan*, grown by the peasants of the *chars* are so suited to their peculiar ecological conditions that their yields are much higher than some of the high yielding varieties of rice.[12] They also cultivate a special variety of *kharif* rice called *bao* which has the ability to grow very quickly and keep pace with the rising water level. Moreover, the use of some simple, but very useful traditional implements such as *bidha* for proper spacing of broadcast rice and removing weeds, is of great significance for good yields. Proper spacing of rice plants by this implement and timely weeding may increase the yield by more than 200 kg per hectare. Thus, the peasants have been able to adapt their farming practice to the sensitive ecology of the *char* areas appreciably.

Similarly, the Nepali immigrants engaged primarily in dairy farming on the seasonal grasslands of the Brahmaputra riverine tracts are well adapted here. The Nepali dairy farmers supply milk to the nearby market and urban centres during the winter months in addition to practising some grain farming to meet their domestic needs.

Mobility and Means of Transport

The mobility of the *char*-dwellers is greatly restricted by physical barriers on one hand and extreme socio-economic backwardness on the other. Till recently, the *chars* had been almost totally cut off from the mainland of Assam resulting in a near-static socio-cultural environment in the *char* areas. However, the last twenty years, especially since the 1983 general elections, which were held at the peak of the foreign national deportation movement costing hundreds of human lives, the *char* areas experienced a remarkable change in the infrastructural facilities. During this period, the road network of the Brahmaputra plain was extended to connect the till then, inaccessible riverine tract. Despite these developments, the *chars* continue to be socio-culturally the most remote and economically the most backward areas of the state.

Inland water transport is provided by the Brahmaputra river with its numerous tributaries and several *beels* and marshes in the active floodplain. Traditionally, the *char*-dwellers with their small country boats navigate the river with almost all the families possessing a country boat to carry out routine outdoor activities. During the summer months, when houses may remain marooned for days, it becomes almost impossible to move out without a boat. The mobility of the dwellers in this season solely depends upon the availability of the country boats. During high floods, many of the families have to move to safer places such as embankments and river-bank roads, for which again, boats are indispensable. Moreover, boats are the only means of transport to cross the network of channels within the islands and the master stream as well. This is adequately reflected in the weekly markets within and outside the islands where a large number of country boats of varying size carrying goods and people congregate. Boats are also invariably used for fishing and carrying riverine grasses for the cattle. Thus, the country boats are an integral part of life in the *char* areas of the Brahmaputra. Recently, however, privately owned motor boats, locally referred to as *bhutbhuti*, ply regularly through the river linking the islands to the river-bank urban and market centres.

In the spacious and relatively permanent *chars*, bufallo-carts are generally used as an important means of land transport. Agricultural produce, manures (cowdung), building materials (bamboo and thatch) and sometime people are also carried by these carts. The main advantage of such carts is that even in the absence of a road they may be easily drawn over the dry fields in any direction during winter.

Undoubtedly locational and ecological constraints restrict movement of people and goods to and from the *char* areas. However, the growing pressure of population in the *char* areas on the one hand, and the ever increasing demand for labour force in the urban and relatively developed rural areas of the state on the other, have encouraged a considerable section of the *char*-dwellers to overcome the limits of their mobility. At present, in most of the towns of Assam, the immigrants from the *char* areas constitute a major proportion of the labour force. Even the female folk who are mostly illiterate and conservative, have now come forward to work as wage labourers and maid servants in the urban areas. Thus, the movement of the *char*-dwellers from the riverine wastelands to the urban best lands during the recent period has assumed immense significance in so far as rural-urban migration and consequent over-crowding in the urban areas of the state are concerned.

Conclusion

The foregoing discussion in a preliminary manner unfolds the process of humanisation of the virgin *chars* of the Brahmaputra and the way the inhabitants have adapted to the very sensitive ecological conditions of these marginal habitats. Being forced by some adverse political and economic factors in their original home regions, the immigrants came to the Brahmaputra Valley with the aim of having a better and easier livelihood. Their own cultural heritage, economic experiences and skills, therefore, obviously get reflected in the mode of their settlement, mobility and economic persuits. The most important aspect of cultural adaptation in the *char* areas of the state is the eco-friendly farming practices. Although apparently very traditional, practically it is found to be the best possible practice in the context of the *char* environment. The locality-based perception and skilful way of adaptation have made their farming practice sustainable, ensuring a positive change in their economy and culture. Thus, the present state of spatial and cultural existence of the *char*-dwellers is primarily the result of rational adjustment between the harsh dictates of nature and the accurate precepts of culture achieved through a relatively short,but mutually helpful evolutionary process.

References and Notes

1. Coleman, J.N. (1969). 'Brahmaputra River, Channel Process and Sedimentation', *Sedimentary Geology*, Vol. 3, No. 2-3, pp. 129-239.
2. Goswami, D.C. (1985). 'Brahmaputra River, Assam, India : Physiography, Basin Denudation and Channel Aggradation', *Water Resource Research*, Vol. 21, No.7, pp. 859-978.
3. Government of Assam, Shillong. (1938). *Report of the Line System Committee*, p. 1.
4. Guha, A. (1977). *Planter Raj to Swaraj*, New Delhi : People's Publishing House.

5. Farmer, B.H. (1974). *Agricultural Colonisation in India Since Independence*, London: Oxford University Press, p. 57.
6. Bhuyan, M.C. (1977). *Immigrant Population of Assam : An Analytico-Synthetic Study with Special Treatment of Darrang District,* Unpublished Ph.D. Thesis, Guwahati University, Guwahati.
7. Bhagabati, A.K. (1990). *Spatial Analysis of Small Scale Agriculture in Assam : A Case Study of Nalbari District*, Unpublished Ph. D. Thesis, Guwahati University, Guwahati.
8. Bhattacharyya, N.N. (1974). 'Rural Settlements in the Brahmaputra Valley A Geographical Analysis', *Journal of the North East India Geographical Society*, Vol. VI, No.1 and 2, pp. 15-19.
9. Bhagabati, A.K. (1994). 'Islands in the River', *Down to Earth*, Vol. 3, No.11, pp. 34-36.
10. Krings, T. (1990). 'Ecofarming in Africa', *Applied Geography and Development*, Vol. 36, pp. 45-60.
11. Rao, C.R.P. and Verghese, S. (1989). 'Land to Lab : Some Issues in Technology Creation in Modern Agriculture', in A.Bananno, (ed.) *Sociology of Agriculture*, New Delhi : Concept Publishing Co., p. 76.
12. Bhagabati, A.K. (1992). 'Prospect of Ecofarming in Assam', *The Assam Tribune*,Guwahati, Nov. 03, 1992.

29

MAJULI : A UNIQUE ECOLOGICAL ENTITY

ANANDA HAZARIKA

Introduction

Majuli, situated between the Brahmaputra and Lohit rivers is the largest fresh water island in the world (Fig. 29.1). Comprising a sub-division under Jorhat district in the state of Assam, it has a significant geographic location. Its geomorphological setting exerts considerable influence on the nature and development of its socio-cultural landscape with the mighty Brahmaputra and its tributaries having extensive impact on the life of the people living on the island. Unlike the other *char, chaporis* found in the Brahmaputra river, its unique character lies in that this island was not formed due to the accumulation of sand and other alluvial materials as were the other islands of the river. There are also some controversies regarding the exact time of formation of the island. This study presents an introduction to this unique ecological entity.

One of the earliest references to Majuli in literature is from A.J. Moffat Mills, who wrote that 'the island is situated between the Brahmaputra and the Lohit rivers, and was formerly well inhabited and cultivated, but is almost entirely over run with deep grass and forest jungle; it contains an area of 2,82,165 acres,... the soil is rich and alluvial, and cultivated for the growth of every kind of crop'.

Historical Background

In about 1189 A.D, Sri Gauri Narayana of the Great Bodo tribes of Sino-Tibetan origin, founded a dynasty with Ratanpur as the capital and assumed the name of Ratnadwarj Pal.[1] Ratanpur was situated near Kherkatia Suti on the south bank of the Lohit River. As two channels flowed on either side of Ratanpur, it was also known as 'Majali' or 'Majuli'.[2] The mighty Brahmaputra with its tributaries flows from east to west for 2,880 kilometres. The Subansiri and its other tributaries have shallow, braided, shifting channels and steep slopes. These tributaries deposit sediments in the Brahmaputra bottom in the northern part, as a result of which the process of erosion in the south bank of the Brahmaputra

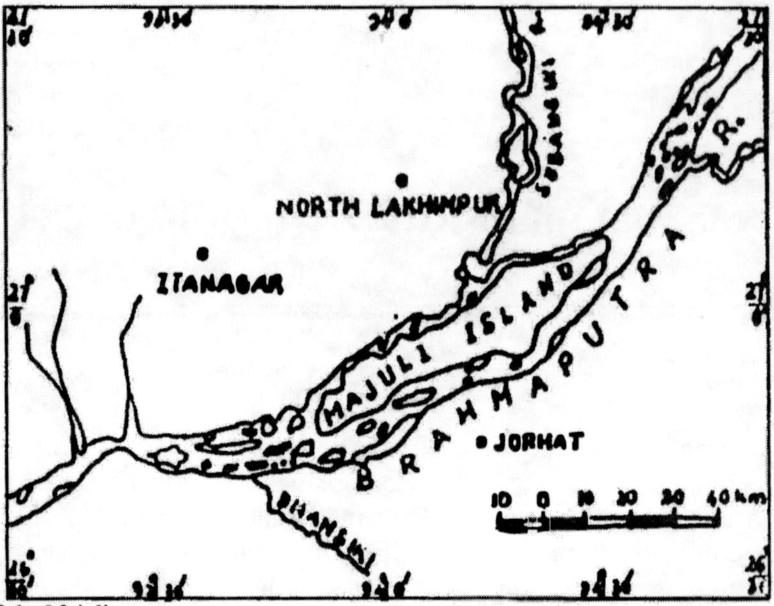

Fig. 29.1 : Majuli

is more active than the north bank. Though Majuli is a part of the Brahmaputra Valley formed of old alluvium and igneous rocks, this southern part is comparatively new from the geological point of view. Burhi-Dihing, Dihing, Dikhow, Namdang, Kokodonga, Jhaji, Bhogdai (Disoi) and Dhansiri are the tributaries of the river Brahmaputra here and form a combined river channel which used to flow parallel to the Brahmaputra. These combined river channels again met the river Brahmaputra at Kaziranga.

According to Edward A. Gait, in 1622 A.D., the Brahmaputra was flowing along the present channel of the Lohit (Lohitya), in the northern part of Majuli, while the Dihing was flowing along the present channel of the Brahmaputra in the south of the island. After being joined by the Dehang and Dikhow, the Dihing joined the Brahmaputra at the western extremity of the island. At a still earlier period the Dihing is believed to have flowed to the Brahmaputra further east than the Burhi Dihing does now. At that time according to native traditions, the Dikhow had an independent course as far as Kajalimukh, part of which still exists in Majuli as the 'Tuni' river and in part of Nagaon as the Kalang.[3] In 1671, the Dihing changed its course to join the Upper Lohit. Later in 1735 the Brahmaputra abandoned its course to flow down the abandoned course of the Dihing. This resulted in a most devastating flood with the Dihing causing widespread devastation in the entire area.[4]

Physical Setting

Majuli, the world's largest river island situated in the upper reaches of the

Brahmaputra river 630 km upstream from the Bangladesh-Bhutan border and 1100 kms from the mouth of the Brahmaputra, has consistently changed dimension (Fig. 29.2). According to the oldest record available in the 17th century, Majuli was 100 kms in length.[5] Later in 1853 the area of Majuli was recorded as 2,82,165 acres. In 1991 the total area of the island was recorded as 1,256 sq km, but as a result of erosion of its banks its size, according to the Brahmaputra Board, is 880 sq. km although according to the official survey, the area of the island is reduced to only 460 sq km.

Majuli experiences the typical temperate humid type of climate associated with the tropical monsoon climate prevailing in this region. Climate exercises an important role in the day-to-day activities of the people of this island.

Natural Resources and Water Bodies

This island is characterised by a rich bio-diversity, the richest perhaps in the North-East Region. The floristic composition includes mainly the three types of flora-herbs, shrubs and trees. Grass sedges are dominant among the herbaceous community, while *Bombax Ceiba* (*Simolo*) and *Azadinacta Indica* (*Azar*) are dominant among trees. The forests of the island are an important source of timber (for furniture), fuel and other products. Bamboo is grown widely on the island and is used extensively for building purposes. The faunistic composition includes amphibia, reptiles, aves, mammals and fish. A survey conducted by Avarde-ne, a voluntary agency, recorded 97 species of birds inhabiting the island.

Water bodies are an important resource of the Majuli. Fishing is one of the important occupations or businesses here. Fish is caught commercially in the Brahmaputra river and its tributaries as well as in other natural and man-made fisheries on the island. According to records of the Agricultural Department, about 33,358 hectares of land is cultivable. Natural production (without chemicals and pesticides) of various agricultural crops still remain due to the natural fertility of the soil. Agriculture here mainly depends on the intensity of floods that occur during the monsoon season each year. In addition to agriculture, the soils of this region are used as raw material for pottery and brick making industries. The sand is also utilised for making buildings and roads. Besides agricultural products, cattle farming, and buffalo raising in the *char* areas of the Brahmaputra river are also very important economic activities. Thus, most of the people are engaged in agriculture and associated activities such as fishing and home crafts.

Roads and Communication

The Majuli road along the east-west direction, from lower (Namuni) Majuli to

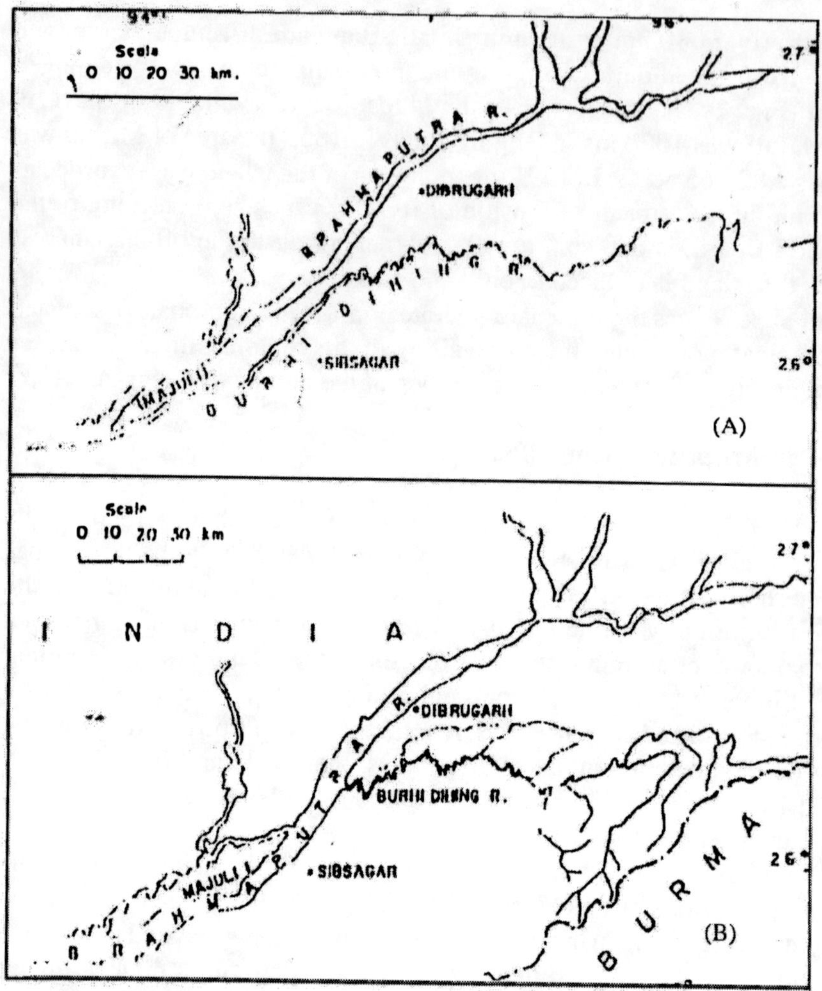

Fig. 29.2 : Majuli

Bongaon is an important line of communication. Another important road known as the NLK road runs from Kamlabari to North Lakhimpur, connecting the Jorhat and Lakhimpur districts. Further, few ferries provided by the Island Water Transport Department are used for crossing the Brahmaputra from different points on the island.

Population

Population, which may be regarded as the major source for the economic development of Majuli sub-division, stood at approximately 1,52,000 in 1996. As per available records, the population of Majuli increased from 36,000 in

1901 to 94,000 in 1971 and as per the Welfare Department to 1,35,378 in 1991. The people of Majuli can be divided into three groups. One, the general castes wherein Hindus, Kalita, Koch, Keot etc. can be included; second, the tribals, wherein plains tribals such as the Mishings, Deuri, Kachari etc, are included and third, the schedule castes wherein the Koibarta, Bonia etc. are included. The distribution of population is quite uneven with the thickly populated areas being limited. The *chars* and *chaporis* are generally sparsely populated. The growth, distribution and pattern of settlement is mainly influenced by ecological factors. Rural settlements to begin with, came up in the alluvial plains of the island.

Religion and Culture

The people of Majuli basically practice Hinduism, which acts here as an important factor in social control. The unique feature here is the dominance of the Hindu society by the *Satra* institutions. The *Satras* are the *Vaishnavite* monastries for religious practices and at the same time, the centres for art and culture. Majuli island is the home of *Vaishnavite* religion and the heart of Assamese culture as well as the *Satra* culture.

Administration

Originally Majuli island was a part of Jorhat district but in 1978, Majuli was declared an independent sub-division except for functions of development, which still vest with the district administration. Majuli itself is an independent (reserved) assembly constituency, though for parliamentary elections its falls under Lakhimpur.

The administrative head in Majuli is the Civil SDO assisted by three Extra Assistant Commissioners and a Sub-Deputy Collector. There is a *Mahukma Parishad*, and two blocks, one with its headquarter at Kamlabari (lower Majuli) and the other with its headquarter at Phuloni, with jurisdiction over Upper Majuli. There are twenty *Gram Panchayats*, and one *Anchalik Panchayat*.

Satra

Satras are institutions for motivating the minds of the people through ritual performances such as *Nam-Prasanga*, prayers, religious festivals, *sanskara*, *gayan-bayan* etc. to realise the existence of a supreme power. They have an ancient and profound tradition.

In 1510, during the Ahom rule, Shri *Shankardev* was forced to leave Nowgong and wanted to live in Bhuyanhat and Belaguri in Majuli. Shankardev evolved a purified form of *Vaishnavism* one which shunned the existing practice of idol worship and expensive propitiatory sacrifices and instead advocated salvation by faith and prayer. In 1523 A.D., *Shankardev* met and became a follower of Madhabdeb, a strict follower of Saktism. The Brahmans at that time were on good terms with the Ahom rulers and were against this kind of religious reform, which they saw as eroding their traditional authority.

As a result, two types of *Satras* came to be established in Assam with separate sects of their own. One, the '*Damodoria*' *Satras* such as *Auniati*, *Dakhinpath* and *Garamur* headed by Deb Damodar, Haridev and Gopaldev and two, the *Satras* established by Shankardev and his followers known as the *Mahapurusia Satras*. The latter included the *Kamlabari*, *Bengenati*, *Samaguri*, *Bhogpur Satras* which have now become the nucleus of *Vaishnav* religion and culture. The chief is called the *Satradhikar*. It may be pointed out that Shankardev and his main followers in Assam were the contemporaries of the famous religious reformers of India such as Guru Nanak, Kabir, Chaitanya Deb, and Tulsidas in the 14th and 15th centuries.

Auniati *Satra*

Although originally there were 65 *Satras* in Majuli, a majority of them have been forced to leave due to floods and erosion. As of now, including both the big and small there are only twenty two *Satras* in Majuli. Of these one of the most well known for its preservation of art and cultural forms is the *Bamunia Satra* (*Damodaria*) at Auniati. When the Auniati *Satra* was first established in Majuli, the land was comparatively high and full of '*Auni*' *pan*. '*Ati*' means highland, and the name came from the conjunction of these two words. Historical accounts of Auniati reveal that King Joyaddhaja known by the Ahom name of Sutamala was responsible for granting land to Sri Sri Niranjan Deva, the first *Satradhikar* of Auniati. For this deed, he is also referred to as the '*sarania*' *Raja* or the one who gave shelter. Joyaddhaja alloted 81,650 *bighas* of land to the Auniati *Satra*, dividing '*Devottara*' and '*Brahmottara*' written on a copper plate.

Satras are usually built along a rectangular pattern, surrounded by walls and houses. There is a '*manikut*' (sanctum sanctorum) in the middle of the *Namghar*. The idol of Krishna, which is referred to as 'Gobinda' at Auniati *Satra* was brought from Jaggannath *khetra* (Puri). Though there are twelve branches of the Auniati *Satra* all over the country, the Uttar Guwahati branch is one of the most important. It was established in 1915 A.D. on the land taken from Aswarlanta Dewalaya by Sri Sri Kamaldeva Goswami. This was reconstructed in 1959 by the then (14th) *Satradhikar* or the *Satra*, known as

Bhagwat Bhaskar Sri Sri Hemachandra Goswami. *Vaishnavas* (*Bhakats*, or disciples) living in the *Satras* can be classified into two classes, one call Udasin (who remain celibate) and the other Bisaya (who can get married). The Auniati *Satriya* is purely monastic, with the *Staradhikar* and *Bhakats* Udasin in nature.

There is a specific system of prayer. The main prayer continues from sunrise to sunset, three times in a day with fourteen *prasangas*. Besides this daily routine, the cultural life of the *Satra* is enriched by regular dramatic performances of *bhaona* and *ankiya*, as well as *geet* and *nritya*.

Erosion

But this unique ecological as also socio-cultural entity is in danger of disappearing into oblivion. After the great Assam earthquake of 1950, Majuli, situated at the confluence of the Subansiri and Brahmaputra rivers, has been rapidly eroding as the river changes its course and undercuts the sandy soil that forms the island. The main reason for this is the age-old tendency to shift on either side and excessive sediment discharge.

The southern side of the island is protected by a 70 km long main dyke stretching from the Kherkotia suti to Kamlabari, the one protection from the wrath of the Brahmaputra , the 'lifeline' of the island. The island is once again divided into two parts at the village Sorapar under Ahataguri *Mouza*, where the Brahmaputra and the Subansari joined together following erosion of their banks. With the braided character of these two rivers, the Ahataguri *Mouza* has virtually disappeared, fragemented into a number of '*chars*' or sandbanked islands. The breach at Kanyajan near Kamlabari where an area of three square kilometres has been eoded since 1989, culminating in a dangerous break of the dyke in 1993, represents a potential threat to Kamlabari, the main township and commerical centre on the island. If Kamlabari and adjacent areas are eroded, a histori-cultural change will take place, leaving Majuli similar to a headless human being.

According to a study[5] based on remote sensing data and interpretation, the Brahmaputra river increased in width from 6.75 kilometres to 8.90 kilometres on the eastern side of Majuli island in the period from1972-1992, eroding away a large tract of land on both banks of the Brahmaputra river.

Flood inundation and loss of land area due to severe river bank erosion have become a burning problem of this island now. It is also important to maintain the perspective that while the inundation has been an annual feature on the island since 1570, the loss of land area that has been taking place since 1950 (and more rapidly in the recent past) is comparitively much more serious, as it displaces people from their roots and occupations. Moreover no efforts have been made by the government to tackle the problem of erosion.

If Majuli is not preserved scientifically and in a permanent way, then in a few years there will be no trace of present Majuli, the largest river island in Assam and in the world.

References and Notes

1. *Sibsagar District Gazetteer* (1967). Shillong: Government of India, pp. 36-37.
2. *Adi Charitra Madhabdev*, p. 61.
3. Gait, E.A. *History of Assam*, Second Edition, p.126.
4. *See* ref. 3.
5. River Bank Erosion of Majuli Island as deciphered from the IRS imagery, by Hussain, Chaudhary and Ghani (all of Brahmaputra Board, Guwahati) and S.D. Naik (of Space Applications Centre, Indian Space Research Organisation, Ahmedabad), in *Proceedings of the National Symposium on Remote Sensing Applications for Resource Management with Special Emphasis on the N.E. Region, Guwahati*, Nov. 25-27, 1993, pp.31-35.

THE CONTRIBUTORS

Nasreen Ahmad is Associate Professor of Geography at the Dhaka University, Bangladesh.

Abani K. Bhagabati is Associate Professor of Geography at the Gauhati University, Guwahati, India. He received his PhD. from Gauhati University and has been specialised on the role of caste structure in agriculture in Assam.

Dirk Bronger is Professor Ordinaire at the Geographical Institute of University of Ruhr-Bochum. He wrote his DLit. thesis on some aspects of regionalisation of Andhra Pradesh and on its caste system in Erdkunde. He has published extensively in German journals and in India on Social Geography and Regional Geography. In recent years he has taken a great deal of interest in the development of Gujarat.

Maurice Carney is research associate at the Department of Geography and Planning, University of Akron, Ohio, U.S.A.

A.K. Chakravarti is Professor of Geography at the University of Saskatchewan, Saskatoon, Canada. He received his PhD. from the University of Wisconsin, U.S.A. He has extensively published on the climatology of U.S.A. and on the Green Revolution and food habits of India.

Anjana Desai is Professor and Head of the Department of Geography at Gujarat University, Ahmedabad, India. She received her PhD. from Calcutta University. She has specialised in urban morphology and has several publications to her credit.

Mamta Desai is presently teaching Geography at Calcutta. She received her doctorate from the University of Calcutta, India.

Ramesh C. Dhussa is Assistant Professor of Geography at the Department of Geography, Drake University, Des Moines, Iowa, U.S.A. He received his PhD. from Kent State University. He specialises in the geographical analysis of regional literature.

Jaymala Diddee is Professor at the Department of Geography, University of Poona, Pune, India. She received her PhD. from Poona University. Her book on Central Places in western Maharashtra has been hailed as the best on this theme. For some years now she has been working on urban development during the British period, urbanisation and some of the nomadic tribes.

Ashok K. Dutt is Professor of Geography at both the Departments of Geography and of Urban Study at the University of Akron, Ohio, U.S.A. He received his PhD. from the University of Patna, India and thereafter worked for sometime as a town planner in the Calcutta Metropolitan Development Organization. He has been publishing on regional and urban planning and Social Geography for more than 30 years. He edited a two-volume research on Regional Planning published by Oxford University Press, England. His book with Allen G. Noble on Cultural Patterns and Processes of India and on Urbanisation in India have brought him wide recognition. He has published widely on many aspects of society in India.

Theodore I. Erski is at the Department of Geography, University of Akron, Akron, Ohio, U.S.A.

Irina Efremova is working as a Researcher in the centre for Indian Studies, Institute of Oriental Studies at the Russian Academy of Sciences, Moscow, Russia. She received her M.A. in Geography and PhD. in History from the University of Moscow. She wrote her doctoral thesis on regionalisation of India on the basis of regional structure of caste. She has done extensive field work with Prof. Jaimala Diddee of the University of Poona on the intermediary and lower administr-ative functionaries.

Neelam Grover is Professor of Geography at Panjab University, Chandigarh, India. She received her PhD. from the Department of Geography, Panjab University, Chandigarh. She wrote her doctoral thesis on the rural settlements of different culture groups of Northern Haryana. She is deeply interested in the evolution and cultural-geographical analysis of the rural landscape and has published on the Rural Settlements of Northern Haryana and on the methodology of Cultural Geography for the study of Indian villages.

Swapna Banerjee Guha is Professor and Head of the Geography Department at Bombay University. She has specialised in the study of the negative role of capitalism in regional development and has recently published a book on this theme.

Sumit Guhathakurta is head of the Foreign Language Section at the Library and Information Services Division of the Bhabha Atomic Research Centre, Bombay.

Ananda Hazarika teaches Geography at the Majuli College, Majuli, Assam, India. He continues to voice his concern at the continuous degradation of the island of Majuli, a fragile ecosystem.

Mumtaz Khan is an Associate Professor of Geography at Jamia Millia Islamia, New Delhi. He is particularly interested in various aspects of Cultural Geography, especially folk regions. In fact both Jaipal Singh and Mumtaz Khan have specialised in the Cultural Geography of parts of Mewat and have recently published a book entitled 'Mythical Space, Cosmology and Lanscape, Towards a Cultural Geography of India.'

Suresh Kumar is also at the Centre for the Study of Regional Development, Jawaharlal Nehru University, New Delhi.

John McKim Malville is Professor at the Department of Astrophysical, Planetary and Atmospheric Sciences, University of Colorado at Boulder, U.S.A.

D.S. Manku retired as a senior lecturer in Geography from the Government College, Hoshiarpur, India. He received both his M.A. and PhD. degrees from the Department of Geography, Panjab University, Chandigarh, India. He wrote his doctoral thesis on the Gujar settlements in the Punjab Kandi. He continues his field work on the Gujars of the Kandi area.

Bhupinder Singh Marh is Reader and Head of the Geography Department in Himachal Pradesh University. He received his M.A. and M. Phil degrees from the Department of Geography, Panjab University, Chandigarh, India. He carried out extensive field work, sketching and photographing of house types in the district of Chamba, Himachal Pradesh, India and submitted this research for his M. Phil degree. He then received a research assistantship at the University of Missourie, Colombia, U.S.A. where he did his second M.A. specializing in aspects of Cultural Geography. He received his PhD. in the field of Fluvial Geomorphology from Indiana State University, U.S.A.

Champa Mitra is presently teaching Geography at Burdwan. She received her PhD. degree from the University of Calcutta, India.

Anupa Mukhopadhyay is at the Maryland Department of Health and Mental Hygiene, AIDS Administration, Baltimore, U.S.A.

Allen G. Noble is Professor and Head of the Geography Department at Akron University, Ohio, U.S.A. He did his doctorate at the University of Illinois. He is well known in the United States for his publications on rural settlements and house types of different regions of U.S.A. Prof. Noble has done intensive field research in parts of Bihar especially on the city of Bhagalpur and has jointly authored with Prof. Ashok K. Dutt a well known book, India : Cultural Patterns and Processes, and a book on Urbanisation in South Asia. He has published several research papers in Journal of Cultural Geography.

William A. Noble retired as Professor of Geography from the University of Missoure, Columbia, U.S.A. He received his PhD. from Louisiana State University. He has specialised in the Cultural Geography of the Nilgiris and its indigenous cultures especially the Todas. He has published extensively on the various aspects of the Cultural Geography of India.

Renuka Patnaik is a research assistant in NCERT at New Delhi. She received her M.Phil in Geography from the University of Poona, Pune, India. She has extensively researched some of the primitive tribes of Maharashtra.

M.H. Qureshi is Professor of Geography at the Centre for the Study of Regional Development, Jawaharlal Nehru University, New Delhi. He received his PhD. from the University of Jodhpur, Rajasthan, India. In recent years he has specialised in the relationship between culture and ecology of different regions.

H.N. Sharma is Professor of Geography at Gauhati University, Guwahati, India. He received his PhD. from Syracuse University, U.S.A. and worked with the late Professor David Sopher. He has specialised in several aspects of Social Geography of Assam and has published several papers on this field.

K.D. Sharma is Professor of Geography at Panjab University, Chandigarh, India. He received his doctorate from Department of Geography, Panjab University, Chandigarh. He wrote his doctoral dissertation on the Urban Geography of Rohtak city. In recent years he has been researching extensively on Himalayan towns and has been publishing on varied aspects of this theme.

Daljit Singh is presently working as Lecturer, Geography at Doaba College, Jallandhar, India. He received both his M.A. and M. Phil degrees from the Geography Department, Panjab University, Chandigarh. He did his M.Phil dissertation on the field pattern of some area of Punjab. A keen observer of the field landscape, he has avidly studied field patterns.

Jaipal Singh has taught Geography at Government Arts College, Alwar, Rajasthan, India. He has been deeply interested in Hindu cosmology, cultural landscape, marital migration and Literary Geography.

Mehar Singh received his M.A. from the Department of Geography, Panjab University, Chandigarh, India. He had a first class first in Diploma in Cartography also from Panjab University. He has been teaching cartography, areal photo interpretation, remote sensing and Rural Settlement Geography. He is a very keen observer of the rural landscape of Punjab.

Rana P.B. Singh is Associate Professor of Geography at Benaras Hindu University, Varanasi, India. He received his PhD also from Benaras Hindu University. He has specialized in the study of clear settlements, pilgrimage centres and the Religious Geography of Varanasi.

Robert H. Stoddard has for a long time been a full Professor of Geography at the University of Nebraska at Lincoln, Iowa, U.S.A. His PhD. thesis was on the location of Hindu Pilgrimages using a very innovative kind of mathematical model to predict such a pattern of locations. He has been interested in Rajasthan on aspects of Cultural Geography of India and is currently supervising Indian PhD students. His special interests have been field techniques, Social Geography and South Asia.

Razia Sultana is Professor of Geography at the Dhaka University, Bangladesh. She received her PhD. from the University of Hawaii, U.S.A.

INDEX